Principles of

SOLID MECHANICS

Rowland Richards, Jr.

CRC Press

Boca Raton London New York Washington, D.C.

FIRST INDIAN REPRINT, 2011

Library of Congress Cataloging-in-Publication Data

Richards, R. (Rowland)
Principles of solid mechanics / R. Richards, Jr.
p. cm. — (Mechanical engineering series)
Includes bibliographical references and index.
ISBN 0-8493-0114-9 (alk. paper)
1. Mechanics, Applied. I. Title. II. Advanced topics in mechanical engineering series.

TA350.R54 2000
620'.1'05—dc21 00-060877

Direct all inquiries to CRC Press LLC, 2000 N.W. Corporate Blvd., Boca Raton, Florida 33431.

International Standard Book Number 0-8493-0114-9
Library of Congress Card Number 00-060877
Printed in India by Nutech Photolithographers, New Delhi

" This book is printed on ECF card and ECF environment-friendly paper manufactured from unconventional and other raw materials sourced from sustainable and identified sources."

Preface[1]

There is no area of applied science more diverse and powerful than the mechanics of deformable solids nor one with a broader and richer history. From Galileo and Hooke through Coulomb, Maxwell, and Kelvin to von Neuman and Einstein, the question of how solids behave for structural applications has been a basic theme for physical research exciting the best minds for over 400 years. From fundamental questions of solid-state physics and material science to the mathematical modeling of instabilities and fracture, the mechanics of solids remains at the forefront of today's research. At the same time, new innovative applications such as composites, prestressing, silicone chips, and materials with memory appear everywhere around us.

To present to a student such a wonderful, multifaceted, mental jewel in a way that maintains the excitement while not compromising elegance and rigor, is a challenge no teacher can resist. It is not too difficult at the undergraduate level where, in a series of courses, the student sees that the simple solutions for bending, torsion, and axial load lead directly to analysis and design of all sorts of aircraft structures, machine parts, buildings, dams, and bridges. However, it is much more difficult to maintain this enthusiasm when, at the graduate level, the next layer of sophistication is necessary to handle all those situations, heretofore glassed over and postponed, where the strength-of-materials approach may be inaccurate or where a true field theory is required immediately.

This book has evolved from over 30 years of teaching advanced seniors and first-term graduate students a core course on the application of the full-range field theory of deformable solids for analysis and design. It is presented to help teachers meet the challenges of leading students in their exciting discovery of the unifying field theories of elasticity and plasticity in a new era of powerful machine computation for students with little experimental experience and no exposure to drawing and graphic analysis. The intention is to concentrate on fundamental concepts, basic applications, simple problems yet unsolved, inverse strategies for optimum design, unanswered questions, and unresolved paradoxes in the hope that the enthusiasm of the past can be recaptured and that our continued fascination with the subject is made contagious.

[1] Since students never read the preface to a textbook, this is written for teachers so they can anticipate the flavor of what follows. Many of the observations in this preface are then repeated in bits and pieces when introducing the various chapters so students cannot actually escape them entirely.

In its evolution this book has, therefore, become quite different from other texts covering essentially the same subject matter at this level.[2] First, by including plastic as well as elastic behavior in terms of a unified theory, this text is wider in scope and more diverse in concepts. I have found that students like to see the full range, nonlinear response of structures and more fully appreciate the importance of their work when they realize that incompetence can lead to sudden death. Moreover, limit analysis by Galileo and Coulomb historically predates elastic solutions and is also becoming the preferred method of analysis for design not only in soil mechanics, where it has always dominated, but now in most codes for concrete and steel structures. Thus in the final chapters, the hyperbolic field equations of plasticity for a general Mohr-Coulomb material and their solution in closed form for special cases is first presented. The more general case requiring slip-line theory for a formal plasticity solution is then developed and applied to the punch problem and others for comparison with approximate upper-bound solutions.

Secondly, while the theory presented in the first three chapters covers familiar ground, the emphasis in its development is more on visualization of the tensor invariants as independent of coordinates and uncoupled in the stress–strain relations. The elastic rotations are included in anticipation of Chapter 4 where they are shown to be the harmonic conjugate function to the first invariant leading to flow nets to describe the isotropic field and closed-form integration of the relative deformation tensor to determine the vector field of displacements.

Although the theory in three dimensions (3D) is presented, the examples and chapter problems concentrate on two-dimensional (2D) cases where the field can be plotted as contour maps and Mohr's circle completely depicts tensors so that the invariants are immediately apparent. Students often find the graphic requirements difficult at first but quickly recognize the heuristic value of field plots and Mohr's circle and eventually realize how important graphic visualization can be when they tackle inverse problems, plasticity, and limit analysis. In addition to the inclusion of elastic rotations as part of the basic field equations, the discussion in Chapter 4 of the properties of field equations and requirements on boundary conditions is normally not included in intermediate texts. However, not only do all the basic types of partial differential equations appear in solid mechanics, but requirements for uniqueness and existence are essential to formulating the inverse problem, understanding the so-called paradox associated with certain wedge solutions, and then

[2] By texts I mean books designed for teaching with commentary, examples and chapter problems. Most can be broadly categorized as either: (a) the presentation of the theory of elasticity with emphasis on generality, mathematical rigor, and analytic solutions to many idealized boundary-value problems; or (b) a more structural mechanics approach of combining elasticity with judicious strength-of-materials type assumptions to develop many advanced solutions for engineering applications. Two representative, recent books of the first type are *Elasticity in Engineering Mechanics* by A.P. Boresi and K.P. Chong, and *Elastic and Inelastic Stress Analysis* by I.H. Shames and F.A. Cozzarelli. Probably *Advanced Mechanics of Materials* by R.D. Cook and W.C. Young and the book by A.P. Boresi, S.J. Schmidt, and O.M. Sidebottom of the same title are the most recent examples of the second type.

appreciating the difficulties with boundary conditions inherent in the governing hyperbolic equations of slip-line theory.

There are only 15 weeks in a standard American academic term for which this text is designed. Therefore the solutions to classic elasticity problems presented in the intermediate chapters have been ruthlessly selected to meet one or more of the following criteria:

a. to best demonstrate fundamental solution techniques particularly in two dimensions,
b. to give insight as to the isotropic and deviatoric field requirements,
c. to present questions, perhaps unanswered, concerning the theory and suggest unsolved problems that might excite student interest,
d. to display particular utility for design,
e. to serve as a benchmark in establishing the range where simpler strength-of-materials type analysis is adequate, or
f. are useful in validating the more complicated numerical or experimental models necessary when closed-form solutions are not feasible.

At one time the term "Rational Mechanics" was considered as part of the title of this text to differentiate it from others that cover much of the same material in much greater detail, but from the perspective of solving boundary value problems rather than visualizing the resulting fields so as to understand "how structures work." The phrase "Rational Mechanics" is now old-fashioned but historically correct for the attitude adopted in this text of combining the elastic and plastic behavior as a continuous visual progression to collapse.

This book makes liberal use of footnotes that are more than just references. While texts in the humanities and sciences often use voluminous footnotes, they are shunned in modern engineering texts. This is, for a book on Rational Mechanics, a mistake. The intention is to excite students to explore this, the richest subject in applied science. Footnotes allow the author to introduce historical vignettes, anecdotes, less than reverent comments, uncertain arguments, ill-considered hypotheses, and parenthetical information, all with a different perspective than is possible in formal exposition. In footnotes the author can speak in a different voice and it is clear to the reader that they should be read with a different eye. Rational Mechanics is more than analysis and should be creative, fun, and even emotional.

To close this preface on an emotional note, I must acknowledge all those professors and students, too numerous to list, at Princeton, Caltech, Delaware, and Buffalo, who have educated me over the years. This effort may serve as a small repayment on their investment. It is love, however, that truly motivates. It is, therefore, my family: my parents, Rowland and Jean; their grandchildren, Rowland, George, Kelvey, and Jean; and my wife, Martha Marcy, to whom this book is dedicated.

Contents

1

Introduction

Solid mechanics deals with the calculation of the displacements of a deformable body subjected to the action of forces in equilibrium for the purpose of designing structures better. Throughout the history of engineering and science from Archimedes to Einstein, this endeavor has occupied many great minds, and the evolution of solid mechanics reflects the revolution of applied science for which no end is in sight.

1.1 Types of Linearity

The development of various concepts of linearity is one central theme in solid mechanics. A brief review of five distinct meanings of "linear analysis" can, therefore, serve to introduce the subject from a historical perspective* setting the stage for the presentation in this text of the field theory of deformable solids for engineering applications. Admittedly, any scheme to introduce such an incredibly rich subject in a few pages with one approach is ridiculously simplistic. However, discussing types of linearity can serve as a useful heuristic fiction.

1.1.1 Linear Shapes—The "Elastic Line"

One fundamental idealization of structures is that, for long slender members, the geometric properties and therefore the stiffness to resist axial torsion and bending deformation, are functions of only the one variable along the length of the rod. This is the so-called elastic line used by Euler in his famous solution for buckling.

---

* In this introduction and succeeding chapters, the history of the subject will appear primarily in footnotes. Most of this information comes from four references: *A History of the Theory of Elasticity and Strength of Materials* by I. Tokhunter and K. Pearson, Cambridge University Press, 1893; *History of Strength of Materials* by S.P. Timoshenko, McGraw Hill, New York 1953; *A Span of Bridges* by H.J. Hopkins, David and Charles, Ltd., 1970; and *An Introduction to the History of Structural Mechanics* by E. Beuvenato, Springer-Verlag, Berlin, 1991.

If the internal stress resultants, moments, torque, shears, and axial force are only dependent on the position, s, along the member, then, too, must be the displacements and stresses.* This, then, is the tacit idealization made in classic structural analysis when we draw line diagrams of the structure itself and plot line diagrams for stress resultants or changes in geometry. Structural analysis for internal forces and moments and, then deformations is, therefore, essentially one-dimensional analysis having disposed of the other two dimensions in geometric properties of the cross-section.**

1.1.2 Linear Displacement (Plane Sections)

The basic problem that preoccupied structural mechanics in the 17th century from Galileo in 1638 onward was the behavior and resistance to failure of beams in bending. The hypothesis by Bernoulli*** that the cross-section of a bent beam remains plane led directly to the result that the resistance to bending is a couple proportional to the curvature. This result, coupled with the concept of linear shape, allowed Euler**** to develop and study the deformation of his "elastic line" under a variety of loadings. Doing this, he was able to derive the fundamental equations of flexure with great generality including initial curvature and large deflections as well as for axial forces causing buckling with or without transverse load. Bernoulli and Euler assumed "elastic" material implicitly lumping the modulus in with their geometric stiffness "constant."

Plane sections is, of course, *the* fundamental idealization of "Strength of Materials" ("simple" solid mechanics) which, for pure bending, is a special

* Actually, for Euler to determine the deformed geometry of loaded bars also required the next three assumptions of linearity and the invention of calculus, which historically predated the concept of the "elastic line."

** The extension to two- and three-dimensional structural analysis is the neutral-surface idealization used for plates and shells.

*** Jacob (1654–1705), the oldest of five Bernoulli-family, applied mathematicians. Like Galileo (1564–1642), Leibnitz (1646–1716), and others, he incorrectly took the neutral axis in bending as the extreme fiber, but by correctly assuming the cross-section plane remained plane, he derived the fundamental equation for bending of a beam, i.e., $\frac{1}{\rho} = \frac{M}{EI}$ (but off by a constant and only for a cantilever without using EI explicitly). Hooke had it right in his drawing of a bent beam (Potentia Restitutive, 1678) 30 years before Jacob got it wrong, but Hooke could not express it mathematically.

**** Euler (1707–1783) was probably the greatest applied mathematician dealing with solid mechanics in the 18th century. He was a student of John Bernoulli (Jacob's brother) at the age of 13 and went in 1727 to Russia with John's two sons, Daniel and Nicholas, as an associate at St. Petersburg. He led an exciting life, and although blind at 60, produced more papers in his last 20 years than ever before. His interest in buckling was generated by a commission to study the failure of tall masts of sailing ships. His famous work on "minimization of energy integrals" came from a suggestion from Daniel Bernoulli (who himself is most noted as the "father of fluid dynamics") that he apply variational calculus to dynamic behavior of elastic curves.

case of the more general elasticity theory in two and three dimensions.* "Strength of Materials" is, in turn, divided into "simple" and "advanced" solutions: simple being when the bar is straight or, if curved, thin enough so all the fibers have approximately the same length. For cases where the axial fibers have the same base length, then linear axial displacements (Bernouilli's Hypothesis) implies linear strains, and therefore linear stresses in the axial direction.

Relatively simple strength of materials solutions are, to the engineer, the most important of solid mechanics. They:

a. may be "exact" (e.g., pure bending, axial loading, or torsion of circular bars);
b. or so close to correct it makes no difference; and
c. are generally a reasonable approximation for preliminary design and useful as a benchmark for more exact analysis.

One important purpose in studying more advanced solid mechanics is, in fact, to appreciate the great power of the plane-section idealization while recognizing its limitations as, for example, in areas of high shear or when the shape is clearly not one-dimensional and therefore an elastic line idealization is dubious or impossible.

1.1.3 Linear Stress Strain Behavior (Hooke's Law)

Of the many fundamental discoveries by Hooke,** linear material behavior is the only one named for him. In his experiments, he loaded a great variety of

* The designation "Strength of Materials" popularized by Timoshenko in a series of outstanding undergraduate texts, has fallen on hard times, and rightly so, since the subject matter has very little to do with strength of materials *per se.* However, it is a useful label for solutions based on the plane-section idealizations (approximation) and will be so used. Titles for introductory solid mechanics texts now in vogue include: *Mechanics of Materials, Mechanics of Solids, Mechanics of Deformable Solids,* and *Statics of Deformable Bodies.* The last is the best, but has never caught on. It is essentially impossible to invent a title to categorize a subject as rich and important to engineers as "Strength of Mechanics."

** Robert Hooke (1635–1703) is probably the most controversial figure in the history of science, perhaps because he was really an engineer—the first great modern engineer. Born in 1635, he died in 1703 which was his greatest mistake for Newton (1642–1727) hated him and had over 20 years to destroy his reputation unopposed. Hooke had anticipated two of Newton's Laws and the inverse square law for gravitation, as well as pointing out errors in Newton's "thought experiments." Also Newton was aloof, dogmatic, religious, and a prude while Hooke was intuitively brilliant, nonmathematical, gregarious, contentious, and lived, apparently "in sin," with his young niece. Newton (and others in the Royal Society) regarded Hooke as inferior by lowly birth and treated him as a technician and servant of the society which paid him to provide "three or four considerable experiments every week." Newton despised Hooke professionally, morally, and socially, and after Newton was knighted and made President of the Royal Society, his vindictive nature was unrestrained. Hooke needs a sympathetic biographer who is an engineer and can appreciate all his amazing achievements—"a thousand inventions"—which included the balance wheel for watches, setting zero as the freezing point of water, the wheel barometer, the air pump or pneumatic engine, not to mention the first, simplest, and most powerful concepts for bending and the correct shape for arches.

materials in tension and found that the elongation was proportional to the load. He did not, however, express the concept of strain as proportional to stress, which required a gestation period of more than a century.

Although often called "elastic behavior" or "elasticity," these terms are misnomers in the sense that "elastic" denotes a material which, when unloaded, returns to its original shape but not necessarily along a linear path. However, elastic, as shorthand for linear elastic, has become so pervasive that linearity is always assumed unless it is specifically stated otherwise.

As already discussed, Hooke's Law combined with the previous idealization of plane sections, leads to the elastic line and the fundamental solutions of "Strength of Materials" and "Structural Mechanics." As we will see, Hooke's Law—that deformation is proportional to load—can be broadly interpreted to include time effects (viscoelasticity) and temperature (thermoelasticity). Moreover, elastic behavior directly implies superposition of any number of elastic effects as long as they add up to less than the proportional limit. Adding the effects of individual loads applied separately is a powerful strategy in engineering analysis.

1.1.4 Geometric Linearity

The basic assumption that changes in base lengths and areas can be disregarded in reducing displacements and forces to strains and stresses is really a first-order or linear approximation. Related to it is the assumption that the overall deformation of the structure is not large enough to significantly affect the equilibrium equations written in terms of the original geometry.*

1.1.5 Linear Tangent Transformation

The fundamental concept of calculus is that, at the limit of an arbitrary small baseline, the change as a nonlinear function can be represented by the slope or tangent. When applied to functions with two or more variables, this basic idea gives us the definition of a total derivative, which when applied to displacements will, as presented in Chapter 2, define strains and rotations.

A profound physical assumption is involved when calculus is used to describe a continuum since, as the limiting process approaches the size of a molecule, we enter the realm of atomic physics where Bohr and Einstein argued about the fundamental nature of the universe. The question: At what

* Neither of these assumptions were made by Euler in his general treatment of the elastic line and buckling shapes where he used the *final* geometry as his reference base. This is now called Eulerian strain or stress (ϵ^E and σ^E) to differentiate it from "engineering" or Lagrangian strain or stress (ϵ^L and σ^L), which we shall use. The "true" value, often called Cauchy strain or stress after Augustan Cauchy (1789–1857), involves the logarithm and is almost never used. Lagrange (1736–1813) was encouraged and supported by Euler. Cauchy, in turn, was "discovered" as a boy by Lagrange who became his mentor. Cauchy was educated as a civil engineer and was doing important work at the port of Cherbourg at the age of 21.

point does the limit process of calculus break down? is also significant in engineering. Even for steel, theoretical calculations of strength or stiffness from solid-state physics are not close to measured values. For concrete, or better yet soil, the idea of a differential base length being arbitrarily small is locally dubious. Yet calculus, even for such discrete materials as sand, works on the average. Advanced analysis, based on, for example, "statistical mechanics" or the "theory of dislocations," is unnecessary for most engineering applications.

1.2 Displacements—Vectors and Tensors

A second basic concept or theme in solid mechanics, is the development of a general method of describing changes in physical quantities within an artificial coordinate system. As we shall see, this involves tensors of various orders.

A tensor is a *physical* quantity which, in its essence, remains unchanged when subject to any admissible transformation of the reference frame. The rules of tensor transformation can be expressed analytically (or graphically), but it is the unchanging aspects of a tensor that verify its existence and are the most interesting physically. Seldom in solid mechanics is a tensor confused with a *matrix* which is simply an operator. A tensor can be written in matrix form, and therefore the two can look alike on paper, but a matrix as an array of numbers has no physical meaning and the transformation of a matrix to a new reference frame is impossible. Matrix notation and matrix algebra can apply to tensors, but few matrices, as such, are found in the study of solid mechanics.

All of physics is a study of tensors of some order. Scalars such as temperature or pressure, where one invariant quantity (perhaps with a sign) describes them, are tensors of order zero while vectors such as force and acceleration are tensors of the first order. Stress, strain, and inertia are second-order tensors, sometimes called dyadics. Since the physical universe is described by tensors and the laws of physics are laws relating them, what we must do in mechanics is learn to deal with tensors whether we bother to call them that or not.

A transformation tensor is the next higher order than the tensor it transforms. A tensor of second order, therefore, changes a vector at some point into another vector while, as we shall see, it takes a fourth-order tensor to transform stress or strain.* A tensor field is simply the spacial (x, y, z) and/or time variation of a tensor. It is this subject, scalar fields, vector fields, and second-order tensor fields that is the primary focus of solid mechanics. More specifically, the goal is to determine the vector field of displacement and second-order stress and strain fields in a "structure," perhaps as a function of time as well as position, for specific material properties (elastic, viscoelastic, plastic) due to loads on the boundary, body forces, imposed displacements, or temperature changes.

* Since we are not concerned with either relativistic speeds or quantum effects, transforming tensors of higher than second order is not of concern. Nor do we deal with tensor calculus since vector calculus is sufficient to express the field equations.

1.3 Finite Linear Transformation

When a body is loaded, the original coordinates of all the points move to a new position. The movement of each point *A*, *B*, or *C* to *A′B′C′* in Figure 1.1 is the movement of the position vectors r_1, r_2, and $(r_1 + r_2)$ where we assume a common origin. A linear transforation is defined as one that transforms vectors according to the rules.*

$$[a](\overline{r}_1 + \overline{r}_2) = [a]\overline{r}_1 + [a]\overline{r}_2 \tag{1.1}$$

where [*a*] is the "finite linear transformation tensor." Also,

$$[a](n\bar{r}) = n[a]\bar{r} \tag{1.2}$$

and:

a. straight lines remain straight
b. parallel lines remain parallel
c. parallel planes remain parallel

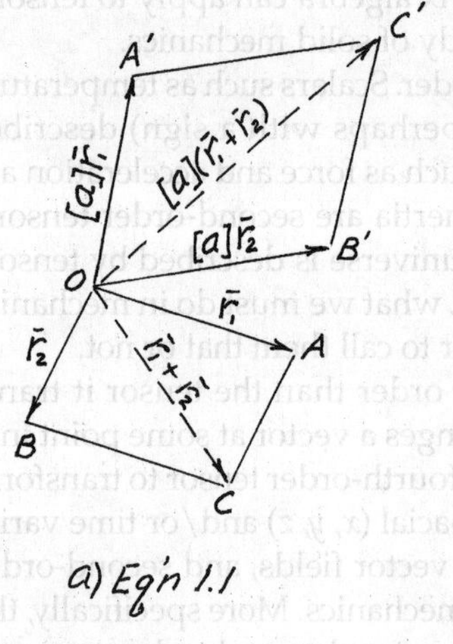

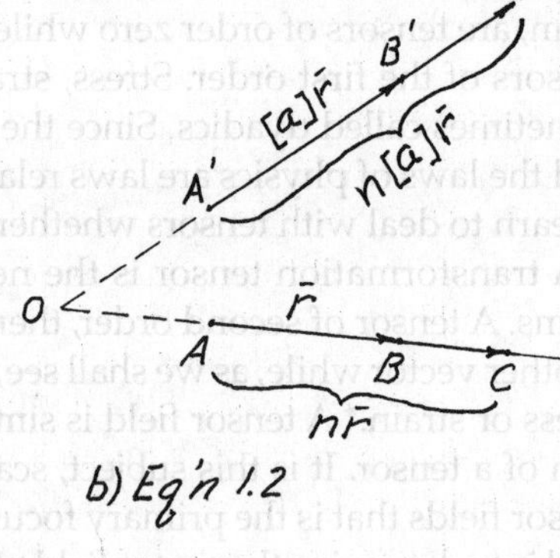

FIGURE 1.1
Linear transformation.

* A much more detailed discussion is given by Saada, A.S., *Elastic, Theory and Applications*, Pergamon Press, 1974, pp. 20–65. This excellent text on mathematical elasticity will be referred to often.

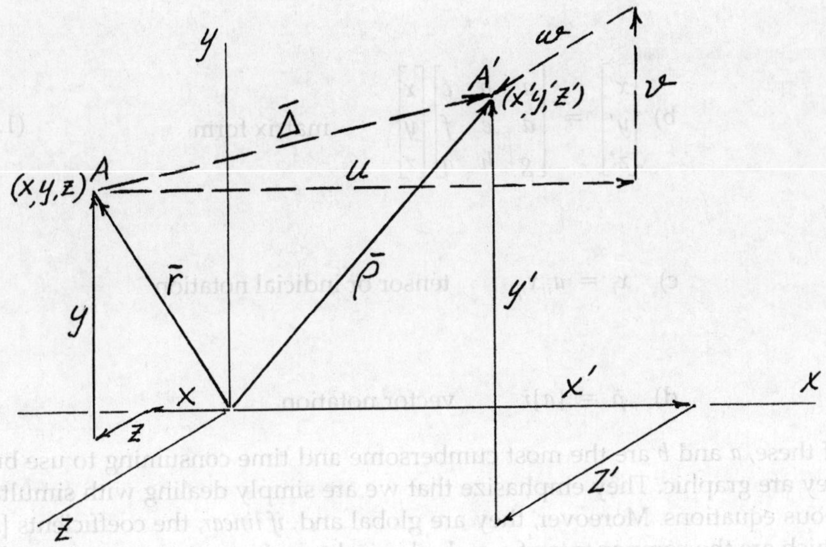

FIGURE 1.2
Displacement of a point.

Equations (1.1) and (1.2) are illustrated in Figures 1.1(a) and (b) where n is some constant.

Consider a point A, with Cartesian coordinates x, y, z, which can be thought of as a position or radius *vector* from the origin as in Figure 1.2. This vector can be written

$$\begin{bmatrix} x \\ y \\ z \end{bmatrix} = \begin{bmatrix} x_1 \\ x_2 \\ x_3 \end{bmatrix} = \bar{r} = x_i$$

matrix representation of a vector	vector	tensor or indicial notation i, j, k take on cyclic values such as x, y, z, or 1, 2, 3, or a, b, c, etc.

On linear transformation, this point moves to a new position A' with new coordinates x', y', z'. This transformation can be written in various forms as

$$\text{a)} \quad \begin{aligned} x' &= ax + by + cz \\ y' &= dx + ey + fz \\ z' &= gx + hy + qz \end{aligned} \qquad \text{explicit}$$

or

$$\text{b)} \quad \begin{bmatrix} x' \\ y' \\ z' \end{bmatrix} = \begin{bmatrix} a & b & c \\ d & e & f \\ g & h & q \end{bmatrix} \begin{bmatrix} x \\ y \\ z \end{bmatrix} \qquad \text{matrix form} \tag{1.3}$$

or

$$\text{c)} \quad x_i = a_{ij} x_{ij} \qquad \text{tensor or indicial notation}$$

or

$$\text{d)} \quad \bar{\rho} = [a]\bar{r} \qquad \text{vector notation}$$

Of these, *a* and *b* are the most cumbersome and time consuming to use but they are graphic. They emphasize that we are simply dealing with simultaneous equations. Moreover, they are global and, *if linear,* the coefficients $[a]$ which are the components of a real, physical transformation tensor must be constant and not themselves functions of x, y, z since if any of the coefficients involve x, y, or z, the equations would have cross products or powers of x, y, z and be nonlinear. Tensor or indicial notation is the most efficient, but only after years of writing out the implied equations does one get a physical feel for this powerful shorthand.

Instead of transforming coordinates of a point (position vectors), it is usually more fruitful in mechanics to talk about how much the coordinates change (i.e., the movement of the tip of the position vector). In fact, it will turn out that if we can find the movement or displacement of all the points of a body, we can easily determine all the strains and usually the stresses causing them. As seen in Figure 1.2, the displacement vector $\bar{\Delta}$ of point A moving to A' has components in the x, y, z directions, which are usually called u, v, w in engineering. That is,

$$\bar{\Delta} = \begin{bmatrix} u \\ v \\ w \end{bmatrix} = \begin{bmatrix} u_x \\ u_y \\ u_z \end{bmatrix} = u_i$$

It is obtained by simply subtracting the original coordinates from the new ones. Therefore,

$$\begin{matrix} \bar{\Delta} & = & \bar{\rho} & - & \bar{r} & = & [\delta] & \bar{r} \\ \begin{bmatrix} u \\ v \\ w \end{bmatrix} & = & \begin{bmatrix} x' \\ y' \\ z' \end{bmatrix} & - & \begin{bmatrix} x \\ y \\ z \end{bmatrix} & = & \begin{bmatrix} (a-1) & b & c \\ d & (e-1) & f \\ g & h & (q-1) \end{bmatrix} & \begin{bmatrix} x \\ y \\ z \end{bmatrix} \end{matrix} \tag{1.4}$$

where $[\delta] = [a] - \begin{bmatrix} 1 \\ 1 \\ 1 \end{bmatrix}$ is the linear displacement tensor associated with the linear coordinate transformation $[a]$.

1.4 Symmetric and Asymmetric Components

Any tensor can be resolved into symmetric and asymmetric components where symmetry or asymmetry is with respect to the diagonal. That is,

$$\begin{bmatrix} a & b & c \\ d & e & f \\ g & h & q \end{bmatrix} = \underbrace{\begin{bmatrix} a & \frac{b+d}{2} & \frac{c+g}{2} \\ \frac{d+b}{2} & e & \frac{h+f}{2} \\ \frac{g+c}{2} & \frac{f+h}{2} & q \end{bmatrix}}_{\text{symmetric}} + \underbrace{\begin{bmatrix} 0 & \frac{b-d}{2} & \frac{c-g}{2} \\ \frac{d-b}{2} & 0 & \frac{f-h}{2} \\ \frac{g-c}{2} & \frac{h-f}{2} & 0 \end{bmatrix}}_{\text{asymmetric}} \tag{1.5}$$

each of which have quite a different physical effect.

1.4.1 Asymmetric Transformation

Consider first the displacement due to an asymmetric tensor such as:

$$\begin{bmatrix} u \\ v \\ w \end{bmatrix} = \begin{bmatrix} 0 & -a & b \\ a & 0 & -c \\ -b & c & 0 \end{bmatrix} + \begin{bmatrix} x \\ y \\ z \end{bmatrix} \tag{1.5a}$$

It can be shown to produce a rotation $\Omega = \tan^{-1}\sqrt{a^2 + b^2 + c^2}$ around an axis whose direction ratios are a, b, and c plus a dilation (a change in size, but not shape). To illustrate, consider the unit square in the x, y plane with $b = c = 0$

$$\begin{bmatrix} u \\ v \\ w \end{bmatrix} = \begin{bmatrix} 0 & -a & 0 \\ a & 0 & 0 \\ 0 & 0 & 0 \end{bmatrix} \begin{bmatrix} x \\ y \\ z \end{bmatrix}$$

As shown in Figure 1.3, all points rotate in the xy plane an angle $\omega = \tan^{-1} a$ around the z axis positive in that the rotation vector is in the positive z direction by the right-hand screw rule. Thus the rotation is independent of orientation of the axes in the xy plane and therefore "invariant." That is $\omega_{xy} = \omega_{\alpha\beta}$ and the symbol ω_z for the rotation vector is sensible although ω_{xy} is more common in the

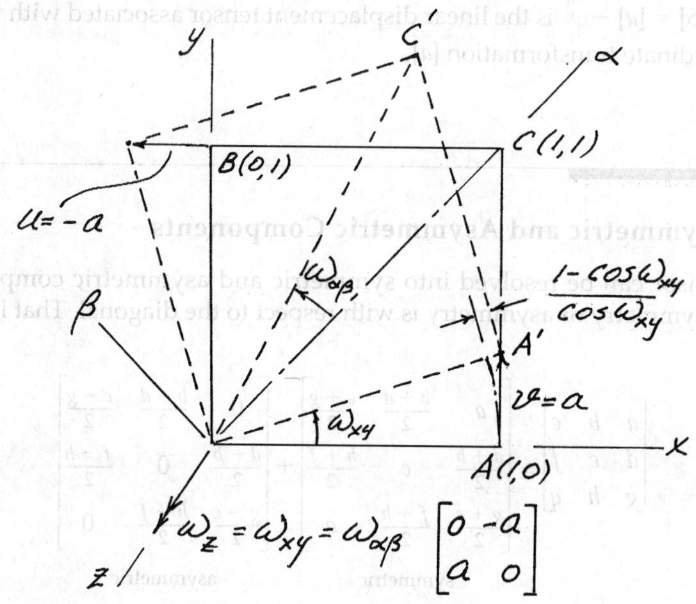

FIGURE 1.3
Asymmetric transformation in 2D.

literature. For "small" rotations, $\omega_z = a$ and the isotropic dilation $(-1 + 1/\cos a)$ is negligible.

Similarly, b and c represent rotation around the y and x axes, respectively. Therefore, the asymmetric displacement tensor can be rewritten

$$\overline{\Omega} = \omega_{ij} = \begin{bmatrix} 0 & -\omega_{xy} & \omega_{xz} \\ \omega_{xy} & 0 & -\omega_{zy} \\ -\omega_{xz} & \omega_{yz} & 0 \end{bmatrix} \quad \text{or} \quad \begin{bmatrix} 0 & -\omega_z & \omega_y \\ \omega_z & 0 & -\omega_x \\ -\omega_y & \omega_x & 0 \end{bmatrix} \tag{1.6}$$

The rotation is not a tensor of second order, but a vector $\overline{\Omega} = \omega_x i + \omega_y j + \omega_z k$ made up of three invariant scalar magnitudes in the subscripted directions.

1.4.2 Symmetric Transformation

The symmetric component of a linear displacement tensor can similarly be understood by simple physical examination of the individual elements. Consider first the diagonal terms a, b, and c in the symmetric tensor:

$$\begin{bmatrix} a & d & e \\ d & b & f \\ e & f & c \end{bmatrix} = \begin{bmatrix} a & & \\ & b & \\ & & c \end{bmatrix} + \begin{bmatrix} 0 & d & e \\ d & 0 & f \\ e & f & 0 \end{bmatrix} \tag{1.7}$$

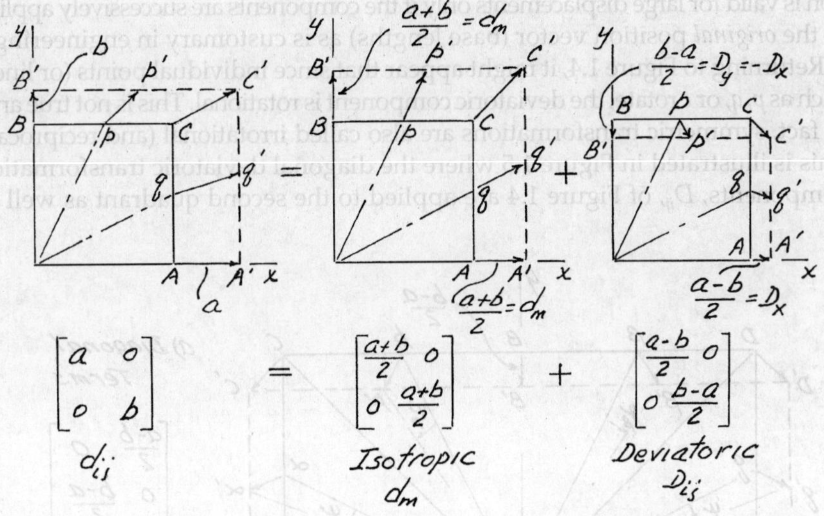

FIGURE 1.4
Diagonalized 2D symmetric tensor "d_{ij}".

Clearly they produce expansion (contraction if negative) in the *xyz* directions proportional to the distance from the origin. The 2D case is shown in Figure 1.4, which also illustrates how the effect of the diagonal terms can be further decomposed into two components that are distinctly different physically. In 3D:

$$\begin{bmatrix} u \\ v \\ w \end{bmatrix} = \underset{d_{ij}}{\begin{bmatrix} a & 0 & 0 \\ 0 & b & 0 \\ 0 & 0 & c \end{bmatrix}} \begin{bmatrix} x \\ y \\ z \end{bmatrix}_o = \underset{\text{isotropic } d_m}{\begin{bmatrix} d_m & & \\ & d_m & \\ & & d_m \end{bmatrix}} \begin{bmatrix} x \\ y \\ z \end{bmatrix}_o + \underset{\text{deviatoric } [D_{ij}]}{\begin{bmatrix} D_{xx} & & \\ & D_{yy} & \\ & & D_{zz} \end{bmatrix}} \begin{bmatrix} x \\ y \\ z \end{bmatrix}_o$$

where the isotropic component producing pure volume change and no distortion is actually a scalar (tensor of order zero) in which $d_m = \frac{a+b+c}{3}$. The so-called deviatoric component terms:

$$D_{xx} = \frac{2a-b-c}{3}; \qquad D_{yy} = \frac{2b-c-a}{3}; \qquad D_{zz} = \frac{2c-a-b}{3} \tag{1.8}$$

sum to zero and produce pure distortion and secondary volume change.* For "small" displacements, the volume change, ΔV, is simply $3d_m$ and the deviatoric volume change is negligible. The subscript, *o*, is to emphasize the linear superposi-

* This decomposition of a symmetric tensor into its isotropic (scalar) and deviatoric components is one of the most important basic concepts in solid mechanics. As we shall see, this uncoupling is profoundly physical as well as mathematical and it permeates every aspect of elasticity, plasticity, and rheology (engineering properties of materials). Considering the two effects separately is an idea that has come to full fruition in the 20th century leading to new and deeper insight into theory and practice.

tion is valid for large displacements only if the components are successively applied to the *original* position vector (base lengths) as is customary in engineering.

Returning to Figure 1.4, it might appear that since individual points (or lines) such as p, q, or c rotate, the deviatoric component is rotational. This is not true and, in fact, symmetric transformations are also called irrotational (and reciprocal). This is illustrated in Figure 1.5 where the diagonal deviatoric transformation components, D_{ij}, of Figure 1.4 are applied to the second quadrant as well as

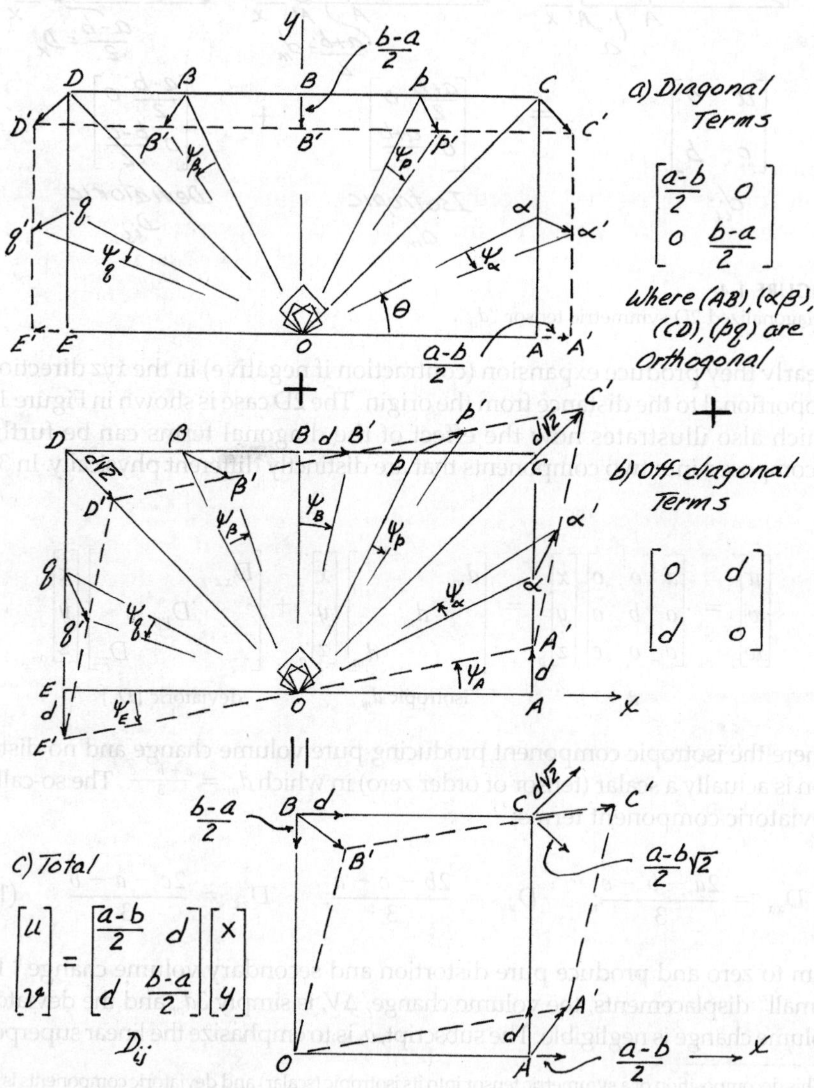

FIGURE 1.5
Deviatoric transformation D_{ij}.

the first (plotted at twice the scale for clarity). The lower two quadrants would be similar.

Thus, line rotations are compensating and there is no net rotation of any pair of orthogonal directions (such as xy, pq, $\alpha\beta$, or CD) or for any reflected pair of lines. *Any* square element, since it is bounded by orthogonal lines, does not rotate either and, for small displacements, its volume will not change. It will undergo pure distortion (change in shape). Certain lines (directions) do not rotate in themselves (in this case, xy) and are termed principal directions while those 45° from them (OC and OD), undergo the maximum compensating angle change.* The angle change itself (in radians) is called the shear and is considered positive if the 90° angle at the origin *decreases*.

Now consider the off-diagonal symmetric terms. The 2D case is shown in Figure 1.5b. Again there is distortion of the shape and, therefore, the off-diagonal symmetric terms are also deviatoric (shear). In this case the lines that do not rotate (principal) are the C–D orthogonal pair and maximum rotation is in the x–y orientation. Thus, again, the maximum and minimum are 45° apart, just as they were for the diagonal deviatoric terms. In fact, it is easy to show that the diagonal and the off-diagonal terms produce the identical physical effect 45° out of phase if $\frac{b-a}{2} = d$. With either set of deviatoric terms, diagonal or off-diagonal, the element expands in one direction while it contracts an equal amount at right angles. If the element boundaries are not in the principal directions, they rotate in compensating fashion and square shapes become rhomboid. The combined effects on an element of the total deviatoric component is shown in Figure 1.5c. Any isotropic component would simply expand or contract the element.

Extension to 3D is straightforward conceptually, but difficult to draw. The components of the general 3D symmetric transformation will be reviewed in terms of strain and fully defined in terms of stress in the next chapter.

1.5 Principal or Eigenvalue Representation

Under a general linear transformation, all points in a body are displaced such that while straight lines remain straight, most rotate while they also change in length. However, as we have seen, certain directions (lines) "transform upon themselves" (or parallel to themselves if not from the origin) without rotation and the general tensor must reduce to the special, simple scalar form:

$$u_i = \begin{bmatrix} u \\ v \\ w \end{bmatrix} = \begin{bmatrix} \lambda & & \\ & \lambda & \\ & & \lambda \end{bmatrix} \begin{bmatrix} x \\ y \\ z \end{bmatrix}$$

* These important observations are best described by Mohr's Circle, presented in the next section.

or $\bar{\Delta} = \lambda\bar{r}$. Thus, in the special direction of $\bar{r}$ (the eigenvector):

$$\begin{bmatrix} \lambda & & \\ & \lambda & \\ & & \lambda \end{bmatrix}\begin{bmatrix} x_1 \\ y_1 \\ z_1 \end{bmatrix} \underset{\text{must}}{=} \begin{bmatrix} \delta_{xx} & \delta_{yx} & \delta_{zx} \\ \delta_{xy} & \delta_{yy} & \delta_{zy} \\ \delta_{xz} & \delta_{yz} & \delta_{zz} \end{bmatrix}\begin{bmatrix} x_1 \\ y_1 \\ z_1 \end{bmatrix}$$

or

$$\begin{aligned} (\delta_{xx} - \lambda)x_1 + \delta_{yx}y_1 + \delta_{zx}z_1 &= 0 \\ \delta_{xy}x_1 + (\delta_{yy} - \lambda)y_1 + \delta_{zy}z_1 &= 0 \\ \delta_{xz}x_1 + \delta_{yz}y_1 + (\delta_{zz} - \lambda)z_1 &= 0 \end{aligned} \tag{1.9}$$

For a nontrivial solution to Equation (1.9), the determinant of the coefficients must equal zero. Expanding this determinant gives *"the characteristic equation"**

$$\begin{aligned} &\lambda^3 - (\delta_{xx} + \delta_{yy} + \delta_{zz})\lambda^2 + (\delta_{xx}\delta_{yy} + \delta_{yy}\delta_{zz} + \delta_{zz}\delta_{xx} - \delta_{xy}\delta_{yx} \\ &- \delta_{yz}\delta_{zy} - \delta_{zx}\delta_{xz})\lambda - (\delta_{xx}\delta_{yy}\delta_{zz} + \delta_{xy}\delta_{yz}\delta_{zx} + \delta_{xz}\delta_{yx}\delta_{zy} - \delta_{xy}\delta_{yx}\delta_{zz} \\ &- \delta_{xx}\delta_{yz}\delta_{zy} - \delta_{xz}\delta_{yy}\delta_{zx}) = 0 \end{aligned} \tag{1.10}$$

The three roots (real or imaginary—some of which may be equal) are called "eigenvalues" each with its own eigenvector or characteristic direction. Once the roots are determined, they each can be substituted back into Equation (1.9) to find the corresponding direction $r_i = x_i i + y_i j + z_i k$.

There are a number of remarkable aspects to this characteristic equation. Perhaps the most important is the invariant nature of the coefficients. The three roots (λ_1, λ_2, λ_3) diagonalize the general transformation for one special orientation of axes. However, the choice of initial coordinate system is completely arbitrary. Thus the coefficients must be *invariant*. That is:

$$\lambda_1 + \lambda_2 + \lambda_3 = \delta_{xx} + \delta_{yy} + \delta_{zz} = (I_\delta)_1 \tag{1.11}$$

$$\begin{aligned} \lambda_1\lambda_2 + \lambda_2\lambda_3 + \lambda_3\lambda_1 &= \delta_{xx}\delta_{yy} + \delta_{yy}\delta_{zz} + \delta_{zz}\delta_{xx} \\ &- \delta_{xy}\delta_{yx} - \delta_{yz}\delta_{zy} - \delta_{zx}\delta_{xz} = (I_\delta)_2 \end{aligned} \tag{1.12}$$

$$\begin{aligned} \lambda_1\lambda_2\lambda_3 = \ &\delta_{xx}\delta_{yy}\delta_{zz} + \delta_{xy}\delta_{yz}\delta_{zx} + \delta_{xz}\delta_{yz}\delta_{zy} - \delta_{xy}\delta_{yx}\delta_{zz} \\ &- \delta_{xx}\delta_{yz}\delta_{zy} - \delta_{xz}\delta_{yy}\delta_{zx} = (I_\delta)_3 \end{aligned} \tag{1.13}$$

* This "eigenvalue problem" appears throughout physics and engineering, which is not surprising given the prevalence of tensors in the mathematical description of the universe. Applied mathematicians enjoy discussing it at great length, often without really appreciating the profound physical implications of the invariant coefficients of Equation (1.10).

The overriding importance of these invariant directions will become apparent when we discuss the strain and stress tensors.

For the general displacement transformation tensor, δ_{ij}, with a rotation component, the invariant directions given by the eigenvectors $\bar{r}_1, \bar{r}_2, \bar{r}_3$ for each root $\lambda_1, \lambda_2, \lambda_3$ need not be orthogonal. However, for the symmetric, d_{ij}, component (nonrotational), the roots are real and the three invariant directions are orthogonal. They are called *principal* (principal values and principal directions perpendicular to principal planes). The search for this principal representation, which diagonalizes a symmetric tensor to the principal values and reduces the invariants to their simplest forms, is crucial to a physical understanding of stress, strain, or any other second-order tensor.

To summarize this brief introduction of linear transformations, we have seen that:

a. Under linear transformation, geometric shapes retain their basic identity: straight lines remain straight, parallel lines remain parallel, ellipses stay elliptic, and so forth.
b. However, such linear transformations may involve:
   (i) Volume change (isotropic effect), d_m,
   (ii) Distortion due to compensating angle change (shear or deviatoric effect), D_{ij}, and
   (iii) Rotation, ω_{ij}.
c. These three effects can be seen separately if the general tensor δ_{ij} is decomposed into component parts: $\delta_{ij} = d_m + D_{ij} + \omega_{ij}$ as illustrated in Example 1.1.
d. The invariant "tensor" quality of a linear transformation is expressed in the coefficients of the characteristic equation, which remain constant for any coordinate system even while the nine individual elements change. Since nature knows no man-made coordinate system, we should expect the fundamental "laws" or phenomena of mechanics (which deals with tensors of various orders) to involve these invariant and not the individual coordinate-dependent elements.
e. The roots of this characteristic equation with their orientation in space (i.e., eigenvalues and eigenvectors, which reduce the tensor to its diagonal form) are called "principal." If the tensor is symmetric (i.e., no rotation), the principal directions are orthogonal.

Example 1.1

Using the general definition [Equation (1.4)], determine the linear displacement tensor that represents the transformation of the triangular shape ABC into $AB'C'$ as shown below and sketch the components.

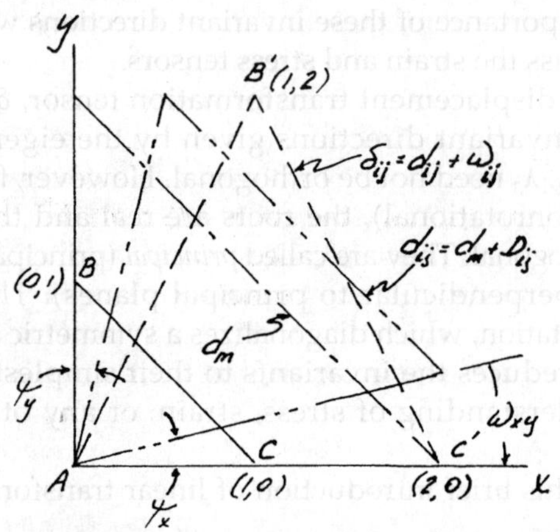

a) Determine δ_{ij} directly

i) pt. C

$$\begin{bmatrix}1\\0\end{bmatrix} = \begin{bmatrix}\delta_{xx} & \delta_{yx}\\ \delta_{xy} & \delta_{yy}\end{bmatrix}\begin{bmatrix}1\\0\end{bmatrix}$$

$$\bar{\Delta} = \delta_{ij}\,\bar{r}_0$$

$$\therefore \delta_{xx} = 1,\ \delta_{xy} = 0$$

ii) pt. B

$$\begin{bmatrix}1\\1\end{bmatrix} = \begin{bmatrix}1 & \delta_{yx}\\ 0 & \delta_{yy}\end{bmatrix}\begin{bmatrix}0\\1\end{bmatrix}$$

$$\therefore \delta_{yx} = 1,\ \delta_{yy} = 1$$

b) Decompose into components

$$\delta_{ij} = \begin{bmatrix}1 & 1\\ 0 & 1\end{bmatrix} = \underbrace{\begin{bmatrix}1 & \frac{1}{2}\\ \frac{1}{2} & 1\end{bmatrix}}_{d_{ij}} + \underbrace{\begin{bmatrix}0 & \frac{1}{2}\\ -\frac{1}{2} & 0\end{bmatrix}}_{\omega_{ij}}$$

$$d_{ij} = \underbrace{\begin{bmatrix}1 & 0\\ 0 & 1\end{bmatrix}}_{d_m} + \underbrace{\begin{bmatrix}0 & \frac{1}{2}\\ \frac{1}{2} & 0\end{bmatrix}}_{D_{ij}}$$

c) Plotted above

$$\psi_x + \psi_y = \tan^{-1}\left(\frac{1}{2}\right) = 26.6°$$

α) Effect of d_m

$$\begin{bmatrix} u \\ v \end{bmatrix} = \begin{bmatrix} 1 & 0 \\ 0 & 1 \end{bmatrix} \begin{bmatrix} x_0 \\ y_0 \end{bmatrix} \therefore \begin{array}{l} \bar{\Delta}_C^\alpha = 1i \\ \bar{\Delta}_B^\alpha = 1j \end{array}$$

∴ All pts move radially outward

β) Effect of D_{ij}

$$\begin{bmatrix} u \\ v \end{bmatrix} = \begin{bmatrix} 0 & \frac{1}{2} \\ \frac{1}{2} & 0 \end{bmatrix} \begin{bmatrix} x_0 \\ y_0 \end{bmatrix} \therefore \begin{array}{l} \bar{\Delta}_C^\beta = \frac{1}{2}j \\ \bar{\Delta}_B^\beta = \frac{1}{2}i \end{array}$$

γ) Effect of $\omega_{ij} = \omega_{xy} = \omega_z$

$$\begin{bmatrix} u \\ v \end{bmatrix} = \begin{bmatrix} 0 & \frac{1}{2} \\ -\frac{1}{2} & 0 \end{bmatrix} \begin{bmatrix} x_0 \\ y_0 \end{bmatrix} \therefore \begin{array}{l} \bar{\Delta}_C^\gamma = -\frac{1}{2}j \\ \bar{\Delta}_B^\gamma = \frac{1}{2}i \end{array}$$

1.6 Field Theory

At the turn of the century, although few realized it, the ingredients were in place for a flowering of the natural sciences with the development of field theory. This was certainly the case in the study of the mechanics of fluids and solids, which led the way for the new physics of electricity, magnetism, and the propagation of light.

An uncharitable observer of the solid-mechanics scene in 1800 might, with the benefit of hindsight, characterize the state of knowledge then as a jumble of incorrect solutions for collapse loads, an incomplete theory of bending, an unclear definition for Young's modulus, a strange discussion of the frictional strength of brittle materials, a semigraphical solution for arches, a theory for the longitudinal vibration of bars that was erroneous when extended to plates or shells, and the wrong equation for torsion. However, this assessment would be wrong. While no general theory was developed, 120 years of research from Galileo to Coulomb had developed the basic mental tools of the scientific method (hypothesis, deduction, and verification) and compiled the necessary ingredients to formulate the modern field theories for strain, stress, and displacement.

The differentiation between shear and normal displacement and the generalization of equilibrium at a point to the cross-section of a beam were both major steps in the logic of solid mechanics as, of course, was Young's insight in relating strain and stress linearly in tension or compression. Newton first proposed bodies made up of small points or "molecules" held together by self-equilibriating forces and the generalization of calculus to two and three dimensions allowed the mathematics of finite linear transformation to be reduced to an arbitrarily small size to describe deformation at a differential scale. Thus the stage was set. The physical concepts and the mathematical tools were available to produce a general field theory of elasticity. Historical events conspired to produce it in France.

The French Revolution destroyed the old order and replaced it with republican chaos. The great number of persons separated at the neck by Dr. Guillotine's invention is symbolic of the beheading of the Royal Society as the leader of an elite class of intellectuals supported by the King's treasury and beholden to imperial dictate.

The new school, L'Ecole Polytechnique founded in 1794, was unlike any seen before. Based on equilitarian principles, entrance was by competitive examination so that boys without privileged birth could be admitted. Moreover, the curriculum was entirely different. Perhaps because there were so may unemployed scientists and mathematicians available, Gospareed Monge (1746–1818), who organized the new school, was able to select a truly remarkable faculty including among others, Lagrange, Fourier, and Poisson. Together they agreed on a new concept of engineering education.They would, for the first two years, concentrate on instruction in the basic sciences of mechanics, physics, and chemistry, all presented with the fundamental language of mathematics as the unifying theme. Only in the third year, once the fundamentals that apply to all branches of engineering were mastered, would the specific training in applications be covered.*

Thus the modern "institute of technology" was born and the consequences were immediate and profound. The basic field theory of mathematical elasticity would appear within 25 years, developed by Navier and Cauchy not only as an intellectual construction but for application to the fundamental

* In fact, engineering at L'Ecole Polytechnique was soon eliminated and students went to "graduate work" at one of the specialized engineering schools such as L'Ecole des Ponts et Chaussees, the military academy, L'Ecole de Marine, and so forth.

problems left by their predecessors.* The first generation of graduates of the Ecole Polytechique such as Navier and Cauchy, became professors and educated many great engineers who would come to dominate structural design in the later half of the 19th century.**

The French idea of amalgamating the fundamental concepts of mathematics and mechanics as expressed by field theory for engineering applications, is the theme of this text. Today, two centuries of history have proven this concept not only as an educational approach, but as a unifying principle in thinking about solid mechanics.*** In bygone days, the term "Rational Mechanics" was popular to differentiate this perspective of visualizing fields graphically with mathematics and experiments so as to understand how structures work rather than just solving specific boundary value problems. The phrase "Rational Mechanics" is now old-fashioned, but historically correct for the attitude adopted in succeeding chapters of combining elastic and plastic behavior as a continuous visual progression to yield and then collapse.

1.7 Problems and Questions

P1.1 Find the surface that transforms into a sphere of unit radius from Equation (1.3). Sketch the shape and discuss the three possible conditions regarding principal directions and principal planes.

* Both Navier and Cauchy, after graduating from L'Ecole Polytechnique, went on to L'Ecole des Ponts et Chaussees and then into the practice of civil engineering where bridges, channels, and waterfront structures were involved. Cauchy quickly turned to the academic life in 1814, but Navier did not join the faculty of L'Ecole Polytechnique until 1830 and always did consulting work, mostly on bridges, until he died.

** The greatest French structural engineer, Gustave Eiffel (1832–1923), actually failed the entrance examination for L'Ecole Polytechnique and graduated from the private L'Ecole Centrale des Arts et Manufactures in 1855 with a chemical engineering degree. However, he always did extensive calculations on each of his structures and fully appreciated the basic idea of combining mathematics with science and aesthetics. Eugene Freyssinet (1879–1962), recognized as the pioneer of prestressed concrete as well as a designer of great bridges, also failed the entrance examination to L'Ecole Polytechnique in 1898. But he persevered being admitted the next year ranking only 161st among the applicants. He graduated 19th in his class and went on to L'Ecole des Ponts et Chaussees where he conceived the idea of prestressing. The obvious moral is to never give up on your most cherished goals.

*** The French approach was not accepted quickly or easily. The great English engineers in iron, such as Teleford and then Stephenson and Brunel, had no use for mathematics or analysis much beyond simple statics. They were products of a class culture with a Royal Society for scientists educated privately and then admitted to Oxford and Cambridge without competitive examination. They considered building bridges, railroads, steamships, and machines a job for workmen. The pioneering English engineers of the first half of the 19th century were primarily entrepreneurs who tacitly agreed with the assessment and conformed to the stereotype. While the industrial revolution was started by the British, they could not maintain the initial technical leadership when France and then the United States began to compete in the second half of the 19th century.

P1.2 Show that symmetric transformations are the only ones to possess the property of reciprocity. (Hint: This can be done by considering any linear transformation that transforms any two vectors $\overline{OP_1}$ $(x_1 y_1 z_1)$ and $\overline{OP_2}(x_2 y_2 z_2)$ into $\overline{OP_1'}$ and $\overline{OP_2'}$. For reciprocity, the dot-product relationship: $\overline{OP_1'} \cdot \overline{OP_2'} = \overline{OP_2'} \cdot \overline{OP_1}$ must hold, which imposes the symmetric condition on the coefficients of the transformation.)

P1.3 See Figure P1.3. Show that the asymmetric transformation [Equation (1.5a)] represents:

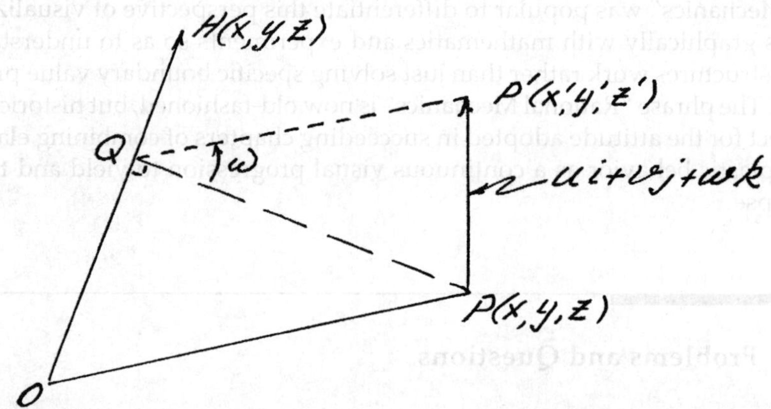

FIGURE P1.3

1. A rotation $\omega = \tan^{-1}\sqrt{a^2 + b^2 + c^2}$ around an axis $\overline{OH}$, and
2. A cylindrical dilation equal to $OP'-QP$ as shown. Hint: Let the coordinates of H be c, b, and a. Then show that, therefore, $\overline{OH}$ is invariant and that therefore, all points on it are fixed. Sum the scalar products $ux + vy + wz$ and $ux' + vy' + wz'$ to show that vector $\overline{PP'}$ is $\perp$ to $\overline{OP}$ and $\overline{OH}$, and thus the plane POH. Finally, show that unit dilation (expansion or contraction)

$$\frac{QP'-QP}{QP} = \epsilon = \sqrt{1 + a^2 + b^2 + c^2} - 1 = \frac{1}{\cos\omega} - 1$$

and use this to prove that, for small ω, it is a second-order effect.

P1.4 Compare the unit dilation (volume change) associated with the off-diagonal terms in a *symmetric* displacement transformation [Equation(1.7)] to that for the asymmetric component (rotation) in P1.3.

P1.5 What is the unit volume due to the isotropic component of the linear displacement transformation, and to what formula does it evolve for small displacements (i.e., small in comparison to the "base-lengths" x, y, z).

P1.6 Derive an expression for the rotation ψ as a function of θ for both the diagonal and off-diagonal terms of the deviatoric component of the symmetric two-dimensional linear displacement tensor in Figures 1.3–1.7. Then:

a. Show that $\psi_\theta = -\psi_{\theta+\pi/2}$ in either case,

b. Plot the two in phase and $\frac{\pi}{4}$ out of phase. What is going on?

c. Write (and plot) the expression for using double angle identities for $\cos 2\theta$, $\sin 2\theta$ and discuss, emphasizing maximums and minimums.

d. Show that you have derived Mohr's Circle (ahead of time) for coordinate transformation of a symmetric 2D linear displacement tensor.

P1.7 Discuss what happens to parabolas, ellipses, hyperbolas, or higher-order shapes under linear transformation. Present a few *simple* examples (graphically) to illustrate.

P1.8 For the (a) unit cube, (b) unit circle, and (c) unit square transformed to the solid position as shown in Figure P1.8:

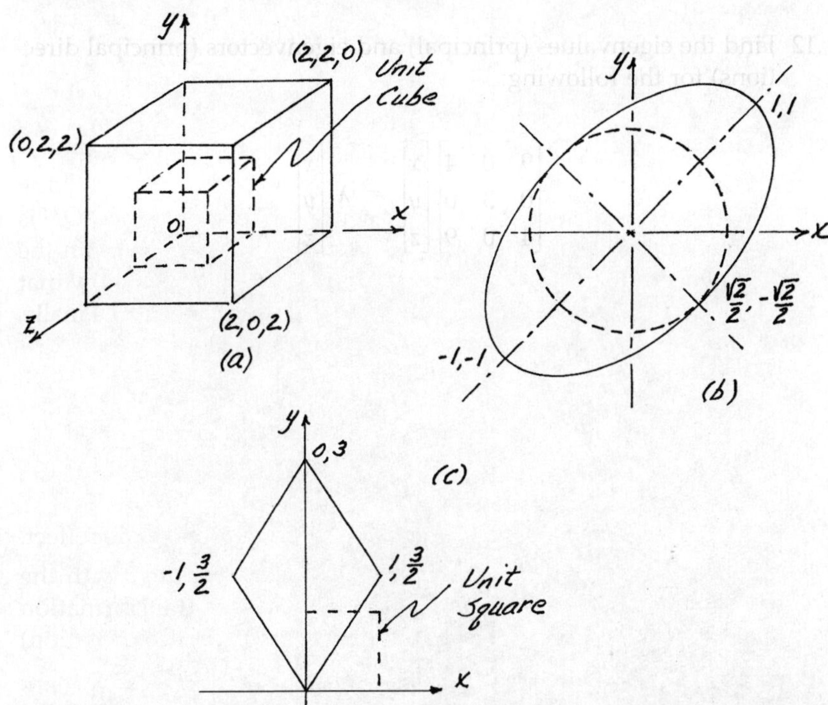

FIGURE P1.8

i. Derive the liner displacement tensor δ_{ij}
ii. Decompose it, if appropriate, into symmetric ($d_m + D_{ij}$) and asymmetric, ω_{xy} components,
iii. Show with a careful sketch, the effect of each as they are superimposed to give the final shape with $\delta_{ij} = d_m + D_{ij} + \omega_{ij}$.

P1.9 Make up a problem to illustrate one or more important concepts in Chapter 2 (and solve it). Elegance and simplicity are of paramount importance. A truly original problem, well posed and presented, is rare and is rewarded.

P1.10 Reconsider the linear displacement transformation in P1.8b and show that your previous answer is not unique. (Hint: Assume point (1, 0) goes to (1, 1) and point (0, 1) goes to $(-\sqrt{2}/2, \sqrt{2}/2)$. By plotting the components of the transformation, reconcile the question.)

P1.11 Find the principal values and their directions for the symmetric linear transformation

$$d_{ij} = \begin{bmatrix} 5 & -1 & 2 \\ -1 & 12 & -1 \\ 2 & -1 & 5 \end{bmatrix}$$

P1.12 Find the eigenvalues (principal) and eigenvectors (principal directions) for the following:

$$\begin{bmatrix} 9 & 0 & 4 \\ 0 & 3 & 0 \\ 4 & 0 & 9 \end{bmatrix} \begin{bmatrix} x \\ y \\ z \end{bmatrix} = \lambda \begin{bmatrix} x \\ y \\ z \end{bmatrix}$$

2

Strain and Stress

2.1 Deformation (Relative Displacement)

Almost all displacement fields induced by boundary loads, support movements, temperature, body forces, or other perturbations to the initial condition are, unfortunately, *nonlinear*; that is: u, v, and w are cross-products or power functions of x, y, z (and perhaps other variables). However, as shown in Figure. 2.1,* the fundamental linear assumption of calculus allows us to directly use the relations of finite linear transformation to depict immediately the relative displacement or deformation du, dv, dw of a differential element dx, dy, dz.

On a differential scale, as long as u, v, and w are continuous, smooth, and small, straight lines remain straight and parallel lines and planes remain parallel. Thus the standard definition of a total derivative:

$$\begin{aligned} du &= \frac{\partial u}{\partial x}dx + \frac{\partial u}{\partial y}dy + \frac{\partial u}{\partial z}dz \\ dv &= \frac{\partial v}{\partial x}dx + \frac{\partial v}{\partial y}dy + \frac{\partial v}{\partial z}dz \\ dw &= \frac{\partial w}{\partial x}dx + \frac{\partial w}{\partial y}dy + \frac{\partial w}{\partial z}dz \end{aligned} \quad \text{or} \quad \begin{bmatrix} du \\ dv \\ dw \end{bmatrix} = \begin{bmatrix} \frac{\partial u}{\partial x} & \frac{\partial u}{\partial y} & \frac{\partial u}{\partial z} \\ \frac{\partial v}{\partial x} & \frac{\partial v}{\partial y} & \frac{\partial v}{\partial z} \\ \frac{\partial w}{\partial x} & \frac{\partial w}{\partial y} & \frac{\partial w}{\partial z} \end{bmatrix} = \begin{bmatrix} dx \\ dy \\ dz \end{bmatrix}$$

$$\overline{d\Delta} = [E_{ij}] \quad \overline{dr} \tag{2.1}$$

is more than a mathematical statement that differential base lengths obey the laws of linear transformation.** The resulting deformation tensor, E_{ij}, also

* This is the "standard blob." It could just as well be a frame, gear, earth dam, shell, or any "structure."

** Displacements due to *rigid body* translation and/or rotation can be added to the displacements due to deformation. Most structures are made stationary by the supports and there are no rigid body displacements. Rigid body mechanics (statics and dynamics) is a special subject as is time-dependent deformation due to vibration or sudden acceleration loads such as stress waves from shock or seismic events.

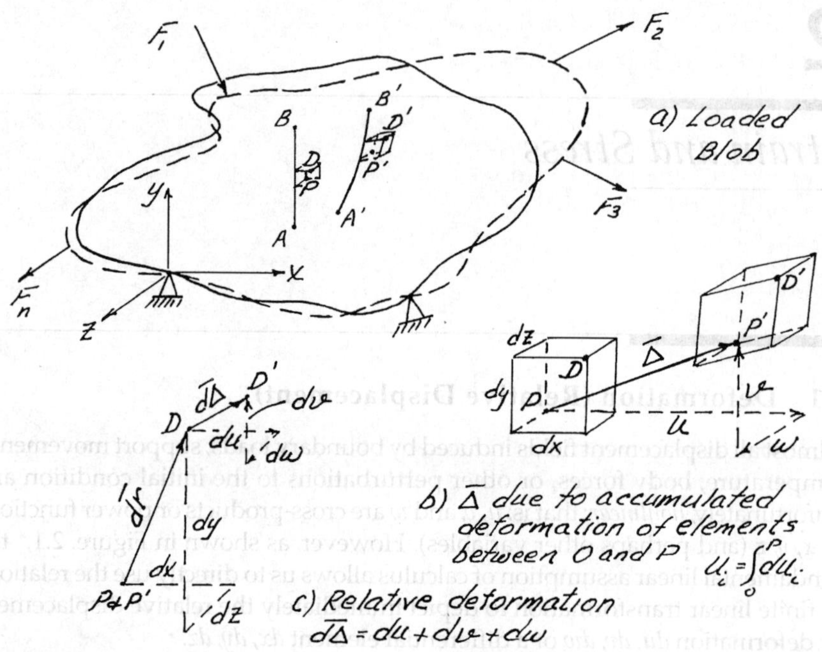

FIGURE 2.1
Nonlinear deformation field, u_i or $\bar{\Delta}$.

called the relative displacement tensor, is directly analogous to the linear displacement tensor, δ_{ij}, of Chapter 1, which transformed finite base-lengths. The elements of E_{ij} (the partial derivatives), although nonlinear functions throughout the field (i.e., the structure), are just numbers when evaluated at any x, y, z. Therefore E_{ij} should be thought of as an average or, in the limit, as "deformation at a point." Displacements u, v, w, due to deformation, are obtained by a line integral of the total derivative from a location where u, v, w have known values; usually a support where one or more are zero. Thus:

$$u = \int_0^P du; \quad v = \int_0^P dv; \quad w = \int_0^P dw \tag{2.2}$$

2.2 The Strain Tensor

As on a finite scale, the deformation tensor can be "dissolved" into its symmetric and asymmetric components:

$$\begin{bmatrix} \frac{\partial u}{\partial x} & \frac{\partial u}{\partial y} & \frac{\partial u}{\partial x} \\ \frac{\partial v}{\partial x} & \frac{\partial v}{\partial y} & \frac{\partial v}{\partial z} \\ \frac{\partial w}{\partial x} & \frac{\partial w}{\partial y} & \frac{\partial w}{\partial z} \end{bmatrix} = \begin{bmatrix} \frac{\partial u}{\partial x} & \frac{1}{2}\left(\frac{\partial u}{\partial y} + \frac{\partial v}{\partial x}\right) & \frac{1}{2}\left(\frac{\partial u}{\partial z} + \frac{\partial w}{\partial x}\right) \\ \frac{1}{2}\left(\frac{\partial v}{\partial x} + \frac{\partial u}{\partial y}\right) & \frac{\partial v}{\partial y} & \frac{1}{2}\left(\frac{\partial v}{\partial z} + \frac{\partial w}{\partial y}\right) \\ \frac{1}{2}\left(\frac{\partial w}{\partial x} + \frac{\partial u}{\partial z}\right) & \frac{1}{2}\left(\frac{\partial w}{\partial y} + \frac{\partial v}{\partial z}\right) & \frac{\partial w}{\partial z} \end{bmatrix}$$

E_{ij} $\qquad$ ϵ_{ij} *(symmetric) Strain Tensor*

$$+ \begin{bmatrix} 0 & \frac{1}{2}\left(\frac{\partial u}{\partial y} + \frac{\partial v}{\partial x}\right) & \frac{1}{2}\left(\frac{\partial u}{\partial z} + \frac{\partial w}{\partial x}\right) \\ \frac{1}{2}\left(\frac{\partial v}{\partial x} + \frac{\partial u}{\partial y}\right) & 0 & \frac{1}{2}\left(\frac{\partial v}{\partial z} + \frac{\partial w}{\partial y}\right) \\ \frac{1}{2}\left(\frac{\partial w}{\partial x} + \frac{\partial u}{\partial z}\right) & \frac{1}{2}\left(\frac{\partial w}{\partial y} + \frac{\partial v}{\partial z}\right) & 0 \end{bmatrix} \tag{2.3}$$

$+\ \omega_{ij}$ *(asymmetric) Rotation Vector*

The individual elements or components of ϵ_{ij} by definition are:*

$$\epsilon_{xx} = \epsilon_x = \frac{\partial u}{\partial x}; \qquad \epsilon_{xy} = \epsilon_{yx} = \frac{\gamma_{xy}}{2} = \frac{1}{2}\left(\frac{\partial u}{\partial y} + \frac{\partial v}{\partial x}\right)$$

$$\epsilon_{yy} = \epsilon_y = \frac{\partial v}{\partial y}; \qquad \epsilon_{yz} = \epsilon_{zy} = \frac{\gamma_{zy}}{2} = \frac{1}{2}\left(\frac{\partial v}{\partial z} + \frac{\partial w}{\partial y}\right) \quad \text{(6 equations, 9 unknowns)}$$

$$\epsilon_{zz} = \epsilon_z = \frac{\partial w}{\partial z}; \qquad \epsilon_{zx} = \epsilon_{xz} = \frac{\gamma_{zx}}{2} = \frac{1}{2}\left(\frac{\partial w}{\partial x} + \frac{\partial u}{\partial z}\right) \tag{2.4}$$

each of which produces a "deformation" of the orthogonal differential baselengths *dx*, *dy*, and *dz*, exactly like its counterpart in the symmetric displacement tensor d_{ij} with respect to finite base lengths x, y, and z when u, v, and w are linear functions. Physically, the diagonal terms are

$$\epsilon_i = \frac{\textit{change in length}}{\textit{original baselength}} \quad \textit{(in the i direction)}.$$

---

* Engineering notation will often be used because it: (i) emphasizes, by using different symbols, the extremely important physical difference between shear (γ) and normal (ϵ) strain behavior and, (ii) the unfortunate original definition (by Navier and Cauchy in 1821) of shear γ as the sum of counteracting angle changes rather than their average is historically significant in that much of the important literature has used it.

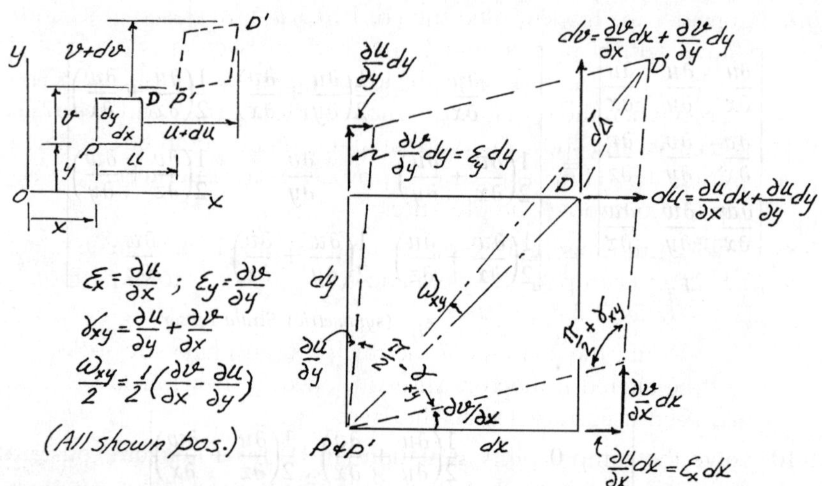

FIGURE 2.2
Deformation of a 2D differential element.

The off-diagonal terms, γ_{ij} = total counteracting angle change between othogonal baselengths (in the i and j directions) and are positive when they increase $\overline{d\Delta}$.

The same symbol is even used for the rotation of the differential element where the vector components of ω_{ij} are defined as:

$$\frac{\omega_{xy}}{2} = \omega_z = \frac{1}{2}\left(\frac{\partial v}{\partial x} - \frac{\partial u}{\partial y}\right)$$
$$\frac{\omega_{yz}}{2} = \omega_x = \frac{1}{2}\left(\frac{\partial w}{\partial y} - \frac{\partial v}{\partial z}\right) \quad (3\ eqns., 3\ unkn's.) \tag{2.5}$$
$$\frac{\omega_{zx}}{2} = \omega_y = \frac{1}{2}\left(\frac{\partial u}{\partial z} - \frac{\partial w}{\partial x}\right)$$

positive by the "right-hand-screw-rule" as in Chapter 1 for finite baselengths. A complete drawing of the deformation of a 3D differential element showing each vector component of strain and rotation adding to give the total derivative is complicated.* The 2D (in-plane) drawing, Figure 2.2, illustrating the basic definitions for ϵ_x, ϵ_y, γ_{xy}, and ω_{xy} is more easily understood.

Generally at this point in the development of the "theory of elasticity," at least in its basic form, these elastic rotations as defined by Equation (2.5) are termed "rigid body" and thereafter dismissed. This is a conceptual mistake and we will not disregard the rotational component of E_{ij}.

* It would be helpful for students, but such figures are seldom shown in texts on mathematics when partial derivatives are defined since "physical feel" is to be discouraged in expanding the mind on the way to n-space. The opposite is the case in engineering where calculus is just a tool. "Physical feel" is the essence of design and developing such figures is a route to expanding the mind for creativity to produce structural art in our 3D world.

Equation (2.4) represents 6 field equations relating the displacement vector field with 3 unknown components to the strain-tensor field with 6 unknown components. With the rotation field we have 3 more equations and unknowns. Since 9 deformation components are defined in terms of 3 displacements, the strains and rotations cannot be taken arbitrarily, but must be related by the so-called compatibility relationships usually given in the form:

$$\frac{\partial^2 \epsilon_x}{\partial y^2} + \frac{\partial^2 \epsilon_y}{\partial x^2} = \frac{\partial^2 \gamma_{xy}}{\partial x \partial y}; \qquad 2\frac{\partial^2 \epsilon_z}{\partial y \partial z} = \frac{\partial}{\partial x}\left(-\frac{\partial \gamma_{yz}}{\partial x} + \frac{\partial \gamma_{xz}}{\partial y} + \frac{\partial \gamma_{xy}}{\partial z}\right)$$

$$\frac{\partial^2 \epsilon_y}{\partial z^2} + \frac{\partial^2 \epsilon_z}{\partial y^2} = \frac{\partial^2 \gamma_{yz}}{\partial y \partial z}; \qquad 2\frac{\partial^2 \epsilon_y}{\partial x \partial z} = \frac{\partial}{\partial y}\left(\frac{\partial \gamma_{yz}}{\partial x} - \frac{\partial \gamma_{xz}}{\partial y} + \frac{\partial \gamma_{xy}}{\partial z}\right) \tag{2.6}$$

$$\frac{\partial^2 \epsilon_z}{\partial x^2} + \frac{\partial^2 \epsilon_x}{\partial z^2} = \frac{\partial^2 \gamma_{xz}}{\partial x \partial z}; \qquad 2\frac{\partial^2 \epsilon_z}{\partial x \partial y} = \frac{\partial}{\partial z}\left(\frac{\partial \gamma_{yz}}{\partial x} + \frac{\partial \gamma_{zx}}{\partial y} - \frac{\partial \gamma_{xy}}{\partial z}\right)$$

These relationships are easily derived.* For example, in the xy plane:

$$\left.\begin{array}{l} \dfrac{\partial^2 \epsilon_x}{\partial y^2} = \dfrac{\partial^3 u}{\partial x \partial y^2} \\[2ex] \dfrac{\partial^2 \epsilon_x}{\partial x^2} = \dfrac{\partial^3 v}{\partial y \partial x^2} \\[2ex] -\dfrac{\partial^2 \gamma_{xy}}{\partial x \partial y} = -\dfrac{\partial^3 u}{\partial x \partial y^2} - \dfrac{\partial^3 v}{\partial y \partial x^2} \end{array}\right\} \quad ADD = \frac{\partial^2 \epsilon_x}{\partial y z} + \frac{\partial^2 \epsilon_y}{\partial x^2} = \frac{\partial^2 \gamma_{xy}}{\partial x \partial y} \tag{2.7}$$

and the first compatibility relationship (for the 2D) case is obtained. Three further compatibility relationships can be found if the rotations are used instead of the shears. For example, in the 2D case (x,y plane) using $\omega_z = \frac{1}{2}\left(\frac{\partial v}{\partial x} - \frac{\partial u}{\partial y}\right)$ and the same procedure used in deriving Equation (2.7):

$$\frac{\partial^2 \epsilon_x}{\partial y^2} - \frac{\partial^2 \epsilon_y}{\partial x^2} = \frac{\partial^2 \omega_z}{\partial x \partial y} \tag{2.8}$$

While these nine compatibility equations are sufficient to ensure that displacement functions u, v, and w will be smooth and continuous, they are never sufficient in themselves to solve for displacements. Thus all structures

---

* Also it might, at this point, be well to reemphasize that our definitions of strain components [Equations (2.4) and (2.5)], as depending *linearly* on the derivatives of displacements *at a point*, relies on the assumption that displacements themselves are small. If they are not (rolled sheet, springs, tall buildings), then *nonlinear* strain definitions such as: $\epsilon_x = \frac{\partial u}{\partial x} + \frac{1}{2}\left[\left(\frac{\partial u^2}{\partial x}\right) + \left(\frac{\partial v}{\partial x}\right)^2 + \left(\frac{\partial w^2}{\partial x}\right)\right]$ *may* be required. In such situations the general field equations, even 2D, become too difficult for a simple solution.

are "geometrically indeterminate" and we must investigate the stress field describing the flow of forces from the boundary to the supports in order to develop additional equations to work with. The tacit "assumption" of Hooke's law validated over years of observation, is that the deformation field (strains and rotations) are caused by stresses (or vice versa) and the two can be related by material properties determined in the laboratory.

2.3 The Stress Tensor

The concept of force coming directly from physical observation of deformation was first quantified in one dimension by Hooke but intuitively recognized by cavemen.* To generalize to 3D and introduce the concept of *internal* force per unit area (stress), requires the idea of equilibrium and the limit process of calculus.

Consider again the standard 3D structural blob in equilibrium from external boundary forces and slice it to expose a plane defined by its normal vector $\bar{n}$ (Figure 2.3a). For equilibrium of the remaining piece, there must be a resultant force and/or couple acting on the cross-section of area A. If they can be calculated from any slice from the six equilibrium equations (in 3D), the structure is statically determinate for internal forces and shear, bending moment, torque, and normal force diagrams can be drawn. However, what the cross-section actually feels is small increments of force $\overline{\Delta F_i}$ at various orientations distributed in some undetermined (and usually difficult-to-determine) way. It is this distribution of force and moment over the area (the stress field) that we hope to discover (stress analysis) and, if we are clever, then change and control (design). As we shall see, this stress distribution is almost never determinate.**

For now let $\bar{n}$ be in the x direction so that the exposed surface is a yz plane. Over a small area $\Delta A_x = \Delta y \Delta z$ surrounding any point 0, there will be a force $\overline{\Delta F}$

* It is an amusing gambit (at a scientist's cocktail party) to argue that there is no such thing as force (or stress) since it cannot be measured directly. We sense it in our muscles, in the lab, and in practice by its effects—the deformation or strains it produces—but there is no independent force meter. Newton and others in the 17th century (such as Descartes with his famous silent tree falling in the forest) agonized over this question for years (and it does have philosophical implications). The practical Hooke simply answered it by calling such essential physical "things," which could not be measured directly, "Primitive Quantities." This neat semantic solution (which is how many philosophical dilemmas are dealt with) is now accepted without qualm or reexamination after 300 years of discovering that many potential functions (e.g., voltage, temp, etc.) are primitive quantities and only their "conjugate pair" can be measured.

** The indeterminate structures we study first as a student are externally so. That is, there are too many unknown reactions to even get started. In this case we use the "extra" geometric conditions on displacements and/or rotations of supports to generate enough extra force equations to solve for the unknown reactions. We then proceed with internal force analysis where we may have to adopt the same strategy all over again (e.g., trusses with extra bars). For stress analysis we use the same strategy still one more time at a differential scale.

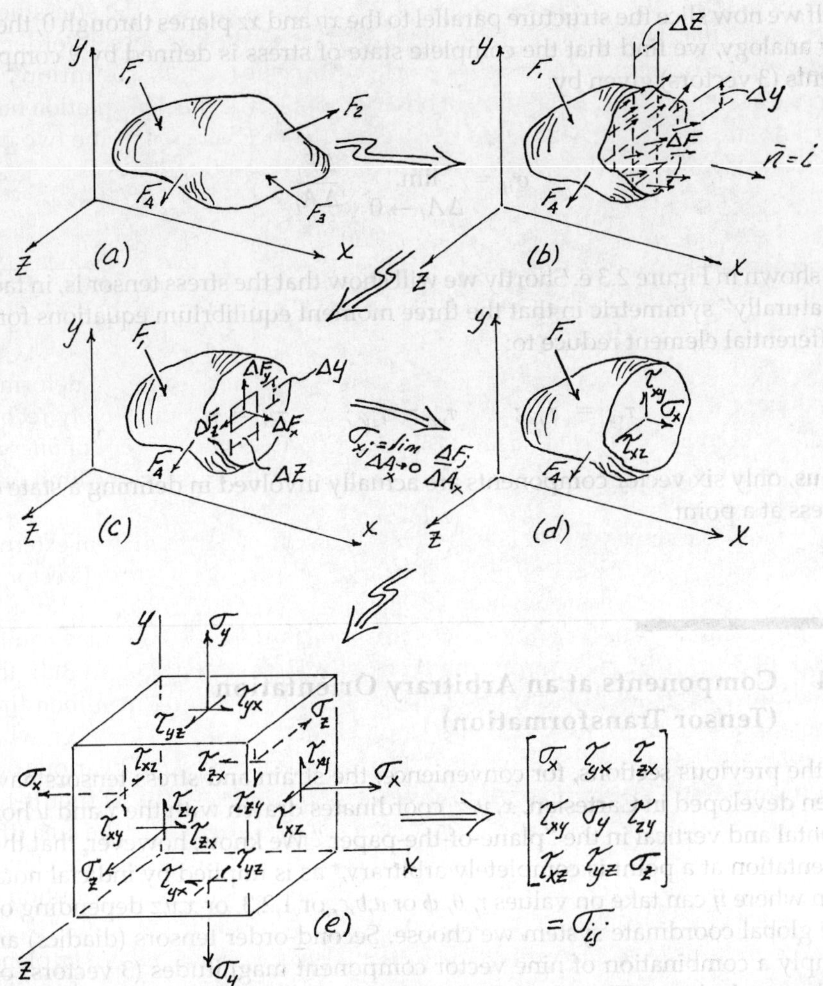

FIGURE 2.3
Stress at a point.

(inclined at an arbitrary angle) with components ΔF_{xx} ΔF_{xy} ΔF_{xz} (Figure 2.3c). The first subscript identifies the plane on which the force acts defined by the direction of its normal, in this case x, and the second the direction of the force component itself. Positive components are those (Figure 2.3e) with both subscripts positive or negative.

The average "intensity of force" or average stress is defined as the ratio $\Delta F/\Delta A$ analogous to "average strain" d_{ij} for *finite* linear displacement. Taking the limit as $\Delta A_x \to 0$, gives x stresses at O (Figure 2.3d). As with strains, the shear components are highlighted by using a different symbol in engineering notation to emphasize the physical difference between normal and in-plane components of stress.

If we now slice the structure parallel to the xy and xz planes through 0, then, by analogy, we find that the complete state of stress is defined by 9 components (3 vectors) given by:

$$\sigma_{ij} = \lim_{\Delta A_i \to 0} \frac{\Delta F_j}{\Delta A_i}$$

as shown in Figure 2.3 e. Shortly we will show that the stress tensor is, in fact, "naturally" symmetric in that the three moment equilibrium equations for a differential element reduce to:

$$\tau_{xy} = \tau_{yx}\ ; \qquad \tau_{yz} = \tau_{zy}\ ; \qquad \tau_{zx} = \tau_{xz}$$

Thus, only six vector components are actually involved in defining a state of stress at a point.

2.4 Components at an Arbitrary Orientation (Tensor Transformation)

In the previous sections, for convenience, the strain and stress tensors have been developed in Cartesian, x, y, z, coordinates drawn with the x and y horizontal and vertical in the "plane-of-the-paper." We know, however, that this orientation at a point is completely arbitrary,* as is implied by indicial notation where ij can take on values r, θ, ϕ or a,b,c, or 1,2,3, or x,y,z depending on the global coordinate system we choose. Second-order tensors (diadics) are simply a combination of nine vector component magnitudes (3 vectors) on orthogonal planes. Thus, as we would expect for vectors, each component transforms into a new value as the coordinate rotates by sine and/or cosine relationships. We will derive them for the stress tensor because most of us find it easiest to visualize forces and use equilibrium arguments. The results are completely general, however, and apply to any symmetric tensor. In particular, they apply to the strain tensor where the normal components ϵ_{ii} replace σ_{ii} and the shear components $\frac{\gamma_{ij}}{2}$ are substituted for τ_{ij}.

Consider a new orientation x',y',z' which would result if the "stress block" were sliced out of the structural blob in a different orientation ($\bar{n}$ corresponding to x' in Figure 2.3e). In other words the new x' coordinate axis is rotated α_x, α_y, α_z, as shown in Figure 2.4. Rather than use the angles directly, let their

* Coordinate systems are an artificial, man-made reference to facilitate the mathematical translation of "nature," which itself knows no such limitation. Basic "truths," or "what nature actually feels," should be expressible independent of a coordinate system and the mathematics should eventually reflect that logic.

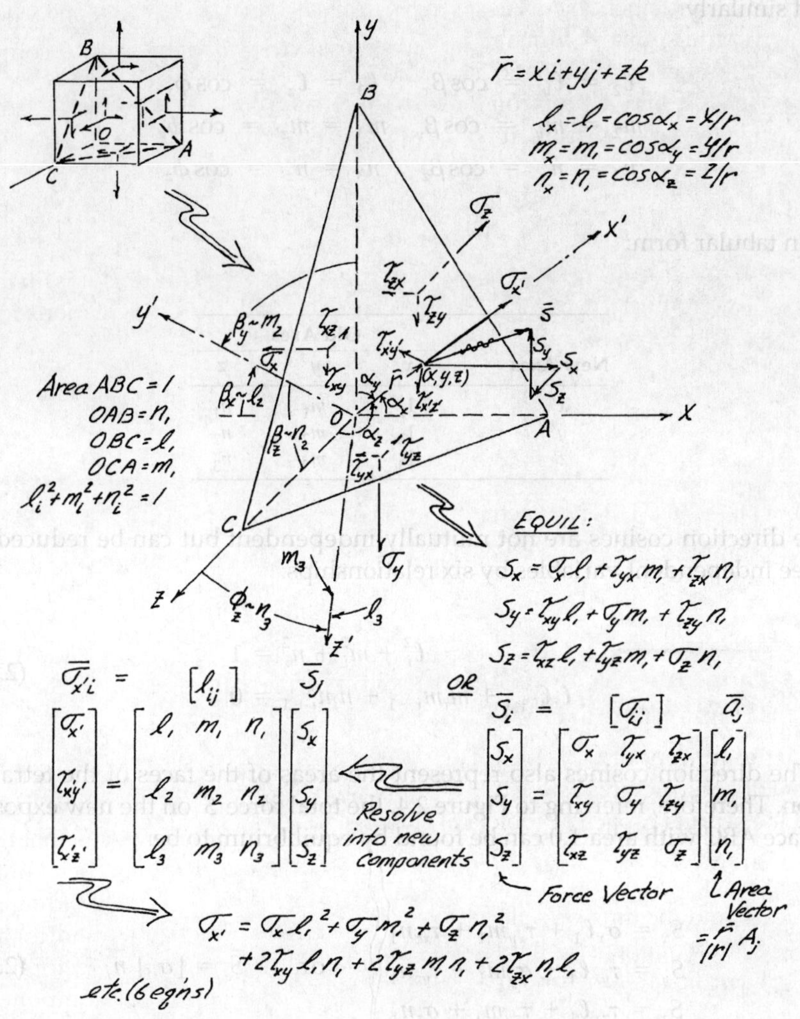

FIGURE 2.4
Stress transformation.

cosines be represented by:

$$\ell_1 = \ell_{x'} = \cos\alpha_x = \frac{x}{r}$$

$$m_1 = m_{x'} = \cos\alpha_y = \frac{y}{r}$$

$$n_1 = n_{x'} = \cos\alpha_z = \frac{z}{r}$$

and similarly

$$\begin{aligned} \ell_2 &= \ell_{y'} = \cos\beta_x & \ell_3 &= \ell_{z'} = \cos\phi_x \\ m_2 &= m_{y'} = \cos\beta_y & m_3 &= m_{z'} = \cos\phi_y \\ n_2 &= n_{y'} = \cos\beta_z & n_3 &= n_{z'} = \cos\phi_z \end{aligned}$$

or in tabular form:

	Old Axes		
New Axes	x	y	z
x'	1_1	m_1	n_1
y'	1_2	m_2	n_2
z'	1_3	m_3	n_3

The direction cosines are not mutually independent but can be reduced to three independent variables by six relationships:

$$\begin{aligned} \ell_i^2 + m_i^2 + n_i^2 &= 1 \\ \ell_i\ell_{i+1} + m_i m_{i+1} + n_i n_{i+1} &= 0 \end{aligned} \tag{2.10}$$

The direction cosines also represent the areas of the faces of the tetrahedron. Therefore, referring to Figure 2.4, the total force $\bar{S}$ on the new exposed x' face ABC with area 1.0 can be found by equilibrium to be:

$$\begin{aligned} S_x &= \sigma_x\ell_1 + \tau_{xy}m_1 + \tau_{xz}n_1 \\ S_y &= \tau_{yx}\ell_1 + \sigma_y m_1 + \tau_{yz}n_1 \qquad \text{or} \qquad \bar{S}_i = [\sigma_{ij}]\,\bar{n}_j \\ S_z &= \tau_{zx}\ell_1 + \tau_{zy}m_1 + \sigma_z n_1 \end{aligned} \tag{2.11}$$

where $\bar{n}_j$ has the direction cosines ℓ_1, m_1, n_1 as components.

As the final step, the force $\bar{S}$ on the x' plane can be "dissolved" back into the new stress components $\sigma_{x'}, \tau_{x'y'}, \tau_{x'z'}$ by again using the direction cosines:

$$\begin{bmatrix} \sigma_{x'} \\ \tau_{x'y'} \\ \tau_{x'z'} \end{bmatrix} = \begin{bmatrix} \ell_1 & m_1 & n_1 \\ \ell_2 & m_2 & n_2 \\ \ell_3 & m_3 & n_3 \end{bmatrix} \begin{bmatrix} S_x \\ S_y \\ S_z \end{bmatrix} \tag{2.12}$$

This same procedure of cutting new tetrahedrons to expose faces with normals in the y' and z' directions will produce the complete set of 3D transformations in Equation (2.13) as follows:

$$
\begin{aligned}
\sigma_{x'} &= \sigma_x l_1^2 + \sigma_y m_1^2 + \sigma_z n_1^2 + 2\tau_{xy} l_1 m_1 + 2\tau_{yz} m_1 n_1 + 2\tau_{zx} n_1 l_1 \\
\sigma_{y'} &= \sigma_x l_2^2 + \sigma_y m_2^2 + \sigma_x n_2^2 + 2\tau_{xy} l_2 m_2 + 2\tau_{yz} m_2 n_2 + 2\tau_{zx} n_2 l_2 \\
\sigma_{z'} &= \sigma_x l_3^2 + \sigma_y m_3^2 + \sigma_z n_3^2 + 2\tau_{xy} l_3 m_3 + 2\tau_{yz} m_3 n_3 + 2\tau_{zx} n_3 l_3 \\
\tau_{x'y'} &= \sigma_x l_1 l_2 + \sigma_y m_1 m_2 + \sigma_z n_1 n_2 + \tau_{xy}(m_1 l_2 + l_1 m_2) \\
&\quad + \tau_{yz}(n_1 m_2 + m_1 n_2) + \tau_{zx}(n_1 l_2 + l_1 n_2) \\
\tau_{y'z'} &= \sigma_x l_2 l_3 + \sigma_y m_2 m_3 + \sigma_z n_2 n_3 + \tau_{xy}(m_2 l_3 + l_2 m_3) \\
&\quad + \tau_{yz}(n_2 m_3 + m_2 n_3) + \tau_{zx}(n_2 l_3 + l_2 n_3) \\
\tau_{z'x'} &= \sigma_x l_1 l_3 + \sigma_y m_1 m_3 + \sigma_z n_1 n_3 + \tau_{xy}(m_1 l_3 + l_1 m_3) \\
&\quad + \tau_{yz}(n_1 m_3 + m_1 n_3) + \tau_{zx}(n_1 l_3 + l_1 n_3)
\end{aligned}
\tag{2.13}
$$

In indicial notation these transformation equations, which provide an analytic test for second-order tensors, can be written:

$$\sigma_{i'j'} = A_{ijkl}\sigma_{kl} \quad \text{or} \quad \sigma_{i'j'} = [A]\sigma_{ij}[A^T]$$

showing that it takes a fourth-order tensor (made up of direction cosines) to transform a second-order tensor. In practice it is more revealing and often easier to accomplish the transformation in two stages rather than use Equation (2.13) directly. First the "principal" state is found and the axes are then rotated, if necessary, from there. Usually the engineer is most interested in the principal state anyway since, in this orientation, the shear stresses are zero and normal components take on their maximum and minimum values.

2.4.1 Invariants and Principal Orientation

A vector can be plotted on paper in its fundamental or principal form as an arrow with magnitude and direction. It can then be decomposed into endless sets of different x', y', z' components analytically or graphically by the parallelogram law (cosines), each orthogonal set representing the *same* vector. Equations (2.13) demonstrate the same essential property for tensors (which are, after all, just a combination of vectors) but there is no neat graphical representation in 3D, or at least it is not directly apparent. At the same time we know that "the point" cannot feel anything different between one set of components or another equivalent set at some new orientation. Thus, although a tensor can be decomposed into endless sets of different vector components at a point, the body must actually feel the same effect regardless of our orientation when looking at it.

This fundamental description of a tensor, not tied to any particular coordinate system, must involve the "invariant" coefficients of the characteristic equation for determining the principal values (or eigenvalues) and their directions as presented in Chapter 1 (for the general linear finite displacement tensor). For the stress tensor, this characteristic Equation (1.10), is:

$$\lambda^3 - (\sigma_x + \sigma_y + \sigma_z)\lambda^2 + (\sigma_x\sigma_y + \sigma_y\sigma_z + \sigma_z\sigma_x - \tau_{xy}^2 - \tau_{yz}^2 - \tau_{zx}^2)\lambda$$
$$-(\sigma_x\sigma_y\sigma_z + 2\tau_{xy}\tau_{yx}\tau_{zx} - \sigma_x\tau_{yz}^2 - \sigma_y\tau_{zx}^2 - \sigma_z\tau_{xy}^2) = 0 \tag{2.14}$$

The three roots (values of λ) are the principal stresses called σ_1, σ_2, σ_3 and they are real since the stress tensor (or the strain tensor) is symmetric. It is in this principal orientation of orthogonal axes 1, 2, 3, that the tensor is diagonalized (no τ or γ) and thus, in a principal direction (say of σ_1), Equation (2.11) takes the form:

$$\begin{aligned}(\sigma_x - \sigma_1)\, l_1 + \tau_{xy}m_1 + \tau_{xz}n_1 &= 0\\ \tau_{xy}l_1 + (\sigma_y - \sigma_1)m_1 + \tau_{yz}n_1 &= 0\\ \tau_{xz}l_1 + \tau_{yz}m_1 + (\sigma_z - \sigma_1)n_1 &= 0\end{aligned} \tag{2.15}$$

which, with the identity $l_1^2 + n_1^2 + m_1^2 = 1$, can be solved for the direction cosines (the eigenvector) giving the orientation of the 1 or major principal axis. The same procedure is followed with σ_2 to find l_2, m_2, n_2 and σ_3 to find l_3, m_3, n_3 giving the directions of the intermediate and minor principal stresses respectively.

But Equation (2.14) is true for any choice of axes and therefore the coefficients must be the same in any coordinate system. These coefficients then are the invariants:

$$\begin{aligned}I_{1_\sigma} &= \sigma_x + \sigma_y + \sigma_z = \sigma_1 + \sigma_2 + \sigma_3\\ I_{2_\sigma} &= \sigma_x\sigma_y + \sigma_y\sigma_z + \sigma_z\sigma_x - \tau_{xy}^2 - \tau_{yz}^2 - \tau_{xz}^2 = \sigma_1\sigma_2 + \sigma_2\sigma_3 + \sigma_3\sigma_1\\ I_{3_\sigma} &= \sigma_x\sigma_y\sigma_z + 2\tau_{xy}\tau_{yz}\tau_{zx} - \sigma_x\tau_{yz}^2 - \sigma_y\tau_{xz}^2 - \sigma_z\tau_{xy}^2 = \sigma_1\sigma_2\sigma_3\end{aligned} \tag{2.16}$$

They are scalers and must be the "fundamental essence" of a stress tensor.

The corresponding invariants of the strain tensor can be deduced by analogy to be:

$$\begin{aligned}I_{1_\epsilon} &= \epsilon_x + \epsilon_y + \epsilon_z = \epsilon_1 + \epsilon_2 + \epsilon_3\\ I_{2_\epsilon} &= \epsilon_x\epsilon_y + \epsilon_y\epsilon_z + \epsilon_z\epsilon_x - \frac{1}{4}(\gamma_{xy}^2 + \gamma_{yz}^2 + \gamma_{zx}^2) = \epsilon_1\epsilon_2 + \epsilon_2\epsilon_3 + \epsilon_3\epsilon_1\\ I_{3_\epsilon} &= \epsilon_x\epsilon_y\epsilon_z + \frac{1}{4}(\gamma_{xy}\gamma_{yz}\gamma_{zx} - \epsilon_x\gamma_{yz}^2 - \epsilon_y\gamma_{zx}^2 - \epsilon_z\gamma_{xy}^2) = \epsilon_1\epsilon_2\epsilon_3\end{aligned} \tag{2.17}$$

or we could write them for moment of inertia, electromagnetic flux, or any second-order tensor.

Example 2.1

A state of stress is $\sigma_x = 3^{ksi}, \sigma_y = 0, \sigma_z = -1^{ksi}, \tau_{xy} = \tau_{zy} = -\sqrt{2}^{ksi}$ $\tau_{xz} = 1^{ksi}$. Determine the principal stresses and their direction.

$$\sigma_{ij} = \begin{bmatrix} 3 & -\sqrt{2} & 1 \\ -\sqrt{2} & 0 & -\sqrt{2} \\ 1 & -\sqrt{2} & -1 \end{bmatrix}$$

∴ The Invariant coefficients for Equation (2.14)

$$I_{1\sigma} = \sigma_x + \sigma_y + \sigma_Z = 2$$

$$I_{2\sigma} = \sigma_x\sigma_y + \sigma_y\sigma_z + \sigma_z\sigma_x - \tau_{xy}^2 - \tau_{yz}^2 - \tau_{zx}^2 = -8$$

and $I_{3\sigma} = \sigma_x\sigma_y\sigma_z + 2\tau_{xy}\tau_{yz}\tau_{zx} - \sigma_x\tau_{yz}^2 - \sigma_y\tau_{xz}^2 - \sigma_z\tau_{xy}^2 = 0$

so the characteristic equation is $\lambda^3 - 2\lambda^2 - 8\lambda - 0 = 0$ (a)

and solving $(\lambda - 0)(\lambda - 4)(\lambda + 2) = 0$

so $\lambda_1 = \sigma_1 = 4^{ksi}$, $\lambda_2 = \sigma_2 = 0$, $\lambda_3 = \sigma_3 = -2^{ksi}$

To solve for the direction cosines using Equation (2.11)
For

$$\sigma_1 = 4: \ (3-4)\,\ell_1 - \sqrt{2}m_1 + 1n_1 = 0 \qquad \therefore n_1 = \ell_1 + \sqrt{2}m_1 \quad \text{(b)}$$

$$-\sqrt{2}\,\ell_1 + (0-4)m_1 - \sqrt{2}n_1 = 0 \quad \text{(c)}$$

$$1\ell_1 - \sqrt{2}m_1 + (-1-4)n_1 = 0 \quad \text{(d)}$$

(b) into (d)

$$\ell_1 - \sqrt{2}m_1 - 5(\ell_1 + \sqrt{2}m_1) = 0 \qquad \text{so} \qquad \ell_1 = -\frac{3}{2}\sqrt{2}m_1, \quad n_1 = -\frac{\sqrt{2}}{2}m_1 \quad \text{(e)}$$

(e) into (c) $3m_1 - 4m_1 + m_1 = 0$ no new information

but $\ell_i^2 + m_i^2 + n_i^2 = 1$ So: $\frac{9}{2}m_1^2 + m_1^2 + \frac{1}{2}m_1^2 = 6m_1^2 = 1$ (f)

$$\therefore m_1 = \sqrt{\frac{1}{6}} = \cos\alpha_y \;\; \therefore \alpha_y = 66° \;; \qquad \ell_1 = -\frac{\sqrt{3}}{2}, \qquad \therefore \alpha_x = -30°$$

$$n_1 = -\frac{1}{2\sqrt{3}}, \qquad \therefore \alpha_z = 73.2°$$

or $$\bar{r}_1 = .866i - .40Tj + .289k$$

Similarly for

$$\sigma_2 = 0 \qquad 3\ell_2 - \sqrt{2}m_2 + n_2 = 0 \tag{g}$$

$$-\sqrt{2}\ell_2 + 0 - \sqrt{2}n_2 = 0 \qquad \therefore \ell_2 = -n_2 \tag{h}$$

$$\ell_2 - \sqrt{2}m_2 - n_2 = 0 \qquad m_2 = -\frac{2}{\sqrt{2}}\ell_2 \tag{i}$$

and

$$\ell_2^2 + m_2^2 + n_2^2 = 1 \qquad \therefore \ell_2^2 + 2\ell_2^2 + \ell_2^2 = 1$$

$$\therefore \ell_2 = \frac{1}{2}, \quad \beta_x = 60°; \quad m_2 = \frac{1}{\sqrt{2}}, \quad \beta_y = 45°; \quad n_2 = -\frac{1}{2}, \quad \beta_z = -60°$$

or $$\bar{r}_2 = .5i + .70Tj - .5k$$

Finally for

$$\sigma_3 = -2 \qquad (3+2)\ell_3 \; - \sqrt{2}m_3 \qquad + n_3 = 0 \tag{j}$$

$$-\sqrt{2}\ell_3 \; + (0+2)m_3 \; - \sqrt{2}n_3 = 0 \tag{k}$$

so $$n_3 = -\ell_3 + \sqrt{2}m_3 \qquad \ell_3 \; - \sqrt{2}m_3 \qquad + (-1+2)n_3 = 0 \tag{l}$$

from (j) $5\ell_3 - \sqrt{2}m_3 - \ell_3 + \sqrt{2}m_3 = 0 \qquad \therefore \ell_3 = 0, \qquad n_3 = \sqrt{2}m_3$

and $\ell_3^2 + m_3^2 + n_3^2 = 1 \qquad \therefore 2m_3^2 + m_3^2 = 1 \qquad$ and $\qquad m_3 = \frac{1}{\sqrt{3}}, \; n_3 = \sqrt{\frac{2}{3}}$

or $\bar{r}_3 = .577j + .817k$

The above is illustrated in Figure E2.1.
Note: as a check $\Sigma\ell_i^2 = \Sigma m_i^2 = \Sigma n_i^2 = 1$

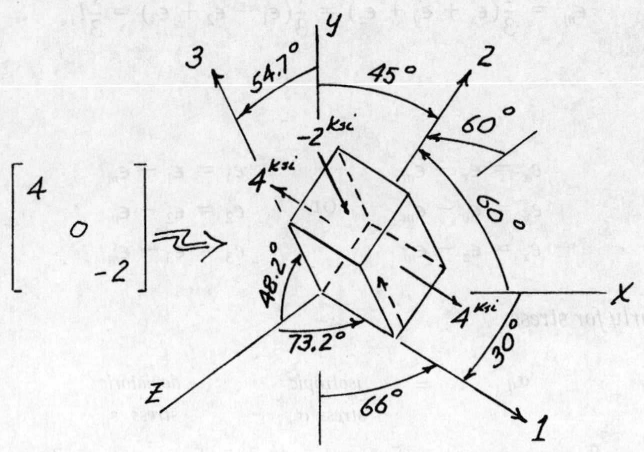

FIGURE E2.1

2.5 Isotropic and Deviatoric Components

As we have seen in Section 2.3, any symmetric tensor can be split into isotropic and deviatoric components as follows:

a) *For strain*:

$$\begin{array}{cccc} \epsilon_{ij} & = \quad \textit{Isotropic strain } \epsilon_m & + & \textit{Deviatoric strain tensor } e_{ij} \end{array}$$

$$\begin{bmatrix} \epsilon_x & \dfrac{\gamma_{xy}}{2} & \dfrac{\gamma_{xz}}{2} \\ \dfrac{\gamma_{xy}}{2} & \epsilon_y & \dfrac{\gamma_{yz}}{2} \\ \dfrac{\gamma_{xz}}{2} & \dfrac{\gamma_{yz}}{2} & \epsilon_z \end{bmatrix} = \begin{bmatrix} \epsilon_m & & \\ & \epsilon_m & \\ & & \epsilon_m \end{bmatrix} + \begin{bmatrix} e_x & \dfrac{\gamma_{xy}}{2} & \dfrac{\gamma_{xz}}{2} \\ \dfrac{\gamma_{xy}}{2} & e_y & \dfrac{\gamma_{yz}}{2} \\ \dfrac{\gamma_{xz}}{2} & \dfrac{\gamma_{yz}}{2} & e_z \end{bmatrix} \tag{2.18}$$

$$\begin{bmatrix} \epsilon_1 & & \\ & \epsilon_2 & \\ & & \epsilon_3 \end{bmatrix} = \underset{\substack{\textit{volume change} \\ \Delta V = 3\epsilon_m V_0}}{\begin{bmatrix} \epsilon_m & & \\ & \epsilon_m & \\ & & \epsilon_m \end{bmatrix}} + \underset{\substack{\textit{distortion} \\ \textit{no } \Delta V}}{\begin{bmatrix} e_1 & & \\ & e_2 & \\ & & e_3 \end{bmatrix}}$$

where:

$$\epsilon_m = \frac{1}{3}(\epsilon_x + \epsilon_y + \epsilon_z) = \frac{1}{3}(\epsilon_1 + \epsilon_2 + \epsilon_3) = \frac{1}{3}I_{1\epsilon} \tag{2.19}$$

and

$$\begin{aligned} e_x &= \epsilon_x - \epsilon_m \\ e_y &= \epsilon_y - \epsilon_m \\ e_z &= \epsilon_z - \epsilon_m \end{aligned} \quad \text{or} \quad \begin{aligned} e_1 &= \epsilon_1 - \epsilon_m \\ e_2 &= \epsilon_2 - \epsilon_m \\ e_3 &= \epsilon_3 - \epsilon_m \end{aligned} \tag{2.20}$$

b) *Similarly for stress:*

$$\sigma_{ij} \quad = \quad \underset{\textit{stress } \sigma_m}{\textit{isotropic}} \qquad \underset{\textit{stress } s_{ij}}{\textit{deviatoric}}$$

$$\begin{bmatrix} \sigma_x & \tau_{xy} & \tau_{xz} \\ \tau_{xy} & \sigma_y & \tau_{yz} \\ \tau_{xz} & \tau_{yz} & \sigma_z \end{bmatrix} = \begin{bmatrix} \sigma_m & & \\ & \sigma_m & \\ & & \sigma_m \end{bmatrix} + \begin{bmatrix} s_x & \tau_{xy} & \tau_{xz} \\ \tau_{xy} & s_y & \tau_{yz} \\ \tau_{xz} & \tau_{yz} & s_z \end{bmatrix} \tag{2.21a}$$

or

$$\begin{bmatrix} \sigma_m & & \\ & \sigma_m & \\ & & \sigma_m \end{bmatrix} = \begin{bmatrix} \sigma_m & & \\ & \sigma_m & \\ & & \sigma_m \end{bmatrix} + \begin{bmatrix} s_1 & & \\ & s_2 & \\ & & s_3 \end{bmatrix} \tag{2.21b}$$

where

$$\sigma_m = \frac{1}{3}(\sigma_x + \sigma_y + \sigma_z) = \frac{1}{3}I_{1\sigma} \tag{2.22}$$

and

$$\begin{aligned} s_x &= \sigma_x - \sigma_m \\ s_y &= \sigma_y - \sigma_m \\ s_z &= \sigma_z - \sigma_m \end{aligned} \quad \text{or} \quad \begin{aligned} s_1 &= \sigma_1 - \sigma_m \\ s_2 &= \sigma_2 - \sigma_m \\ s_3 &= \sigma_3 - \sigma_m \end{aligned} \tag{2.23}$$

The isotropic component, ϵ_m, σ_m is essentially a scalar (tensor of order zero) having only magnitude with all directions principal. It is sometimes called the spherical component. The mean stress, σ_m, is hydrostatic pressure if negative, or suction if positive, and causes only volume change, $3\epsilon_m$, per unit volume. The deviatoric component is just the opposite in that it causes only distortion or shear with no volume change. The invariants of the deviatoric tensors are:

$$I_{1s} = I_{1e} = 0 \tag{2.24a}$$

$$I_{2s} = S_xS_y + S_yS_z + S_zS_x - \tau_{xy}^2 - \tau_{yz}^2 - \tau_{zx}^2 = S_1S_2 + S_2S_3 + S_3S_1$$

$$= -\frac{2}{3}\left[\left(\frac{\sigma_1 - \sigma_2}{2}\right)^2 + \left(\frac{\sigma_2 - \sigma_3}{2}\right)^2 + \left(\frac{\sigma_3 - \sigma_1}{2}\right)^2\right] = I_{2\sigma} - \frac{1}{3}I_{1\sigma}^2 \tag{2.24b}$$

$$\text{or } I_{2e} = \ldots = e_1e_2 + e_2e_3 + e_3e_1 = \ldots = I_{2\epsilon} - \frac{1}{3}I_{1\epsilon}^2$$

$$I_{3s} = S_xS_yS_z + 2\tau_{xy}\tau_{yz}\tau_{zx} - S_x\tau_{yz}^2 - S_y\tau_{zx}^2 - S_z\tau_{xy}^2 = S_1S_2S_3$$

$$= \frac{1}{27}[(2\sigma_1 - \sigma_2 - \sigma_3)(2\sigma_2 - \sigma_3 - \sigma_1)(2\sigma_3 - \sigma_1 - \sigma_2)] \tag{2.24c}$$

$$= I_{3\sigma} - \frac{1}{3}\left[I_{1\sigma}I_{2\sigma} - \frac{2}{9}I_{1\sigma}^3\right]$$

$$\text{or } I_{3e} = \ldots e_1e_2e_3 = \ldots = I_{3\epsilon} - \frac{1}{3}\left[I_{1\epsilon}I_{2\epsilon} - \frac{2}{9}I_{1\epsilon}^3\right]$$

This uncoupling of the tensor into isotropic and deviatoric parts will turn out to be a fundamental physical reality that is reflected both in the way materials behave under load, and how the stress, strain, and displacements "flow" through a body. It also turns out that there is a particular orientation of axes (of viewing the tensor) where, in fact, this fundamental uncoupling occurs naturally and the tensor components become the invariants themselves. This is the so-called octahedral state.

2.6 Principal Space and the Octahedral Representation

Since the principal stresses are orthogonal, we can, once we calculate their directions by the procedure in Section 2.4, imagine that now the x, y, z, axes of Figure 2.4 are oriented in the principal directions. Thus plane ABC, as redrawn in Figure 2.5a, is normal to a straight line x' from the origin with

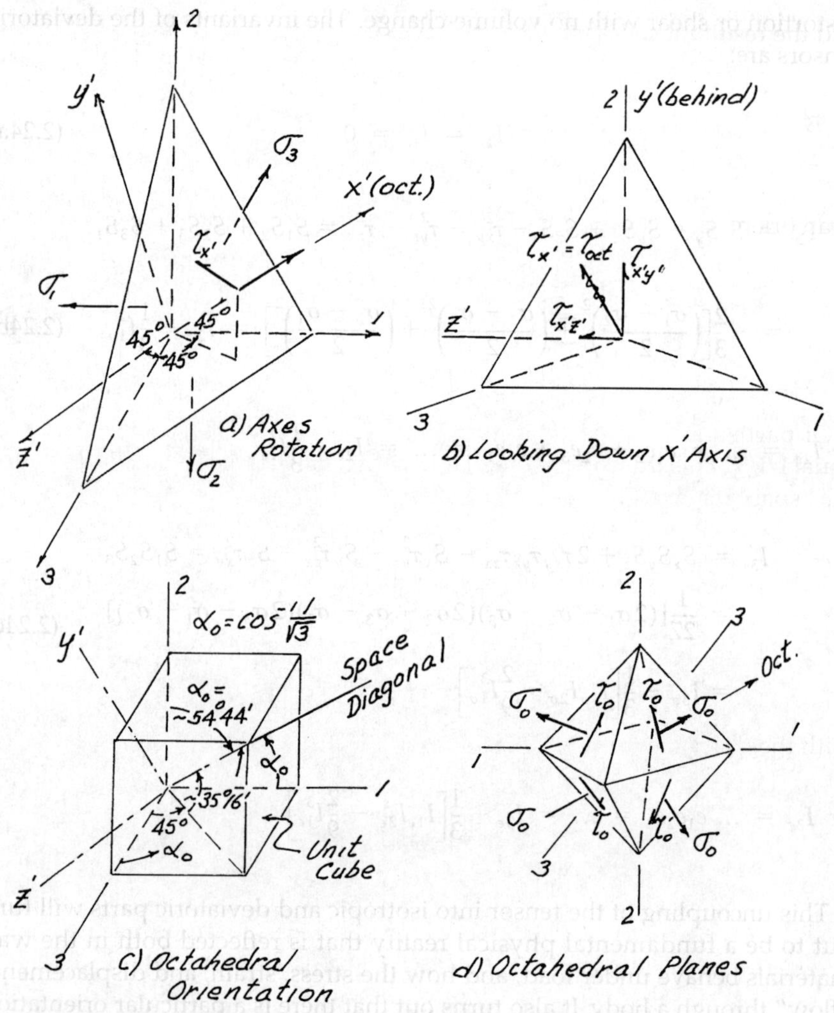

FIGURE 2.5
Transformation from principal to octahedral space.

direction cosines with respect to the 1, 2, 3 axes. For convenience let the z' axis lie in the 1–3 plane. From Equation (2.12) then:

$$\sigma_{x'} = \sigma\ell^2 + \sigma_2 m^2 + \sigma_3 n^2$$
$$\tau_{x'z'} = \frac{\ell n}{\sqrt{1-m^2}}(\sigma_3 - \sigma_1) \tag{2.25}$$
$$\tau_{x'y'} = -\sigma_1 \frac{\ell^2 m}{\sqrt{1-m^2}} - \sigma_3 \frac{n^2 m}{\sqrt{1-m^2}} + \sigma_2 m\sqrt{1-m^2}$$

and the resultant shear stress on x' will be:

$$\tau_{x'}^2 = \tau_{x'y'}^2 + \tau_{x'z'}^2 = (\sigma_1 - \sigma_2)^2\ell^2 m^2 + (\sigma_2 - \sigma_3)^2 m^2 n^2 + (\sigma_3 - \sigma_1)^2 n^2 \ell^2 \tag{2.26}$$

at an orientation (Figure 2.5b):

$$\tan\alpha = \frac{\tau_{x'z'}}{\tau_{x'y'}} = \frac{n}{\ell m}\left[\frac{(\sigma_1 - \sigma_3)\ell^2 + (\sigma_2 - \sigma_3)m^2}{(\sigma_1 - \sigma_2)}\right] \tag{2.27}$$

Of particular interest is the special case in Figure 2.5a where ℓ, m, n all equal $1/\sqrt{3}$. This direction is the diagonal of a unit cube and thus often called the "space diagonal." For this orientation:

$$\begin{aligned} \sigma_{x'} &= \sigma_m = \sigma_{oct} = \frac{1}{3}(\sigma_1 + \sigma_2 + \sigma_3) = \frac{1}{3}I_{1\sigma} \\ \tau_{x'} &= \tau_{oct} = \frac{1}{3}\sqrt{(\sigma_1 - \sigma_2)^2 + (\sigma_2 - \sigma_3)^2 + (\sigma_3 - \sigma_1)^2} = \sqrt{-\frac{2}{3}I_{2S}} \end{aligned} \tag{2.28}$$

with the components

$$\begin{aligned} \tau_{oz'} &= \frac{1}{\sqrt{6}}(\sigma_2 - \sigma_1) \\ \tau_{oy'} &= \frac{1}{3\sqrt{2}}(2\sigma_3 - \sigma_1 - \sigma_2) \end{aligned} \tag{2.29}$$

at

$$\tan\alpha = \frac{1}{\sqrt{3}}\frac{(\sigma_1 + \sigma_2 - 2\sigma_3)}{(\sigma_1 - \sigma_2)} \tag{2.30}$$

The subscript designation "oct" or o is used since there are eight such faces normal to the space diagonals forming a regular octahedron as in Figue 2.5d. The total "stress resultant" on an octahedral plane is:

$$p_{x'} = p_{oct} = p_o = \sqrt{\sigma_o^2 + \tau_o^2} \tag{2.31}$$

As already mentioned, the remarkable property of this octahedral orientation is, as shown by Equation (2.28), that the stress or strain tensors uncouple naturally* into invariant isotropic and deviatoric components when viewed in this perspective in principal space. Moreover, the angle α in Equation (2.30) is related to $I_{\sigma 3}$ or more specifically to $I_{s3}/\sqrt{I_{s2}^3}$ so that all three invariants appear on the octahedral plane. Thus, in this one special orientation, the tensor can be reduced (visualized) in its most physical form as three numbers (scalar invariants) σ_{oct}, τ_{oct}, α and three direction cosines relating the direction of the space diagonal (the octahedral plane) to the coordinate system.**

The corresponding octahedral strain components are, by analogy:

$$\begin{aligned} \epsilon_{oct} &= \epsilon_o = \frac{1}{3}(\epsilon_1 + \epsilon_2 + \epsilon_3) = \frac{1}{3}I_{\epsilon 1} \\ \gamma_{oct} &= \frac{2}{3}\sqrt{(\epsilon_1 - \epsilon_2)^2 + (\epsilon_2 - \epsilon_3)^2 + (\epsilon_3 - \epsilon_1)^2} = 2\sqrt{-\frac{2}{3}I_{e2}} \end{aligned} \tag{2.32}$$

with components corresponding to Equations (2.29) and (2.30).

2.7 Two-Dimensional Stress or Strain

In the majority of important practical situations, the general three-dimensional stress-strain array (Figure 2.3e) can, to an acceptable degree of accuracy, be simplified to a 4-component symmetric tensor by assuming either no stress or no strain components on one set of planes (say the z faces). The first is called "plane stress" corresponding to free surfaces normal to the z axis so that $\sigma_z = \tau_{xx} = \tau_{zy} = 0$ and is the most common and the most easily achieved in the laboratory. For plane stress there will generally be strain normal to the unrestrained z planes ($e_z \neq 0$) but no shearing distortion (i.e., $\gamma_{zy} = \gamma_{zx} = 0$).

* There is no sign that this uncoupling was recognized by those who originally formulated the theory of elasticity. Among other complications, this has led to the current confusion in defining the engineering properties of materials. The attempt to formulate yield criteria and plasticity theory seems to mark the start of the gradual realization that nature feels isotropic and deviatoric effects separately and independently. Tresca and St. Venant in 1868 began this development by relating plastic flow to maximum shear stress and noting there was no volume change in the process. Maxwell shortly afterwards, in his usual brilliant fashion, correctly related yield to the "energy of distortion." However, it was not until the 1900s when the theory was formulized by Huber and others in terms of the second deviatoric invariant, that it fully dawned on everyone that this uncoupling extended throughout the elastic–plastic range making experiments, theory, solutions, and calculation clearer and easier.

** We could, at this point, further discuss the properties of stress and strain in terms of Cartesian "Invariant Space" with coordinate axes I_1, I_2, I_3. This is done in detail with very interesting results by V.K. Stokes in his paper "On the Space of Stress Invariants," G.E. Corp. (1980).

"Plane strain" arises physically in situations such as dams, pipelines, embankments, long shafts, etc. where there are no displacements in the z direction nor distortion in the z plane. Therefore $\epsilon_z = \gamma_{zx} = \gamma_{zy} = 0$. In general for plane strain $\sigma_z \neq 0$ but $\tau_{zx} = \tau_{zy} = 0$.

Thus, in either case, the in-plane tensor reduces to:

$$\sigma_{xy} = \begin{bmatrix} \sigma_x & \tau_{xy} \\ \tau_{xy} & \sigma_y \end{bmatrix} \quad \text{or in the principal orientation,} \quad \sigma_{12} = \begin{bmatrix} \sigma_1 & \\ & \sigma_2 \end{bmatrix}$$

$$\underbrace{\begin{bmatrix} \frac{\sigma_x + \sigma_y}{2} & \\ & \frac{\sigma_x + \sigma_y}{2} \end{bmatrix}}_{\text{isotropic } \sigma_m} + \underbrace{\begin{bmatrix} \frac{\sigma_x - \sigma_y}{2} & \tau_{xy} \\ \tau_{xy} & -\frac{\sigma_x - \sigma_y}{2} \end{bmatrix}}_{\text{deviatoric } S_{xy}} \quad \text{or} \quad \underbrace{\begin{bmatrix} \frac{\sigma_1 + \sigma_2}{2} & \\ & \frac{\sigma_1 + \sigma_2}{2} \end{bmatrix}}_{\sigma_m} + \underbrace{\begin{bmatrix} \frac{\sigma_1 - \sigma_2}{2} & \\ & -\frac{\sigma_1 - \sigma_2}{2} \end{bmatrix}}_{S_{12}}$$

σ_z and ϵ_Z are principal and either $I_{3\sigma}$ or $I_{3\epsilon}$ is zero.*

The 2D transformation (rotation of coordinate axes) from the z, y, z to a new x, y, z axes can be deduced from the 3D equations [Equations (2.13)]. The direction cosines in terms of the positive counterclockwise angle of axis rotation θ are shown in Figure 2.6.

Therefore,

$$\begin{aligned} \sigma_{x'} &= \sigma_x \cos^2\theta + \sigma_y \sin^2\theta + 2\tau_{xy} \sin\theta\cos\theta \\ \sigma_{y'} &= \sigma_x \sin^2\theta + \sigma_y \cos^2\theta - 2\tau_{xy} \sin\theta\cos\theta \\ \tau_{x'y'} &= -(\sigma_x - \sigma_y)\sin\theta\cos\theta + \tau_{xy}(\cos^2\theta - \sin^2\theta) \end{aligned} \tag{2.33}$$

or in terms of double angles:

$$\begin{aligned} \sigma_{x'} &= \frac{\sigma_x + \sigma_y}{2} + \frac{\sigma_x - \sigma_y}{2}\cos 2\theta + \tau_{xy}\sin 2\theta \\ \sigma_{y'} &= \frac{\sigma_x + \sigma_y}{2} - \frac{\sigma_x - \sigma_y}{2}\cos 2\theta - \tau_{xy}\sin 2\theta \\ \sigma_{x'y'} &= -\frac{\sigma_x - \sigma_y}{2}\sin 2\theta + \tau_{xy}\cos 2\theta \end{aligned} \tag{2.34}$$

* A more exact definition of a two-dimensional tensor. If both I_2 and I_3 are zero, the stress or strain tensor is one-dimensional (a scalar magnitude). If I_1 and I_3 are zero, the stress is still two-dimensional. I_3 is thus a measure of the three-dimensionality and I_2 the two-dimensionality of a tensor.

	x'	y'	z'
x	$l_1 = \cos\theta$	$m_1 = \sin\theta$	0
y	$l_2 = -\sin\theta$	$m_2 = \cos\theta$	0
z	$l_3 = 0$	0	1

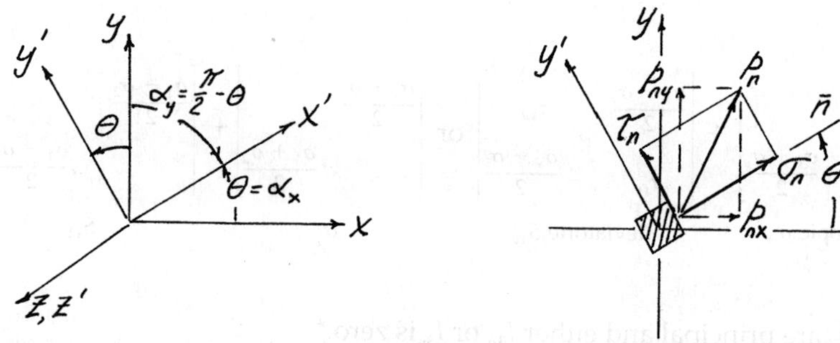

FIGURE 2.6
2D Stress transformation.

The transformation of the 2D strain tensor is the same by analogy, with ϵ_x, ϵ_y, $\gamma_{xy}/2$, replacing $\sigma_x \sigma_y \tau_{xy}$.

The 2D characteristic equation becomes

$$\lambda^2 - (\sigma_x + \sigma_y)\lambda + (\sigma_x \sigma_y - \tau_{xy}^2) = 0 \tag{2.35}$$

The roots being the principal stresses

$$\sigma_{1,2} = \frac{\sigma_x + \sigma_y}{2} \pm \sqrt{\left(\frac{\sigma_x - \sigma_y}{2}\right)^2 + \tau_{xy}^2} \tag{2.36}$$

at an angle of axis rotation

$$\theta_{1,2} = \tan^{-1}\left(\tau_{xy} \Big/ \frac{\sigma_x - \sigma_y}{2}\right) \tag{2.37}$$

The coefficients of the characteristic equation must again be the same in any orientation so the 2D invariants are:

$$\begin{aligned} P_{\sigma 1} &= \sigma_x + \sigma_y = \sigma_1 + \sigma_2 \\ P_{\sigma 2} &= \sigma_x \sigma_y - \tau_{xy}^2 = \sigma_1 \sigma_2 \end{aligned} \tag{2.38}$$

or considering the isotropic and deviatoric components separately

$$\sigma_m = \frac{\sigma_x + \sigma_y}{2} = \frac{P_{\sigma 1}}{2} \tag{2.39}$$

and

$$\begin{aligned} P_{s1} &= 0 \\ P_{s2} &= S_x S_y - \tau_{xy}^2 = -\left[\left(\frac{\sigma_x - \sigma_y}{2}\right)^2 + \tau_{xy}^2\right] = S_1 S_2 = -\left(\frac{\sigma_1 - \sigma_2}{2}\right)^2 \end{aligned} \tag{2.40}$$

In principal space the "octahedral" orientation along the "space diagonal" becomes the "quadrahedral" orientation along the "square diagonal." In this orientation the quadrahedral stress components are:

$$\begin{aligned} \sigma_{\text{quad}} &= \sigma_q = \sigma_m = \frac{\sigma_1 + \sigma_2}{2} = \frac{\sigma_x + \sigma_y}{2} = \frac{P_{\sigma 1}}{2} \\ \tau_{\text{quad}} &= \tau_q = \tau_{12} = \tau_{\max} = \sqrt{\left(\frac{\sigma_x - \sigma_y}{2}\right)^2 + \tau_{xy}^2} = \sqrt{-P_{s2}} \end{aligned} \tag{2.41}$$

at 45° to the principal directions as shown in Figure 2.7. Their resultant is:

$$p_{\text{quad}} = p_q = \sqrt{\sigma_q + \tau_q} = \sqrt{\frac{1}{2}(\sigma_1^{\,2} + \sigma_2^{\,2})} \tag{2.42}$$

Thus in the quadrahedral orientation, we again see all the invariants as individual, uncoupled stress components in themselves. There can be no doubt that this is what nature actually feels at a point in a 2D stress field. Again the analogous equations for the 2D strain tensor in the plane [Equations (2.35)–(2.42)] are obtained by substituting ϵ_x for σ_x, ϵ_y, for σ_y, $\gamma_{xy}/2$ for τ_{xy}.

Perhaps it should be noted that the elastic rotation around the z axis in two dimensions $\omega_{xy}/2$ or ω_z of Equation (2.5) (the asymmetric component of the relative deformation tensor) is already an invariant.* While in 3D the elastic rotation is a vector with three components, in 2D, $\omega_z = \omega_q$ at a point is actually a scalar quantity involving only magnitude since the direction is known. There is no comparable asymmetric component of stress, but elastic rotations

* Clearly correct intuitively, but the formal proof is left as a homework exercise.

occur due to stress fields and must not be forgotten while we are preoccupied with symmetric stress and strain tensors.

2.8 Mohr's Circle for a Plane Tensor

A complete graphical representation of any symmetric 2D tensor (stress, strain, inertia, etc.) in all orientations simultaneously is given by Mohr's Circle. This construction, therefore, provides an invaluable tool for visualizing all the equations derived in the previous section* and, when plotted accurately, can be used to obtain values for engineering computations. Mohr's Circle is also very useful in describing yield behavior, slip surfaces, strain hardening, and failure in plasticity. Thus, it is extremely important in engineering mechanics that Mohr's Circle be mastered in all its nuances and totally understood. In this and the next section, Mohr's Circle will be presented in terms of stress, knowing the results can be translated immediately to any other tensor quantity by analogy.

For the two-dimensional case, Mohr's Circle can be plotted from the values of the components of the planar tensor in any orientation as shown in Figure 2.7. This construction should be thought of as a mapping of the normal and shear components on any face to a point in stress space (the $\sigma - \tau$ plane) where:

1. σ is positive in tension and is plotted to the right of the origin; compression is negative to the left.
2. τ is plotted positively if it rotates the stress-block clockwise. Thus, positive τ_{xy} is plotted downward for the x face and upward for the y face even though "in real life" they have the same sign.
3. Angles θ from one axis to another around the origin are in the same direction but twice those "in real life."

Thus any orthogonal set of axes are 180° to one another on Mohr's Circle and therefore a diameter of the circle. With these mapping rules, the circle is a direct representation of Equation (2.34) for 2D coordinate transformation in terms of double angles.

Every possible orientation of σ_{ij} is represented by some diameter of the circle. Of particular interest are, first, the principal 1–2 orientation when τ

* Applied mathematicians therefore argue that Mohr's Circle is unnecessary and is simply a heuristic device since the equations alone are sufficient. This is true pedagogically, but a nightmare for engineers who want to understand tensor relationships physically. Historically from Hooke onwards, mental pictures, real and thought experiments, graphical analysis, plots, diagrams, etc. have led to profound science and great engineering. Mathematics as an art form is beautiful and develops the brain, perhaps the soul, but pictures can be worth a thousand symbols.

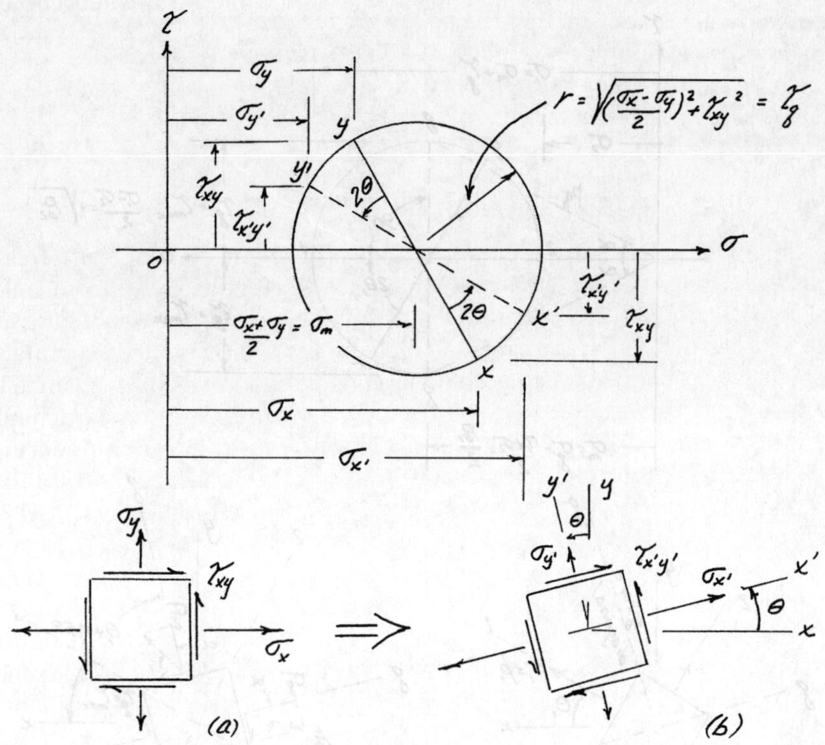

FIGURE 2.7
2D Mohr's Circle.

disappears and, secondly the quadrahedral orientation q–q where τ is a maximum and where the scalar invariants occur as the stress components themselves. These two states are shown in Figure 2.8 for the same stress as in Figure 2.7. Since each quadrahedral face has the same normal and shear stress on it, there is no need to differentiate between axes. This property of the quadrahedral orientation along the two square principal-plane diagonals is unique as it was in 3D for the octahedral orientation along the eight space diagonals.

By providing a graphic (geometric) picture of a 2D tensor at a point in all orientations simultaneously, Mohr's Circle is the ultimate representation of a 2D tensor. It is far superior to a matrix or block representation, which can only depict one array of vector components. Moreover, as the center of Mohr's Circle and the radius, the invariants are self-evident. So, too, is the separation into isotropic and deviatoric components shown in Figure 2.9. The two effects are unlinked graphically just as they are physically. Changes in the isotropic stress σ_m (causing isotropic strain ϵ_m and pure volume change) simply moves the center of Mohr's Circle as a point back and forth along the σ axis without affecting the circle itself. It is the deviatoric component (pure shear causing

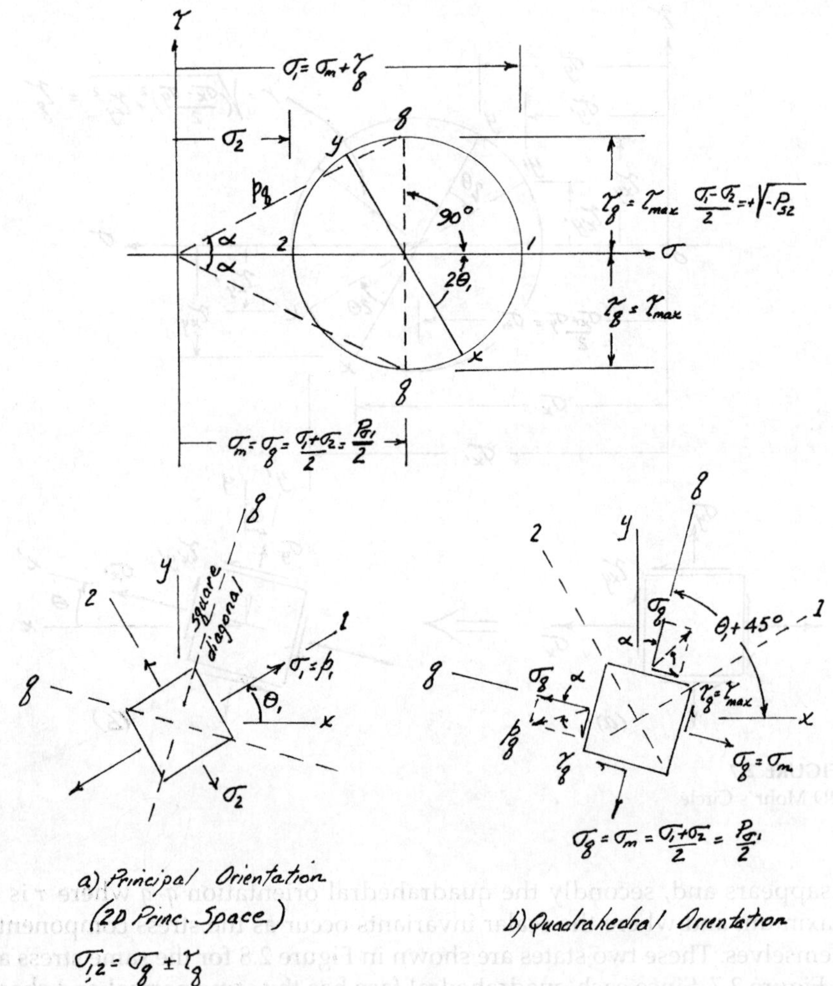

FIGURE 2.8
Principal and quadrahedral orientations.

only distortion) which, as it increases, makes the point grow into larger and larger circles and establishes the orientation of the tensor.*

The limit to how far the center of Mohr's Circle can move to the right in isotropic tension or left in isotropic (hydrostatic) compression, or how large the circle can grow before the onset of serious nonlinearity depend on the strength

* Isotropic stress (the 1st invariant σ_m) is one of the few important fundamental scalar quantities pervasive throughout the universe for which no absolute scale has been established. Zero is gauge pressure (actually 15 psi) at sea level on Earth. Thus, like temperature when we use Fahrenheit or Centigrade, we are stuck with positive and negative stress. If we made absolute zero the hydrostatic pressure in a black hole, we would live in a positive universe of tension. The deviatoric scalar invariant (as the radius of Mohr's Circle) is always positive naturally.

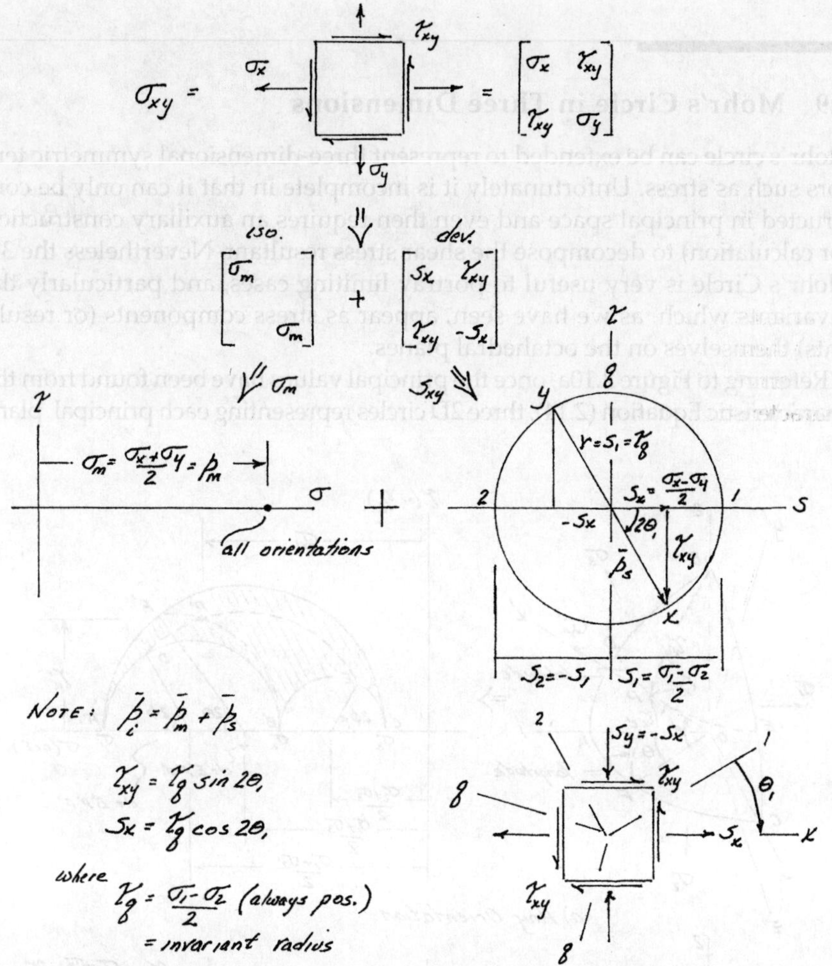

FIGURE 2.9
Separation into isotropic and deviatoric components.

of the material in which the stress or strain (at the point) exists.* These material properties (isotropic and deviatoric "strength" and associated flow rules), which will be introduced in the next chapter, are also best visualized in terms of Mohr's Circle and, in fact, led Mohr to develop his construction in the first place.

* These strength properties also differentiate the specialties within continuum mechanics from ideal fluids, which have no shear strength, to ideal plastic solids with constant shear strength independent of σ_m.

2.9 Mohr's Circle in Three Dimensions

Mohr's circle can be extended to represent three-dimensional symmetric tensors such as stress. Unfortunately it is incomplete in that it can only be constructed in principal space and even then requires an auxiliary construction (or calculation) to decompose the shear stress resultant. Nevertheless the 3D Mohr's Circle is very useful to portray limiting cases, and particularly the invariants which, as we have seen, appear as stress components (or resultants) themselves on the octahedral planes.

Referring to Figure 2.10a, once the principal values have been found from the characteristic Equation (2.14), three 2D circles representing each principal plane

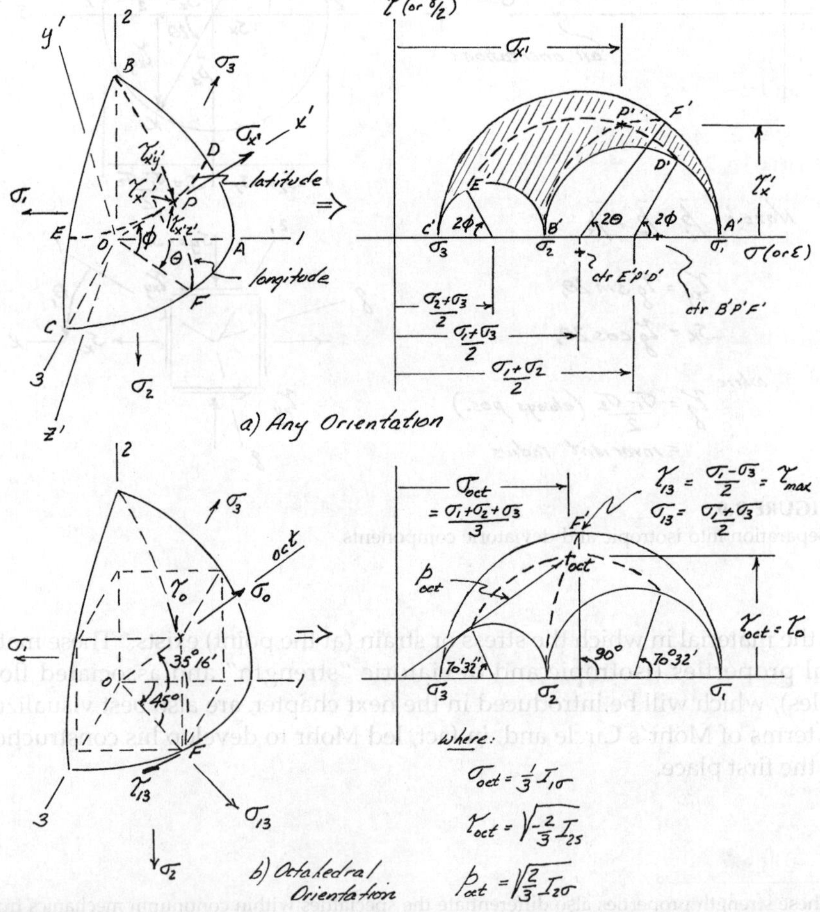

FIGURE 2.10
Mohr's Circle in 3D principal space.

can be drawn defining the tensor in principal space. Since the principal values are limit values, all other orientations for the tensor must be in the area between these circles (shown slashed).*

There are many ways to then locate a point (say point P) in this principal space, which represents a new orientation of axes. The transformation equations use the direction cosines, but point P can be described more graphically using just two angles: latitude, ϕ, and longitude, θ, as shown in Figure 2.10a. As we have already seen in 2D, these will be mapped as double angles. Circle BPF of longitude is a "great circle" (center at 0) so its center must lie along the σ (or ϵ) axis and arc B′F′ can be drawn. While circle DPE of latitude is not a great circle, all circles of latitude must have their centers along the same line as the equator AFC at $\frac{\sigma_1 - \sigma_3}{2}$ in stress space. That is, circles of latitude are only seen as circles when looking down the σ_2 axis. Thus, again by trial-and-error, arc $E'D'$ can be drawn locating, point P′ uniquely in the $\sigma - \tau$ (or $\epsilon - \gamma/2$) space to determine the normal and shear-resultant components normal and tangential to the new orientation x'. The octahedral orientation along the space diagonal for the same set of stress is shown in Figure 2.10b. Also shown is the maximum shear stress orientation which, in principal space, is actually the quadrahedral state in the 1–3 plane. There are corresponding 1–2 and 2–3 states which are intermediate quadrahedral orientations.

Example 2.2

For the state of stress of Example 2.1 ($\sigma_1 = 4^{ksi}, \sigma_2 = 0, \sigma_3 = -2^{ksi}$), determine using Mohr's Circle (and check by formula)

(a) the normal and total shear stress on an x′ face given by direction cosines $\ell_1 = 0.433$, $m_1 = 0.866$, $n_1 = 0.25$ and

(b) the octahedral orientation.

(a)Referring to Figure E2.2a:

$$\cos^{-1}\alpha'_y = .866 \qquad \therefore \alpha'_y = 30° \qquad \therefore \phi' = 60°$$

$$\cos 60° \cos\theta' = .433 \qquad \therefore \cos\theta' = .866 \qquad \therefore \theta' = 30°$$

$$\therefore \text{ From Mohr's Circle (Figure E2.2c) } \sigma_{x'} \cong 0.6^{ksi}, \qquad \tau_{x'} \cong 1.7^{ksi}$$

$$\text{check: } \ell_2 = .753, \qquad m_2 = -.5, \qquad n_2 = .433$$

$$\ell_3 = -.5, \qquad m_3 = 0, \qquad n_3 = .866$$

---

* There are other graphical representations of stress at a point (i.e., any two- or three-dimensional symmetric tensor). The two primary ones are:

1. The Stress Ellipsoid of Lamé
2. The Stress Quadric of Cauchy

Although interesting, neither is as general, more complete, or as useful in practice as Mohr's Circle in either two or three dimensions. A full discussion of these and other geometric constructions is presented in *Analysis of Stress and Strain* by Durelli, Phillips, and Tsao, McGraw Hill, New York, 1958.

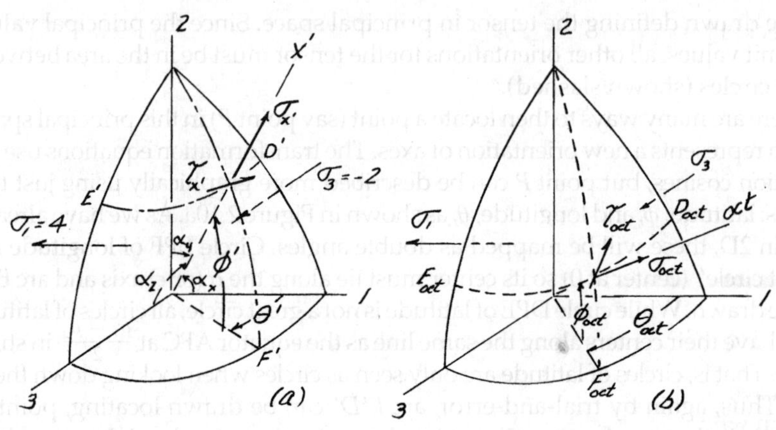

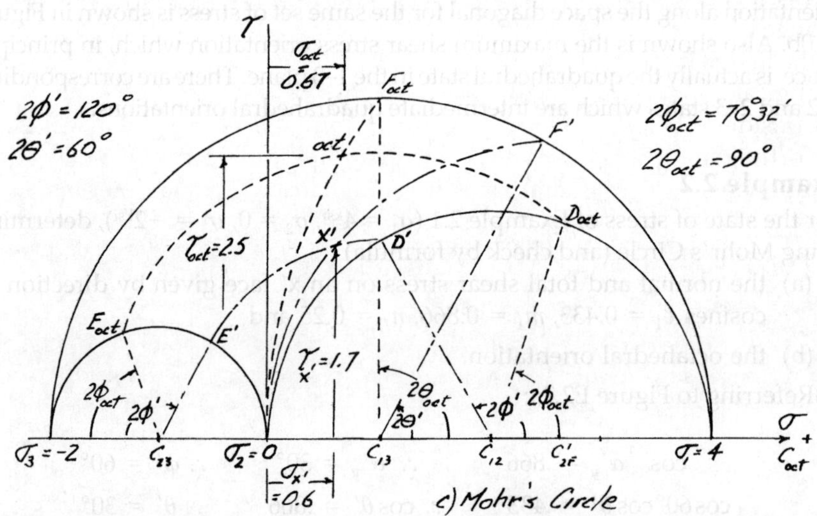

FIGURE E2.2

$\therefore$ From Equations (2.13)

$$\sigma_{x'} = \sigma_1 \ell_1^2 + \sigma_2 m_1^2 + \sigma_3 n_1^2 = 4(.433^2) + 0 - 2(.25^2) = 0.625^{\text{ksi}}$$

$$\tau_{x'y'} = \sigma_1 \ell_1 \ell_2 + \sigma_2 m_1 m_2 + \sigma_3 n_1 n_2 = 4(.326) + 0 - 2(.108) = 1.088^{\text{ksi}}$$

$$\tau_{x'y'} = \sigma_1 \ell_1 \ell_3 + \sigma_2 m_1 m_3 + \sigma_3 n_1 n_3 = 4(-.2165) + 0 - 2(.215) = -1.3^{\text{ksi}}$$

$$\therefore \tau_{x'} = \sqrt{\tau_{x'y'}^2 + \tau_{x'z'}^2} = 1.695^{ksi}$$

b)Referring to Figure E2.2b $\quad \phi_{\text{oct}} = 37°16', \theta_{\text{oct}} = 45°$

$\therefore$ From Mohr's Circle (Figure E2.2c) $\quad \sigma_{\text{oct}} = 0.67^{\text{ksi}}$

$$\tau_{\text{oct}} = 2.5^{\text{ksi}}$$

Check: From Equations (2.28)

$$\sigma_{oct} = \sigma_m = \frac{1}{3}(\sigma_1 + \sigma_2 + \sigma_3) = \frac{2}{3} = .67^{ksi}$$

$$\tau_{oct} = \frac{1}{3}\sqrt{(\sigma_1 - \sigma_2)^2 + (\sigma_2 - \sigma_3)^2 + (\sigma_3 - \sigma_1)^2} = \frac{1}{3}\sqrt{56} = 2.494^{ksi}$$

2.10 Equilibrium of a Differential Element

So far we have discussed the properties of the stress tensor and its transformation with various orientations of orthogonal axes by visualizing a "state of stress at a point" (a mathematical abstraction?). In general, in the loaded blob, stresses change over differential base lengths and to discover how they change is a primary reason that we study the mechanics of deformable solids in the first place. For design, we would like to control how stress components change along with their magnitudes, which is more difficult yet.

Consider first the free-body diagram of a differential element in rectangular orthogonal coordinates x,y,z as shown in Figure 2.11. The body-force components B_x, B_y, B_z (force per unit volume) include gravity, inertial effects

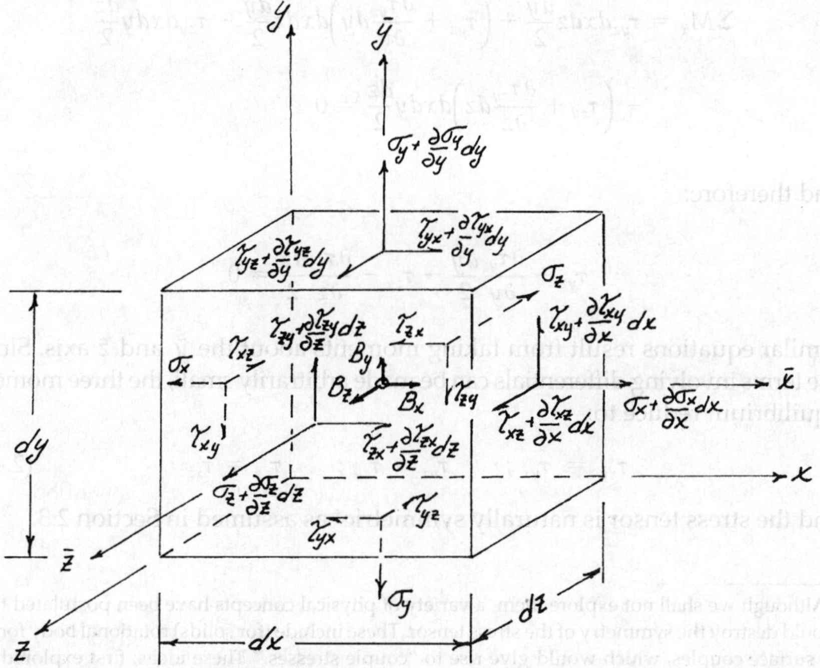

FIGURE 2.11
Free body diagram of a differential element in a changing stress field.

due to acceleration, magnetic fields, etc. Summing forces in the x direction:

$$\Sigma F_x = \left(\sigma_x + \frac{\partial \sigma_x}{\partial x}\,dx\right)dydz - \sigma_x dydz + \left(\tau_{yx} + \frac{\partial \tau_{yx}}{\partial y}dy\right)dzdx$$

$$- \tau_{yx}dzdx + \left(\tau_{zx} + \frac{\partial \tau_{zx}}{\partial z}\,dz\right)dxdy - \tau_{zx}\,dxdy + B_x\,dxdydz = 0$$

Summation of forces in the y and z directions give similar results and the three equations of equilibrium are thus:

$$\begin{aligned} \frac{\partial \sigma_x}{\partial x} + \frac{\partial \tau_{yx}}{\partial y} + \frac{\partial \tau_{zx}}{\partial z} + B_x = 0 \\ \frac{\partial \tau_{xy}}{\partial x} + \frac{\partial \sigma_y}{\partial y} + \frac{\partial \tau_{zy}}{\partial z} + B_y = 0 \\ \frac{\partial \tau_{xz}}{\partial x} + \frac{\partial \tau_{yz}}{\partial y} + \frac{\partial \sigma_z}{\partial z} + B_z = 0 \end{aligned} \tag{2.43}$$

For equilibrium of the element, the sum of the moments with respect to any axis is also zero. If the centroidal axis $\bar{x}$ parallel to x is considered first:

$$\Sigma M_{\bar{x}} = \tau_{yz}dxdz\frac{dy}{2} + \left(\tau_{yz} + \frac{\partial \tau_{yz}}{\partial y}dy\right)dxdz\frac{dy}{2} - \tau_{zy}dxdy\frac{dz}{2}$$

$$- \left(\tau_{zy} + \frac{\partial \tau_{zy}}{\partial z}dz\right)dxdy\frac{dz}{2} = 0$$

and therefore:

$$\tau_{yz} + \frac{\partial \tau_{yz}}{\partial y}\frac{dy}{2} - \tau_{zy} - \frac{\partial \tau_{zy}}{\partial z}\frac{dz}{2} = 0$$

Similar equations result from taking moments about the $\bar{y}$ and $\bar{z}$ axis. Since the terms involving differentials can be made arbitrarily small, the three moment equilibrium reduce to:

$$\tau_{xy} = \tau_{yx}\ ; \qquad \tau_{yz} = \tau_{zy}\ ; \qquad \tau_{zx} = \tau_{xz} \tag{2.44}$$

and the stress tensor is naturally symmetric* as assumed in Section 2.3.

---

* Although we shall not explore them, a variety of physical concepts have been postulated that would destroy the symmetry of the stress tensor. These include (for solids) rotational body forces or surface couples, which would give rise to "couple stresses." These ideas, first explored by Voigt and Cosserat, were reintroduced by Mindlin in 1960 in connection with his pioneering work for Bell Labs on pure quartz crystals. Couple stresses have never been shown to be significant in the laboratory or in practice for macroscopic structures.

Equations (2.43), the statement of $F = ma$ on a differential scale, are perhaps the most important ingredients in the theory of the mechanics of deformable solids. They are *field equations* that apply throughout the structure, whatever its geometry and independent of what material it is built of, the load level, or whether it is elastic, plastic, viscoelastic, or whatever. They are the starting points for all solutions or for design.

Unfortunately the differential equations of equilibrium are not sufficient in themselves to solve for a unique stress field for a given boundary loading. For three dimensions there are [with Equations (2.44)], three equations and six unknowns. In the plane two-dimensional case, the equilibrium equations reduce to:

$$\begin{aligned} \frac{\partial \sigma_x}{\partial x} + \frac{\partial \tau_{xy}}{\partial y} + B_x = 0 \\ \frac{\partial \tau_{xy}}{\partial x} + \frac{\partial \sigma_y}{\partial y} + B_y = 0 \end{aligned} \tag{2.45a}$$

which, with two equations and three unknowns, are still insufficient. Only in the rare case of one dimension can the remaining one equation of differential equilibrium with one unknown: $d\sigma_x/dx + B_x = 0$ be integrated to give a solution directly. Thus determining the stress field is inherently a statically indeterminate problem.*

2.11 Other Orthogonal Coordinate Systems

The laws of tensor transformation [Equation (2.13) or Mohr's Circle] allow us to describe stress or strain in any orientation of coordinate axes once we have solved for the tensor in one particular orthogonal set. In many physical situations, the field equations, which have been derived in a global, rectangular coordinate system, can be more easily solved in some other coordinate system.

Rather than re-derive them from consideration of stress and geometry changes for a new differential element in each coordinate system, it is easier to present the most general form and show how it can be reduced in each specific case.** The differential equations of equilibrium relative to general

* The student must be careful not to confuse *structural analysis* with *stress analysis* (although they are related and may depend on one another). Structural analysis is concerned with finding force and moment *resultants* (vectors) acting externally (reactions) and internally (bending moments, torque, shear, and normal forces on cross-sections). Stress analysis is concerned with determining how these forces and moments are distributed (what the cross-section actually "feels"). They are both part of solid mechanics, but distinct conceptually. The vector field of structural analysis may be statically determinate if there are as many vector equations of equilibrium as there are force resultants. The tensor field of stress analysis never is.

** Derived in detail in *Elasticity in Engineering Mechanics* by Boresi and Lynn, Prentice-Hall, Englewood Cliffs, NJ, 1974.

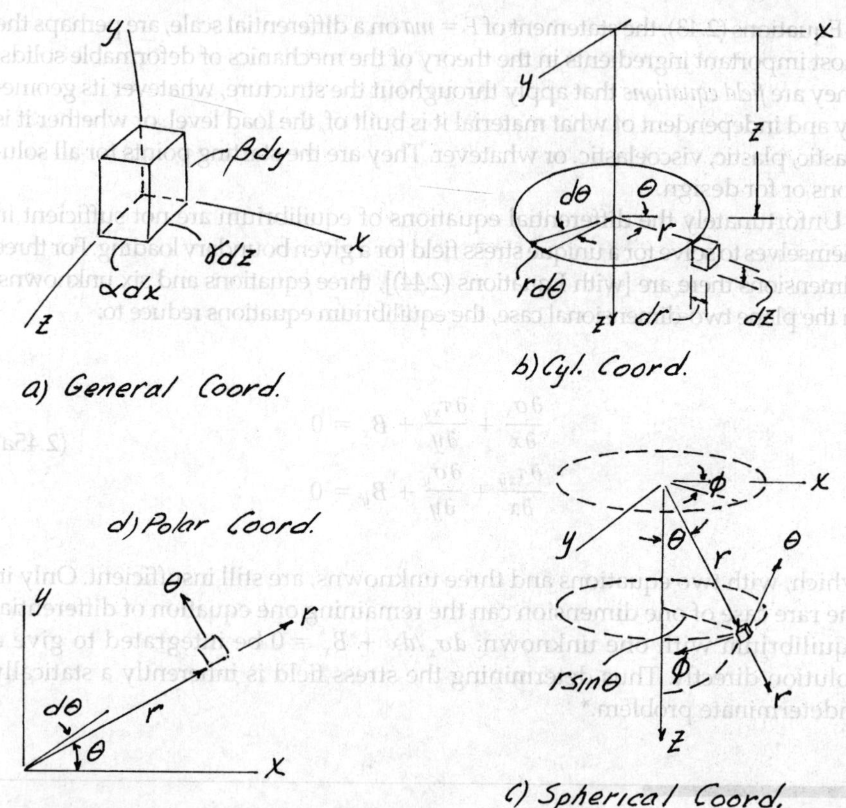

FIGURE 2.12
Orthogonal/curvilinear coordinates.

curvilinear coordinates (x, y, z) shown in Figure 2.12a are:

$$\frac{\partial(\beta\gamma\sigma_x)}{\partial x} + \frac{\partial(\gamma\alpha\tau_{xy})}{\partial y} + \frac{\partial(\alpha\beta\tau_{xz})}{\partial z} + \gamma\tau_{xy}\frac{\partial\alpha}{\partial y} + \beta\tau_{xz}\frac{\partial\alpha}{\partial z} - \gamma\sigma_y\frac{\partial\beta}{\partial x} - \beta\sigma_z\frac{\partial\gamma}{\partial x} + \alpha\beta\gamma B_x = 0$$

$$\frac{\partial(\beta\gamma\tau_{yx})}{\partial x} + \frac{\partial(\gamma\alpha\sigma_y)}{\partial y} + \frac{\partial(\alpha\beta\tau_{yz})}{\partial z} + \alpha\tau_{yz}\frac{\partial\beta}{\partial z} + \gamma\tau_{yx}\frac{\partial\beta}{\partial x} - \gamma\sigma_z\frac{\partial\gamma}{\partial y} - \gamma\sigma_x\frac{\partial\alpha}{\partial y} + \alpha\beta\gamma B_y = 0 \qquad (2.45b)$$

$$\frac{\partial(\beta\gamma\tau_{zx})}{\partial x} + \frac{\partial(\gamma\alpha\tau_{zy})}{\partial y} + \frac{\partial(\alpha\beta\sigma_z)}{\partial z} + \beta\tau_{zx}\frac{\partial\gamma}{\partial x} + \alpha\tau_{zy}\frac{\partial\gamma}{\partial y} - \beta\sigma_x\frac{\partial\beta}{\partial z} - \beta\sigma_x\frac{\partial\alpha}{\partial z} + \alpha\beta\gamma B_z = 0$$

in which (α, β, γ) are metric coefficients which are functions of the coordinates (x, y, z) defined by

$$ds^2 = \alpha^2 dx^2 + \beta^2 dy^2 + \gamma^2 dz^2 \tag{2.46}$$

where ds is the differential diagonal of the element (Figure 2.12a) with edge lengths αdx, βdy, γdz. In Cartesian coordinates $\alpha = \beta = \gamma = 1$ and Equations (2.45) reduce to Equation (2.43).

The strain-displacement equations for the element are:

$$\begin{aligned}
\epsilon_x &= \frac{1}{\alpha}\left[\frac{\partial u}{\partial x} + \frac{v}{\beta}\frac{\partial \alpha}{\partial y} + \frac{w}{\gamma}\frac{\partial \alpha}{\partial z}\right] \\
\epsilon_y &= \frac{1}{\beta}\left[\frac{\partial v}{\partial y} + \frac{w}{\gamma}\frac{\partial \beta}{\partial z} + \frac{u}{\alpha}\frac{\partial \beta}{\partial x}\right] \\
\epsilon_z &= \frac{1}{\gamma}\left[\frac{\partial w}{\partial z} + \frac{u}{\alpha}\frac{\partial \gamma}{\partial x} + \frac{v}{\beta}\frac{\partial \gamma}{\partial y}\right] \\
\gamma_{xy} &= \frac{1}{\beta}\frac{\partial u}{\partial y} + \frac{1}{\alpha}\frac{\partial v}{\partial x} - \frac{v d\beta}{\alpha\beta dx} - \frac{u}{\alpha\beta}\frac{\partial \alpha}{\partial y} \\
\gamma_{xz} &= \frac{1}{\alpha}\frac{\partial w}{\partial x} + \frac{1}{\gamma}\frac{\partial u}{\partial z} - \frac{u}{\alpha\gamma}\frac{\partial \alpha}{\partial z} - \frac{w}{\alpha\gamma}\frac{\partial \gamma}{\partial x} \\
\gamma_{yz} &= \frac{1}{\beta}\frac{\partial w}{\partial y} + \frac{1}{\gamma}\frac{\partial v}{\partial z} - \frac{w}{\beta\gamma}\frac{\partial \gamma}{\partial y} - \frac{v}{\beta\gamma}\frac{\partial \beta}{\partial z}
\end{aligned} \tag{2.47}$$

where u,v,w are the projections of the displacement vector of point x,y,z on the tangents to the coordinate lines at the point.

2.11.1 Cylindrical Coordinates (r, θ, z)

Let $x = r$, $y = \theta$, $z = z$ as shown in Figure 2.12b. Then:

$$ds^2 = dr^2 + r^2 d\theta^2 + dz^2 \tag{2.48}$$

and comparing to Equation (2.46)

$$\alpha = 1, \qquad \beta = r, \qquad \gamma = 1$$

Thus the equilibrium equations become:

$$\begin{aligned}
&\frac{\partial \sigma_r}{\partial r} + \frac{1}{r}\frac{\partial \tau_{r\theta}}{\partial \theta} + \frac{\partial \tau_{rz}}{\partial z} + \frac{\sigma_r - \sigma_\theta}{r} + B_r = 0 \\
&\frac{\partial \tau_{r\theta}}{\partial r} + \frac{1}{r}\frac{\partial \sigma_\theta}{\partial \theta} + \frac{\partial \tau_{\theta z}}{\partial z} + \frac{2\tau_{r\theta}}{r} + B_\theta = 0 \\
&\frac{\partial \tau_{rz}}{\partial r} + \frac{1}{r}\frac{\partial \tau_{\theta z}}{\partial \theta} + \frac{\partial \sigma_z}{\partial z} + \frac{\tau_{rz}}{r} + B_z = 0
\end{aligned} \tag{2.49}$$

and the strain-displacement relations are:

$$
\begin{aligned}
\epsilon_r &= \frac{\partial u}{\partial r}; \qquad \epsilon_\theta = \frac{u}{r} + \frac{1}{r}\frac{\partial v}{\partial \theta}; \qquad \epsilon_z = \frac{\partial w}{\partial z} \\
\gamma_{r\theta} &= \frac{1}{r}\frac{\partial u}{\partial \theta} + \frac{\partial v}{\partial r} - \frac{v}{r}; \qquad \gamma_{rz} = \frac{\partial u}{\partial z} + \frac{\partial w}{\partial r} \\
\gamma_{\theta z} &= \frac{\partial v}{\partial z} + \frac{1}{r}\frac{\partial w}{\partial \theta}
\end{aligned}
\tag{2.50}
$$

2.11.2 Spherical Coordinates (r, θ, ϕ)

Let $x = r, y = \theta, z = \phi$ where θ is the latitude and ϕ the longitude as shown in Figure 2.12c. The equilibrium equations become:

$$
\begin{aligned}
&\frac{\partial \sigma_r}{\partial r} + \frac{1}{r}\frac{\partial \tau_{r\theta}}{\partial \theta} + \frac{1}{r\sin\theta}\frac{\partial \tau_{r\phi}}{\partial \phi} + \frac{1}{r}(2\sigma_r - \sigma_\theta - \sigma_\phi + \tau_{r\theta}\cot\theta) + B_r = 0 \\
&\frac{\partial \tau_{r\theta}}{\partial r} + \frac{1}{r}\frac{\partial \sigma_\theta}{\partial \theta} + \frac{1}{r\sin\theta}\frac{\partial \tau_{\theta\phi}}{\partial \phi} + \frac{1}{r}[(\sigma_\theta - \sigma_\phi)\cot\theta + 3\tau_{r\theta}] + B_\theta = 0 \\
&\frac{\partial \tau_{r\phi}}{\partial r} + \frac{1}{r}\frac{\partial \tau_{\theta\phi}}{\partial \theta} + \frac{1}{r\sin\theta}\frac{\partial \sigma_\phi}{\partial \phi} + \frac{1}{r}(3\tau_{r\phi} + 2\tau_{\phi\theta}\cot\theta) + B_\phi = 0
\end{aligned}
\tag{2.51}
$$

The strain-displacement relations in spherical coordinates from Equation (2.47) become:

$$
\begin{aligned}
&\epsilon_r + \frac{\partial u}{\partial r}; \qquad \epsilon_\theta = \frac{u}{r} + \frac{1}{r}\frac{\partial v}{\partial \theta}; \qquad \epsilon_\phi = \frac{u}{r} + \frac{v}{r}\cot\theta + \frac{1}{r\sin\theta}\frac{\partial w}{\partial \phi} \\
&\gamma_{r\theta} = \frac{1}{r}\frac{\partial u}{\partial \theta} + \frac{\partial v}{\partial r} - \frac{v}{r}; \qquad \gamma_{r\phi} = \frac{1}{r\sin\theta}\frac{\partial u}{\partial \phi} + \frac{\partial w}{\partial r} - \frac{w}{r} \\
&\gamma_{\theta\phi} = \frac{1}{r}\left(\frac{\partial w}{\partial \theta} - w\cot\theta\right) + \frac{1}{r\sin\theta}\frac{\partial v}{\partial \phi}
\end{aligned}
\tag{2.52}
$$

2.11.3 Plane Polar Coordinates (r, θ)

Polar coordinates for plane problems are, it will turn out, equally as useful as Cartesian coordinates. They can be considered a special case of cylindrical coordinates where $\tau_{rz} = \tau_{\theta z} = (\partial/\partial z) = 0$ and from Equation (2.49), equilibrium requires

$$
\begin{aligned}
&\frac{\partial \sigma_r}{\partial r} + \frac{1}{r}\frac{\partial \tau_{r\theta}}{\partial \theta} + \frac{\sigma_r - \sigma_\theta}{r} + B_r = 0 \\
&\frac{\partial \tau_{r\theta}}{\partial r} + \frac{1}{r}\frac{\partial \sigma_\theta}{\partial \theta} + \frac{2\tau_{r\theta}}{r} + B_\theta = 0
\end{aligned}
\tag{2.53}
$$

The strain-displacement relations are:

$$\epsilon_r + \frac{\partial u}{\partial r}; \qquad \epsilon_\theta = \frac{u}{r} + \frac{1}{r}\frac{\partial v}{\partial \theta}$$
$$\gamma_{r\theta} = \frac{1}{r}\frac{\partial u}{\partial \theta} + \frac{\partial v}{\partial r} - \frac{v}{r} \tag{2.54}$$

2.12 Summary

There are two basic concepts that are discussed and developed in this chapter beyond the introductory level. The first of these, as reviewed to some extent in Table 2.1, is to examine the physical and analytic properties of tensors, and in particular relative deformation, strain, and stress. We see that the components in different orientations are completely different but related by the equations of tensor transformation. The mathematics reflect fundamental physical meaning through the invariant coefficients in the characteristic equation. Alternatively, Mohr's Circle gives a complete graphic picture of a second-order tensor in 2D in all orientations simultaneously and, as such, is an invaluable tool for visualizing stress and strain for analysis and design. Mohr's Circle in 3D serves the same purpose, but only in principal space where the three invariants occur on the octahedral plane. Mohr's Circle also emphasizes graphically the decoupling of isotropic and deviatoric components of symmetric tensors that represent the separate effects of volume change and distortion.

The second major theme is the examination of the deformation and the equilibrium of a differential element to develop the field equations of deformable solids. These are summarized in Table 2.2 using tensor notation for brevity where the comma between subscripts signifies the partial derivative with respect to the index following the comma , e.g.: $u_{i,i} = \frac{\partial u_i}{\partial x_i}$. It is important to remember that these equations are general in the sense that they are independent of material properties. However, the assumption of small and smooth u, v, w implies no yield, in that no kinks are allowed in the strain field.

Thus we discover that so far, from at least an analytic standpoint, the field analysis of deformable solids is intractable. From a geometric view there are 12 unknowns, u, v, w, ϵ_x, ϵ_y, ϵ_z, γ_{xy}, γ_{yz}, γ_{zx}, ω_x, ω_y, and ω_z which can change throughout the blob as functions of position, temperature, material properties, etc., and only 9 equations. From an equilibrium point of view we are again deficient by 3 equations. Even in two dimensions the situation remains indeterminate. We must therefore, like Mr. Hooke, turn to the laboratory to relate stress to strain by experiment.

TABLE 2.1

Types and Properties of Tensors

	Zero Order	1st Order		2nd Order			
	Scaler	Vector		Tensor (Diadic)			
Examples	Pressure	Displacement		Relative Displacement			
Notation	p	u_i or $\bar{u}$		δ_{ij} or $[\delta_{ij}]$			
Components	$3^0 = 1$	2D $2^1 = 2$	3D $3^1 = 3$	2D $2^2 = 4$		3D $3^2 = 9$	
				Symmetric	Asymmetric	Symmetric	Asymmetric
Invariant scalar magnitudes + directional quant.	1 0	1 1	1 2	2 1	1 0	3 3	1 2
Matrix representation	p	$\begin{bmatrix} u_x \\ u_y \end{bmatrix}$ or $\begin{bmatrix} u \\ v \end{bmatrix}$	$\begin{bmatrix} u_x \\ u_y \\ u_z \end{bmatrix}$ or $\begin{bmatrix} u \\ v \\ w \end{bmatrix}$	$\begin{bmatrix} \epsilon_x & \frac{\gamma_{xy}}{2} \\ \frac{\gamma_{xy}}{2} & \epsilon_y \end{bmatrix}$	ω	$\begin{bmatrix} \epsilon_x & \frac{\gamma_{xy}}{2} & \frac{\gamma_{xz}}{2} \\ \frac{\gamma_{xy}}{2} & \epsilon_y & \frac{\gamma_{yz}}{2} \\ \frac{\gamma_{xz}}{2} & \frac{\gamma_{yz}}{2} & \epsilon_z \end{bmatrix}$	$\begin{bmatrix} 0 & -\omega_z & \omega_y \\ \omega_z & 0 & -\omega_x \\ -\omega_y & \omega_x & 0 \end{bmatrix}$
Along principal space diagonal	p	$\bar{\delta}_{quad}$	$\bar{\delta}_{oct}$	$\epsilon_m = \epsilon_{quad}$ $\gamma_{quad}/2$ θ	ω_z	$\epsilon_m = \epsilon_{oct}$ $\gamma_{oct}/2$ α_{oct} + 3 angles	$\bar{\omega}$ (magnitude + 2 angles)

TABLE 2.2

Field Equations of Solids

Unknown Tensor Components	Field Equations: 3 Dimensions	Field Equations: 2 Dimensions
Displacements u_i	Very smooth and continuous	Very smooth and continuous
Strains ϵ_{ij}	$\epsilon_i = u_{i,j}$	$\epsilon_x = \frac{\partial u}{\partial x}, \quad \epsilon_y = \frac{\partial v}{\partial y}$
	$\frac{\gamma_{ij}}{2} = \frac{1}{2}(u_{j,i} + u_{i,j})$	$\frac{\gamma_{xy}}{2} = \frac{1}{2}\left(\frac{\partial v}{\partial x} + \frac{\partial u}{\partial y}\right)$
Rotations ω_{ij}	$\omega_{ij} = \frac{1}{2}(u_{j,i} - u_{i,j})$	$\frac{\omega_{xy}}{2} = \omega_z = \frac{1}{2}\left(\frac{\partial v}{\partial x} - \frac{\partial u}{\partial y}\right)$
Stresses σ_{ij}	$\sigma_{ij,j} + B_i = 0$	$\frac{\partial \sigma_x}{\partial x} + \frac{\partial \tau_{xy}}{\partial y} + B_x = 0$
		$\frac{\partial \tau_{xy}}{\partial x} + \frac{\partial \sigma_y}{\partial y} + B_x = 0$

2.13 Problems and Questions

P2.1 Concerning the footnote in Section 2.3, list four "primitive quantities" and the associated laws of measurement (such as Darcy's or Ohm's law).

P2.2 As mentioned in the footnote in Section 2.7, prove that the 2D elastic rotation ω_{xy} or ω_z is a scalar invariant.

P2.3 What are the invariants of the general (nonsymmetric) deformation tensor, E_{ij}, given by Equation (2.1) in terms of strain and rotation components. It is easiest to do this in 2D and extend to 3D in principal space. The result of P2.2 above may also be useful.

P2.4 Where does the third invariant or some measure of it appear on the 3D Mohr's Circle?

P2.5 Find the principal stresses and the direction cosines corresponding to the state of stress: $\sigma_x = 0.5$, $\sigma_y = 0.7$, $\sigma_z = 0.6$, $\tau_{xy} = 1$, $\tau_{yz} = 1.2$, $\tau_{zx} = 0.8$. Illustrate with a sketch showing the spacial angles of this transformation from x, y, z to 1, 2, 3 orientation.

P2.6 Find the octahedral normal and shear stress and their resultant corresponding to the state of stress given in P2.5 above and illustrate on Mohr's Circle. Also find the orientation from the x, y, z axes to the space diagonal in the first octant of principal space.

P2.7 A plate is stretched as shown in Figure P2.7. Determine ϵ_1, ϵ_2, γ_{max}, ω_{12}.

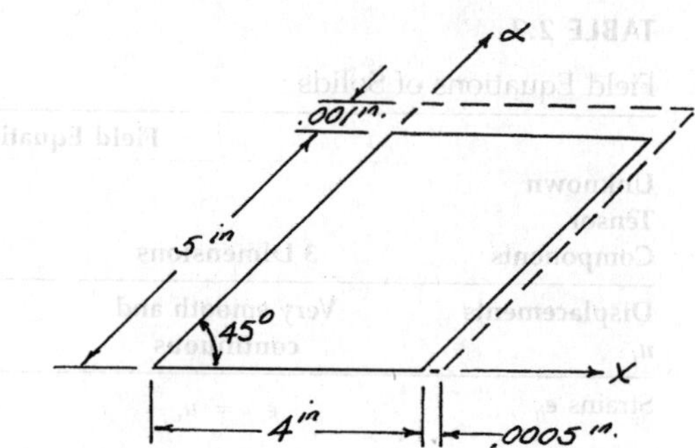

FIGURE P2.7

P2.8 The stresses at a point are given as

$$\sigma_x = -500 \text{ psi}, \quad \sigma_y = 700 \text{ psi}, \quad \sigma_z = 1000 \text{ psi}$$
$$\tau_{xy} = 1200 \text{ psi}, \quad \tau_{yz} = 800 \text{ psi}, \quad \tau_{zx} = -300 \text{ psi}$$

a. Determine the principal stresses.
b. Using Mohr's Circle, determine the normal and shearing stress on a plane normal to a line inclined 30° clockwise from the major principal axis.
c. Resolve both the original and the principal tensor into their isotropic and deviatoric components.
d. Show the isotropic and deviatoric components of normal stress (i.e., σ_m, S_1, S_2, S_3) on the Mohr's Circle (carefully drawn to scale).
e. Calculate the invariants of the general stress tensor σ_{ij} and its isotropic and deviatoric components for both the original x, y, z orientation and the principal 1, 2, 3 orientation (checking to make sure they are actually the same in either system).
f. Calculate the octahedral normal and shear stress (σ_{oct} and τ_{oct}) and show them on the Mohr's Circle.

P2.9 The strains in a body vary linearly as:

$$\epsilon_x = \epsilon_y = \epsilon_z = \gamma_{xy} = 0; \qquad \gamma_{xz} = -cy; \qquad \gamma_{yz} = cx$$

Determine ϵ_1, ϵ_2, ϵ_3 at the point 1, 2, 4 and show their directions on a 3D sketch. What is the physical situation that gives rise to this strain field?

P2.10 Determine by tensor transformation the area moments of inertia I_{xx}, I_{yy}, I_{xy} of the cross-sections shown and illustrate by a Mohr's Circle (use Mohr's Circle directly if you wish).

P2.11 What are the engineering terms for the two invariants of the plane moment of inertia tensor? Determine what they are for the cross-sections in Figure P2.10.

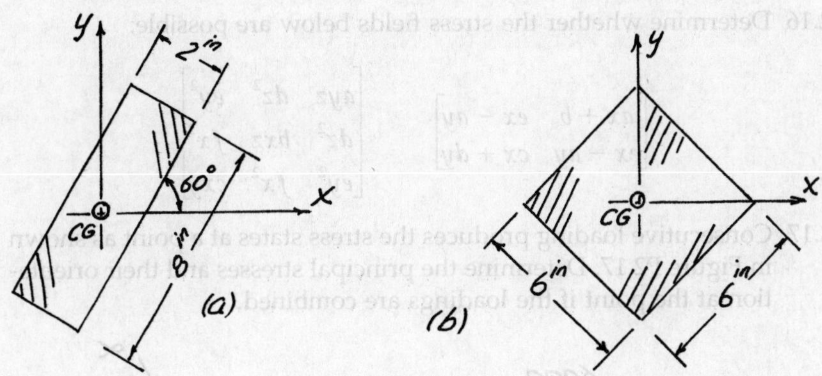

FIGURE P2.10

P2.12 Discuss a 3D moment of inertia tensor. Where does it appear in engineering analysis (if at all)?

P2.13 For a state of stress given by $\sigma_x = 12$, $\sigma_y = 6$, $\tau_{xy} = 3$, $\sigma_z = \tau_{xz} = \tau_{yz} = 0$, compare σ_{oct} and τ_{oct} to σ_{quad} and τ_{quad} to $\frac{\sigma_1 - \sigma_3}{2}$ and τ_{max} and show on a sketch the orientation of each with respect to the x, y, z and 1, 2, 3 axes (in the first octant).

P2.14 Compute τ for equilibrium and discuss uniqueness (Fig. P2.14).

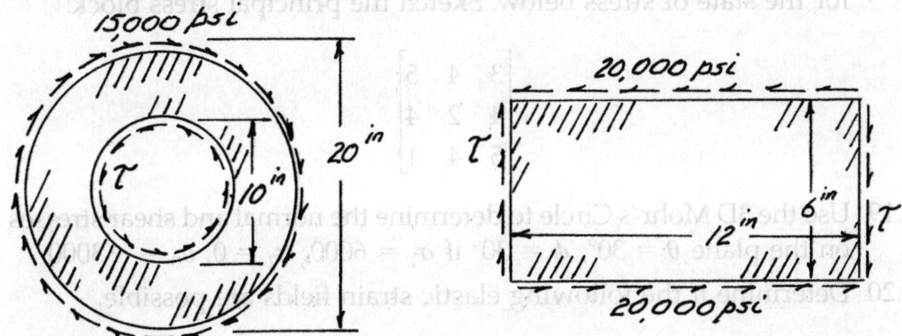

FIGURE P2.14

P2.15 Indicate whether the following statements are true or false or poorly defined:

a. Strain theory depends upon the material being discussed.
b. Stress theory depends upon strain theory.
c. The mathematical theories of stress and strain are equivalent.
d. The correct strains of a strained continuum must be compatible.
e. If the stress components are directly proportional to the corresponding strain components, the values of the principal stresses are related to the values of the principal strains by the same proportional factor.

P2.16 Determine whether the stress fields below are possible:

$$\begin{bmatrix} ax+b & ex-ay \\ ex-ay & cx+dy \end{bmatrix} \qquad \begin{bmatrix} ayz & dz^2 & ey^2 \\ dz^2 & bxz & fx^2 \\ ey^2 & fx^2 & cxy \end{bmatrix}$$

P2.17 Consecutive loading produces the stress states at a point as shown in Figure P2.17. Determine the principal stresses and their orientation at the point if the loadings are combined.

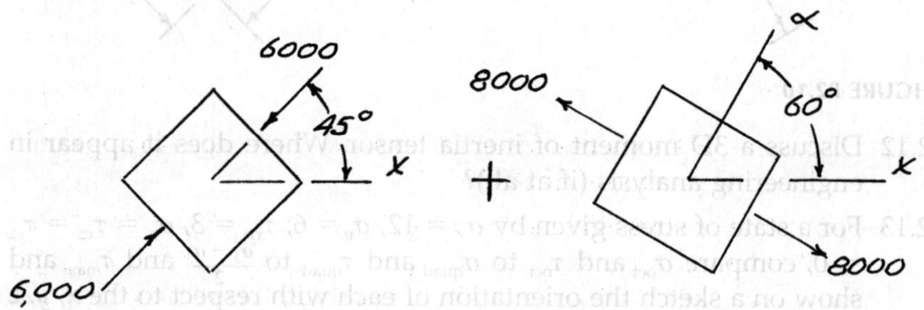

FIGURE P2.17

P2.18 Determine the principal stresses and the related directional cosines for the state of stress below. Sketch the principal stress block:

$$\begin{bmatrix} 3 & 4 & 5 \\ 4 & 2 & 4 \\ 5 & 4 & 1 \end{bmatrix}$$

P2.19 Use the 3D Mohr's Circle to determine the normal and shear stresses on the plane $\theta = 30°$, $\phi = 30°$ if $\sigma_1 = 6000$, $\sigma_2 = 0$, $\sigma_3 = -3000$.

P2.20 Determine if the following elastic strain fields are possible.

$$\begin{bmatrix} x^2 & bxy \\ bxy & a(x^2+y^2) \end{bmatrix} \qquad \begin{bmatrix} x^2z & bxyz & 0 \\ bxyz & az(x^2+y^2) & 0 \\ 0 & 0 & 0 \end{bmatrix}$$

P2.21 Is the displacement field:

$$\begin{bmatrix} u \\ v \\ w \end{bmatrix} = \begin{bmatrix} .002 & 0 & -.002 \\ .006 & .002 & 0 \\ 0 & .001 & -.006 \end{bmatrix} \begin{bmatrix} x \\ y \\ z \end{bmatrix}$$

permissible in a continuous body and, if so, what is the relative deformation between points P (1,0,0) and D(0,1,2). Sketch and resolve into strain and rigid body components.

3

Stress–Strain Relationships (Rheology)

The goal of rheology is to relate, by laboratory testing, each component of stress to each component of strain while, in the process, introducing as few material properties as is necessary to capture the fundamental behavior with sufficient accuracy for engineering purposes. Such "constitutive relationships" for engineering analysis and design are only a small part of material science which, using chemistry and physics, tries to answer the question "why" in addition to "how" materials behave. As yet, material science is not capable of deriving with sufficient accuracy for engineers the load–deformation response of any "solid" material from fundamental principles (such as bond strength, crystal structure, etc). Thus, while such considerations can often lead to a better understanding of the behavior of materials under load, decisions as to what might be done to make materials better, or even the design of new materials for special uses, the material properties from standard tests remain the basis for engineering analysis and design.

3.1 Linear Elastic Behavior

A material specimen put under load deforms and with sensitive instruments we can measure, at a given time and temperature, its new shape. The first and simplest response idealization is linear-elastic behavior. If, in our test, we increase the load and the deformations increase proportionately, the material is linear. If we unload the specimen and it returns to its original shape, it is also elastic. Hooke was the first to do experiments of this sort and to recognize that this fundamental concept, now called Hooke's Law,* defining a linear load–deformation relationship could lead to revolution in structural mechanics.

The uniaxial tension test used by Hooke has become standard because it is so simple. If we measure and plot the deflection, Δ, of bars with different

* Hooke tested all sorts of springs and solid materials in both tension and compression including "metal, wood, stones, baked earth, hair, horns, silk, bones, sinews, glass, and the like," and applied his proportional load–deformation concept directly to tuning forks, his invention of the watch spring (stolen by the president of the Royal Society), and unsuccessful attempts to measure a decrease in gravitational attraction with elevation.

lengths and cross-sectional areas under increasing load, P, we obtain a variety of straight lines for P vs. Δ. Young* was apparently the first to divide the load by the area and the deflection by the length to make all these lines coincide for a given material. Although he may not have quite realized it at least originally, Young thereby related normal stress to normal strain in the same direction by the portionality constant, E, now named after him.** In the simple tension test shown in Figure 3.1, the lateral contraction ΔW can be measured and is found to be proportional to the longitudinal elongation ΔL. Dividing by the original base lengths, W_0 and L_0, to

* Thomas Young (1773–1829) was the prototypical English intellectual adventurer of remarkable genius. He turned his transcendent ability to every type of intricate problem from deciphering the first eleven scripts of the Rosetta Stone to hydraulics, tides, annuities, carpentry, life preservers, steam engines, and the double refraction of light (among others). Fluent at the age of 14 in a dozen languages, he went to London, Edinburgh, and Gottingen to get his medical degree before going to study science at Cambridge. At Cambridge he seemed totally bored. In the words of one of his colleagues:

> "When the Master introduced Young to his tutors, he jocularly said: 'I have brought you a pupil qualified to read lectures to his tutors.' This, however, he did not attempt… He never obtruded his various learning in conversation; but if appealed to on the most difficult subject, he answered in a quick, flippant, decisive way, as if he was speaking of the most easy; and in this mode of talking he differed from all the clever men that I ever saw. His reply never seemed to cost him an effort, and he did not appear to think there was any credit in being able to make it. He did not assert any superiority, or seem to suppose that he possessed it; but spoke as if he took it for granted that we all understood the matter as well as he did… His language was correct, his utterance rapid, and his sentences, though without any affectation, never left unfinished. But his words were not those in familiar use, and the arrangement of his ideas seldom the same as those he conversed with. He was, therefore, worse calculated than any man I ever knew for the communication of knowledge… It was difficult to say how he employed himself; he read little, and though he had access to the College and University libraries, he was seldom seen in them. There were no books piled on his floor, no papers scattered on his table, and his room had all the appearance of belonging to an idle man… He seldom gave an opinion, and never volunteered one. A philosophical fact, a difficult calculation, an ingenious instrument, or a new invention, would engage his attention; but he never spoke of morals, of metaphysics, or of religion."

This quotation is taken from a biography of Young by G. Peacock, London, 1855.

** Actually Young talked about "weight of the modulus," AE, and the "height of the modulus," LE, which is typical of his penchant for insight expressed with confusion (in this case confusing the modulus, E, which relates stress to strain and the stiffness or spring constant AE/L relating load to deformation). This inability to express his ideas to persons of lesser intelligence (practically everyone) led to his failure as a lecturer at the Royal Institute where, in 1801, he was appointed professor after his famous discovery of the interference of light and prompt election as a member of the Royal Society. Thus Young was forced to resign his professorship after two years but continued his research on all fronts. He not only considered axial loads but suggested a proportional relationship between shear stress and angle of twist in torsion (the shear modulus), recognized the Poisson ratio effect, and noted that the onset of yielding and permanent set was the logical limit for design. He explained in detail the combined bending and buckling of beams and columns solving all sorts of practical problems, but his work was unrecognized until Lord Rayleigh reviewed it 100 years later when he gave similar lectures at the Royal Institute. For example, Young's estimate of the molecular diameter as 2 to 10×10^{-9} inches anticipated Kelvin's similar estimate by 50 years, but it went unnoticed.

reduce the data to normal engineering strain, the proportionality factor, ν, relating them is called Poisson's ratio.* Thus in simple tension (or compression):

$$\epsilon_x = \frac{\sigma_x}{E}; \qquad \epsilon_y = \epsilon_z = -\frac{\nu}{E}\sigma_x = -\nu\epsilon_x; \qquad \sigma_y = \sigma_z = 0 \tag{3.1}$$

with x, y, z principal as there are no shears. Since a uniaxial test includes both isotropic and deviatoric components, the data, as shown in Figure 3.1, can also be used to relate both the individual distortional components e_{ij} to s_{ij} term by term** and the volumetric behavior, ϵ_m to σ_m.

Finally, effects of temperature can be included. If the "blob" is free to expand or contract and the temperature change is uniform, only the volume changes and no additional stresses result. However, if there are redundant external restraints or thermal gradients, the stress field will also be affected. On a differential scale, temperature changes, ΔT, do not directly affect the deviatoric components of stress since there is no distortion of a differential element by ΔT. Thus the term $\epsilon_i = \alpha\Delta T$ can be incorporated directly in the relationship between normal strains and stresses*** (α being the coefficient of thermal expansion).

Implicit in our discussion of a simple tension test (Figure 3.1) is the assumption that the specimen represents a material that is isotropic (the same in all directions) and also homogeneous (the same at all points). If this were not the case, the proportionality constants would need subscripts and as many as 36 such subscripted "moduli" (not including α) might be necessary to define linear elastic behavior. However, for isotropic materials, many nice

---

Thus Young like Hooke (whom he called the "greatest of all philosophical mechanicers") was not appreciated by his contemporaries and while he may have had major impact on the development of structural design and the development of elasticity through personal interaction with good friends such as John Rennie or conversations with Navier (who came to England seeking his help more than once), there was little immediate recognition. Nor did he look for scientific acclaim. Devoid of arrogance, even ambition, he was unable and uninterested in promoting his ideas developed primarily for the self-gratifying pleasure of satisfying his prodigious mind.

* S. D. Poisson (1781–1840), while not a pioneer in the formulation of the Theory of Elasticity to the extent that his contemporaries Navier and Cauchy were, was the first to recognize the fundamental phenomenon of dilitational and deviatoric stress waves as well as formulate the basic equation for plate bending. Poisson also worked on the vibration of plates where he seems to have introduced the concept of using trigonometric series. He takes his ratio $\nu = -\epsilon_y/\epsilon_x$ from the tension test as $1/4$ in most of this work.

** Actually the shear modulus, G, is usually found from a separate, purely deviatoric experiment such as simple torsion or a biaxial test with equal tension and compression. Similarly a purely isotropic test to determine the bulk modulus, K, can be run in a triaxial cell by measuring the volume change with increasing increments of hydrostatic stress. However, this is a very difficult test in practice and usually K is computed from E and ν.

*** It is often easier in linear analysis to solve the thermal stress–strain field separately and superimpose with the field due to the applied loads.

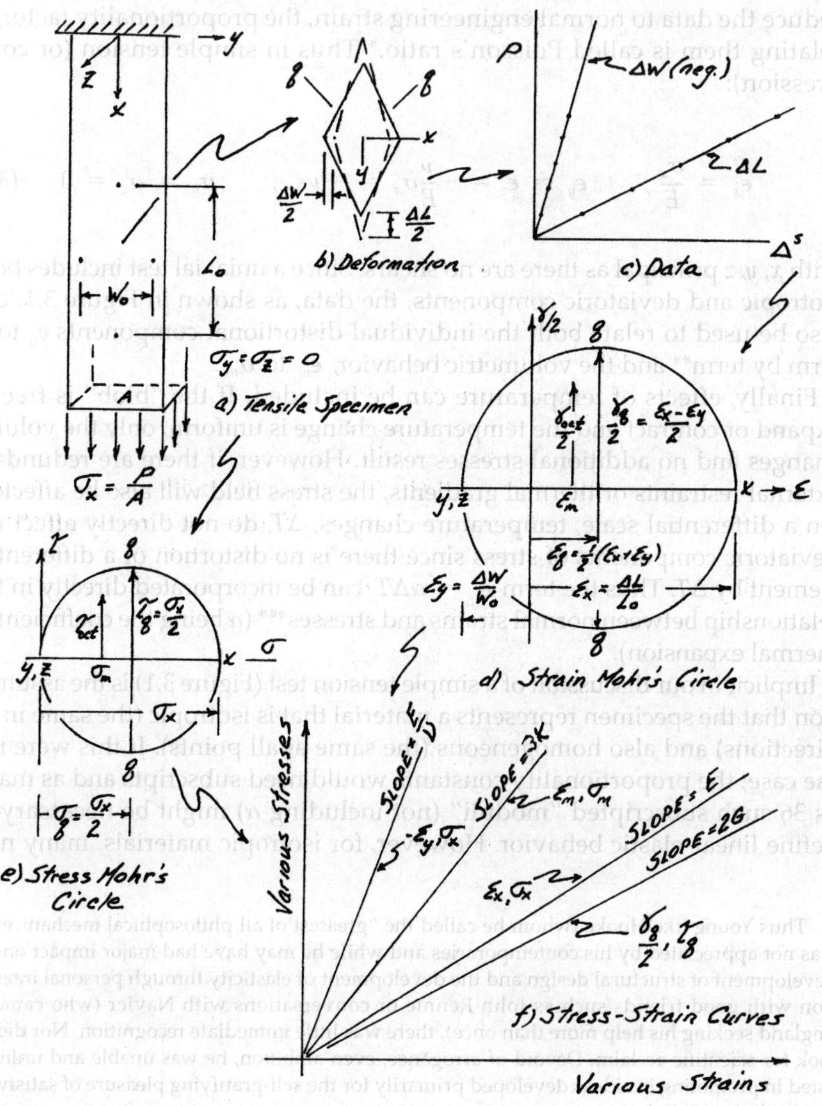

FIGURE 3.1
Simple tension test.

things happen. Any state of stress or strain can be transformed to its principal representation (without shear components). Considering this orientation, the principal directions for stress and strain must coincide and therefore the same is true for any other orientation of orthogonal axes. Thus we can add together three simple tension tests [Equation (3.1)] in orthogonal directions to derive the general 3D constitutive relationships:

$$\epsilon_x = \frac{1}{E}(\sigma_x - \nu\sigma_y - \nu\sigma_z) + \alpha\Delta T \qquad \sigma_x = 2G\left(\epsilon_x + \frac{3\nu}{1-2\nu}\epsilon_m\right) - 3K\alpha\Delta T$$

$$\epsilon_y = \frac{1}{E}(\sigma_y - \nu\sigma_z - \nu\sigma_x) + \alpha\Delta T \quad \text{or} \quad \sigma_y = 2G\left(\epsilon_y + \frac{3\nu}{1-2\nu}\epsilon_m\right) - 3K\alpha\Delta T$$

$$\epsilon_z = \frac{1}{E}(\sigma_z - \nu\sigma_x - \nu\sigma_y) + \alpha\Delta T \qquad \sigma_z = 2G\left(\epsilon_z + \frac{3\nu}{1-2\nu}\epsilon_m\right) - 3K\alpha\Delta T \tag{3.2a}$$

with the relationship between shear components:

$$\gamma_{xy} = \frac{\tau_{xy}}{G} \qquad \tau_{xy} = G\gamma_{xy}$$

$$\gamma_{yz} = \frac{\tau_{yz}}{G} \quad \text{or} \quad \tau_{yz} = G\gamma_{yz} \tag{3.2b}$$

$$\gamma_{zx} = \frac{\tau_{zx}}{G} \qquad \tau_{zx} = G\gamma_{zx}$$

where x, y, z could just as well be r, θ, z or a, b, c, or any set of orthogonal coordinates. Superposition used at the stress–strain level with isotropy applies as well for the load–deformation behavior of a homogeneous body. The inverse equations for stresses in terms of strain components, also given in Equations (3.2), are very useful for converting experimental data and for expressing displacement boundary conditions in terms of stresses.

These "coupled," linear-elastic constitutive relationships between vector components of the *total* stress and strain tensors are often the easiest to use. Conceptually it is much better, however, to think in terms of the separate isotropic and deviatoric components. A clue to this insight is that the off-diagonal terms are already uncoupled in that only one modulus, G, is involved. It is easy to show using Equation (3.2) that the deviatoric terms, $e_i = \epsilon_i - \epsilon_m$ and $s_i = \sigma_i - \sigma_m$, are related in the same way and derive the relationship:

$$G = \frac{E}{2(1+\nu)} \tag{3.3}$$

Similarly, the relationship between the scalar isotropic components

$$\frac{\epsilon_\nu}{3} = \epsilon_m = \frac{\sigma_m}{3K} + \alpha\Delta T \qquad \text{where} \qquad K = \frac{E}{3(1-2\nu)} \tag{3.4}$$

is obtained by adding together the three normal stress–strain relationships in terms of E and ν. Thus Equation (3.2) can be written:

$$\epsilon_m = \frac{1}{3K}\sigma_m + \alpha\Delta T \qquad \text{or} \qquad \sigma_m = 3K\epsilon_m - 3K\alpha\Delta T \tag{3.5a}$$

and

$$
\begin{aligned}
\epsilon_x - \epsilon_m &= e_x = \frac{1}{2G}(\sigma_x - \sigma_m) = \frac{1}{2G}s_x \\
\epsilon_y - \epsilon_m &= \frac{1}{2G}(\sigma_y - \sigma_m) \\
\epsilon_z - \epsilon_m &= \frac{1}{2G}(\sigma_z - \sigma_m) \qquad \text{or} \qquad s_{ij} = 2Ge_{ij} \\
\gamma_{xy}/2 &= \frac{1}{2G}\tau_{xy} \\
\gamma_{yz}/2 &= \frac{1}{2G}\tau_{yz} \\
\gamma_{zx}/2 &= \frac{1}{2G}\tau_{zx}
\end{aligned}
\tag{3.5b}
$$

The plane-stress and plane-strain relationships are easily extracted from Equations (3.2). Without ΔT they are

a) For Plane Stress:

$$
\begin{aligned}
\epsilon_x &= \frac{1}{E}(\sigma_x - \nu\sigma_y) \quad &\text{or} \quad \sigma_x &= \frac{E}{1-\nu^2}(\epsilon_x + \nu\epsilon_y) \\
\epsilon_y &= \frac{1}{E}(\sigma_y - \nu\sigma_x) \quad &\text{or} \quad \sigma_y &= \frac{E}{1-\nu^2}(\epsilon_y + \nu\epsilon_x) \\
\epsilon_z &= -\frac{\nu}{E}(\sigma_x + \sigma_y) \quad &\text{or} \quad \sigma_z &= 0 \\
\gamma_{xy} &= \frac{1}{G}\tau_{xy} \quad &\text{or} \quad \tau_{xy} &= G\gamma_{xy} \\
\gamma_{yz} &= \gamma_{zx} = 0 \quad &\text{or} \quad \tau_{yz} &= \tau_{zx} = 0
\end{aligned}
\tag{3.6}
$$

b) and for Plane Strain:

$$
\begin{aligned}
\epsilon_x &= \frac{1}{E'}(\sigma_x - \nu'\sigma_y) \quad &\text{or} \quad \sigma_x &= \frac{E'}{1-\nu'^2}(\epsilon_x + \nu'\epsilon_y) \\
\epsilon_y &= \frac{1}{E'}(\sigma_y - \nu'\sigma_x) \quad &\text{or} \quad \sigma_y &= \frac{E'}{1-\nu'^2}(\epsilon_y + \nu'\epsilon_x) \\
\epsilon_z &= 0 \quad &\text{or} \quad \sigma_z &= \nu(\sigma_x + \sigma_y) \\
\gamma_{xy} &= \frac{1}{G}\tau_{xy} \quad &\text{or} \quad \tau_{xy} &= G\gamma_{xy} \\
\gamma_{yz} &= \gamma_{zx} = 0 \quad &\text{or} \quad \tau_{yz} &= \tau_{zx} = 0
\end{aligned}
\tag{3.7}
$$

where the fictitious plane-strain moduli are defined as:

$$E' = \frac{E}{1-\nu^2}; \qquad \nu' = \frac{\nu}{1-\nu}$$

These equations show that the *in-plane* equations are the same form in both cases which is, as we will see, very significant when it comes to finding 2D fields either analytically or experimentally. It is also interesting to note that G and K are not affected by the assumption of plane stress or plane strain giving the same result whether E and ν or E' and ν' are substituted into Equations (3.3) and (3.4).

It seems from Equation (3.5) that materials respond to volumetric (hydrostatic) tension or compression quite independent of their distortional (shearing) behavior. As noted previously, nature seems to favor simplicity and we should not be surprised that isotropic and deviatoric behavior are uncoupled. Each component is modeled separately by a simple spring and materials can have totally dissimilar isotropic and deviatoric stiffness. At one extreme is the perfectly elastic solid where both K and G values are generally of the same order of magnitude, and at the other extreme are liquids and gasses where K can still be large under hydrostatic pressure while G is nearly zero. For fluids, the only way significant shear stresses can develop is if significant velocities of shear deformation are imposed.*

Example 3.1

An air chamber in a circular pipe is formed between two circular plates connected by an aluminum rod as shown. The rigid circular plates are assumed to fit the rigid pipe exactly so that no air can escape from the chamber but they are free to slide along the pipe. Suppose the temperature has been raised 150 °F and air pressure is increased by 20,000 psi. Radius of the pipe $R = 20$ in., radius of the rod $r = 10$ in., initial length of the rod $L = 100$ in., and the material properties of the rod are $E = 10 \times 10^6$ psi, $\alpha = 20 \times 10^{-6}/\text{F}°$, $v = 0.25$. Find the length change of the aluminum rod.

* And thus we come to the fundamental division of fluid mechanics ($G = 0$) from solid mechanics. If $G = 0$ there is no direct resistance to deviatoric stress (no shear stiffness) and the material must flow until equilibrium is achieved through viscous resistance and/or inertial forces. Although, for fluids, conservation of momentum, conservation of mass, and some equation of state also must be considered separately, the basic field equations of equilibrium and compatibility still apply. Thus one might think it would be economic to maintain this correspondence and learn both subjects simultaneously. Books and courses on "continuum mechanics" do in fact periodically appear trying to accomplish this but they never catch on. Some scholars after years of effort do eventually achieve this synthesis and can visualize and enjoy roving around in the unified world of mechanics. It is worthwhile to recognize this goal and strive for it, but to present the unified theory in all its complexity first and then work down to specific cases later, does not seem to be a fruitful pedagogical strategy. Presenting "the big picture" first tends to destroy motivation and, more importantly, is not in tune with the way the normal human mind works which, like a good computer routine, debugs and perfects subroutines before assembling them into a general program.

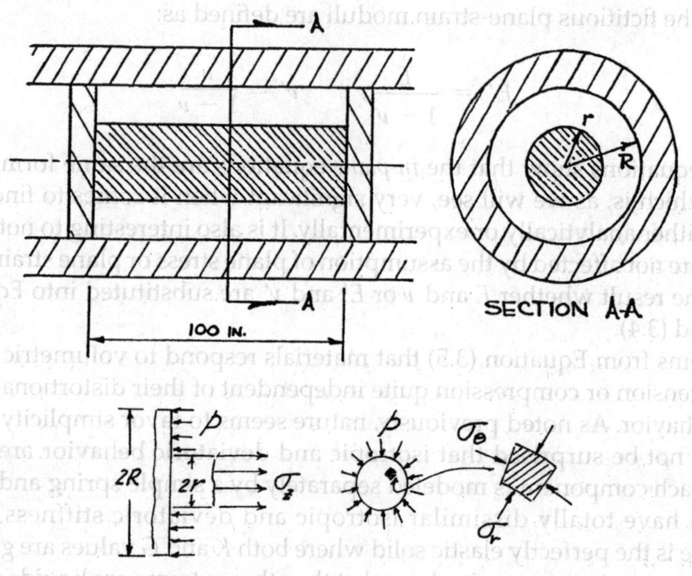

$$\sigma_r = \sigma_\theta = -p$$

$$\Sigma F_z = 0 \qquad \therefore \; \sigma_z \pi r^2 = p\pi(R^2 - r^2)$$

$\therefore$ the stress Field is: $\quad \sigma_r = \sigma_\theta = -p = -20{,}000$ psi

$$\sigma_z = 60{,}000 \text{ psi}$$

Axial Strain

$$\epsilon_z = \frac{1}{E}(\sigma_z - \nu\sigma_r - \nu\sigma_\theta) + \alpha\Delta T$$

$$\Delta L = \epsilon_z L = 100\left\{\frac{1}{10^7}\left[60{,}000 + \frac{1}{4}(40{,}000)\right] + 20 \times 10^6(150)\right\}$$

$$\Delta L = 100[10^{-7}(70{,}000 + 30{,}000)] = 1 \text{ in.}$$

3.2 Linear Viscous Behavior

Such time-dependent stiffness (the resistance through induced stresses to the time-rate-of-change of strain) is called viscosity. If this relationship between stress and strain rate is proportional, the material exhibits Neutonion or linear viscosity. Such ideal behavior, shown in Figure 3.2, is generally observed in the laboratory to a reasonable degree at least at moderate strain rates for fluids just as linear elastic response is so common for solids at moderate strains.

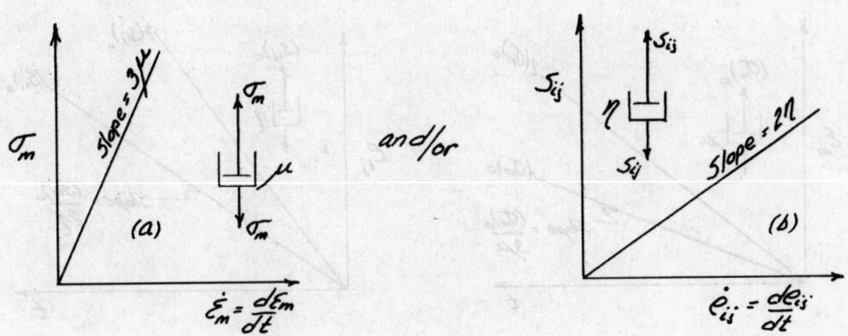

FIGURE 3.2
Dashpot models for linear viscosity: (a) isotropic and (b) deviatoric (both either positive or negative).

The linear viscous stiffness element called a "dashpot," physically represents a piston in a sleeve where the surrounding viscous fluid is expelled (in compression) or sucked into the chamber at a rate dependent on the size of the gap between the piston and the sleeve (through which the fluid must flow). Real dashpots, often used as shock absorbers, are usually filled with oil as the working fluid (silicone oil is a good choice because it is relatively inert and temperature insensitive) and the viscous stiffness is controlled by the size of holes drilled through the piston as well as the gap around it.

Two dashpots, one for the isotropic viscosity, μ, (corresponding to the elastic bulk molulus, K) and one for the deviatoric viscosity, η, (corresponding to the elastic shear modulus, G) are all that are necessary to represent linear viscosity. By analogy:

$$\sigma_m = 3\mu\frac{d\epsilon_m}{dt} = 3\mu\dot{\epsilon}_m \quad \text{or} \quad \sigma_m = \mu\dot{\epsilon}_v \tag{3.8a}$$

and

$$\begin{aligned} \sigma_x - \sigma_m &= 2\eta(\dot{\epsilon}_x - \dot{\epsilon}_m) \\ \sigma_y - \sigma_m &= 2\eta(\dot{\epsilon}_y - \dot{\epsilon}_m) \\ \sigma_z - \sigma_m &= 2\eta(\dot{\epsilon}_z - \dot{\epsilon}_m) \quad \text{or} \quad S_{ij} = 2\eta\dot{e}_{ij} \\ \tau_{xy} &= \eta\dot{\gamma}_{xy} \\ \tau_{yz} &= \eta\dot{\gamma}_{yz} \\ \tau_{zx} &= \eta\dot{\gamma}_{zx} \end{aligned} \tag{3.8b}$$

where the dot over each strain component indicates the time derivative.*

* Equation (3.8) substituted into the equilibrium equations in terms of the time derivatives of strain along with conservation of mass and energy lead to the most general equations of fluid mechanics (Navier-Stokes equations). If there is no shear stiffness ($\eta = 0$), these, in turn, reduce to the Euler equations (compressible but inviscid fluids). They further reduce to potential flow (or the Bernoulli equation in 1D) for incompressible, irrotational flow.

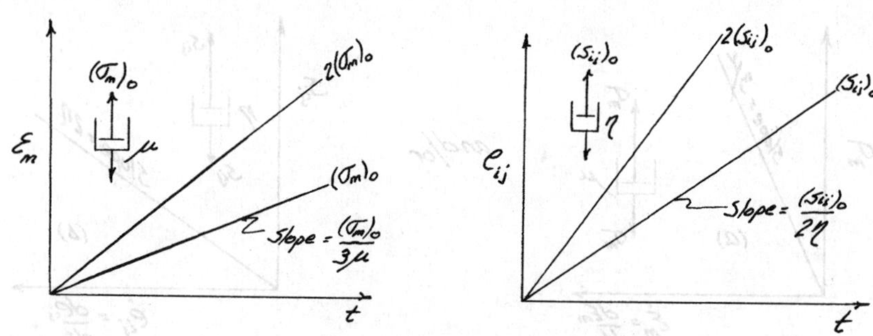

FIGURE 3.3
Linear viscous creep behavior (stress = constant).

A special and very important case called *creep* arises if the dashpot is loaded by a constant stress $(\sigma_m)_o$ or $(S_{ij})_o$. For creep the stress–strain relationship can be integrated directly and, as shown in Figure 3.3, the strain *at any time* is proportional to the stress since the strain rate is constant. Thus a highly viscous fluid can be thought of as having time-dependent "elastic" moduli $K(t) = \mu/t$ and $G(t) = \eta/t$ in creep.

3.3 Simple Viscoelastic Behavior

All solids are to some extent "fluid" in that they will flow, even if only a minuscule amount, at working stress levels if enough time passes. Thus stress–strain relationships are actually functions of time (and temperature) and it can be dangerous, even disastrous, to neglect creep effects in such materials as concrete, plastics, wood, or soil when dimensional tolerances must be maintained over time.* Moreover, as we might expect, an increase in temperature accelerates viscous flow. All solids turn into liquids or (gases) at their melting point, but long before that, many materials such as metals (which appear elastic at room temperature) will loose stiffness and become viscous** at higher temperatures. Such viscous flow, which can occur under either isotropic or deviatoric stress, is easily confused with plastic flow which, as we will see, is a purely deviatoric phenomenon. However, in areas of plastic flow,

* For example, time-dependent creep distortion of the concrete in Hoover dam (where the load is essentially constant) eventually (after forty years), necessitated replacement of the steel penstock liners at great expense. Creep of concrete and wood can reduce effective prestress significantly (or destroy shallow arches) and the long-term deviatoric creep sometimes thought of as secondary consolidation can, even after hundreds of years, lead to foundation failure as witnessed in the Tower of Pisa.

** The melting point of iron, for example, is 1535°C but at only 300°C it begins to flow under load. That is why unprotected steel buildings have a lower fire rating (higher insurance premiums) than concrete or even wooden ones.

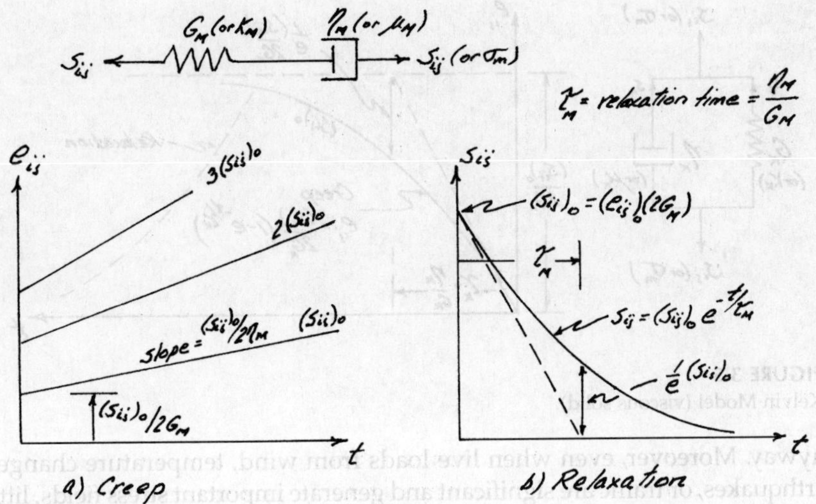

FIGURE 3.4
Maxwell Model (elastic viscous fluid).

all distortional resistance is gone* and there is therefore no deviatoric stiffness. An increase in temperature will reduce both the deviatoric stiffness and the yield strength, but it is important to keep the two properties, stiffness and strength, separate in our minds.

Elastic behavior can be modified to include viscous effects by combining dashpots with springs to any degree of complexity deemed necessary to capture the essential stiffness response observed in the laboratory. The simplest such "models" for viscoelastic behavior combine a single spring and a single dashpot either in series or in parallel. The corresponding "Maxwell Model" and "Kelvin Model" are shown in Figures 3.4 and 3.5. In general a separate model may be necessary to depict the observed isotropic and deviatoric behavior. This is true, for example, in saturated clays where the volumetric behavior is well represented by a Kelvin model and the deviatoric response by the Maxwell model. The viscoelastic deformation of most solids, however, exhibits much more flow under shear than mean stress. Thus we will discuss the various viscoelastic models in terms of deviatoric stress and the corresponding distortional strain, recognizing that similar models can apply to volumetric behavior with σ_m, ϵ_m, $3K$, and 3μ substituted in place of S_{ij}, e_{ij}, $2G$, and 2η.

Since in this book we consider only static loads, the viscoelastic creep (and relaxation) behavior of our models is of primary importance. For many structures, particularly large ones like bridges or the foundations on which they sit, the major part of the stress field in service is due to constant "dead" loads

* This is only true for ideal plastic materials, which yield without strain hardening. For such materials it will turn out also that after yield, $\nu \cong \frac{1}{2}$ (so $K \to \infty$) and the material behaves like a perfect incompressible fluid. Thus plasticized areas are areas of potential (irrotational) flow governed by the Laplace equation.

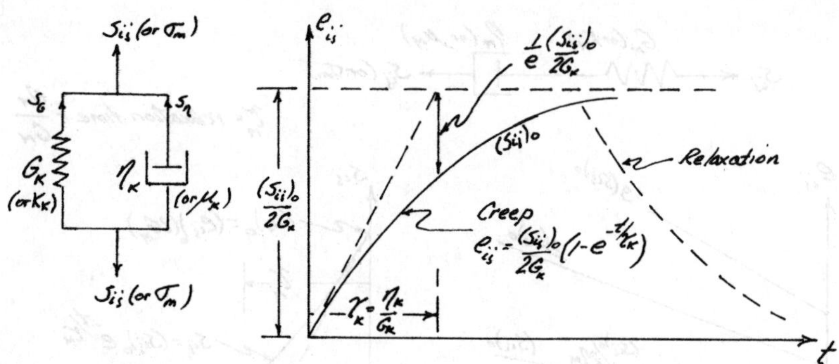

FIGURE 3.5
Kelvin Model (viscous solid).

anyway. Moreover, even when live loads from wind, temperature changes, earthquakes, or traffic are significant and generate important stress fields, little viscous flow will occur since dashpots cannot react to sudden loads of short duration.* Thus, it is generally a reasonable engineering assumption for viscoelastic solids to calculate live-load stresses from the purely elastic response of the material and the viscous strains from the dead-load part of the stress field neglecting the comparatively transitory effects.**

The creep curves for the Maxwell and Kelvin models are shown in Figures 3.4 and 3.5. In each case the derivation of the fundamental stress–strain equation from equilibrium and compatibility is straightforward. For the Maxwell model with the spring and dashpot in series, the stress in each component equals the total applied stress, S_{ij}, and the total deviatoric strain is the sum of that in each. Therefore, the general equation for the deviatoric Maxwell model is

$$\dot{e}_{ij} = \frac{1}{2G_M}\dot{S}_{ij} + \frac{1}{2\eta_M}S_{ij} \tag{3.9a}$$

with the solution for creep, S_{ij} = constant = $(S_{ij})_o$:

$$e_{ij} = \frac{(S_{ij})_o}{2G_M} + \frac{1}{2\eta_M}(S_{ij})_o t = \frac{(S_{ij})_o}{2G_M}\left[1 + \frac{t}{\tau_M}\right] \tag{3.9b}$$

---

* Actually the viscosity of so-called "solids" as a part of structural damping has important effects on resonance or shock waves but this is a subject covered in courses on vibrations or stress waves or in earthquake engineering (where both subjects are combined).

** One must keep in mind the relative magnitudes of the elastic and viscous components we are talking about. Creep of *solids* used to build structures should take years to become significant (or else the solid is not solid). Also within the constraints of simple linear theory, the viscous flow must never become very large or the assumption of "small" strains (less than, say, 5%) and negligible geometric nonlinearity are violated. Thus for wood, concrete, plastics, soil, rock, metals at elevated temperature, or viscoelastic solids used in durable structures, viscosities are in units such as Ksi-yrs.

and for relaxation, e_{ij} = constant = $(e_{ij})_o$

$$S_{ij} = 2G_M(e_{ij})_o e^{-t/\tau_M} \tag{3.9c}$$

where

$$\tau_M = \frac{\eta_M}{G_M}$$

As can be seen in Figure 3.4, the Maxwell model is essentially a fluid. Although it responds elastically and appears to be solid, it flows indefinitely under sustained stress. Silly putty is a perfect example of such a material.

The Kelvin* model where the components are in parallel (Figure 3.5) is, on the other hand, a solid with the dashpot only serving to delay the elastic response. In this case the stress in each component depends on time with the dashpot taking all the stress at $t = 0$ and the spring only able to accept the stress given up by the dashpot as it strains. Equilibrium and compatibility require that $S_G + S_\eta = S_{ij}$ and $e_G = e_\eta = e_{ij}$. The general equation is therefore:

$$S_{ij} = 2G_K e_{ij} + 2\eta_K \dot{e}_{ij} \tag{3.10a}$$

Solving for creep: S_{ij} = constant = $(S_{ij})_o$

$$e_{ij} = \frac{(S_{ij})_o}{2G_K}(1 - e^{t/\tau_K}) \tag{3.10b}$$

where

$$\tau_K = \frac{\eta_K}{G_K}$$

More complicated models can be developed to depict viscoelastic behavior to any degree of sophistication. To illustrate the procedure for determining the basic equations for such hybrid models, the derivation for the "standard solid" is shown as an example in Figure 3.6 while others are left for homework problems.

---

* William Thompson (1814–1907), later Lord Kelvin, maintained a lifelong involvement in all aspects of physics and thermodynamics relating to stress–strain relationships. He was knighted for extensive work on the Atlantic telegraph in addition to his reputation as an inventor, research engineer, and the leading physicist in England.

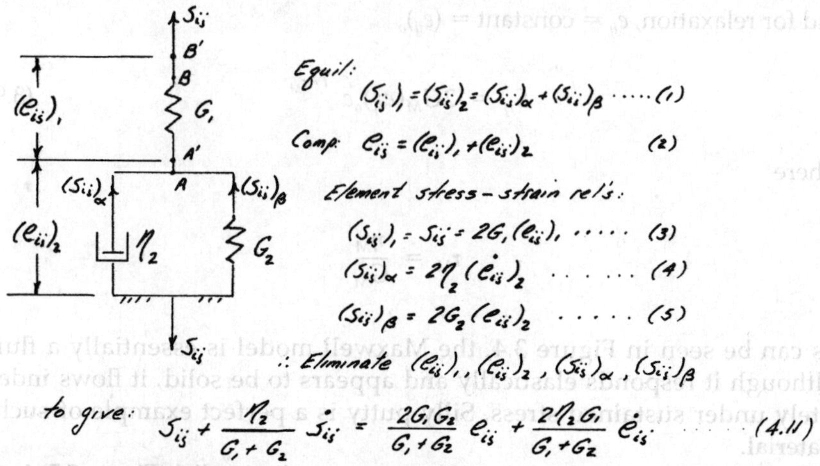

Equil:
$(S_{ij})_1 = (S_{ij})_2 = (S_{ij})_\alpha + (S_{ij})_\beta$ ....(1)
Comp: $e_{ij} = (e_{ij})_1 + (e_{ij})_2$ (2)
Element stress–strain rel's:
$(S_{ij})_1 = S_{ij} = 2G_1 (e_{ij})_1$ ..... (3)
$(S_{ij})_\alpha = 2\eta_2 (\dot{e}_{ij})_2$ ...... (4)
$(S_{ij})_\beta = 2G_2 (e_{ij})_2$ ..... (5)
∴ Eliminate $(e_{ij})_1, (e_{ij})_2, (S_{ij})_\alpha, (S_{ij})_\beta$
to give: $S_{ij} + \frac{\eta_2}{G_1 + G_2}\dot{S}_{ij} = \frac{2G_1G_2}{G_1+G_2} e_{ij} + \frac{2\eta_2 G_1}{G_1+G_2}\dot{e}_{ij}$ ...... (4.11)
S_{ij}
B'
B
A'
A
G_1
G_2
η_2
$(e_{ij})_1$
$(e_{ij})_2$
$(S_{ij})_\alpha$
$(S_{ij})_\beta$

FIGURE 3.6
Derivation of governing equation—"Standard Solid."

Example 3.2

A six-inch thick wooden bridge deck is post-tensioned with 1 in.^{2} Kevlar tendons with a 1 ft spacing jacked to an initial prestressing force $F = 21{,}600$ lb as shown. Kevlar can be considered elastic ($E = 7.2 \times 10^6$ psi), but wood might be an elastic material in isotropic stress and a Kelvin material in deviatoric stress with properties as shown.

Calculate and plot the stress in the wood and the change in length ΔL as a function of time.

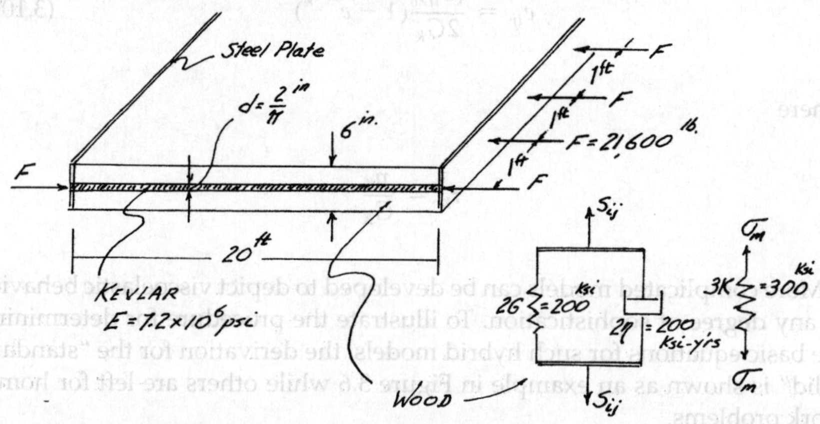

Note:
a) We are concerned only with stress and strain in x direction.
b) Assume both wood and Kevlar are free to expand in y and z directions.

1. EQUIL:

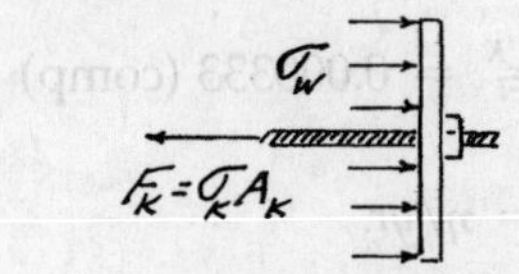

GEOM:

$A_K = 1\ in^2$

$A_W = 72\ in^2$

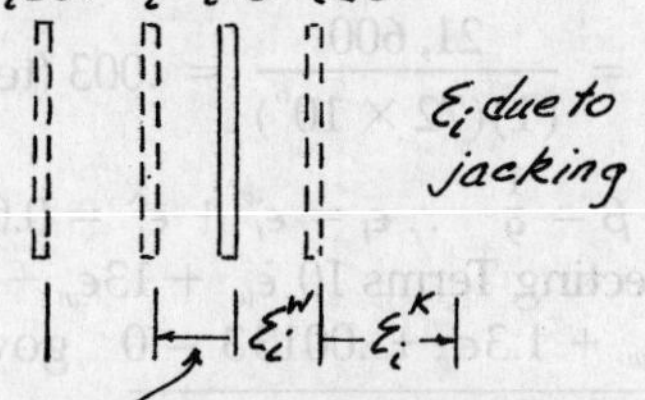

$$F_W = A_W \sigma_W = F_K = A_K \sigma_K \cdots (1)$$

$$\text{for } t > 0 \quad \epsilon_W = \epsilon_K = \epsilon_x \cdot\cdot (2)$$

$$\& \ \epsilon_i = \epsilon_i^W + \epsilon_i^K$$

$$\text{or } \epsilon_K = \epsilon_W + \epsilon_i \quad (2')$$

$$\sigma_W = \frac{A_K}{A_W}\sigma_K = \frac{A_K}{A_W}\epsilon_K E_K$$

$$= \frac{1.0}{72}(7.2 \times 10^6)\epsilon_K = \epsilon_K \times 10^5 \ldots\ldots (3)$$

$$\therefore @ t = 0 \quad F_K = 21{,}600 \ \therefore \ \epsilon_K = \frac{21{,}600}{7.2 \times 10^6} = 0.003 \ \therefore \ \sigma_W^i = -300\ psi$$

But

$$[\sigma_w] = \begin{bmatrix} \sigma_m & & \\ & \sigma_m & \\ & & \sigma_m \end{bmatrix} + \begin{bmatrix} S_x & & \\ & S_y & \\ & & S_z \end{bmatrix}$$

$$\updownarrow 3K \qquad\qquad \updownarrow 2G \ \& \ 2\eta$$

$$[\epsilon_w] = \begin{bmatrix} \epsilon_m & & \\ & \epsilon_m & \\ & & \epsilon_m \end{bmatrix} + \begin{bmatrix} e_x & & \\ & e_y & \\ & & e_z \end{bmatrix}$$

Where $\sigma_m = \dfrac{\sigma_x + \sigma_y + \sigma_z}{3} = \dfrac{\sigma_W}{3}$

& $S_x = \sigma_w - \sigma_m = \frac{2}{3}\sigma_w$

$\therefore \sigma_m = 3K\epsilon_m = \frac{\sigma_w}{3}$ or

$\sigma_w = 9K\epsilon_m$ (a)

and for a Kelvin Body

$S_x = 2Ge_x + 2\eta\dot{e}_x = \frac{2}{3}\sigma_w$

or $\sigma_w = 3Ge_x + 3\eta\dot{e}_x$ (b)

So equating (2a) & (2b) $\quad 9K\epsilon_m = 3Ge_x + 3\eta\dot{e}_x$ (4)

where: $\epsilon_x = \epsilon_w = \epsilon_m + \epsilon_x$ (5)
and we can think of the isotropic and deviatoric elements as in series

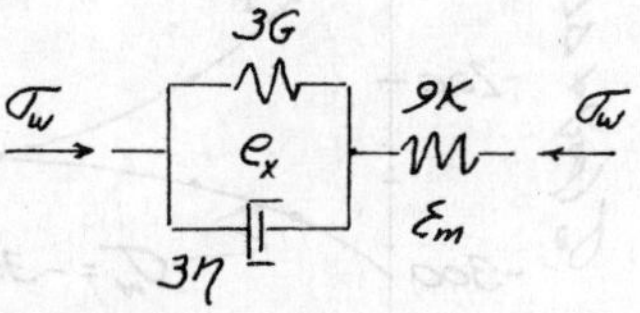

From (3) $\epsilon_m = \dfrac{\sigma_W}{9K} = \dfrac{\epsilon_x \times 10^5}{9K} = -\beta\epsilon_x$ where $\beta = \dfrac{10^5}{9K} = \dfrac{1}{9}$ (6)

Combining (4) & (5) $9K\epsilon_m = 3G(\epsilon_x - \epsilon_m) + 3\eta(\dot{\epsilon}_x - \dot{\epsilon}_m)$

substituting (6) $\quad -9K\beta\epsilon_x = 3G(\epsilon_w + \beta\epsilon_K) + 3\eta(\dot{\epsilon}_w + \beta\dot{\epsilon}_K)$

but from (2′) $\epsilon_K = \epsilon_w + \epsilon_i$ and $\dot{\epsilon}_K = \dot{\epsilon}_w$

$$\therefore -9K\beta(\epsilon_w + \epsilon_i) = 3G(\epsilon_w + \beta\epsilon_w + \beta\epsilon_i) + 3\eta(\dot{\epsilon}_w + \beta\dot{\epsilon}_w)$$

$$\therefore 3\eta(1 + \beta)\dot{\epsilon}_w + [3G(1 + \beta) + 9K\beta]\epsilon_w + [3G\beta + 9K\beta]\epsilon_i = 0$$

Substituting the numerical values

$$\epsilon_i^K = \frac{21,600}{(1)(72\times 10^6)} = .003\text{ (tens.)} \qquad \epsilon_i^w = \beta\epsilon_i^K = 0.000333\text{ (comp)}$$

where $\beta = \frac{1}{9}$ $\therefore \epsilon_i = \epsilon_i^w + \epsilon_i^K = 0.0033\bar{3},\quad K = G = \eta/yr.$

Collecting Terms $10\,\dot{\epsilon}_w + 13\epsilon_w + 4\epsilon_i = 0$

or $\dot{\epsilon}_w + 1.3\epsilon_w + .00133 = 0$ gov.d.e.

This equation is in the form

$$\dot{\epsilon}_w + \alpha_i\epsilon_w + \alpha_2 = 0 \quad \therefore \epsilon_w(t) = Ce^{-\alpha_1 t} - \frac{\alpha_2}{\alpha_1}$$

$$\therefore \epsilon_w(0) = C - \frac{\alpha_2}{\alpha_1} = \epsilon_i^w \quad \text{giving} \quad C = \epsilon_i^w + \frac{\alpha_2}{\alpha_1}$$

$$\therefore \epsilon_w(t) = \left(\epsilon_i^w + \frac{\alpha_2}{\alpha_1}\right)e^{-\alpha_1 t} - \frac{\alpha_2}{\alpha_1} = \left(-.000333 + \frac{.00133}{1.3}\right)e^{-1.3t} - \frac{.00133}{1.3}$$

$$\epsilon_w(t) = 0.0006923e^{-1.3t} - 0.001026$$

For $t \to \infty$, $\epsilon_w(\infty) = -.001026$

$\sigma_w = -300 + \Delta\epsilon_w \times 10^5$

$\therefore \sigma_w(\infty) = -300 + .000693 \times 10^5$

$= -300 + 69.3 = -230.7$ psi

$F(\infty) = \sigma_w A_w = 16{,}610$ lb.

t(yr.)	ϵ_w	ΔL(in)	σ_w(psi)
0	$-.00033\bar{3}$	$-.08$	-300
.5	$-.000665$	$-.16$	-267
1	$-.000837$	$-.20$	-250
2	$-.000975$	$-.23$	-236
4	$-.001022$	$-.245$	-231
∞	$-.001026$	$-.246$	-230.7

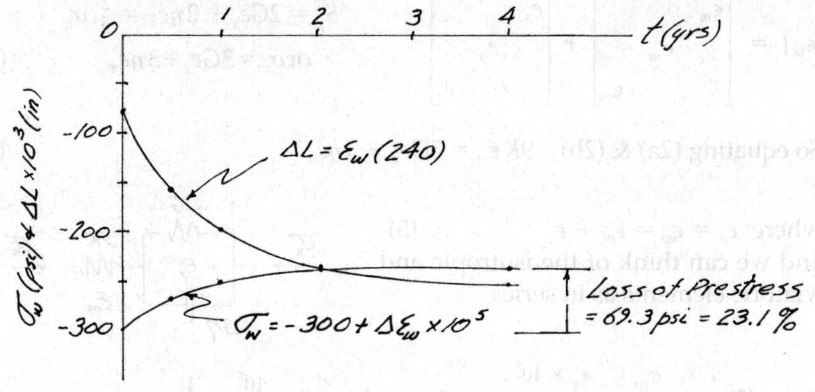

3.4 Fitting Laboratory Data with Viscoelastic Models

Properties of viscoelastic materials are usually determined in the laboratory with a creep test run at constant temperature. Often a tensile test is used with measurements of axial and transverse deformations as a function of time to determine $E(t)$ and $\nu(t)$. It is better (often necessary) to run separate creep tests in pure

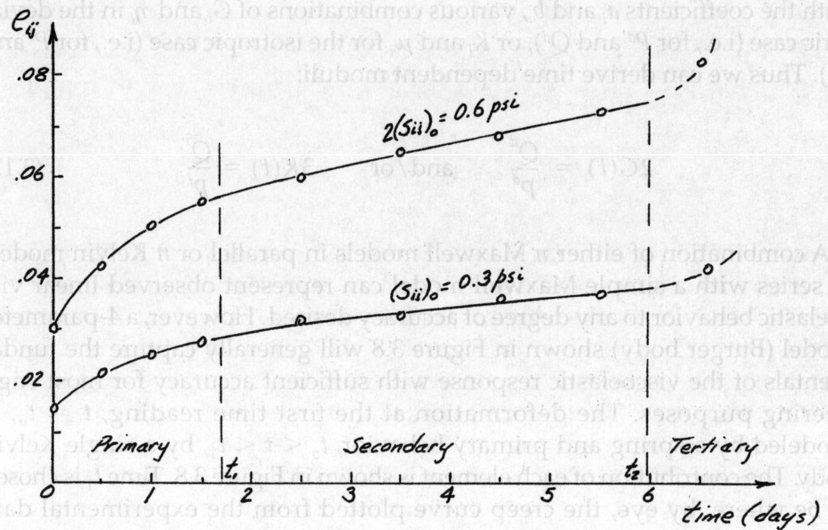

FIGURE 3.7
Torsional creep test of gelatin.

states of constant isotropic and deviatoric stress to determine $K(t)$ and $G(t)$ individually, since for some linear viscoelastic materials, the elastic relationships between E, ν, G, and K do not translate directly when they are functions of time.

Typical results that were obtained in the laboratory for the creep of gelatin* (in this case a mixture of 10% gelatin, 50% glycerin, and 40% water) in pure shear at room temperature are shown in Figure 3.7. One might decide to fit such curves with a least-squares polynomial, but it is much better, since it is physical and leads to easier mathematics, to combine linear springs and dashpots. As we have seen for the Maxwell, Kelvin, and standard solid models, this leads to a differential equation for the behavior of a particular arrangement of springs and dashpots of the form:

$$P^d S_{ij} = Q^d e_{ij} \qquad \text{or} \qquad P^i \sigma_m = Q^i \epsilon_m \tag{3.11}$$

where P and Q are partial differential operators:

$$P = a_o + a_1 \frac{\partial}{\partial t} + a_2 \frac{\partial^2}{\partial t^2} \cdots a_m \frac{\partial^m}{\partial t^m}$$

$$Q = b_o + b_1 \frac{\partial}{\partial t} + b_2 \frac{\partial^2}{\partial t^2} \cdots b_m \frac{\partial^m}{\partial t^m}$$

---

* Gelatin, like most transparent materials, is birefringent under stress and is so sensitive that it can be used for photoelastic analysis of gravity stress fields with small-scale models. It is incompressible ($\nu = 1/2$) and not only is the shear modulus linearly viscoelastic, but so too are the stress and strain fringe coefficients. In his youth, Maxwell pioneered the use of gelatin for photoelasticity and named the fringes of constant τ_{max} isochromatics since they are of different colors in white light. Maxwell's work with polarized light for stress analysis led eventually to his famous equations for electromagnetic fields.

with the coefficients a_i and b_i, various combinations of G_i and η_i in the deviatoric case (i.e., for P^d and Q^d), or K_i and μ_i for the isotropic case (i.e., for P^i and Q^i). Thus we can derive time dependent moduli:

$$2G(t) = \frac{Q^d}{P^d} \qquad \text{and/or} \qquad 3K(t) = \frac{Q^i}{P^i} \tag{3.12}$$

A combination of either n Maxwell models in parallel or n Kelvin models in series with a simple Maxwell model can represent observed linear viscoelastic behavior to any degree of accuracy desired. However, a 4-parameter model (Burger body) shown in Figure 3.8 will generally capture the fundamentals of the viscoelastic response with sufficient accuracy for most engineering purposes. The deformation at the first time reading, $t = t_o$, is modeled by a spring and primary behavior, $t_o < t < t_1$, by a single Kelvin body. The contribution of each element is shown in Figure 3.8. Time t_1 is chosen to be where, by eye, the creep curve plotted from the experimental data appears to become a straight line at the angle ϕ_M. The initial tangent to the curve at $t = t_o$ is drawn and τ_2 determined as shown.

The basic equation for the 4-parameter model is:

$$\dot{e}_{ij} + \frac{\eta_2}{G_2}\ddot{e}_{ij} = \frac{s_{ij}}{2\eta_1} + \frac{\dot{s}_{ij}}{2}\frac{\eta_1 G_2 + \eta_1 G_1 + \eta_2 G_1}{\eta_1 G_2 G_1} + \frac{\ddot{s}_{ij}}{2}\frac{\eta_2}{G_1 G_2} \tag{3.13}$$

which is, again, seen to be an equation of the form $Q^d e_{ij} = P^d s_{ij}$ with P and Q differential operators of order 2.

For creep with $s_{ij} = (s_{ij})_o$

$$e_{ij} = (s_{ij})_o\left[\frac{1}{2G_1} + \frac{1}{2G_2}(1 - e^{-t/\tau_2}) + \frac{t}{2\eta_1}\right] \tag{3.14}$$

or

$$G(t) = \frac{G_1 G_2 \eta_1}{G_2\eta_1 + G_1\eta_1(1 - e^{-t/\tau_2}) + G_1 G_2 t}$$

where

$$\tau_2 = \eta_2 / G_2$$

From t_1, τ_2, and ϕ_M as found from the experimental curve drawn by eye to best fit the laboratory data, the element parameters G_1, G_2, η, η_2 can be computed. If, as in Figure 3.6, the experiment indicates that bonds begin to weaken at large strains, this tertiary behavior can be represented by plastic elements such as those discussed later.

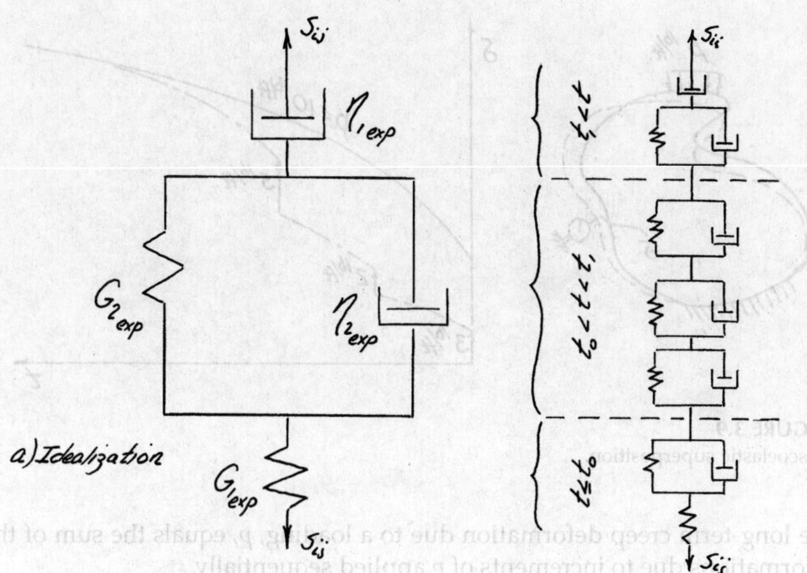

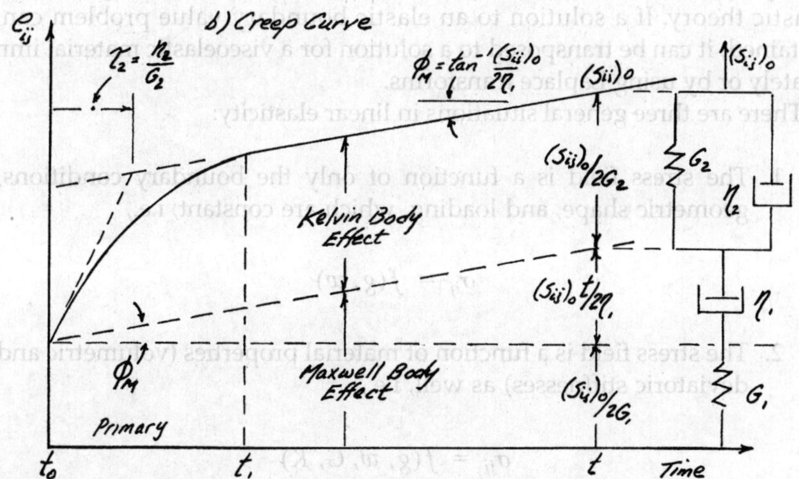

FIGURE 3.8
4-parameter model to best fit experimental data.

3.5 Elastic–Viscoelastic Analogy

One of the most important characteristics of linear viscoelastic "mechanical" models used to represent time-dependent material behavior is that such models obey Boltzmann's superposition principle illustrated in Figure 3.9.

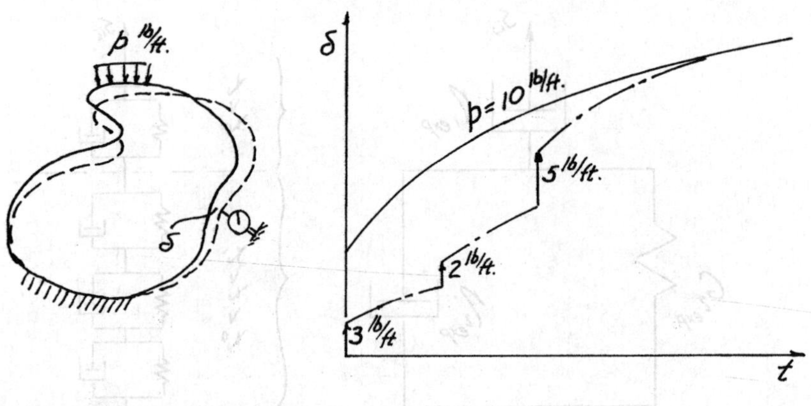

FIGURE 3.9
Viscoelastic superposition.

The long-term creep deformation due to a loading, p, equals the sum of the deformations due to increments of p applied sequentially.

Superposition (with linearity), in turn, allows direct correspondence to elastic theory. If a solution to an elastic boundary value problem can be obtained, it can be transposed to a solution for a viscoelastic material immediately or by using Laplace transforms.

There are three general situations in linear elasticity:

1. The stress field is a function of only the boundary conditions, geometric shape, and loading, which are constant, i.e.,

$$\sigma_{ij} = f(g, w)$$

2. The stress field is a function of material properties (volumetric and deviatoric stiffnesses) as well, i.e.,

$$\sigma_{ij} = f(g, w, G, K)$$

3. The loads are time dependent, i.e.,

$$\sigma_{ij} = f[g, w(t)] \qquad \text{or} \qquad \sigma_{ij} = f[g, w(t), G, K]$$

If the elastic solution is of the first type (as it is in the two-dimensional, homogeneous, isotropic situation) then there really is no problem; viscoelastic stresses will be the same as elastic stresses and time-dependent strains and displacements are obtained directly using $G(t)$ and $K(t)$ in the stress–strain relationships. In addition, Laplace transforms can sometimes be avoided in the second case for special fields, but in general it is more satisfactory to apply transform methods.* In the third situation when the loading is dynamic, Laplace transforms must be used. This may be straightforward theoretically,** but very difficult in practice since the solution in the transformed domain, where time is eliminated as a variable in both the load and the moduli, must be referred back to the real world. The inverse mapping function to do this is seldom found in standard tables, except in the simplest cases, and may be very difficult to determine from complex variables.

The main point in all this, however, is that there is a *general elastic–viscoelastic analogy.* Thus the simple elastic idealization is all important. When we talk about "elastic" field theory, requirements for uniqueness and/or existence, strategy and tactics of elastic analysis, and procedures for closed-form and numerical solution for elastic problems, we are talking about far more than is implied by the word, "elastic." The introduction of time as a variable through linear viscosity (and/or loads that change with time) makes the mathematics much more involved, but the formulation directly corresponds to elasticity. Thus there is no real need to differentiate linearly viscoelastic materials from linear elastic materials in a fundamental way, or consider the two separately hereafter.

Moreover, simple elastic analysis is the basic ingredient for nonlinear analysis by numerical methods. Nonlinear or temperature-dependent material properties (and geometric nonlinearity due to large finite strains***) introduce difficulties so severe that a closed-form solution becomes intractable.

Field equations incorporating each or all of these nonlinear effects can be formulated, but they are so complicated that it is, today, a waste of time to even try to solve them. Instead, all types of nonlinearity can be handled satisfactorily to any degree of accuracy with a piecewise linear approximation if a powerful enough computer is available. The loads are increased incrementally and the material properties adjusted. If necessary, the geometry is revised after each step before the elastic solution is again used for the next increment. ****

---

* This is covered in some detail in Chapter 9 of *Elastic and Inelastic Stress Analysis* by I.H. Shames and F.A. Cozzarelli, Prentice Hall, Englewood Cliffs, NJ, 1992. This book presents the fundamentals of viscoelasticity with elegance and gives simple illustrative examples. Moreover, there is some discussion of an analogy between nonlinear viscoelasticity and plastic flow which, in turn, establishes limits useful for engineering estimates of errors introduced by neglecting nonlinear viscous effects.

** See for example, *The Theory of Linear Viscoelasticity,* by D.R. Bland, Pergamon Press, NY, 1960.

*** Euler's "elastic line" as a true one-dimensional "beam" (and therefore imaginary) is a notable exception.

**** The finite element structural idealization is ideal for this. Canned programs are available for purchase with a variety of element material properties to choose from. However, modifying the structural geometry and/or material stiffnesses of each element after each load increment is usually not included in standard "software" and must be built in as a separate subroutine. This is not difficult since the finite element method computes nodal point displacements and average isotropic and deviatoric stress and strain in each element as a matter of course. Thus the new geometry is the old nodal point coordinates plus the incremental displacements and the new K and G for each element is proscribed by whatever the slope of the nonlinear stress–strain relationship is corresponding to the new level of σ_m and s_{ij} in the element.

Again, therefore, linear elasticity with the help of the computer is the cornerstone even for nonlinear analysis. However, there is an unfortunate consequence to the advent of computer solutions. The uninitiated, when they discover the tremendous power of the finite element (or

3.6 Elasticity and Plasticity

The fundamental assumption, which allows us to develop the relative deformation tensor $E_{ij} = \epsilon_{ij} + \omega_{ij}$ in Chapter 2, is that displacements (u, v, w) are small, continuous, and very smooth. Thus, over differential base lengths, deformations are linear. When this is no longer true (kinks develop in the displacement field) or is a bad approximation, we define that condition as *yield*. If we couple this with the assumption of a linear stress–strain behavior* we have completed our development of a "basic" or simple "Theory of Elasticity." The yield condition provides the end point or boundary of the "elasticity domain." The 9 field equations (in 3D) relating strain and rotation to displacement and the 6 stress–strain relations can still be used in regions of the blob that have not yielded, but to describe the overall behavior, the "Theory of Plasticity" must be introduced.** The "field problem" of determining u, v, w throughout a structure after the "loads" first cause yield becomes nonlinear and is very difficult in closed form (and thus one of the prime areas where numerical methods thrive). We will discuss post-yield behavior and derive approximate plasticity solutions (and a few exact ones) in later chapters.

It is very important to emphasize that engineers generally do no want their structures to yield*** since once yield has occurred the structure is permanently deformed, residual stresses are usually introduced, and fatigue life is drastically reduced. Thus designers apply some factor of safety to the yield condition so that under "working loads" the structure operates in the elastic range. Thus a definition of yield and its laboratory determination is as important an engineering property as E and ν (or G and K) in using elastic analysis for design.

---

boundary element) method coupled with super computing, leap to the erroneous conclusion that elasticity (or plasticity) is irrelevant today; that by making the linear assumptions of elasticity at a finite scale, the computer replaces calculus at some sacrifice in accuracy, but with a tremendous increase in versatility. They see that, as computers have become more and more powerful, the difference between the calculus and results from finite models evaluated numerically is often insignificant and this trend toward greater and greater accuracy will continue. After all, calculus is only linearity taken to the limit of infinitely-small base lengths, which is completely unnecessary for engineering analysis.

The refutation of this myopic view of mechanics is so easy that it is left for a homework problem.

* Remember that the word "elasticity" is jargon for "linear stress–strain relationship," which is too long a phrase for repeated use.

** The equilibrium equations are, however, not invalidated by yielding and, being independent of material properties, are valid in plasticity.

*** Engineers sometimes design for "permanent set" as in hot or cold working metals. Another classic example was the second ECHO satellite designed to deform plastically under internal pressure so that (unlike ECHO 1) it would retain its spherical shape even after meteorite puncture released the inflating gas.

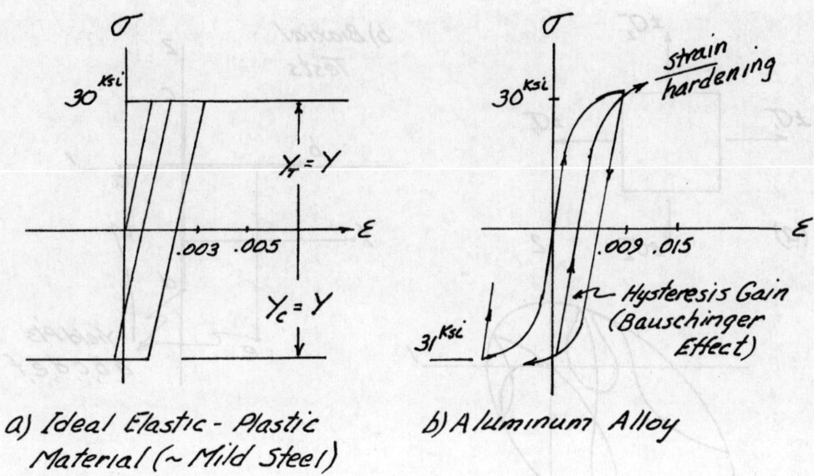

FIGURE 3.10
Uniaxial test.

3.7 Yield of Ductile Materials

Mild steel is very nearly an ideal elastic–perfectly plastic* solid for which we will use the acronym "EPS." A uniaxial test of an EPS would give results as shown in Figure 3.10a as compared to a less ideal material such as an aluminum alloy in Figure 3.10b. If the yield stress in tension, Y_T, is the same as compression, Y_C, and there is a yield plateau, then the material is ductile and the uniaxial yield strength is characterized by a single value, Y.

Many early yield theories were based on EPS tests. These include:

a. Max Stress Theory (Lamé-Navier-Rankine)

$$Y_C \leq \sigma \leq Y_T$$

b. Max Strain Theory (St. Venant)

$$\epsilon_{YC} \leq \epsilon_1 \leq \epsilon_{YT}$$

c. Max Potential Energy Theory (Beltrami)

$$W = \epsilon_1\sigma_1 + \epsilon_2\sigma_2 + \epsilon_3\sigma_3 \leq \epsilon_Y Y$$

* Perfectly plastic is jargon for a "prolonged horizontal yield plateau."

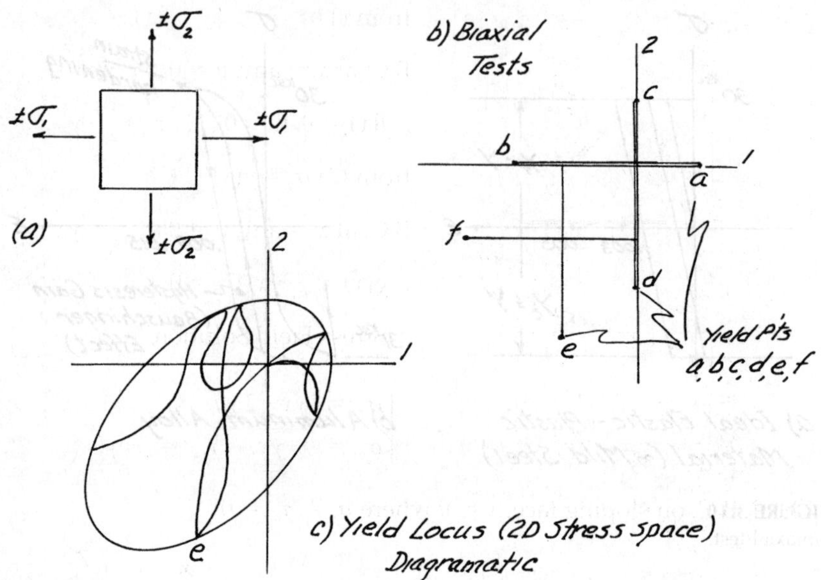

FIGURE 3.11
Hypothetical biaxial test.

However, these ideas have been shown to be incorrect (or not general enough) since materials do not yield under isotropic compression (except perhaps at extreme values). What is actually observed in multidimensional tests is that *shear* causes yield.

Consider first a two-dimensional, imaginary biaxial test on a solid. Simple tension and compression tests give points *a* and *b* in Figure 3.11b in the 1 direction and *c* and *d* in the 2 direction (this is not a block of EPS). However, now imagine we apply some value of $\sigma_1 < Y_T$ and then increase σ_2. Yield might occur at point *e*. Doing this for various combinations of σ_1 and σ_2 we can develop a *"yield Locus"* in 2D stress space (actually looking down the 3 axis), which separates elastic from plastic behavior for this hypothetical material. Since behavior is linear in the elastic domain, this locus will be "independent of path" as diagrammatically indicated in Figure 3.11c.

For the general, three-dimensional situation (test), the yield locus becomes a yield surface. Since we anticipate for our ideal EPS that the isotropic pressure or mean stress will not affect yield, any point on this yield surface should be parallel to the line $\sigma_1 + \sigma_2 + \sigma_3 = \text{const} = 3\sigma_m = 3\sigma_{\text{oct}}$ which we know is the principal-space diagonal. Looking down this orientation we would see an octahedral plane on which a yield locus would represent only the deviatoric component or shear at yield. Thus, as shown in Figure 3.12a, the yield surface is a prism generated by sliding a yield locus in the octahedral plane along the space diagonal perpendicular to it.

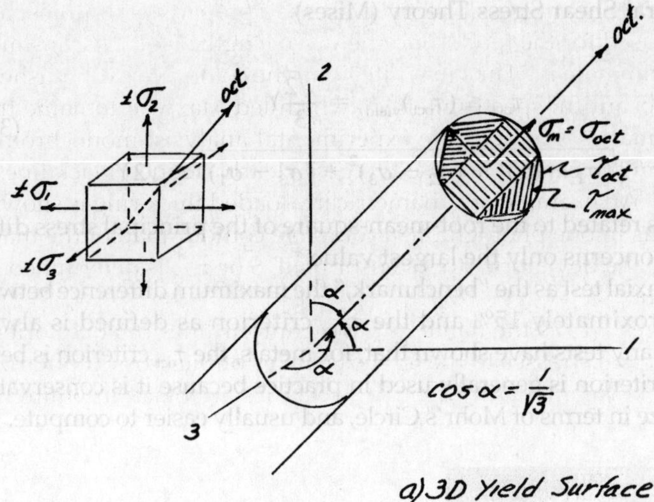

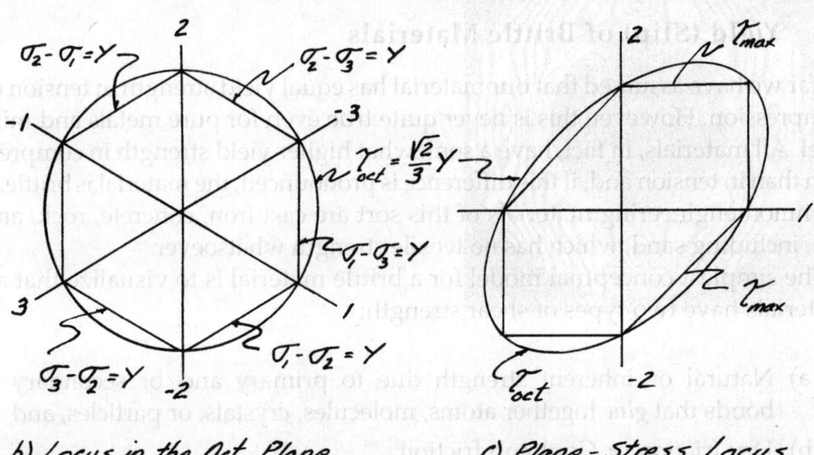

FIGURE 3.12
Tresca and von Mises yield criteria.

Two theories or yield conditions based on test results, which are used to describe the shape and size of this shear locus* (or cross-section of the prism) in the octahedral plane, are compared in Figure 3.12. They are:

a) Maximum Shear Stress Theory (Tresca)

$$\tau_{max} \leq \frac{\sigma_{max} - \sigma_{min}}{2} \tau_{yield} = \frac{Y}{2}$$

* Actually, it is only the "initial" yield locus for a material with strain hardening where we would expect the shape to grow and perhaps change with repeated cycles of loading.

b) Octahedral Shear Stress Theory (Mises)

$$\tau_{oct} \leq (\tau_{oct})_{yield} = \frac{\sqrt{2}}{3}Y$$
$$\text{or} \quad \sqrt{(\sigma_1 - \sigma_2)^2 + (\sigma_2 - \sigma_3)^2 + (\sigma_3 - \sigma_1)^2} \leq \sqrt{2}Y \qquad (3.15)$$

Essentially τ_{oct} is related to the root-mean-square of the principal stress differences and τ_{max} concerns only the largest value.*

Using the uniaxial test as the "benchmark," the maximum difference between the two is approximately 15% and the τ_{max} criterion as defined is always *conservative*. Many tests have shown that, for metals, the τ_{oct} criterion is better, but the Tresca criterion is generally used in practice because it is conservative, easier to visualize in terms of Mohr's Circle, and usually easier to compute.

3.8 Yield (Slip) of Brittle Materials

So far we have assumed that our material has equal yield strength in tension or compression. However, this is never quite true even for pure metals and mild steel. All materials, in fact, have a somewhat higher yield strength in compression than in tension and, if this difference is pronounced, the material is brittle.** Common engineering materials of this sort are cast iron, concrete, rock, and soil, including sand, which has no tensile strength whatsoever.

The simplest conceptual model for a brittle material is to visualize that all materials have two types of shear strength:

a) Natural or inherent strength due to primary and/or secondary bonds that *glue* together atoms, molecules, crystals, or particles, and
b) Dry friction or Coulomb friction.

Coulomb recognized and incorporated them both in his more powerful version of the Tresca theory. Because it is best described in terms of Mohr's Circle, this elegant and unifying concept is today called the Mohr-Coulomb Theory.

The simplest version is shown in Figure 3.13. It will be discussed in much more detail later since it represents all engineering materials to a reasonable

* Both shear criterion have long histories of development. An outgrowth of the experimental observation that materials yield as a result of slip along crystalline planes, the τ_{max} theory was proposed and used by Coulomb (in a more general form) but is now referred to as the Tresca (or Guest) theory because of the many examples worked out by them. τ_{oct} theory was proposed by Maxwell and later by Huber in 1904 in terms of maximum distortional energy. It was further developed by von Mises (1913) and Hencky (1925) and now usually goes by their names.

** In the laboratory we tend to think of brittleness as a description of sudden, often explosive failure of a specimen, which may not happen with confining pressure.

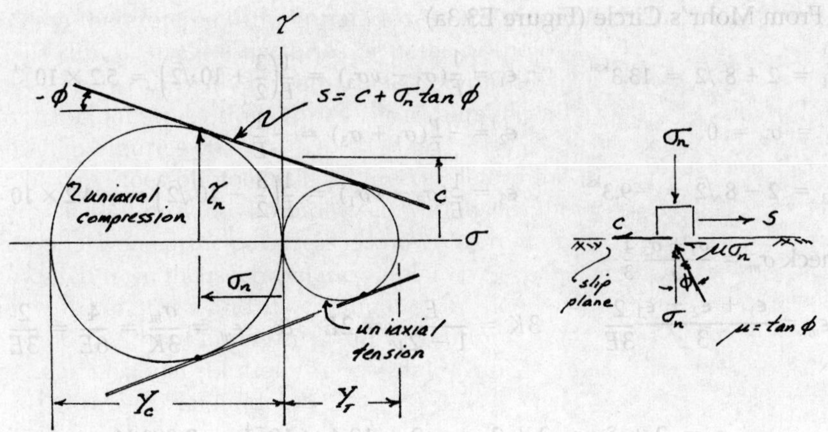

FIGURE 3.13
Mohr-Coulomb theory.

approximation. In terms of a rheological model, the simple EPS must now include the normal stress σ_n on the slip plane, which adds a component to the strength proportional to the coefficient of friction. Thus the Mohr-Coulomb criterion is written:

$$\tau_n \leq s = c + \sigma_n \tan\phi \tag{3.16}$$

which is the envelope of the Mohr Circles for each test in stress space. The "cohesion," c, represents the inherent or natural shear strength which, for a ductile EPS, would be $Y/2$. The Mohr-Coulomb criterion is, therefore, non-linear. For easier computations it can be stated in terms of $\tau_{\max}$ or τ_{oct} rather than the shear stress on the actual slip surface, τ_n.

Example 3.3

A plate (plane stress) is subjected to the constant stress field:

$$\sigma_x = 10^{\text{ksi}}, \qquad \sigma y = -6^{\text{ksi}}, \qquad \tau_{xy} = 8^{\text{ksi}}$$

Assume $E = 30 \times 10^3$ ksi, $\nu = \frac{1}{4}$ and determine:

a) the lengths and directions of the semi-axes of the ellipse formed from a circle whose equation before straining is: $x^2 + y^2 = 4$.

b) the safety factor by the Tresca ($\tau_{\max}$) and von Mises (τ_{oct}) yield criteria if $Y = 30^{\text{ksi}}$

a) From Mohr's Circle (Figure E3.3a)

$$\sigma_1 = 2 + 8\sqrt{2} = 13.3^{\text{ksi}} \qquad \therefore \epsilon_1 = \frac{1}{E}(\sigma_1 - \nu\sigma_3) = \frac{1}{E}\left(\frac{3}{2} + 10\sqrt{2}\right) = 5.2 \times 10^{-4}$$

$$\sigma_2 = \sigma_z = 0 \qquad \epsilon_2 = -\frac{\nu}{E}(\sigma_1 + \sigma_3) = -\frac{1}{E}$$

$$\sigma_3 = 2 - 8\sqrt{2} = -9.3^{\text{ksi}} \qquad \epsilon_3 = \frac{1}{E}(\sigma_3 - \nu\sigma_1) = \frac{1}{E}\left(\frac{3}{2} - 10\sqrt{2}\right) = -4.2 \times 10^{-4}$$

Check $\sigma_m = \frac{\sigma_1 + \sigma_3}{3} \frac{4}{3}$

$$\epsilon_m = \frac{\epsilon_1 + \epsilon_2 + \epsilon_3}{3} \frac{2}{3E} \qquad 3K = \frac{E}{1 - 2\nu} = 2E \quad \& \quad \epsilon_m = \frac{\sigma_m}{3K} = \frac{4}{6E} = \frac{2}{3E}$$

so

$$a = 2 + \delta_1 = 2 + 2\epsilon_1 = 2 + 10.4 \times 10^{-4} = \underline{\underline{2.00104}}$$

$$b = 2 + \delta_2 = 2 + 2\epsilon_2 = 2 - 8.4 \times 10^{-4} = \underline{\underline{1.99958}}$$

@$22\frac{1}{2}^{\circ}$ from x, y as shown in Figure E3.3b

b)Tresca

$$\tau_{\max} = \frac{\sigma_1 - \sigma_3}{2} = 11.3^{\text{ksi}} \qquad \therefore F.S. = \frac{Y/2}{8\sqrt{2}} = \underline{\underline{1.33}}$$

von Mises

$$\tau_{\text{oct}} = \frac{1}{3}\sqrt{\frac{1}{2}[(\sigma_1 - \sigma_2)^2 + (\sigma_2 - \sigma_3)^2 + (\sigma_3 - \sigma_1)^2]}$$

$$= \frac{1}{3}\sqrt{132 + 132 + 512} = 9.29^{\text{ksi}}$$

$$\therefore F.S. = \frac{\frac{\sqrt{2}}{3}(Y)}{\tau_{\text{oct}}} = \underline{\underline{1.52}}$$

$\therefore$ Tresca conservative by ~13%

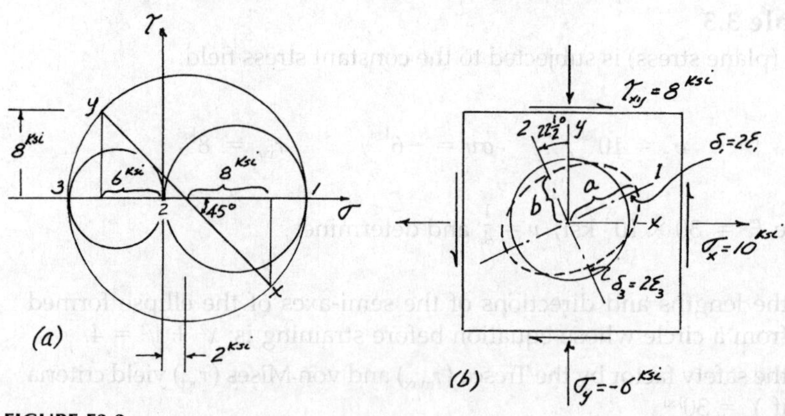

FIGURE E3.3

3.9 Problems and Questions

P3.1 Show that the uncoupled elastic stress–strain relations [Equations (3.5)] in terms of G and K are the same as those expressed in terms of E and ν in Equation (3.2).

P3.2 What kind of material and yield theory might produce an initial yield locus such as shown in Figure 3.13.

P.3.3 For the elastic–perfectly plastic material with uniaxial yield stress, Y, and $\nu = \frac{1}{3}$, plot the yield locus in 1–2 stress space for plane stress ($\sigma_3 = 0$) for:

a. Max Stress Theory
b. Max Strain Theory ($\nu = \frac{1}{3}$)
c. Max Potential Energy Theory ($\nu = \frac{1}{3}$)
d. Max Shear Stress Theory
e. Octahedral Shear Stress Theory

Plot them together so they can be compared.

P3.4 Redo Problem 3.3 for plane-strain ($\nu = \frac{1}{3}$).

P3.5 For cast iron $Y_C = 3Y_T$. Plot the yield locus in 1–2 stress space for the Mohr-Coulomb Theory for plane stress ($\sigma_3 = 0$).

P3.6 A strain rosette is mounted on a steel specimen in plane stress. The measured strains are: $\epsilon_{0°} = 50 \times 10^{-6}$, $\epsilon_{60°} = 50 \times 10^{-6}$, $\epsilon_{90°} = 150 \times 10^{-6}$. Determine the principal stresses and maximum shear stress at the gaged point ($E = 30 \times 10^{-6}$ psi, $\nu = \frac{1}{3}$).

P3.7 A *rigid* horizontal bar AC is supported by a pin and two EPS rods of equal cross-sectional area and length as shown in Figure P3.7. The load, P, is increased until both rods become plastic and is then removed. Find:

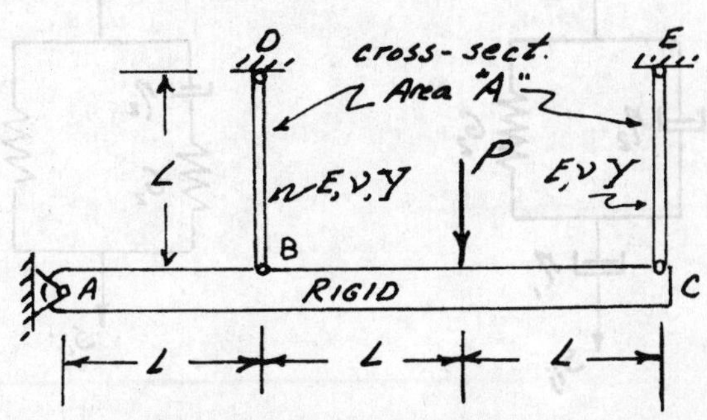

FIGURE P3.7

a. Load at first yield, P_Y

b. Load at collapse, P_L

c. Residual stresses and deformation after P_L is removed.

P3.8 Referring to Figure P3.8, the state of strain at a point "0" in a flat steel plate is the sum of two states, namely: strains as indicated in Figure P3.8a, and strains produced by the stresses as illustrated in Figure P3.8b. The material constants for steel are $E = 30 \times 10^6$ psi, $\nu = \frac{1}{3}$.

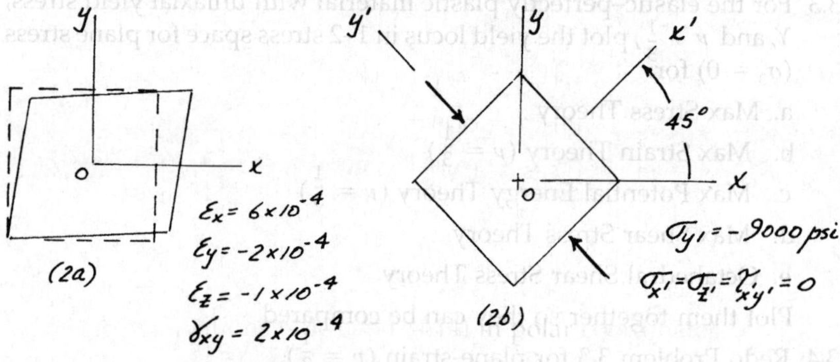

FIGURE P3.8

a. Find the total strain on x, y planes at point 0.

b. Find the magnitudes of the principal strains at 0 in the x, y plane and draw a sketch showing the planes on which they act.

P3.9 Derive the basic equations for the viscoelastic models shown in Figure P3.9 for the deviatoric component of stress and strain. Solve for a constant applied stress $(S_{ij})_o$ to obtain the functional relationship for $G(t)$ (i.e., G in terms of η_i, G_i, t) and plot a representative creep curve.

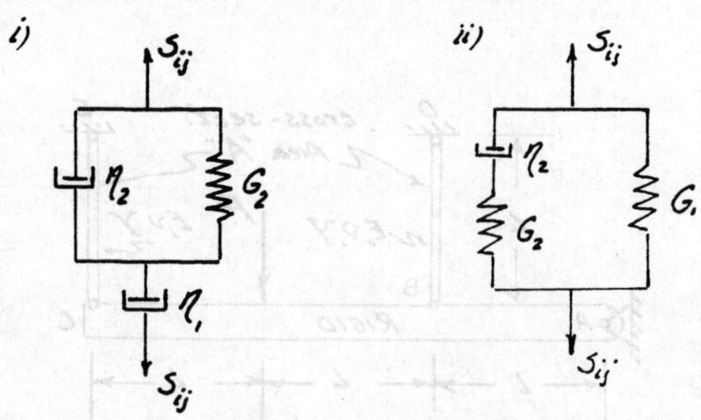

FIGURE P3.9

P3.10 Strain gauge investigation reveals the following data at one point of a sheet in plane stress:
$\epsilon_x = 0.001$; $\epsilon_y = -0.001$; $\gamma_{xy} = \frac{2\sqrt{3}}{3} \times 10^{-3}$. The metal of the sheet is steel with $\nu = 0.25$, $E = 30 \times 10^6$ psi.

a. Find the longitudinal strain at the point in a direction 30° counterclockwise from the x-axis.

b. If the yield stress of the steel is $\sigma_Y = 50{,}000$ psi in a tensile text, will the given state of strain be above or below the yield state according to the Mises criterion?

c. What conclusion would be drawn about yielding if the maximum shear stress criterion were used.

P3.11 A wing panel of a supersonic jet (Figure P.3.11) is prestressed so that the skin is stressed in panel A as shown in (a). During wind tunnel testing, the strains were measured in roll at this point as shown in (b). Assume uniform states of stress and strain throughout the panel:

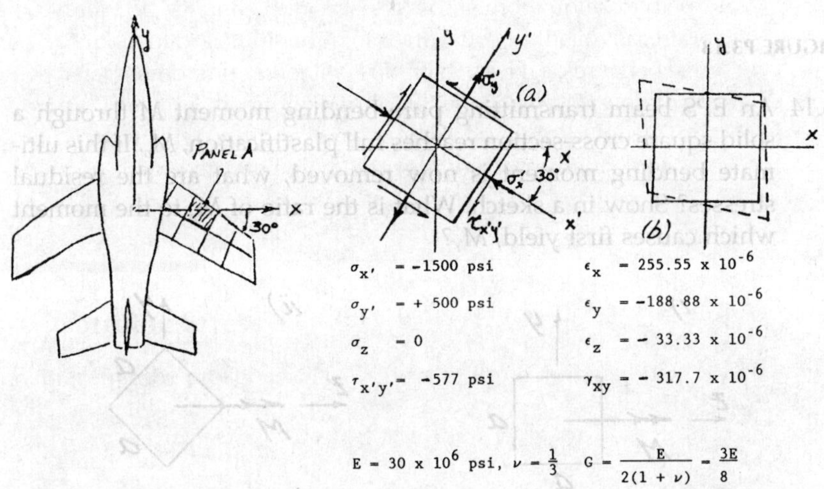

PANEL A
30°
(a)
(b)
$\sigma_{x'}$ = -1500 psi
$\sigma_{y'}$ = + 500 psi
σ_z = 0
$\tau_{x'y'}$ = -577 psi
ϵ_x = 255.55 x 10^{-6}
ϵ_y = -188.88 x 10^{-6}
ϵ_z = - 33.33 x 10^{-6}
γ_{xy} = - 317.7 x 10^{-6}
E = 30 x 10^6 psi, $\nu = \frac{1}{3}$ G = $\frac{E}{2(1+\nu)} = \frac{3E}{8}$

FIGURE P3.11

a. Determine the total stress in the x, y plane

b. Find the magnitude of directions of the principal stresses and their orientation with the x and y axes.

P3.12 An EPS rod transmitting torque through a solid circular cross-section is loaded until the cross-section reaches full plastification, T_P. If this ultimate torque is now removed, what are the residual stresses? Show in a sketch. What is the ratio of T_P to the torque which causes first yield, T_Y?

P3.13 A rectangular steel plate 10 inches wide, 15 inches long, and 1 inch thick is supported between two rigid walls as shown in Figure P3.13. Neglect any friction between the plate and the walls and assume the plate is initially unstressed. If the temperature of the plate is increased by 100°F, find the magnitude and the direction of the principal strains and stresses at any point in the plate.

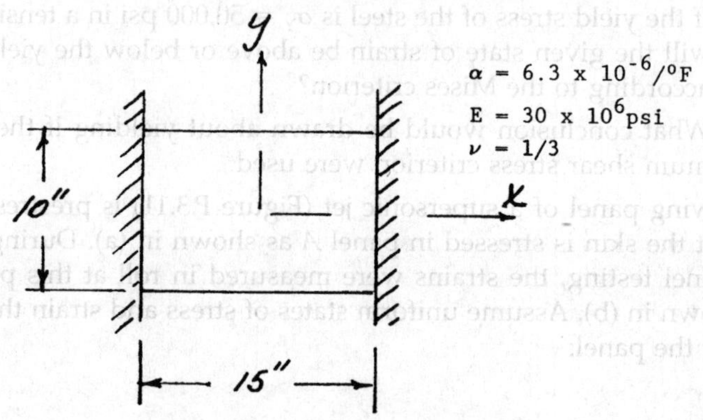

FIGURE P3.13

P3.14 An EPS beam transmitting pure bending moment M through a solid square cross-section reaches full plastification, M_p. If this ultimate bending moment is now removed, what are the residual stresses? Show in a sketch. What is the ratio of M_p to the moment which causes first yield, M_Y?

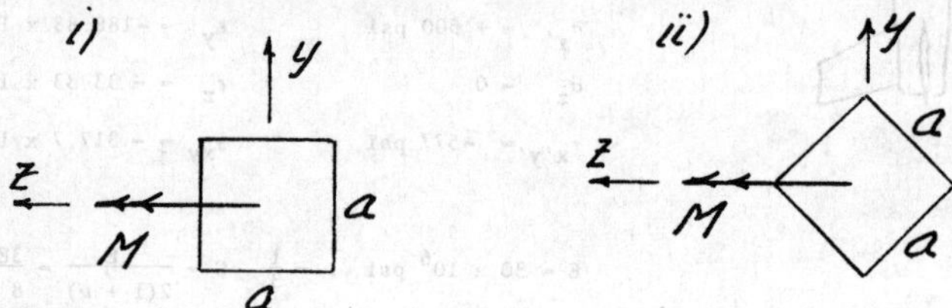

FIGURE P3.14

P3.15 A cube (6 inches on each side) of aluminum ($E = 10^7$ psi, $\nu = \frac{1}{3}$) is subjected to a uniform pressure of 24,000 psi in the x, y plane and restricted to a total deformation of 0.0009 inches in the z direction. Determine $\sigma_z(\tau_{zx} = \tau_{zy} = \tau_{xy} = 0)$ and the change in length of the diagonal, i.e., point (6, 6, 6).

P3.16 The cube of P3.15 is at a temperature such that it behaves like a Maxwell model where $\eta = \mu = 1000$ ksi · days. (E and ν unchanged). Plot σ_z and the diagonal length $0 < t < 5$ days.

P3.17 The 2D stress field in an elastic solid is given by

$$\begin{bmatrix} -ky^2 & 0 \\ 0 & kx^2 \end{bmatrix}$$

where k is constant. Derive the displacements $u(x, y)$ and $v(x, y)$.

P3.18 Referring to the strain field of Problem P2.9:

a. Derive the associated stress field.

b. Derive the associated displacement field (assume at $z = 0$, $u = v = 0$)

c. For a circular bar, $x^2 + y^2 = 9$ of steel ($G = 10 \times 10^6$ psi), what is the physical situation this solution represents?

P3.19 Consider the tensile test for a linearly elastic material shown in Figure 3.1 where $A = 10\,\text{in.}^2$, $L_0 = 6$ in., and $W_0 = 3$ in. At a load of $P = 120{,}000$ lb, $\Delta L = .0024$ in. and $\Delta W = -.0003$ in. Determine E, ν, G, K for this material from this data and plot the corresponding stress–strain curves [as in Figure 3.1(f)]. Do not use the relationships between the moduli except as a check.

P3.20 Derive the stress–strain relationships for the two-dimensional idealizations of plane stress and plane strain including a change in temperature ΔT [corresponding to Equations (3.6) and (3.7)]. Thereby determine the fictitious coefficient of thermal expansion α' such that, along with E' and ν', the in-plane 2D relationships for plane strain have the same form as plane stress.

P3.21 Derive the strain for the creep behavior of the "standard solid" shown in Figure 3.6. Plot and derive the expression for the time dependent shear modulus $G(t)$ relating $(S_{ij})_o$ to e_{ij}.

P3.22 From the experimental torsion test for gelatine (Figure 3.7), determine the appropriate 4-parameter model with values for the two springs and two dashpots.

P3.23 Write a short (one page) essay in rebuttal to the argument outlined in the footnote in Section 3.6 that computer solutions using finite models will (should) replace elasticity (or plasticity) theory based on calculus. You may, if you wish, marshall your points in outline form and then elaborate, but the result should be short, powerful, and persuasive. The best arguments are not entirely technical but have "philosophic" overtones.

4

Strategies for Elastic Analysis and Design

4.1 Rational Mechanics

Design and analysis are intertwined concepts for the engineer. The Theory of Elasticity is now complete with 15 equations and 15 unknowns and a pure mathematician, after some discussion of uniqueness and existence, would at this point go on to other things. To the engineer, the fun has just begun.

How can we solve these equations for specific structures—the classic "boundary-value problem"? Better yet, can we control the internal stresses flowing through a structure and its displacements by adjusting its shape or composition or support conditions? This is the "inverse problem" or "design problem" and is fundamentally much more obscure, difficult, and rewarding.

For all intents and purposes, textbooks and the literature today deal with the first question—finding solutions to the boundary value problem.* It is more manageable, particularly with the increasing power of the computer with refined numerical models. Compared to "the design problem" it may not be as much fun, but it is easy to do and easy to teach. Moreover, for most engineers, design is thought of exclusively as a process of trial and error using successive boundary-value solutions for different shapes, materials, or support conditions.

To imaginative engineers, the great designers, solving boundary value problems is a prelude necessary but not sufficient for creating new and better structures. For them the object is to understand solid mechanics at the more profound level of discovering "how structures work." Solving boundary-value problems can produce this fundamental understanding from which creativity springs but only if the mathematics is thought of as a means to explore the physical truths hidden in the equations and only if the results are completed through

* Historically, however, this dichotomy did not appear for nearly 200 years. In Galileo's *Discoursi*, published in 1638, the analysis and design perspectives are maintained simultaneously. Galileo's solution for the collapse load for an end-loaded cantilever beam goes directly from limit analysis (not elastic) to the inverse problem of determining the shape of such a cantilever having minimum weight. Thus he poses the fundamental *design* problem of structural optimization through shape change, which he calls the problem of "solids of equal resistance," finding that a parabolic depth variation minimizes the weight of a cantilever beam of a given length to resist a given load at its end.

plots of the most revealing field variables with sensitivity studies of the material properties and boundary parameters. All this is then coupled with comparison to simple cases, approximate solutions from strength of materials, and evidence from structural models tested in the laboratory or at full scale in the field.

This sort of visual approach is sometimes called "Rational Mechanics." In this and succeeding chapters, the approach of rational mechanics* will be emphasized. This theme differentiates this text from others that cover the same material. We will not present an in-depth development of all the closed-form solution techniques for the field equations or a compilation of known results. There are many wonderful books on elasticity with hundreds of solutions and references to hundreds more.** The field of plasticity and limit analysis is nearly as rich. Nor will we do more than comment on the various numerical methods such as Finite Difference solutions for differential equations or finite modeling techniques such as the Boundary Element and Finite Element methods, which dominate the literature today. These are already extremely powerful tools for analysis and becoming more so, but they require a clear understanding of the fundamentals of elasticity and plasticity to be appreciated and applied properly.

Similarly the entire subject of experimental analysis must be omitted even though observation of real behavior is of paramount importance in developing a physical feel for the flow of stress and strain through structures. Photographic results from models using such experimental methods as Moire, holographic interferometry, and photoelasticity will, on occasion, be given for comparison to analytic solutions. All these techniques produce optical, full-field interference patterns, which are actually contour maps of the various displacement or stress components (more precisely, strain components). Unfortunately, however, no real explanation of these wonderful methods is possible.***

Instead we will concentrate on a few classical structures such as beams, rings, and wedges primarily in their two-dimensional idealizations. We will explore their behavior in both the elastic and plastic range from both an analysis and a design perspective. Because the regular geometry of such structures can be described by simple coordinate systems, closed-form solutions can often be found that best illustrate strategies at a level where the student can appreciate the intricacy of the concepts involved and their interrelationships. Subsequent courses on numerical or physical modeling to consider structures with irregular geometry, time-dependent boundary or thermal

* This is as opposed to the irrational mechanics practiced by most applied mathematicians who could care less about the visualization of results and their implication for design. Most young engineers, particularly with doctorates and access to high-powered computers, also do irrational mechanics. However with proper guidance, they can eventually outgrow this fixation on solving with more and more sophistication less and less important boundary-value problems, and begin to think about the inverse problem of design.

** The classic is *A Treatise on the Mathematical Theory of Elasticity* by A.E.H. Love first published in 1892 and last revised in 1926. *Theory of Elasticity,* by Timoshenko & Goodier, 3rd Ed. McGraw Hill, 1970 is probably still the best for the engineer. It brings to the reader tremendous scope, physical interpretation of solutions, some commentary, and exhaustive references.

*** Perhaps the most inclusive modern text on experimental solid mechanics is *Experimental Stress Analysis*, by J.W. Dally and W.F. Riley, McGraw Hill, 2nd Ed. 1978.

loading, built perhaps of anisotropic, inhomogeneous, or nonlinear materials will then be based on a firm foundation.

4.2 Boundary Conditions

There are then, within our structural blob, four interrelated fields flowing between the boundaries according to certain rules (the field equations) but controlled by conditions specified on the boundaries. Besides the shape of the boundary, the conditions may include:

a. Displacements at "points" or surfaces, and
b. Applied loads, often called "tractions," at "points" or distributed over an area.

This includes unloaded sections of the boundary where the normal and tangential tractions are zero. Traction boundary conditions are given by Equation (2.11) in 3D which, in two dimensions, reduce to:

$$\begin{aligned} \sigma_x \sin\theta - \tau_{xy}\cos\theta &= S_x \\ \tau_{xy}\sin\theta - \sigma_y \cos\theta &= S_y \end{aligned} \quad \text{or} \quad S_i = \sigma_{ij}\begin{bmatrix} \sin\theta \\ -\cos\theta \end{bmatrix} \tag{4.1}$$

which can be seen to be simply the equilibrium requirement.

Thus there are basically three types of boundary value problems:

1. Only tractions specified
2. Only displacements specified
3. Mixed

The last is the most common in practice and, if the conditions are specified "properly," there will be a unique solution. For the first type, tractions only, the solution for the stress field will involve an arbitrary constant translating to arbitrary linear displacements. The question of what is a "proper" specification of boundary conditions for a unique solution will be discussed later. Obviously a solution for real physical problems must exist, but in rare cases uniqueness is questionable. From the design perspective however, existence of an inverse solution is certainly not guaranteed.

The yield condition is also, in a sense, an "internal" boundary condition separating, along surfaces of unknown geometry, the sections of the structure that have "gone plastic" from those that remain elastic. Determining the geometry of the plastic zones is difficult and is obviously related to the inverse problem of design. If the plastic zones can be isolated, however, then the elastic field theory can be used outside plastic zones to determine structural behavior even past first yield. This difficult challenge is the subject of the concluding chapters of this text.

4.3 Tactics for Analysis

There are two clear plans of attack: a) determine the vector field for displacements and then differentiate to find the strains and stresses, or b) determine the tensor field and integrate to find the displacements. In either case, we can try all sorts of educated guesses or other tricks to shorten the process which, with either approach, is sometimes then called the semi-inverse method.*

4.3.1 Direct Determination of Displacements

Direct determination of displacements is clearly the more obvious strategy and intuitively satisfying since this is what we actually see and can measure. Using the definitions of strains in terms of the relative displacements, and substituting the strain relationships Equation (3.2) becomes:

$$\sigma_x = 2G\left(\frac{\partial u}{\partial x} + \frac{\nu}{1-2\nu}\epsilon_\nu\right), \qquad \tau_{xy} = G\left(\frac{\partial v}{\partial x} + \frac{\partial u}{\partial y}\right)$$

$$\sigma_y = 2G\left(\frac{\partial v}{\partial y} + \frac{\nu}{1-2\nu}\epsilon_\nu\right), \qquad \tau_{yz} = G\left(\frac{\partial w}{\partial y} + \frac{\partial v}{\partial z}\right)$$

$$\sigma_z = 2G\left(\frac{\partial w}{\partial z} + \frac{\nu}{1-2\nu}\epsilon_\nu\right), \qquad \tau_{zx} = G\left(\frac{\partial u}{\partial z} + \frac{\partial w}{\partial x}\right)$$

where $\epsilon_\nu = \epsilon_x + \epsilon_y + \epsilon_z = \frac{\partial u}{\partial x} + \frac{\partial v}{\partial y} + \frac{\partial w}{\partial z}$

Substituting into the first of the equilibrium equations (2.43):

$$2G\left(\frac{\partial^2 u}{\partial x^2} + \frac{\nu}{1-2\nu}\frac{\partial \epsilon_\nu}{\partial x}\right) + G\left(\frac{\partial^2 v}{\partial x \partial y} + \frac{\partial^2 u}{\partial y^2}\right) + G\left(\frac{\partial^2 u}{\partial z^2} + \frac{\partial^2 w}{\partial x \partial z}\right) + B_x = 0$$

and therefore:

$$\nabla^2 u + \frac{1}{1-2\nu}\frac{\partial \epsilon_\nu}{\partial x} + \frac{B_x}{G} = 0$$

Similarly

$$\nabla^2 v + \frac{1}{1-2\nu}\frac{\partial \epsilon_\nu}{\partial y} + \frac{B_y}{G} = 0 \qquad (4.2)$$

$$\nabla^2 w + \frac{1}{1-2\nu}\frac{\partial \epsilon_\nu}{\partial y} + \frac{B_y}{G} = 0$$

* An early historical survey of the development of the theory and solutions to elastic boundary value problems is given by I. Todhunter and K. Person, *A History of the Theory of Elasticity* Vol. 1, Cambridge, 1886.

These are the Navier Field Equations* and they are very nice indeed. Fifteen equations have been reduced to only three second-order differential equations in terms of the components of the displacement vector field, which must be smooth and continuous. Moreover, if there is no volume change (naturally, or if $\nu = 0.5$) and no body force, the displacement field is harmonic. Even if ϵ_v is linear and the body forces constant, the equations are manageable. Unfortunately, the complications introduced through the boundary conditions expressed in terms of displacements more than counterbalance the simplicity of the field equations and usually make the direct closed-form determination of the displacement field intractable.

Transforming the traction BCs [Equation (2.11)] and using the stress–strain and strain-displacement relations with $\lambda = \frac{E\nu}{(1+\nu)(1-2\nu)}$,

$$
\begin{aligned}
S_x &= \lambda\epsilon_v\ell + G\left(\frac{\partial u}{\partial x}\ell + \frac{\partial u}{\partial y}m + \frac{\partial u}{\partial z}n\right) + G\left(\frac{\partial u}{\partial x}\ell + \frac{\partial v}{\partial x}m + \frac{\partial w}{\partial x}n\right)\\
S_y &= \lambda\epsilon_v m + G\left(\frac{\partial v}{\partial y}m + \frac{\partial v}{\partial z}n + \frac{\partial v}{\partial x}\ell\right) + G\left(\frac{\partial v}{\partial y}m + \frac{\partial w}{\partial y}n + \frac{\partial u}{\partial y}\ell\right) \qquad (4.3)\\
S_z &= \lambda\epsilon_v n + G\left(\frac{\partial w}{\partial z}n + \frac{\partial w}{\partial x}\ell + \frac{\partial w}{\partial y}m\right) + G\left(\frac{\partial w}{\partial z}n + \frac{\partial u}{\partial z}\ell + \frac{\partial v}{\partial z}m\right)
\end{aligned}
$$

Seldom are the displacements or their derivatives known on the boundary and Equations (4.3) end up as integral equations which can only be evaluated numerically (the basis of the boundary integral method).

4.3.2 Direct Determination of Stresses

Direct determination of stresses is an alternative tactic for analysis. Finding a tensor field involving six unknowns rather than three would appear to be a

---

* Their publication in 1821 is considered as the initiation date for the formal field theory of elasticity. Navier's father, Claude Louis Marie Henri, died when the boy was only fourteen years old and his uncle, Gauthey, a famous engineer, directed his education, illuminating the theory he received at L'Ecole Polytechnique and L'Ecole des Ponts et Chaussees with practical examples from bridge and channel construction. Gauthey died in 1807 leaving Navier the job of editing the three volumes of Gauthey's treatise on civil engineering. Although others before him alluded to the disappearance of indeterminacy at a differential scale, Navier, according to Todhunter and Pearson, "seems to have been the first to notice that problems relating to reactions, for the determination of which elementary statics does not provide sufficient equations, are perfectly determinate when account is taken of the elasticity of the reacting bodies." Thus Navier ushered in modern structural analysis as well as elasticity (not to mention his fundamental work on fluid mechanics). It is interesting to note, however, that when he applied the theory to bending, Navier, like so many before him, misplaced the neutral axis until in 1826, after two trips to England to converse with the indecipherable, polymath Thomas Young, he finally got it right.

step backwards, but it is not. Shortly after introducing the concept of stress in its general tensor representation, Cauchy* recognized its field properties and the advantages of expressing boundary conditions in terms of equilibrium. Combining the stress–strain relations with the compatibility Equation (3.6) and the equilibrium Equation (3.43), and letting:

$$\sigma_v = \sigma_x + \sigma_y + \sigma_z = I_{1\sigma} = 3\sigma_m \quad \text{and} \quad \text{div}\,\bar{B} = \frac{\partial B_x}{\partial x} + \frac{\partial B_y}{\partial y} + \frac{\partial B_z}{\partial z}$$

then

$$\begin{aligned}
(1+\nu)\nabla^2\sigma_x + \frac{\partial^2\sigma_v}{\partial x^2} &= -\frac{\nu(1+\nu)}{1-\nu}\text{div}\,\bar{B} - 2(1+\nu)\frac{\partial B_x}{\partial x} \\
(1+\nu)\nabla^2\sigma_y + \frac{\partial^2\sigma_v}{\partial y^2} &= -\frac{\nu(1+\nu)}{1-\nu}\text{div}\,\bar{B} - 2(1+\nu)\frac{\partial B_y}{\partial y} \\
(1+\nu)\nabla^2\sigma_z + \frac{\partial^2\sigma_v}{\partial z^2} &= -\frac{\nu(1+\nu)}{1-\nu}\text{div}\,\bar{B} - 2(1+\nu)\frac{\partial B_z}{\partial z} \\
(1+\nu)\nabla^2\tau_{xy} + \frac{\partial^2\sigma_v}{\partial x\,\partial y} &= -(1+\nu)\left(\frac{\partial B_x}{\partial y} + \frac{\partial B_y}{\partial x}\right) \\
(1+\nu)\nabla^2\tau_{yz} + \frac{\partial^2\sigma_v}{\partial y\,\partial z} &= -(1+\nu)\left(\frac{\partial B_y}{\partial z} + \frac{\partial B_z}{\partial y}\right) \\
(1+\nu)\nabla^2\tau_{zx} + \frac{\partial\sigma_v}{\partial z\,\partial x} &= -(1+\nu)\left(\frac{\partial B_z}{\partial x} + \frac{\partial B_x}{\partial z}\right)
\end{aligned} \tag{4.4}$$

From these it can be shown that:

$$\nabla^2\sigma_v = \nabla^2 I_{1\sigma} = -\frac{1+\nu}{1-\nu}\text{div}\,\bar{B} \tag{4.5}$$

In this case, σ_v is harmonic if the body force is constant or zero. Even some of the individual stress components can be harmonic in certain circumstances.

* In its final form, the stress tensor, as we have presented it, is due to St. Venant and the general 3D field equations in terms of stress are usually called the Beltrami-Mitchell compatibility equations. However, Cauchy is primarily responsible for developing stress as a fundamental field variable. Navier considered only forces acting between individual molecules in his derivation and not planes with shear as well as normal "pressure."

In summary, then,* we have developed two fundamental and unified descriptions of the intertwined scalar (ϵ_v and σ_v), vector ($\bar{\Delta} = ui + vj + wk$), and tensor ($\sigma_{ij} \leftrightarrow \epsilon_{ij}$) fields of elasticity. We can try to either:

1. Determine the vector displacement field and differentiate to obtain the tensor field, or
2. Determine the tensor stress field, convert to strains and integrate, if possible, to obtain the vector field.

The first tactic is preferable in every way except in practice. We therefore will concentrate on the second approach for closed-form solutions although reference will often be made to the first, particularly in the chapter on torsion.

4.4 St. Venant's Principle

One of the most useful arguments in solid mechanics is one for which there is no precise analytic statement. St. Venant,** in discussion of beams, notes that:

> "If a system of forces acting over 'small' areas of a solid are replaced by a statically equivalent system (same force and moment) then the stresses change 'significantly' only in the 'neighborhood' of the loaded region."

* The implications of the principles and unified equations of elasticity developed in the 1820s by Navier, Cauchy, and other Frenchmen were profound, producing immediate consequences. The unifying concept of force fields on a molecular level transmitted as stresses by a universal law of interatomic attraction and repulsion excited imaginations everywhere, but particularly in Britain. Although George Green's attempt to unite all interactions such as: heat, temperature, electricity, light, and magnetism with an all-inclusive mathematical theory was unsuccessful, his vision led the way. In 1831, the genius of Michael Faraday for superb experiments to demonstrate his concept of the electromagnetic field was augmented with insights from Stokes, Kelvin, Hamilton, and others to eventually culminate in Maxwell's famous equations. These remarkable field equations, which include light as electromagnetic radiation, have needed no revision since Maxwell, against the vehement advice of his colleagues, included relativistic effects for "the sake of beauty" as dictated by his aesthetic intuition.

** Another engineer with great mathematical ability, Barre de St. Venant was admitted to L'Ecole Polytechnique in 1813 at the age of 16. He quickly became first in his freshman class. When the students were mobilized to defend Paris in 1814, St. Venant refused to "fight for the usurper" (Napoleon I) and was subsequently declared a deserter and denied readmission to L'Ecole Polytechnique (Napoleon's exile to Elba probably saved St. Venant from being shot). Only 8 years later he was allowed to enter L'Ecole des Ponts et Chaussees, where he was still shunned by other students (as a coward no less!).

St. Venant worked as an engineer on channel design (his first paper was on fluid dynamics), municipal projects for Paris (where he quit in disgust), and applications of hydraulics in agriculture (for which he was awarded the gold medal by the French Agricultural Society for a series of papers on the subject). However, his favorite work was on the theory of elasticity, which he could indulge in full-time only after he was elected to the Academy of Sciences in 1868.

One should appreciate that this ill-defined St. Venant principle is quite profound.* It is a physical statement of a fundamental property of the governing field equations of elasticity and the proper specification of boundary conditions for solution. It is certainly important since in the hands of an engineer "with good judgement," it can be used to great effect to simplify and give insight to design. St. Venant's principle, for example, is the reason why simple strength-of-materials type analysis and design is so powerful in practice, and conversely, why more advanced elasticity solutions are often just localized. We will refer to the St. Venant principle often and the student can see it in evidence even more frequently if on the lookout for it.

4.5 Two-Dimensional Stress Formulation

In 2D it is easy to follow the steps only outlined in 3D for Equations (4.4) and (4.5) for the direct determination of stress formulation. Substituting the stress–strain relationships [Equation (2.4)] into the strain compatibility [Equation (2.7)] gives:

$$\frac{\partial^2}{\partial y^2}(\sigma_x - \nu\sigma_y) + \frac{\partial^2}{\partial x^2}(\sigma_y - \nu\sigma_x) = 2(1+\nu)\frac{\partial^2 \tau_{xy}}{\partial x\,\partial y} \qquad \text{(a)}$$

where, in plane strain, ν would be ν'. Now, differentiating the first equilibrium Equation (3.45) with respect to x and the second with respect to y and adding:

$$2\frac{\partial^2 \tau_{xy}}{\partial x\,\partial y} + \frac{\partial^2 \sigma_y}{\partial x^2} + \frac{\partial^2 \sigma_x}{\partial y^2} + \frac{\partial B_x}{\partial x} + \frac{\partial B_y}{\partial y} = 0 \qquad \text{(b)}$$

Finally, substituting (b) into (a) we get:

$$\left(\frac{\partial^2}{\partial x^2} + \frac{\partial^2}{\partial y^2}\right)(\sigma_x + \sigma_y) = -(1+\nu)\left(\frac{\partial B_x}{\partial x} + \frac{\partial B_y}{\partial y}\right) \qquad (4.6)$$

He became the world's expert on mathematical elasticity. He never published his own book, but edited Navier's "Resume des legons..." (1864), in which St. Venant's added notes make up 90% of the total volume! Among many important discoveries, he is responsible for solutions to the Torsion Problem summarized in his famous memoir of 1853, where he used a semi-inverse method and the displacement formulation: all new concepts. His last paper appeared four days before his death in January 1886.

* See, for example, "On St. Venant's Principle," by E. Sternberg, *Quarterly of Applied Mathematics*, Vol. 11, 1954, p. 393. While there are no precise definitions of the words "small" and "neighborhood," the approximation is generally considered entirely satisfactory in the region beyond 1.5 times the dimension "small." One can also argue that St. Venant's principle applies to any type of "local" irregularity such as geometric or material imperfections, small inclusions, or holes.

Equation (4.6) for nonconstant body forces is known as the "compatibility equation in terms of stress," but it actually combines compatibility, equilibrium, and stress–strain relations.

This development is remarkable. For a linear stress–strain material and constant body forces, the 2D *field* equations

$$\begin{aligned} \frac{\partial \sigma_x}{\partial x} + \frac{\partial \tau_{xy}}{\partial y} + B_x &= 0 \\ \frac{\partial \tau_{xy}}{\partial x} + \frac{\partial \sigma_y}{\partial y} + B_y &= 0 \end{aligned} \tag{4.7}$$

and

$$\nabla^2(\sigma_x + \sigma_y) = 0$$

are independent of material properties!
Thus:

a. In-plane stresses are identical for any given geometry, loading, and BCs (i.e., stresses in similar steel, concrete, wood structures are the same).
b. This, in turn, allows the experimental determination of stress fields using plastic or other materials.
c. The in-plane stresses are the same for plane strain or plane stress (which again is very nice for experimental work since plane strain is difficult to achieve in the laboratory).
d. Linear viscoelastic stress fields are the same for constant load (creep) and do not change with time.

Two-dimensional solutions are, therefore, very general and powerful.* Conceptually it is usually easier to think in terms of these stress equations rather than their displacement counterparts and we will concentrate on this approach.

It is also remarkable that the mathematics and our previous discussion of the physical importance of the invariants (independent of coordinate system) seem to be converging. All of a sudden, the first invariant (the mean stress as the center of Mohr's Circle) has appeared and in the form of the most common, most studied, best understood differential equation in the universe:** the Laplace or harmonic equation. The isotropic component seems to have an

* Uniqueness is assured only for a simply-connected region or, if there are holes, only if loads on each hole boundary are self-equilibriating.

** A constant theme in the history of science and engineering is that for some unknown reason, mathematics, no matter how abstract, eventually becomes useful. Mathematics seems to be the language of applied science or at least provides the rules of grammar.

equation of its own, uncoupled from the deviatoric component, which must be described by the equilibrium Equations (2.45).

4.6 Types of Partial Differential Field Equations

The world of Continuum Mechanics is very much a study of partial differential field equations (p.d.e.s) describing how vector and scalar components of tensors vary with position and how that field may vary with time. To make it easy, consider the general second order, linear p.d.e. in two variables:

$$A\frac{\partial^2 \psi}{\partial x^2} + B\frac{\partial^2 \psi}{\partial x \partial y} + C\frac{\partial^2 \psi}{\partial y^2} + D\frac{\partial \psi}{\partial x} + E\frac{\partial \psi}{\partial y} + F\psi + G = 0 \tag{4.8}$$

where ψ is a general symbol for the field quantity and the two coordinates may be either space coordinates or one space coordinate plus time.

There are an infinite number of possible solutions to this general field equation. The boundary conditions determine which particular solution is appropriate. The solution $\psi(x,y)$ is a surface above and/or below the xy plane and the boundary is a specified curve in the xy plane. On this boundary, the boundary conditions are then, either:

a. The height ψ "above" the boundary curve (the value of ψ) called the Dirichlet conditions, or
b. The slope of the ψ surface normal to the boundary curve ($\bar{n} \cdot \text{Grad}\, \psi$) called the Neumann condition, or
c. Both, which is called the Cauchy boundary condition.

Three types of p.d.e.s are subsets of Equation (4.8) which are classified as shown in Table 4.1 along with their simplest examples. A unique and stable solution for each type of equation is possible only if the boundary conditions

TABLE 4.1
Types of Partial Differential Equations

Classification	Type	Example
$B^2 < 4AC$	Elliptic	Laplace: $\frac{\partial^2 \phi}{\partial x^2} + \frac{\partial^2 \phi}{\partial y^2} = 0$
$B^2 = 4AC$	Parabolic	Diffusion: $\frac{\partial^2 T}{\partial x^2} = \frac{1}{\alpha^2}\frac{\partial T}{\partial t}$
$B^2 > 4AC$	Hyperbolic	Wave: $\frac{\partial^2 u}{\partial x^2} = \frac{1}{a^2}\frac{\partial^2 u}{\partial t^2}$

TABLE 4.2
Required Boundary Conditions

Conditions	Type of Boundary	Hyperbolic	Elliptic	Parabolic
Dirichlet or Neumann (value *or* slope)	Open	Insufficient	Insufficient	Unique and Stable Solution in + Direction Unstable in Direction
	Closed	Solution not unique	Unique and Stable Solution	Over-specified
Cauchy (value *and* slope)	Open	Unique and Stable Solution	Solution Unstable	Over-specified
	Closed	Over-specified	Over-specified	Over-specified

are specified properly* as shown in Table 4.2. For example, it is not possible to look backward in time** with a parabolic equation, which can describe such processes as the diffusion of a pollutant into a stream, thermal diffusion of temperature, or consolidation of saturated soil. The wave equation, on the other hand, which is hyperbolic, works equally well with negative or positive time.

4.7 Properties of Elliptic Equations

Since we will deal primarily with elliptic equations, let us list a few important properties of these field equations and the associated boundary conditions:

a. Boundary conditions must be Dirichlet or Neumann over a *closed* boundary. Part of the boundary may be at infinity (where the slope must be zero).
b. The scalar function, ψ, can have no maxima or minima inside the region where $\nabla^2\psi = 0$. In other words, all maximums and minimums occur *on* the boundary of a Laplacian field. This is obviously a very important property for engineers who are particularly interested in maximums and minimums.
c. As a consequence, if the boundary conditions start the field increasing in a given direction, it has no choice but to continue increasing until it either goes to infinity (not a maximum, but a singularity) or the opposite boundary intervenes in time to stop it. That is why Cauchy conditions on an open boundary are not suitable.
d. However, elliptic equations are strongly damping (St. Venant's Principle). In fact, Laplace's equation is really a statement that the function

* A complete discussion is given in *Methods of Theoretical Physics* by Morse, P.M. and Feshback, H., Part I, Chap. 6, McGraw-Hill, 1953.

** Such fundamental concepts can have profound practical implications. For example, there is no way, from measuring a concentration distribution "downstream," to identify which source of pollution upstream to take to court.

ψ and its slope at any point is the average of the values at neighboring points.

e. Elliptic equations represent steady-state conditions. The field is what remains after the transient or particular solution in time disappears. Thus the other types of p.d.e.s (diffusion or wave equations) reduce to Laplace's equation when changes due to time damp out. Therefore, it is not surprising, perhaps, that elliptic equations are so common in the universe.*

f. Solutions to the Laplace equation come in pairs called *conjugate harmonic functions*. They obey the first-order Cauchy-Riemann conditions:

$$\frac{\partial \phi}{\partial x} = \pm \frac{\partial \psi}{\partial y}$$
$$\frac{\partial \phi}{\partial y} = \mp \frac{\partial \psi}{\partial x} \tag{4.9}$$

Thus taking the derivative of the first with respect to x and the second with respect to y and adding, gives $\nabla^2\psi = 0$, *or* taking the derivative of the first with respect to y and the second with respect to x and subtracting gives $\nabla^2\phi = 0$. One of the conjugate pair is often called the potential function and the other the stream function.

g. The solution to Laplace's equation is a *flow net* where lines of equal ϕ (contours of ϕ) intersect lines of equal ψ (contours of ψ) at right angles. The Cauchy-Riemann conditions [Equation (4.9)] also require that the ratio of the adjacent sides of the resulting "rectangles" remain constant (usually taken as 1.0 so that the flow net becomes a system of squares**). Actually ϕ contours can be flow lines "down" the Ψ potential "hill" (the surface of Ψ plotted above the xy plane) or Ψ contours can be flow paths down the ϕ potential hill.*** Constructing the flow net, which is an acquired skill is, in fact, a full graphical solution to Laplace's equation.****

h. The contours of equal $(\sigma_1 + \sigma_2)$ or mean stress σ_m, which we know now is either a potential or a stream function if the body force is zero or constant, are called *isopachics*.

---

* By the laws of thermodynamics, forcing functions eventually dissipate and everything runs down (without supernatural renewal?). Thus the universe will finally become a gigantic Laplacian Field of unavailable energy in the form of waste heat. The depressing implications of such speculation are covered in philosophy courses.

** Curvilinear squares, which may not have four sides (often three or five), may result. Finer and finer division, however, produces squares.

*** Physically we usually know which conjugate function is the "potential" driving the "flow," but mathematically the two are interchangeable.

**** Drawing flow nets to solve Laplace's equation should be familiar from undergraduate courses in fluid dynamics, heat transfer, or ground water flow.

Another elliptic equation which plays a major role in static elasticity is the nonhomogeneous Poisson's equation:

$$\nabla^2 \psi = q \tag{4.10}$$

which arises when there is a source or sink $q(x,y)$ in the field. For example, electric charge density, p, causes a negative concentration of the electric potential, V, so that $\nabla^2 V = -cp$ or a source of heat, Q, causes a concentration in temperature T so that $\nabla^2 T = KQ$.

One way to visualize the Laplace and Poisson equations is to consider a perfectly elastic membrane in equilibrium under a uniform tension applied around its edge. If the edge lies entirely in a plane (perhaps tilted), the trivial solution $\psi = ax + by + c$ results. If the boundary edge is now distorted into a nonlinear shape, the membrane is distorted also, but its height above the plane $\Psi(x,y)$ still satisfies $\nabla^2\psi = 0$. The tension straightens out all the "bulges" introduced at the boundary and the displacement at any point equals the average of the neighboring points. One can also say that the membrane has arranged itself so that the average slope is a minimum. Thus the field condition $\nabla^2\psi = 0$ damps out all the edge values (St. Venant's principle) so that the membrane assumes a shape with the minimum stretching possible.

If now a distributed force, $q(x,y)$, normal to the stretched membrane (which already satisfies $\nabla^2\psi = 0$) is applied, the membrane will bulge. The Laplacian ψ is now proportional to this load per unit area, q, at any point. Shown in Figure 4.1 is a circular membrane loaded uniformly over the area $0 \le r \le 0.5$, and unloaded $0.5 \le r \le 1.0$. In the center region we have Poisson's equation

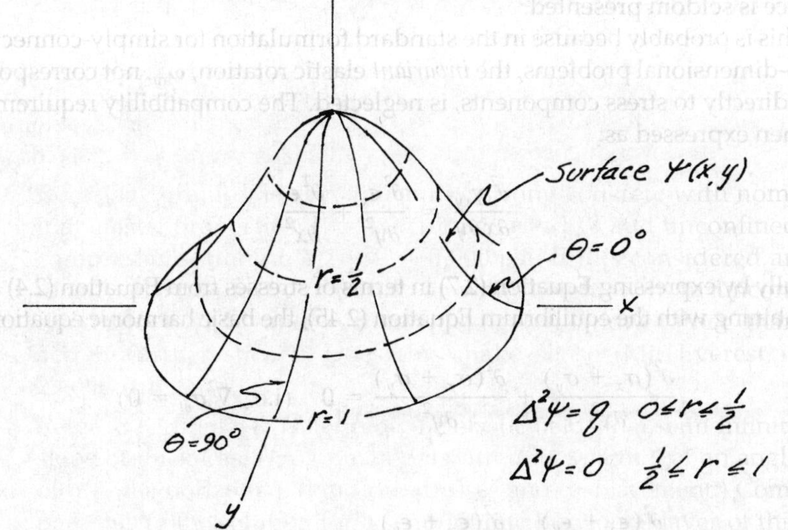

FIGURE 4.1
Shape of a circular membrane clamped at $r = 1.0$ loaded uniformly (q = const.) over the area $0 \le r \le \frac{1}{2}$.

$\nabla^2\psi = q =$ const (a local maximum at $r = 0$) and in the unloaded annulus $0.5 \le r \le 1.0$, $\nabla^2\psi = 0$ (with maximum slope and magnitude on the "boundary" at $r = 0.5$).

It is important to recognize that superposition holds for the Laplacian. If the boundary of the membrane in Figure 4.1 is now distorted, the new membrane shape will be simply $\psi + \alpha$ where α is the solution of $\nabla^2\alpha = 0$ for the membrane satisfying the distorted boundary. This is very useful in matching boundary conditions in that any amount of any solution of Laplace's equation $\nabla^2\alpha = 0$ (homogeneous) can be added to Ψ and still have a solution of $\nabla^2\psi = q$ for the same particular q.

4.8 The Conjugate Relationship Between Mean Stress and Rotation

The first invariant of the stress or strain tensor, the sum of the normal components, is harmonic for a simply-connected domain and constant body forces. This in turn implies the existence of a conjugate function, which must similarly be harmonic and also invariant to linear tangent transformation.

From experience with the significance of conjugate pairs in other potential fields, we might expect a similar importance to be attached to the conjugate harmonic function in the elastic continuum. This is not so. Occasionally passing reference is made to a conjugate relationship between elastic rotations and the mean stress or strain, but the explicit statement of this correspondence is seldom presented.

This is probably because in the standard formulation for simply-connected, two-dimensional problems, the *invariant* elastic rotation, ω_{xy}, not corresponding directly to stress components, is neglected. The compatibility requirement is then expressed as:

$$\frac{\partial^2 \gamma_{xy}}{\partial x \partial y} = \frac{\partial^2 \epsilon_x}{\partial y^2} + \frac{\partial^2 \epsilon_y}{\partial x^2}$$

Finally by expressing Equation (2.7) in terms of stresses from Equation (2.4) and combining with the equilibrium Equation (2.45), the basic harmonic equations:

$$\frac{\partial^2(\sigma_x + \sigma_y)}{\partial x^2} + \frac{\partial^2(\sigma_x + \sigma_y)}{\partial y^2} = 0 \quad (\text{i.e., } \nabla^2\sigma_m = 0)$$

or

$$\frac{\partial^2(\epsilon_x + \epsilon_y)}{\partial x^2} + \frac{\partial^2(\epsilon_x + \epsilon_y)}{\partial y^2} = 0 \quad (\text{i.e., } \nabla^2\epsilon_m = 0)$$

result whenever the body forces B_x and B_y are negligible or constant. The mean stress in the plane $\sigma_m = \frac{\sigma_x + \sigma_y}{2}$ or the mean strain $\epsilon_m = \frac{\epsilon_x + \epsilon_y}{2}$ are invariant, scalar, harmonic functions related by the material constants (i.e., $\epsilon_m = \frac{1-\nu}{E}\sigma_m$).

Consider instead the equilibrium equations rewritten in terms of strain:

$$\begin{aligned} &\frac{E}{1-\nu^2}\frac{\partial}{\partial x}(\epsilon_x + \nu\epsilon_y) + G\frac{\partial \gamma_{xy}}{\partial y} + B_x = 0 \\ &\frac{E}{1-\nu^2}\frac{\partial}{\partial y}(\epsilon_y + \nu\epsilon_x) + G\frac{\partial \gamma_{xy}}{\partial x} + B_y = 0 \end{aligned} \tag{4.11}$$

and the compatibility requirement in the form:*

$$\begin{aligned} \frac{\partial \epsilon_x}{\partial y} &= \frac{\partial}{\partial x}\left(\frac{\gamma_{xy}}{2} - \frac{\omega_{xy}}{2}\right) \quad \text{or} \quad \frac{1}{2}\frac{\partial \gamma_{xy}}{\partial x} = \frac{\partial \epsilon_x}{\partial y} + \frac{\partial \omega_{xy}}{2\partial x} \\ \frac{\partial \epsilon_y}{\partial x} &= \frac{\partial}{\partial y}\left(\frac{\gamma_{xy}}{2} + \frac{\omega_{xy}}{2}\right) \quad \text{or} \quad \frac{1}{2}\frac{\partial \gamma_{xy}}{\partial y} = \frac{\partial \epsilon_y}{\partial x} - \frac{\partial \omega_{xy}}{2\partial y} \end{aligned} \tag{4.12}$$

Substituting:

$$\begin{aligned} \frac{1}{E}\frac{\partial(\sigma_x + \sigma_y)}{\partial x} &= \frac{1}{1-\nu}\frac{\partial}{\partial x}(\epsilon_x + \epsilon_y) = \frac{\partial \omega_{xy}}{2\partial y} - \frac{B_x}{2G} \\ \frac{1}{E}\frac{\partial(\sigma_x + \sigma_y)}{\partial y} &= \frac{1}{1-\nu}\frac{\partial}{\partial y}(\epsilon_x + \epsilon_y) = -\frac{\partial \omega_{xy}}{2\partial x} - \frac{B_y}{2G} \end{aligned} \tag{4.13}$$

Thus, neglecting body forces, we have the Cauchy-Riemann conditions directly and the conjugate relationship between the elastic rotation of a differential element and the first invariant of the 2D stress tensor is established.

To determine elastic rotations around the x and y axes associated with plane stress and plane strain remember that:

$$\frac{\partial w}{\partial x} = \frac{\omega_{zx}}{2} \qquad \frac{\partial w}{\partial y} = \frac{\omega_{zy}}{2} \qquad \text{and} \qquad \frac{\partial w}{\partial z} = \epsilon_z$$

since $\gamma_{zx} = \gamma_{zy} = 0$ in either case. Therefore,

$$\frac{\partial \epsilon_z}{\partial x} = \frac{\partial}{\partial z}\left(-\frac{\omega_{zx}}{2}\right); \qquad \frac{\partial \epsilon_z}{\partial y} = \frac{\partial}{\partial z}\left(\frac{\omega_{zy}}{2}\right)$$

* Remember that $\frac{\partial u}{\partial x} = \frac{1}{2}(\gamma_{xy} + \omega_{xy})$ and $\frac{\partial u}{\partial y} = \frac{1}{2}(\gamma_{xy} - \omega_{xy})$.

For plane strain then ω_x *and* ω_y are zero. However, for plane stress where $\epsilon_z = -\frac{\nu}{E}(\sigma_x + \sigma_y)$:

$$\frac{\partial}{\partial z}\left(\frac{\omega_{zx}}{2}\right) = +\frac{\nu}{E}\frac{\partial(\sigma_x + \sigma_y)}{\partial x} = +\nu\frac{\partial}{\partial y}\left(\frac{\omega_{xy}}{2}\right)$$
$$\frac{\partial}{\partial z}\left(\frac{\omega_{zy}}{2}\right) = -\frac{\nu}{E}\frac{\partial(\sigma_x + \sigma_y)}{\partial y} = +\nu\frac{\partial}{\partial x}\left(\frac{\omega_{xy}}{2}\right) \tag{4.14}$$

Some general observations follow directly from these equations:

1. For an incompressible material ($\nu = 0.5$) in plane strain, ϵ_m is zero and the displacements u and v themselves become conjugate harmonic functions.
2. From Equation (4.12) we can obtain the compatibility equation in its standard form [Equation (2.7)] and derive a second independent relationship:

$$\frac{\partial^2 \epsilon_x}{\partial y^2} - \frac{\partial^2 \epsilon_y}{\partial x^2} = \frac{\partial^2 \omega_{xy}}{\partial x\, \partial y} \tag{4.15}$$

3. Both the elliptic equations associated with Equation (4.13) are of similar form:

$$\nabla^2(\epsilon_x + \epsilon_y) = \frac{(1-\nu)}{2G}\left(\frac{\partial B_x}{\partial x} + \frac{\partial B_y}{\partial y}\right)$$
$$\nabla^2\frac{\omega_{xy}}{2} = \frac{1}{2G}\left(\frac{\partial B_x}{\partial x} - \frac{\partial B_y}{\partial y}\right) \tag{4.16}$$

and are Laplacian whenever the body forces are negligible or constant. In the event the body forces are constant such as in a gravity field, Equation (4.13) can be reduced to the Cauchy-Riemann conditions and direct solution achieved by a suitable transformation. For example, if we define:

$$A = \frac{\sigma_x + \sigma_y}{E} \tag{a}$$

$$B = \frac{\omega_{xy}}{2} + \frac{1}{2G}(B_y x - B_x y) \tag{b}$$

then from Equation (4.13)

$$\frac{\partial A}{\partial x} = \frac{\partial B}{\partial y}; \quad \frac{\partial A}{\partial y} = -\frac{\partial B}{\partial x}$$

and we can again solve for A and B by standard methods.

4. Direct integration to determine displacements for plane strain (or plane stress in the plane $z = 0$) is greatly facilitated in that:

$$\begin{aligned} du &= \epsilon_x dx + \frac{\gamma_{xy}}{2} dy - \frac{\omega_{xy}}{2} dy \\ dv &= \epsilon_y dy + \frac{\gamma_{xy}}{2} dx + \frac{\omega_{xy}}{2} dx \end{aligned} \tag{4.17}$$

where now we can determine the rotation from the stress field with Equation 4.13.

5. Finally for the general 3D case, both the first invariant of the stress or strain tensor:

$$\sigma_m = \frac{\sigma_x + \sigma_y + \sigma_z}{3} \quad \text{or} \quad \epsilon_m = \frac{\epsilon_x + \epsilon_y + \epsilon_z}{3}$$

and the invariant mean rotation of a volume element:

$$\Omega = |\Omega| = \sqrt{\omega_{yz}^2 + \omega_{zx}^2 + \omega_{xy}^2}$$

will also clearly be Laplacian.

In general the conjugate relationships are more complicated in three-dimensional situations. Often, however, as is the case of plane stress [Equation (4.14)] or where there is a high degree of symmetry, the components of rotation can be determined easily, in which case displacements can be found by direct integration of the relative displacement equations:

$$\begin{aligned} du &= \epsilon_x dx + \frac{\gamma_{xy}}{2} dy + \frac{\gamma_{zx}}{2} dz - \frac{\omega_{xy}}{2} dy + \frac{\omega_{zx}}{2} dz \\ dv &= \frac{\gamma_{xy}}{2} dx + \epsilon_y dy + \frac{\gamma_{yz}}{2} dz - \frac{\omega_{yz}}{2} dz + \frac{\omega_{xy}}{2} dx \\ dw &= \frac{\gamma_{zx}}{2} dx + \frac{\gamma_{yz}}{2} dy + \epsilon_z dz - \frac{\omega_{xz}}{2} dx + \frac{\omega_{yz}}{2} dy \end{aligned} \tag{4.18}$$

Thus we see that, without body forces, the elastic rotations times E are the conjugate harmonic function to twice the mean stress and that these two scalar invariants go "hand-in-hand" throughout a structure, one controlling the other in accordance with Laplace's equation. If we can solve for one, we have solved for the other, and in that sense it is correct to neglect the rotations in that they will come out of the solution for the stress (or strain) field of their own accord.* This scalar part of the total tensor field giving the conjugate harmonic invariants σ_q and ω_{xy} (or σ_m and Ω in 3D) we shall call the *"isotropic field."* Only the center of Mohr's Circle or, physically, pure volume change is involved.

Unfortunately it is seldom possible to specify a complete set of proper boundary conditions in terms of σ_q or ω_{xy} for problems in solid mechanics. For example, at most boundaries the tangential normal stress is unknown, and only if the boundary is clamped will there be zero rotation. Thus it is *usually* not possible to solve for the isotropic field independent of the deviatoric field.

Nevertheless, flow nets are extremely useful:

1. To depict the isotropic field so it can be understood once known.
2. Determine the isotropic field *if* the boundary conditions can be found by other methods such as photoelasticity which gives σ_m on the boundary directly.
3. As a conceptual tool for optimum design.

A number of isotropic flow nets for important stress fields will be plotted in later chapters and will be called for in chapter problems. Two examples are shown in Figures 4.2 and 4.3 which will be referred to later when the solutions to these important problems are presented. In both these structures, symmetry requires that certain lines do not rotate, which simplifies the construction.

Interferometry provides the best optical full-field method to see, in the laboratory, the isopachics—contour lines of "equal thickness"—which, by Equation (4.6), are proportional to the mean stress. This is to say that, for plane stress, the normal strain $\epsilon_z = -\frac{\nu}{E}(\sigma_x + \sigma_y)$. Thus when a model is loaded, the change in the thickness, t, will be

$$w = t - t_o = \epsilon_z t_0 = -\frac{\nu t_o}{E}(\sigma_x + \sigma_y) \tag{4.19}$$

By interfering with the wave front of light through or reflected off the unstressed model with the wavefront from the stressed model, the change in

* A further indication that "Couple-Stress" theory is unnecessary and perhaps spurious. Gradients in the rotation field are entirely accounted for by the gradient of the isotropic mean stress (or strain).

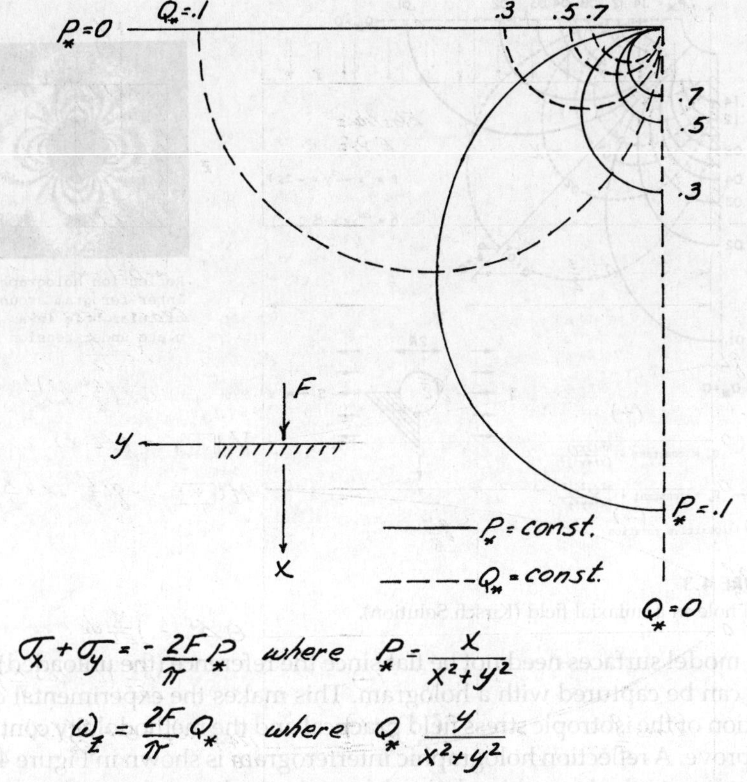

FIGURE 4.2
Line load on the half space (Flamant Solution).

thickness produces "fringes" of integer order N, proportional to w, with precision in multiples of the wavelength of light, λ. More specifically

$$\sigma_x + \sigma_y = c\frac{NE\lambda}{t_o\,\nu} = \frac{Nf}{t}$$

in which c is 2, 1, or 1/2 depending on whether transmission or reflective interferometry is employed. Thus, for example, with a helium-neon laser ($\lambda = 250 \times 10^{-9}$ in.), a Plexiglas model ($E = \frac{1}{3} \times 10^6$ psi, $\nu = \frac{1}{3}$) of thickness, $t_o = \frac{1}{4}$ in., with reflection off the back surface ($c = 1$), $f = 25$ psi/fringe, and

$$\sigma_x + \sigma_y = 100N$$

Until the advent of the laser, interferometry was extremely expensive and not very successful even in the hands of the best experimentalist. However, laser light is almost perfectly monochromatic and, most importantly, coherent.

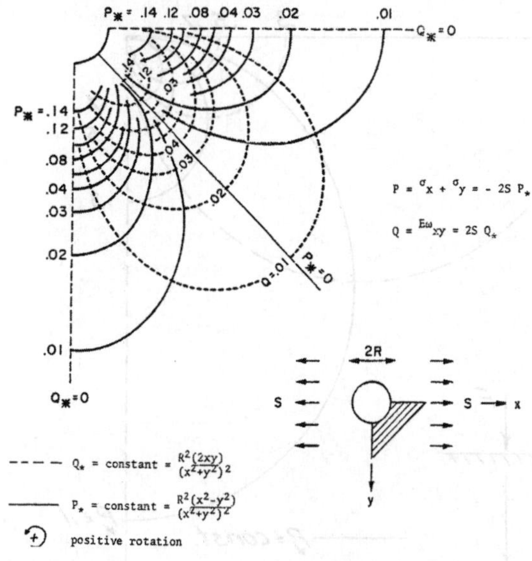

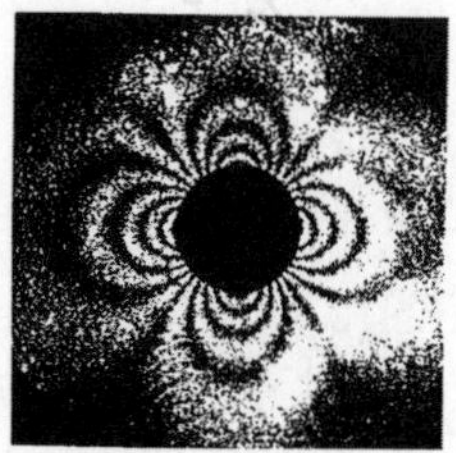
Reflection holographic inter-ferogram around a circular hole in a plate under tension

FIGURE 4.3
Small hole in a uniaxial field (Kirsch Solution).

Thus, model surfaces need not be flat since the reference (the unloaded) light beam can be captured with a hologram. This makes the experimental determination of the isotropic stress field practical and the methodology continues to improve. A reflection holographic interferogram is shown in Figure 4.3 for comparison to the isopachics derived from the analytic solution.

Example 4.1

A beam (plane stress) tapered at 45° has a linearly increasing load $p = \gamma x$ lb/ft^2 on its vertical face as shown below. Neglect body forces and assume a linear distribution of $\sigma_x = My/I$, i.e., $\sigma_x = 2\gamma(x/2 - y)$

a. Derive a solution (i.e., expressions for σ_y and τ_{xy} as functions of x and y) to satisfy equilibrium and BCs. Plot your results on a horizontal cross-section at $x = H$.
b. Is this the elasticity solution? Discuss.
c. Plot the flow net for $(\sigma_1 + \sigma_2)$ vs. $\omega_z E$ and calculate the displacement at the top ($x = y = 0$) relative to a point at a distance $x = H$, $y = 0$ on the loaded face.

Equilibrium:

$$\frac{\partial \sigma_x}{\partial x} + \frac{\partial \tau_{xy}}{\partial y} = \gamma + \frac{\partial \tau_{xy}}{\partial y} = 0 \tag{1}$$

$$\frac{\partial \tau_{xy}}{\partial x} + \frac{\partial \sigma_y}{\partial y} = 0 \tag{2}$$

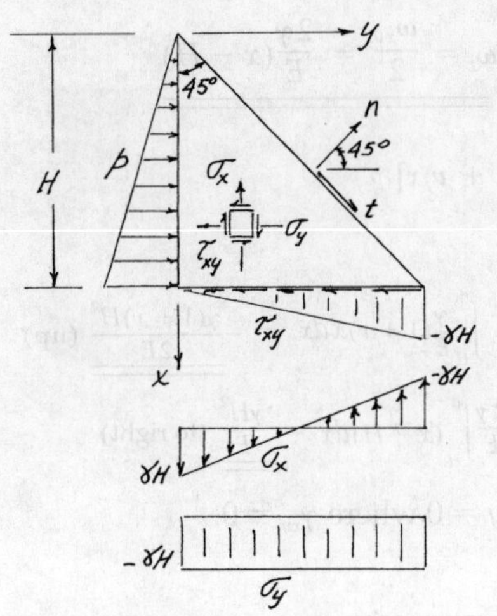

from (1) $\tau_{xy} = -\gamma_y + f(x) + C$

B.C. at $x =$ any, $y = 0 \quad \tau_{xy} = 0$

$\therefore f(x) = 0, C = 0 \quad \therefore \tau_{xy} = -\gamma_y$

from (2) $\sigma_y = g(x) + C'$

B.C. at $x =$ any, $y = 0$ $\sigma_y = -\gamma x$

$\therefore g(x) = -\gamma x, C' = 0$

$\therefore$ Stress Field Solution

$$\begin{cases} \sigma_x = \underline{\underline{\gamma x - 2\gamma y}} \\ \sigma_y = \underline{\underline{-\gamma x}}, \quad \tau_{xy} = \underline{\underline{-\gamma y}} \end{cases}$$

b) Check B.C. on sloping face, $x = y$ where $\sigma_n = \tau_{nt} = 0$

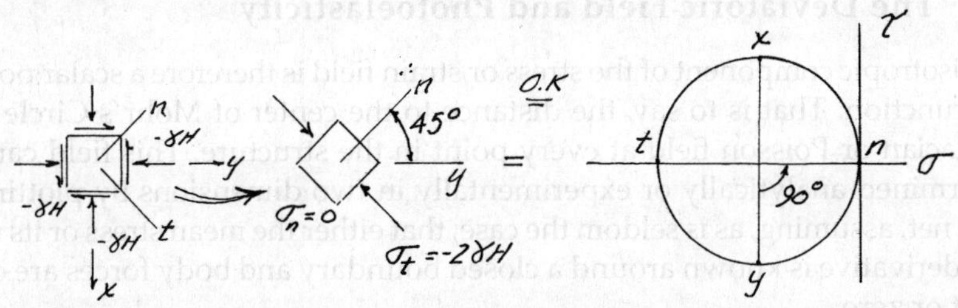

Note: Although B.C. on the face $x =$ const. not specified, overall equilibrium is satisfied and the linear increase as $x \to \infty$ is reasonable.

Check compatibility $\nabla^2(\sigma_x + \sigma_y) = 0 = \nabla^2(-2\gamma y) = 0$

$\therefore$ *This is the elasticity Solution*

Satisfies Equilibrium, Compatibility, and Boundary Conditions (which are specified properly)

c) $\sigma_x + \sigma_y = -2\gamma_y \quad$ Let $\omega_z = \dfrac{\omega_{xy}}{2}$

1st C.R. cond.: $\dfrac{\partial\phi}{\partial x} = \dfrac{\partial\psi}{\partial y}$

$$\therefore \frac{\partial(\sigma_x + \sigma_y)}{\partial x} = \frac{E\,\partial\omega_z}{\partial y} = -2$$

$$\therefore \omega_z = f(x) + C$$

2nd C.R. cond.

$$\frac{\partial(\sigma_x + \sigma_y)}{\partial y} = -\frac{E\,\partial\omega_z}{\partial x} = -2\gamma$$

$$\omega_z = \frac{2\gamma x}{E} + C$$

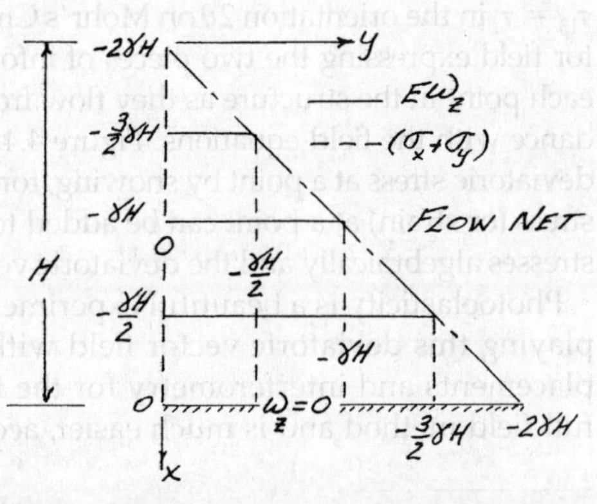

Say $\omega_z = 0$ at $x = \text{H} \therefore C = -\frac{2\gamma H}{E}$ and $\therefore \omega_z = \frac{\omega_{xy}}{2} = \frac{2\gamma}{E}(x - H)$

Strains: $\epsilon_x = \frac{1}{E}(\sigma_x - \nu\sigma_y) = \frac{\gamma}{E}[-2y + (1 + \nu)x]$

$\epsilon_y = \frac{1}{E}(\sigma_y - \nu\sigma_x) = \frac{\gamma}{E}[2\nu y - (1 + \nu)x]$

$$\therefore u^{\text{top}} = \int_H^0 du = \int \epsilon_x dx + \int \frac{\gamma_{xy}}{2} dy - \int \frac{\omega_{xy}}{2} dy = \int_H^0 \frac{\gamma}{E}(1 + \nu)x\,dx = -\frac{\gamma(1 + \nu)H^2}{2E} \text{ (up)}$$

$$v^{\text{top}} = \int_H^0 dv = \int \epsilon_y dy + \int \frac{\gamma_{xy}}{2} dx + \int \omega_z dx = \frac{2\gamma}{E}\int_H^0 (x - H)dx = \frac{\gamma H^2}{E} \text{ (to right)}$$

where we are integrating along the line $y = 0$ where $\gamma_{xy} = 0$.

4.9 The Deviatoric Field and Photoelasticity

The isotropic component of the stress or strain field is therefore a scalar potential function. That is to say, the distance to the center of Mohr's Circle is a Laplacian or Poisson field at every point in the structure. This field can be determined analytically or experimentally in two dimensions by plotting a flow net, assuming, as is seldom the case, that either the mean stress or its normal derivative is known around a closed boundary and body forces are constant or zero.

The deviatoric component of the stress tensor is more complicated in that vector fields are involved rather than a simple scalar field. In three dimensions, the deviatoric tensor involves three magnitudes, S_i, (principal values) and associated angles, α_i, (orientation to the reference frame) and can, therefore, be thought of as three vectors, $\bar{S}_i$, describing the size and aspect ratios of the Mohr's Circles.

In the planar care, there is just one deviatoric vector, $\bar{q}$, of magnitude $\tau_{max} = \tau_{13} = \tau_q$ in the orientation 2θ on Mohr's Circle. Thus the deviatoric field is a vector field expressing the two pieces of information, magnitude and direction, at each point in the structure as they flow from one boundary to another in accordance with the field equations. Figure 4.4 illustrates this vector property of the deviatoric stress at a point by showing, for a specific example, how two states of stress (or strain) at a point can be added together by adding the isotropic mean stresses algebraically and the deviatoric vectors by the parallelogram rule.*

Photoelasticity is a beautiful experimental method of discovering and displaying this deviatoric vector field with photographs. Like Moiré for displacements and interferometry for the isotropic stress, photoelasticity is a full-field method and is much easier, accurate, and more versatile. Two sets

* In contrast to the procedure of rotating one state into the same orientation and adding components (as called for in Prob. 2.19 for the same example).

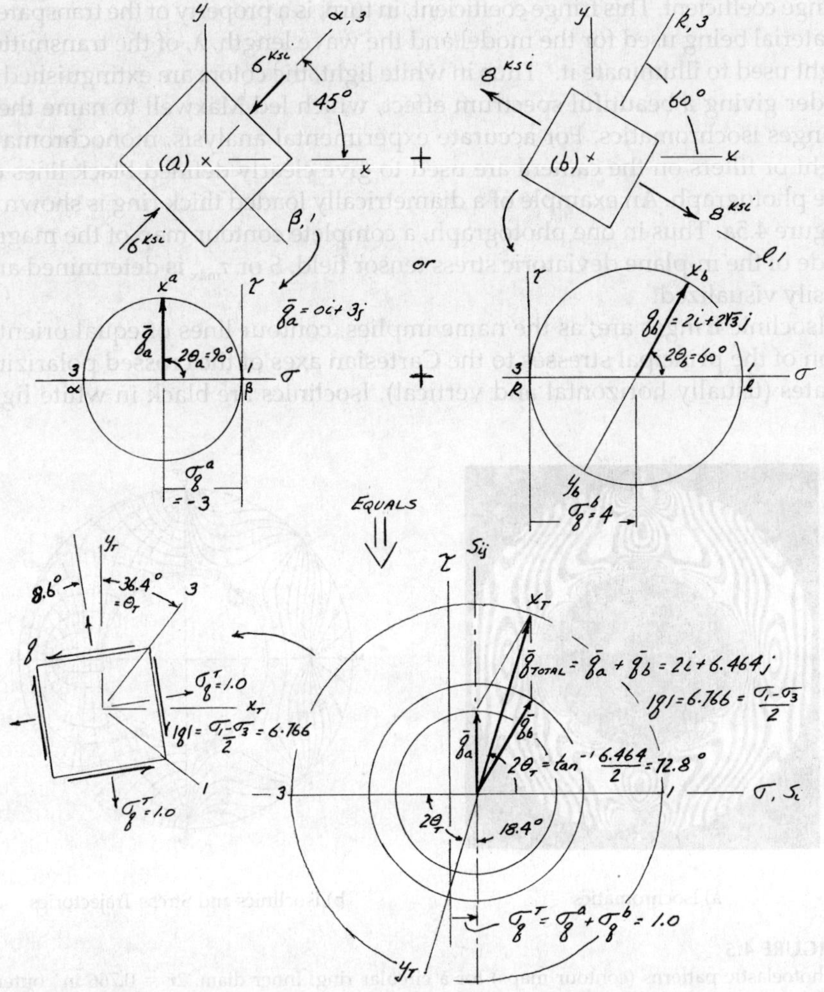

FIGURE 4.4
Vectorial addition of deviatoric stress states.

of interference fringes are observed when plane, polarized light is transmitted through a stressed sheet (a model) of almost any transparent material in a crossed polariscope. One of these sets is the isochromatic fringe pattern given by the fundamental relationship or stress-optic law:

$$(\sigma_1 - \sigma_2) = 2\tau_{\max} = \frac{Nf}{t} \tag{4.20}$$

where the stresses are in the plane of the model, t is the thickness of the model, N are either integer, 0, 1, 2 ... or half-order (1/2, 3/2, 5/2...) depending on whether the polariscope is crossed or aligned, and f is the stress-optic fringe coefficient. This fringe coefficient, in turn, is a property of the transparent material being used for the model and the wave length, λ, of the transmitted light used to illuminate it.* Thus in white light, the colors are extinguished in order giving a beautiful spectrum effect, which led Maxwell to name these fringes isochromatics. For accurate experimental analysis, monochromatic light or filters on the camera are used to give clearly defined black lines on the photograph. An example of a diametrically loaded thick ring is shown in Figure 4.5a. Thus in one photograph, a complete contour map of the magnitude of the in-plane deviatoric stress tensor field, S or τ_{max} is determined and easily visualized.

Isoclinic fringes are, as the name implies, contour lines of equal orientation of the principal stresses to the Cartesian axes of the crossed polarizing plates (usually horizontal and vertical). Isoclinics are black in white light

a) Isochromatics

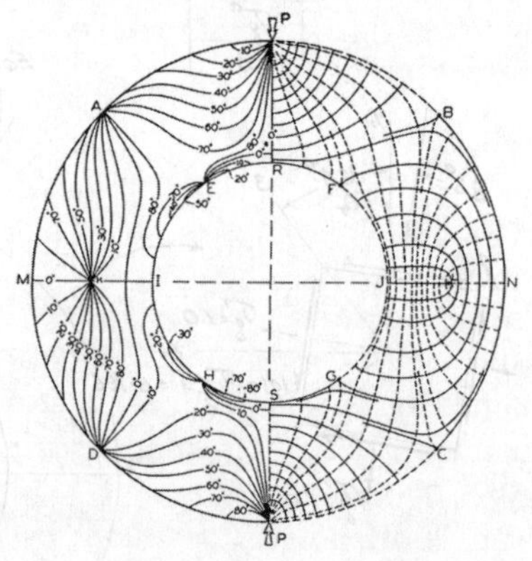

b) Isoclinics and Stress Trajectories

FIGURE 4.5
Photoelastic patterns (contour maps) for a circular ring. Inner diam. $2r$ = 0.766 in.; outer diam. D = 1.532 in.; thickness t = 0.1795; model fringe coeff. f = 240 psi/fr. in. (Frocht)

* Transparent materials have a wide range of stress sensitivity expressed by f, which has the units lb per fringe inch. For example, in order of *increasing* sensitivity in red light:

Material	Tensile Strength Y (psi)	Modulus, E psi	ν	f lb/fringe. in
Glass	10,000	10×10^6	.4	1100
Plexiglas	7,000	0.5×10^6	.33	800
Epoxy (Rm.temp)	12,000	0.5×10^6	.36	60
Lexon		0.33×10^6	.33	30
Urethane Rubber		425	.48	1
Gelatin		1	.5	0.1

and can therefore be differentiated from the colored isochromatics.* From the isoclinics, stress-trajectories or paths of the principal stresses or maximum shears can easily be constructed graphically. Observed isoclinics and the principal stress trajectories for a diametrically loaded thick ring are shown in Figure 4.5b.

Not only does photoelasticity directly determine the complete deviatoric vector field, it provides boundary conditions for the isotropic field. Since on a boundary one principal stress is known (zero if unloaded), the other can be computed from the isochromatic value for $\sigma_1 - \sigma_2$. The flow-net for the harmonic mean stress and its conjugate can then be drawn or calculated by computer without ambiguity and the entire stress field determined. Even three-dimensional solutions are possible with photoelastic analysis of slices from a stress-frozen model.

4.10 Solutions by Potentials

Not too surprisingly since the field is steady state, the deviatoric vector field, like the isotropic scalar field, can be derived from potential functions. These potentials may, under certain conditions, be harmonic themselves or, more generally, can be related to harmonic functions. Most of the classic analytic solutions were obtained by inverse methods where one guesses a part of the solution to the field equations leaving enough freedom in terms of unspecified coefficients or functions so that both the differential equations and specific boundary conditions can be satisfied. However, a direct approach is possible for the linear fields of classic strength-of-materials or certain half-space situations, which is very important. This direct approach is the subject of the next chapter.

For more general problems, various potential functions have been developed that lead to solutions. Related to displacements are the scalar and vector potentials, the Galukin vectors, and the Papkovich-Neuber functions. However, as we have noted, boundary conditions in this case are seldom simple. Potentials that generate solutions for stress fields are, in three dimensions, those of Maxwell and Morera and for two dimensions the Airy stress function. Both Maxwell's and Morera's stress functions can be reduced to Prandtl's stress function for the torsion problem. Airy's stress function and the solution of plane problems will be presented in Chapter 6. Other important solutions, which can be obtained by direct solution of the field equations using semi-inverse reasoning, will be given as we go along or in the homework problems.

* Actually the isochromatics are viewed and photographed separately by inserting quarter-wave plates in the polariscope; the first spinning the light (circular polarization) through the model at greater than 10^{14} cycles/sec and the second despinning it. The isoclinics, which are extinction contours related to orientation of the principal stresses rather than the magnitude of their difference, are "washed out" visually just as the blade of a propeller is at only a few hundred rpm.

4.11 Problems and Questions

P4.1 Derive Equation (4.3).

P4.2 What do the Navier Equation (4.2) and BCs [Equations (4.3)] reduce to in 2D (plane stress and plane strain)?

P4.3 What stresses would you anticipate (magnitude and distribution) on the horizontal and vertical boundaries of the square plate shown in Figure P4.3 with a small hole in it loaded by uniform pressure? What would they be if the pressure varied $p = p_o \sin^2\theta$?

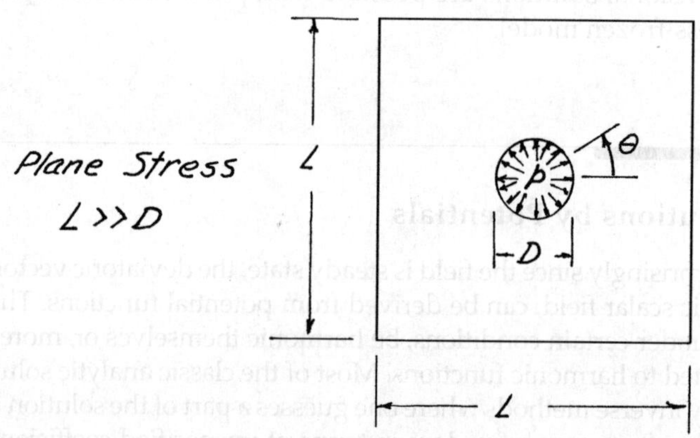

FIGURE P4.3

P4.4 For Laplace's equation with Neumann conditions on a closed boundary, is the choice of slopes at the boundary arbitrary? Remember $\nabla^2\psi = 0$ can have no internal maximums or minimums.

P4.5 An elliptic tent $\frac{x^2}{a^2} + \frac{y^2}{b^2} = 1$ (where $a = 20$ ft, $b = 10$ ft) is supported by a single pole at $x = \frac{a}{2}$, $y = 0$. Plot the flow net and sketch the resulting shape for ψ in isometric projection. What additional information would you need to know to calculate the height ψ at $x = 0$, $y = 0$. What might be the conjugate function to ψ?

P4.6 Repeat P4.5 for a rectangular tent 2a by 2b (a and b the same).

P4.7 Compare the steady-state flow into a well (6 in. radius) being pumped at a constant rate Q in a homogeneous isotropic soil of permeability k to a circular tent. Extend the analogy to other physical phenomena. Illustrate with a flow net and specify boundary conditions in terms of the conjugate functions.

P4.8 Present an argument to support the statement that the Neumann condition for the Laplace equation at infinity is that the slope be zero.

P4.9 Plot the isotropic field (σ_m and ω_{xy}) for a rectangular strip in uniaxial tension (i.e., $\sigma_x = P/A$, $\sigma_y = \sigma_z = \tau_{xy} = \tau_{zx} = 0$ both as a flow net and in isometric projection above the xy plane.

P4.10 Plot the isotropic field (σ_m and ω_{xy}) for a rectangular strip loaded by its own weight γ lbs/ft³ as both a flow net and in isometric projection above the xy plane. Assume the ideal support condition so that $\sigma_y = \sigma_z = \tau_{xy} = \tau_{yz} = \tau_{zx} = 0$.

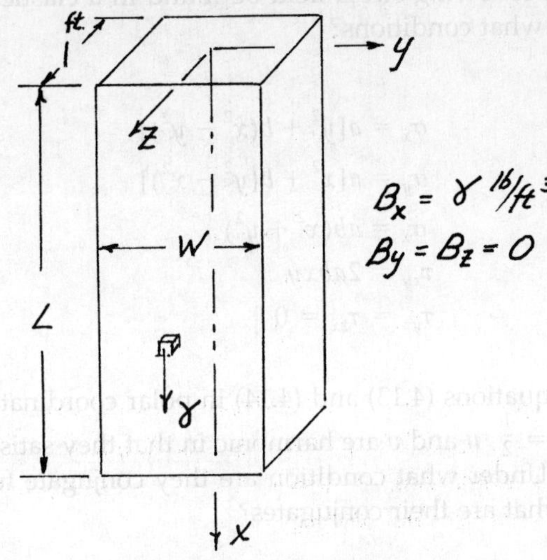

FIGURE P4.10

P4.11 Plot the isotropic field (σ_m and ω_{xy}) for a rectangular cantilever beam ($I_z = I$) in pure bending as both a flow net and in isometric projection above (and/or below) the xy plane. From this flow net:

a. Which is the potential driving the flow of what?
b. Suggest a counterpart to Darcy's Law $Q = kiA$ or Ohm's Law for pure bending.
c. Relate the flow net to the standard equations $\frac{1}{\rho} = \frac{M}{EI}$, $\theta = \int \frac{M}{EI} dx$

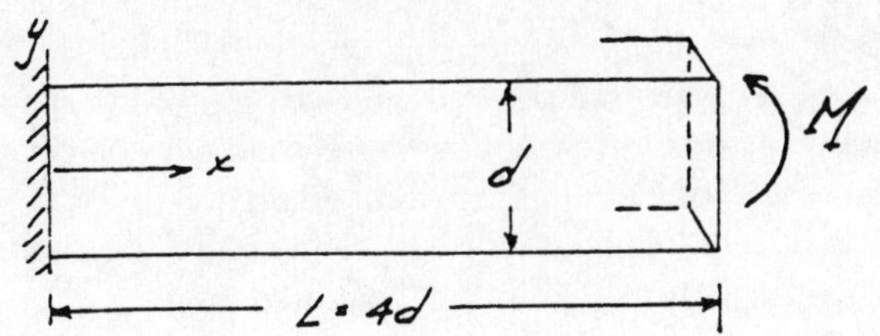

FIGURE P4.11

P4.12 Sketch the isotropic flow net that would be predicted by the standard strength of materials solution for the cantilever beam of P4.11 loaded by a vertical force at its end rather than a moment.

P4.13 What are the boundary conditions at infinity in terms of σ_m for the isotropic fields shown in Figures 4.2 and 4.3? What are they in terms of ω_{xy}? What does the boundary condition $\bar{n} \cdot grad\ \sigma_m =$ const. correspond to terms of ω_{xy}, and vice versa.

P4.14 Could the following stress field be found in a elastic field and, if so, under what conditions?

$$\sigma_x = a[y^2 + b(x^2 - y^2)]$$
$$\sigma_y = a[x^2 + b(y^2 - x^2)]$$
$$\sigma_z = ab(x^2 + y^2)$$
$$\tau_{xy} = 2abxy$$
$$\tau_{yz} = \tau_{zx} = 0$$

P4.15 Rewrite Equations (4.13) and (4.14) in polar coordinates.

P4.16 In 2D if $\nu = \frac{1}{2}$, u and v are harmonic in that they satisfy Laplace's equation. Under what condition are they conjugate functions? In general, what are their conjugates?

5

Linear Free Fields

The second-order, partial differential field equations in terms of stress [Equations (4.4)] can obviously be satisfied by constant or linear functions of position if body forces are also constant or linear functions. If the equilibrium equations are satisfied, then such a constant or linear stress field is the correct solution to the problem specified by the appropriate boundary conditions.

While such elementary solutions may appear trivial in a mathematical sense, they are extremely important in practice since they correspond to simple "strength-of-materials-type" analysis on which the great majority of design decisions are based. Moreover, they are self-sustaining or "free fields" that do not damp out by St. Venant's Principle into some more uniform distribution.* Each stress component or combination of them, such as the invariants, can be visualized as a flat membrane, either level or tilted, which is stretched between a closed boundary. They are, therefore, harmonic functions satisfying the Laplace equation. To achieve this condition, the boundary traction obviously must be constant or linear on shapes dictated by the membrane of constant slope.**

5.1 Isotropic Stress

Consider first the most basic stress field in three dimensions:

$$\sigma_x = \sigma_y = \sigma_z = \sigma_m = \pm p; \qquad \tau_{xy} = \tau_{yz} = \tau_{zx} = 0 \tag{5.1}$$

Everywhere Mohr's Circle is a point and all directions are principal (i.e., the membrane is flat at the elevation, p, above any cross-section). The boundary surface must also be loaded with uniform tension (vacuum) or compression (hydrostatic pressure). The normal strains are also isotropic and equal $p/3K$. Since there are no deviatoric stress components or elastic rotations, the displacements are a linear function of the distance from the origin, the location of which is arbitrary.

---

* Rather, constant or linear stress fields are the distribution to which more complicated fields near irregular boundaries or near loaded areas damp into. (As Churchill pointed out, ending a sentence with a preposition is acceptable when used for emphasis.)

** Linear fields in polar coordinates (torsion and special ring solutions) are left for later chapters.

5.2 Uniform Stress

If, in some orientation (x,y,z), shear stresses vanish throughout the body and the normal stresses are constant everywhere,

$$\sigma_x = a, \quad \sigma_y = b \quad \sigma_z = c; \qquad \tau_{xy} = \tau_{yz} = \tau_{zx} = 0 \tag{5.2a}$$

and all the field equations are satisfied. Strains are also constant giving the displacements by *finite* linear transformation as discussed in Chapter 1. If one stress is zero, the general case reduces to plane stress or, if say $\sigma_y = \nu(\sigma_x + \sigma_z)$, there is uniform plane strain ($\epsilon_y = 0$).

Another common situation is the triaxial state where two of the stresses are equal, say:

$$\sigma_z = c; \qquad \sigma_x = \sigma_y = -p \tag{5.2b}$$

Unlike plane stress or plane strain, the triaxial field really is two-dimensional, in that, because the stresses are isotropic in the x-y plane, there is only one Mohr's Circle throughout the body.

We create this situation in the laboratory in a triaxial cell by first pressurizing the sample in a chamber to some confining pressure, $-p$, and then increasing or decreasing the vertical stress with weights or with a hydraulic piston (Figure 5.1). This device is particularly useful for testing granular materials, which have little or no tensile strength. Both invariants can be controlled independently. A pure deviatoric test can be devised to determine the shear modulus, G, and

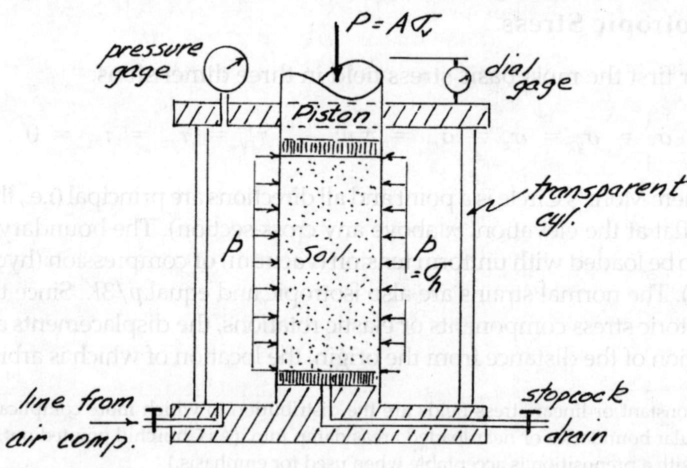

FIGURE 5.1
Triaxial cell (diagrammatic).

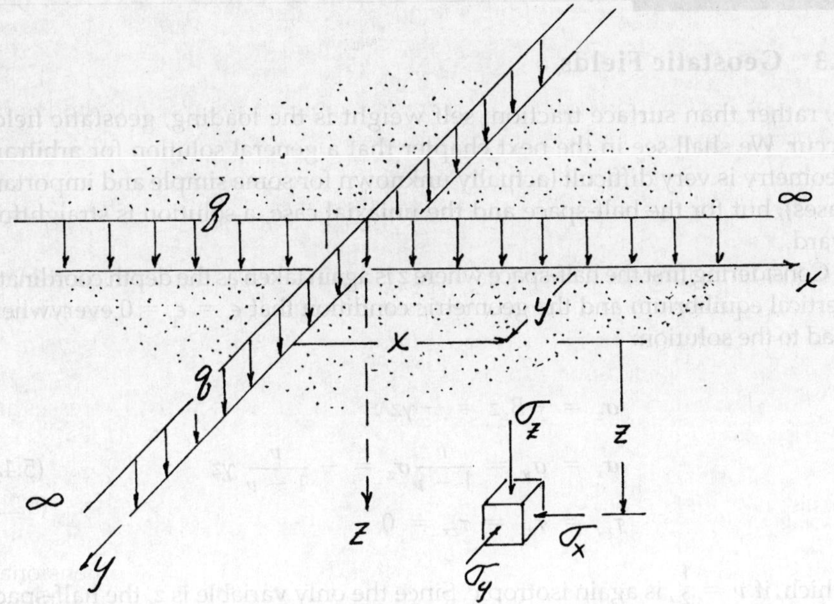

FIGURE 5.2
Uniform load on the half-space.

yield point. For example, the confining pressure can be decreased while the vertical stress is increased such that the radius of Mohr's Circle grows while the mean stress remains constant. Conversely, we can determine the bulk modulus, K, by keeping $\sigma_z = -p$ throughout the test. A uniaxial state of uniform stress is achieved corresponding to an unconfined compression test if $p = 0$.

Depending on the boundary conditions and material properties, all of these uniform stress fields can occur in the half-space loaded by a constant surface load, q (Figure 5.2). Because of the semi-infinite geometric condition, there can be no shear strains and $\epsilon_x = \epsilon_y = 0$. Therefore

$$\sigma_z = -q; \qquad \sigma_x = \sigma_y = \frac{\upsilon}{1-\upsilon}\sigma_z = -\frac{\upsilon}{1-\upsilon}q \tag{5.3a}$$

which is a triaxial field. For plane stress, say $\sigma_y = 0$

$$\sigma_z = -q; \qquad \sigma_x = \nu\sigma_z = -\nu q \tag{5.3b}$$

If $\nu = 0$, then the solution reduces to a simple uniaxial field of unconfined compression or if $\nu = 0.5$ (incompressible), it becomes the isotropic field of hydrostatic compression ($\sigma_x = \sigma_y = \sigma_z = -q$).

5.3 Geostatic Fields

If, rather than surface traction, self weight is the loading, geostatic fields occur. We shall see in the next chapter that a general solution for arbitrary geometry is very difficult (actually unknown for some simple and important cases), but for the half-space and the uniaxial case, a solution is straightforward.

Considering first the half-space where z is again taken as the depth coordinate, vertical equilibrium and the geometric condition that $\epsilon_x = \epsilon_y = 0$ everywhere lead to the solution:

$$\begin{aligned} \sigma_z &= -B_z z = -\gamma z \\ \sigma_x = \sigma_y &= \frac{\nu}{1-\nu}\sigma_z = -\frac{\nu}{1-\nu}\gamma z \\ \tau_{xy} = \tau_{yz} &= \tau_{zx} = 0 \end{aligned} \tag{5.4a}$$

which, if $\nu = \frac{1}{2}$, is again isotropic. Since the only variable is z, the half-space can be considered of finite depth, H, supported on a rigid base as shown in Figure 5.3. The vertical displacement is therefore:

$$u = v = 0; \qquad w = -\left(\frac{1-\nu-2\nu^2}{1-\nu}\right)\frac{\gamma}{2E}(H^2 - z^2) \tag{5.4b}$$

Determining the support condition necessary to achieve the linear geostatic stress field in the uniaxial case provides a simple demonstration of the physical importance of elastic rotations and the efficiency of determining them as a prelude to calculating the displacement field. Consider the block shown in Figure 5.4 standing under its own self weight and free now to

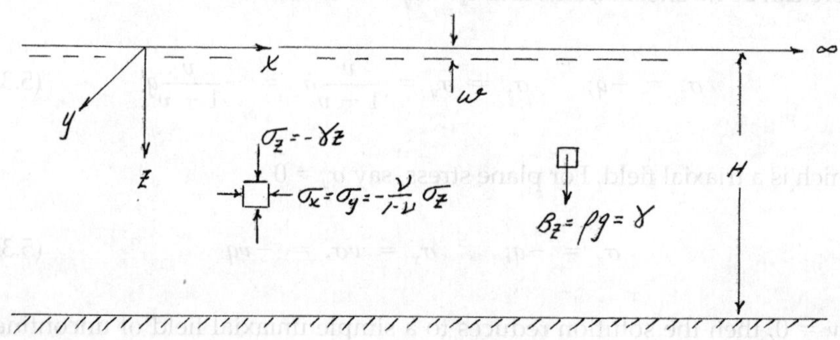

FIGURE 5.3
Geostatic stresses in the half-space of finite depth.

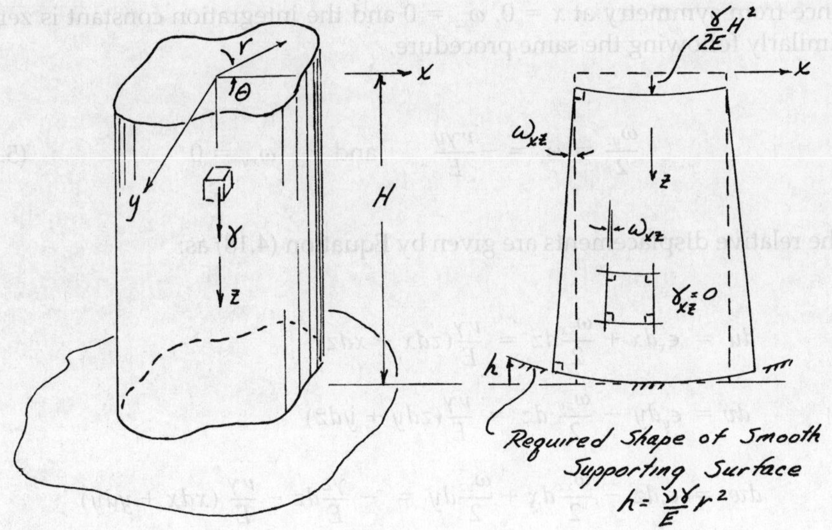

FIGURE 5.4
Uniaxial geostatic stresses.

expand laterally. If we postulate the uniaxial geostatic stress field:

$$\sigma_z = -\gamma z, \qquad \sigma_x = \sigma_y = \tau_{xy} = \tau_{yz} = \tau_{zx} = 0 \tag{5.5}$$

then:

$$\epsilon_z = -\frac{\gamma z}{E}, \; \epsilon_x = \epsilon_y = \epsilon_r = \epsilon_\theta = \frac{\nu\gamma z}{E}, \; \gamma_{xy} = \gamma_{yz} = \gamma_{zx} = 0$$

The elastic rotations can be determined from the conjugate relationships. First in the *xz* plane

$$\frac{1}{E}\frac{\partial\sigma_z}{\partial x} = -\frac{\partial}{\partial z}\left(\frac{\omega_{xz}}{2}\right) = 0; \qquad \frac{1}{E}\frac{\partial\sigma_z}{\partial z} = \frac{\partial}{\partial x}\left(\frac{\omega_{xz}}{2}\right) - \frac{\gamma}{2G}$$

and therefore:

$$\frac{d}{dx}\left(\frac{\omega_{xz}}{2}\right) = -\gamma\left(\frac{1}{E} - \frac{1}{2G}\right) = \frac{\nu\gamma}{E}$$

to give:

$$\frac{\omega_{xz}}{2} = \omega_y = \frac{\nu\gamma x}{E} \tag{5.6}$$

since from symmetry at $x = 0$, $\omega_{xz} = 0$ and the integration constant is zero. Similarly following the same procedure,

$$\frac{\omega_{yz}}{2} = \omega_x = -\frac{\nu\gamma y}{E} \qquad \text{and} \qquad \omega_{xy} = 0 \tag{5.7}$$

The relative displacements are given by Equation (4.18) as:

$$du = \epsilon_x dx + \frac{\omega_{zx}}{2}dz = \frac{\nu\gamma}{E}(zdx + xdz)$$

$$dv = \epsilon_y dy - \frac{\omega_{zy}}{2}dz = \frac{\nu\gamma}{E}(zdy + ydz)$$

$$dw = \epsilon_z dz - \frac{\omega_{xz}}{2}dx + \frac{\omega_{yz}}{2}dy = -\frac{\gamma z}{E}dz - \frac{\nu\gamma}{E}(xdx + ydy)$$

Integrating

$$u = \frac{\nu\gamma}{E}xz + C_1, \qquad v = \frac{\nu\gamma}{E}yz + C_2$$

and

$$w = -\frac{\gamma z^2}{2E} - \frac{\nu\gamma}{2E}(x^2 + y^2) + C_3 \tag{5.8}$$

On the centerline ($x = y = 0$), u and v are zero everywhere, and $w = 0$ at $z = H$. Therefore

$$C_1 = C_2 = 0; \qquad C_3 = \frac{\gamma}{2E}H^2$$

and the displacement field is:

$$u = \frac{\nu\gamma}{E}xz; \qquad v = \frac{\nu\gamma}{E}yz; \qquad w = \frac{\gamma}{2E}(H^2 - z^2) - \frac{\nu\gamma}{2E}(x^2 + y^2) \tag{5.9}$$

Thus to achieve a uniform contact pressure, $-\gamma H$, at $z = H$, the supporting surface (Figure 5.4) must be a smooth parabolic dish satisfying the displacement boundary condition dictated by the required vertical displacement. If, however, the supporting surface were flat, this geostatic solution, by St. Venant's principle, would still be correct some distance from the bottom since the actual distribution of the contact stress must, by vertical equilibrium, be statically equivalent to a uniform distribution.

5.4 Uniform Acceleration of the Half-space

With a horizontal component of body force, the behavior of the half-space is quite different. Although such body forces may arise in many ways (seepage being a common source in soils), acceleration, as for example in an earthquake, is probably the most important.

Consider the half-space of finite depth H of material with unit weight γ. From the previous section, the geostatic field is:

$$\sigma_z = -\gamma z; \qquad \sigma_x = K_o\sigma_z = -K\gamma z; \qquad \tau_{xy} = 0$$

where K, the coefficient of lateral pressure, is on the order of $\frac{1}{2}$ for sand. If the layer is thought of as elastic, this corresponds to $\nu = \frac{1}{3}$ since $K_o = \nu/(1-\nu)$.

If uniform horizontal* and vertical accelerations, a_x and a_z, are now applied to the layer as in Figure 5.5, and the two-dimensional differential equations of equilibrium become:

$$\begin{aligned} \frac{\partial\sigma_x}{\partial x} + \frac{\partial\tau_{zx}}{\partial z} &= k_h\gamma \\ \frac{\partial\tau_{zx}}{\partial x} + \frac{\partial\sigma_z}{\partial z} &= -(1-k_v)\gamma \end{aligned} \tag{5.10}$$

As before, the geometric condition for the half-space is that

$$\epsilon_x = \epsilon_y = 0 \tag{5.11}$$

and the field can be thought of as "one-dimensional plane strain" (i.e., $\sigma_x = \sigma_y$ and *nothing* can be a function of x or y). Thus, with the free-surface boundary condition at $z = 0$, Equations (5.10) can be integrated directly to give the solution:

$$\begin{aligned} \sigma_z &= -(1-k_v)\gamma z \\ \sigma_x &= K\sigma_z = -K(1-k_v)\gamma z \\ \tau_{zx} &= k_h\gamma z \end{aligned} \tag{5.12}$$

Essentially this dynamic (actually pseudostatic) solution for the free field is the geostatic free field with a unit weight of $(1-k_v)\gamma$ superimposed with shear

* Horizontal accelerations in an earthquake are primarily due to a shear (S) wave propagating upward through the earth's crust. In soil near the surface, a S wave velocity of perhaps 150 meters/sec at a frequency of 3 cycles/sec. is typical. Thus a half wave length of 25 meters (80 ft.) is a representative value for H. The horizontal acceleration coefficient, k_h, can be taken as the average for approximate analysis or as the peak value for a conservative design.

Vertical accelerations are associated with compressive (P) waves which travel roughly twice as fast at the same sort of frequency and therefore have longer wavelengths. P and S waves from an earthquake are generally not correlated (may have any phase shift).

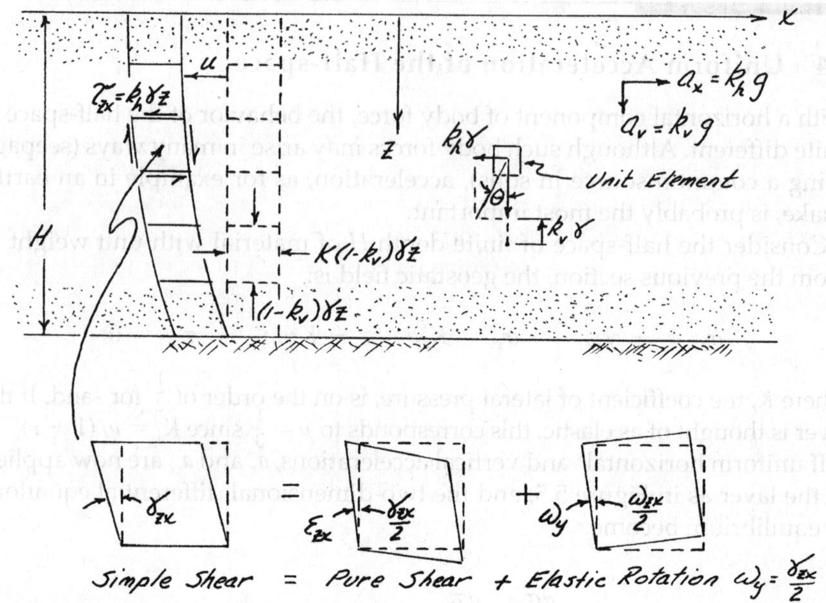

FIGURE 5.5
Seismic free field.

introduced by the horizontal acceleration component. However, considering the deformation, this is not "pure shear," but a case of "simple shear" since there must be an elastic rotation field that keeps horizontal planes from rotating. From Equation (4.13) or by inspection

$$\frac{\gamma_{xz}}{2} = \epsilon_{xz} = \frac{\omega_{xz}}{2} = \omega_y = \frac{\tau_{xz}}{2G} = k_h\frac{\gamma z}{2G} \tag{5.13}$$

The relative deformation then, due to the horizontal acceleration, is:

$$du = \frac{\gamma_{zx}}{2}dz + \frac{\omega_{zx}}{2}dz$$

which can be integrated directly. Assuming the shear modulus, G, is constant, the horizontal deflection is parabolic:*

$$u = -k_h\frac{\gamma(H^2 - z^2)}{2G} \tag{5.14}$$

* In soils the shear modulus, G,(~shear stiffness) increases with confining pressure and therefore with depth. If this increase were linear (often assumed), then the shear strain and rotation would be constant and the horizontal displacement linear. Even nonhomogeneous or anisotropic materials pose no problems. The layered half-space common in most soil deposits can be integrated piecewise.

This simple result is remarkable. Both the geometric field condition $\epsilon_x = \epsilon_y = 0$ (nothing is a function of x or y) and the equilibrium equations hold no matter what the distribution of acceleration or material properties. There is really no restriction on the distribution of k_h with depth, and assuming it constant is for convenience, not necessity. Moreover, the equations and their solution for the stress field must apply throughout the acceleration pulse past yield in the plastic range to eventual failure, which can only occur by base sliding. This exact solution does require that, if there are finite lateral boundaries, then the boundary traction on them and their displacement be distributed as the field equations dictate. However, by St. Venant's principle, perturbations to the free field will be insignificant away from the lateral boundaries as long as the resultant lateral force is equivalent. We will return to the seismic postyield behavior in later chapters on plasticity and limit analysis.

5.5 Pure Bending of Prismatic Bars

Elementary beam theory, where axial strains and therefore stresses are proportional to the bending moment and vary linearly from a centroidal axis, is an exact solution for pure bending of a prismatic bar if body forces are neglected.* Consider a beam such as that shown in Figure 5.6 with the y, z coordinates at the centroid of the rectangular cross-section oriented with the principal directions ($I_{yz} = 0$). The internal moment, M_z, is constant and if the axial strain is linear with the distance from thecentroidal z axis, then the beam is bent into an arc of a circle with radius, ρ. Thus:

$$\epsilon_x = -\frac{y}{\rho}; \qquad \therefore \sigma_x = -\frac{Ey}{\rho} \tag{5.15}$$

and

$$\sigma_y = \sigma_z = \tau_{xy} = \tau_{yz} = \tau_{zx} = 0$$

This stress field satisfies the equilibrium and compatibility equations in three dimensions and the boundary conditions require that at any cross-section, say at $x = a$

$$-M_z = \int \sigma_x y dA = \int \frac{Ey^2 dA}{\rho} = -\frac{EI_z}{\rho} \tag{5.16}$$

or

$$\frac{1}{\rho} = \frac{M_z}{EI_z} \quad \text{and} \quad \sigma_x = -\frac{M_z y}{I_z}$$

* See Section 1.1.2. Often called the Bernoulli Hypothesis, it is only true for pure bending, but it will be shown in Chapter 6 to be nearly exact for all but very short beams.

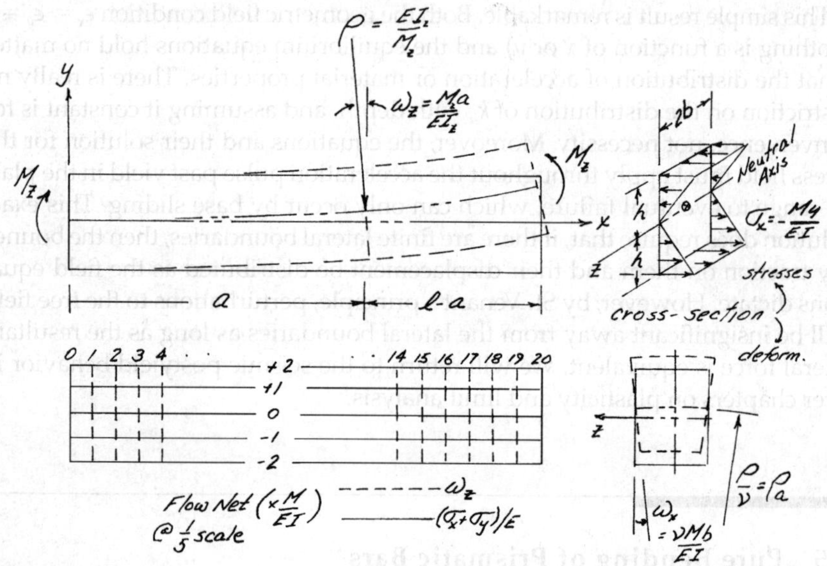

FIGURE 5.6
Pure bending.

These are the standard formulae of elementary bending theory and are the exact free field if the surface tractions are distributed over the ends $x = 0, \ell$ in the same way as the internal stresses σ_x.

To determine the displacements for pure bending, the relative displacements must be integrated. From the stress–strain relationships:

$$\epsilon_x = -\frac{y}{\rho}; \qquad \epsilon_z = \epsilon_y = \frac{\nu y}{\rho}; \quad \gamma_{xy} = \gamma_{yz} = \gamma_{zx} = 0 \tag{5.17}$$

The elastic rotations can be found by drawing a flow net as in Figure 5.6 or from the conjugate relationships. From Equation (4.13):

$$\frac{\partial \omega_z}{\partial x} = \frac{\partial}{\partial x}\left(\frac{\omega_{xy}}{2}\right) = -\frac{1}{E}\frac{\partial(\sigma_x + \sigma_y)}{\partial y} = \frac{1}{E}\frac{\partial}{\partial y}\left(-\frac{My}{I}\right) \tag{5.18}$$

Therefore $\quad d\omega_z = \dfrac{M}{EI}\,dx \qquad$ or $\qquad \omega_z = \dfrac{\omega_{xy}}{2} = \displaystyle\int \frac{M}{EI}\,dx$

which are the basic differential and integral relationships between the "slope diagram" and the "moment diagram" (derived differently in a beginning strength-of-materials course). In this case where the moment is constant:

$$\omega_z = \frac{\omega_{xy}}{2} = \frac{x}{\rho} + C_1 = \frac{Mx}{EI} + C_1 \tag{5.19}$$

Since this is a case of Plane Stress, Equation (4.14) applies and

$$\omega_x = \frac{\omega_{zy}}{2} = \frac{\nu z}{\rho} + C_2 = \frac{\nu M z}{EI}; \qquad \omega_y = 0 \tag{5.20}$$

where from symmetry, $C_2 = 0$.

The arbitrary constants of integration must be determined from conditions on the geometry of deformations dictated by the support. Assuming no rotation, ω_z, of the cross-section at $x = 0$ and no displacement of the origin, then C_1 is zero as are all integration constants for u, v, and w. Thus from Equation (4.18):

$$du = \epsilon_x dx - \frac{\omega_{xy}}{2}\,dy = -\frac{1}{\rho}(ydx + xdy) \tag{5.21}$$

Therefore $\quad u = \int du = -\frac{yx}{\rho} = -\frac{Mxy}{EI}$

and

$$dv = \epsilon_y\,dy - \frac{\omega_{yz}}{2}\,dz + \frac{\omega_{xy}}{2}\,dx = \frac{\nu y}{\rho}\,dy - \frac{\nu z}{\rho}\,dz + \frac{x}{\rho}\,dx \tag{5.22}$$

Therefore $\quad v = \int dv = \frac{\nu y^2}{2\rho} - \frac{\nu z^2}{2\rho} + \frac{x^2}{2\rho} = \frac{M}{2EI}[x^2 - \nu(z^2 - y^2)]$

Finally $\quad dw = \epsilon_z dz + \frac{\omega_{yz}}{2}\,dy = \frac{\nu y}{\rho}dz + \frac{\nu z}{\rho}dy \qquad (5.23)$

Therefore $\quad w = \int dw = \frac{\nu yz}{\rho} = \nu\frac{M}{EI}yz$

The deflection of the centroidal axis ($y = z = 0$) is, therefore:

$$v = \frac{Mx^2}{2EI} = \int_0^x \omega_z\,dx \tag{5.24}$$

which is simply the integral of the slope (rotation) diagram as given by the elementary theory.

Considering the deformation of any cross-section, say at $x = a$, points on the yz plane end up at:

$$x = x_0 + u = a\left(1 - \frac{My}{EI}\right)$$

and, in fact, plane sections do remain plane as was assumed. The in-plane deformation of the cross-section is shown in Figure 5.6. After bending, the position of the vertical edges at $z = \pm b$ become

$$z = \pm b + w = \pm b\left(1 + \frac{\nu y}{\rho}\right)$$

and the sides are rotated through the angle $\omega_x = bv/\rho$.

The top and bottom sides at $y = \pm h$ are bent into parabolas:

$$y = \pm h + v = \pm h + \frac{1}{2\rho}\left[a^2 - \nu(z^2 - h^2)\right]$$

If the deformations are small and $h \ll \ell$, the parabolic curves are closely approximated by a circle with a radius of "anticlastic curvature" $\rho_a = \rho/\nu$.

Example 5.1

An element in a structure built of an elastic material displaces as shown below. Determine (if possible) a compatible elastic field (i.e., σ_{ij}, ϵ_{ij}, ω_{ij}, u_i). Assume plane stress and assume $\sigma_x = 0$. Can you determine elastic properties of the material from this test? If so, do so.

NOTE: Remember the simple bending relationship: $\frac{1}{\rho} = \frac{M}{EI}$ and observe that there are, obviously, elastic rotations involved. Use the engineering definition of strain and neglect body forces and any nonlinearity due to large deformations.

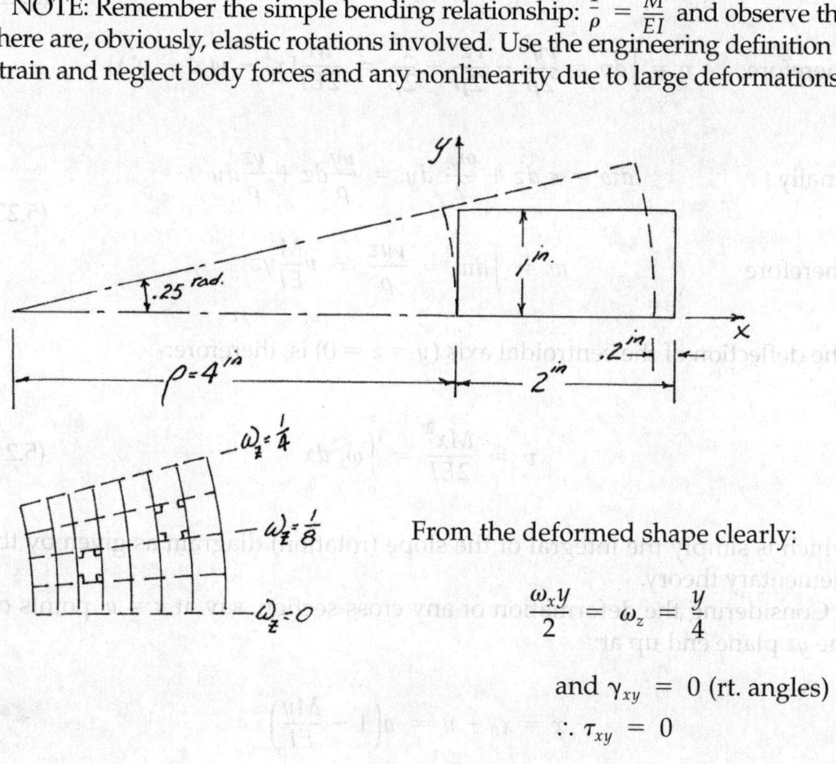

From the deformed shape clearly:

$$\frac{\omega_x y}{2} = \omega_z = \frac{y}{4}$$

and $\gamma_{xy} = 0$ (rt. angles)

$$\therefore \tau_{xy} = 0$$

From equil: $$\frac{\partial \tau_{xy}}{\partial x}^{\nearrow o} + \frac{\partial \sigma_y}{\partial y} = 0$$

$$\therefore \sigma_y = f(x) + C^{\nearrow o}$$

Also $\nabla^2(\sigma_x + \sigma_y) = 0 \quad \therefore f(x)$ must be linear

C.R cond: $$\frac{\partial(\sigma_x + \sigma_y)}{\partial x} = \frac{\partial \sigma_y}{\partial x} = E\frac{\partial \omega_z}{\partial y} = \frac{E}{4} \quad \therefore \sigma_y = \frac{Ex}{4}$$

$\therefore$ Stress Field: $\sigma_x = 0 \; , \sigma_y = \frac{Ex}{4} \; , \quad \tau_{xy} = 0$

Strains

$$\epsilon_x = \frac{1}{E}(\sigma_x - \nu\sigma_y) = -\frac{\nu x}{4}$$

$$\epsilon_y = \frac{1}{E}(\sigma_y - \nu\sigma_x) = \frac{x}{4}$$

or $$\epsilon_{ij} = \begin{bmatrix} -\frac{\nu x}{4} & 0 \\ 0 & \frac{x}{4} \end{bmatrix}$$

Therefore

$$\qquad \epsilon_{ij} \qquad + \qquad \omega$$

$$\begin{bmatrix} du \\ dv \end{bmatrix} = \left\{ \begin{bmatrix} -\frac{\nu x}{4} & 0 \\ 0 & \frac{x}{4} \end{bmatrix} + \begin{bmatrix} 0 & -\frac{y}{4} \\ \frac{y}{4} & 0 \end{bmatrix} \right\} \begin{bmatrix} dx \\ dy \end{bmatrix} = \begin{bmatrix} \frac{\partial u}{\partial x} & \frac{\partial u}{\partial y} \\ \frac{\partial v}{\partial x} & \frac{\partial v}{\partial y} \end{bmatrix} \begin{bmatrix} dx \\ dy \end{bmatrix}$$

and integrating

$$u = \int du = \int \epsilon_x dx - \int \omega_z dy = -\int \nu\frac{x}{4}\, dx + \int -\frac{y}{4}\, dy$$

$$v = \int dv = \int \epsilon_y dy + \int \omega_z dx = \int \frac{x}{4}\, dy + \int \frac{y}{4}\, dx$$

$$\therefore u = -\nu\,\frac{x^2}{8} - \frac{y^2}{8}, \quad v = \frac{xy}{4}$$

at $x = 2, y = 0 \quad u = -.2 \quad \therefore -\nu/2 = -.2$ and $\nu = 0.4$ but it is not possible to find E (no loads given). Also σ_x & σ_y may include an unknown constant.

5.6 Pure Bending of Plates

This result for pure, one-way bending can, by superposition, be used to generate the solution for combined bending of prismatic beams where a moment, M_y, is applied in addition to M_z. The result from the previous section can also be used to derive a solution for pure bending of plates which, as for beams, is the starting point for the study of their flexural behavior under more complicated loading and support conditions.

First consider a plate of uniform thickness, h, in pure bending around only the y axis due to a uniformly distributed moment, M_y, per unit length. The top and bottom of the plate will become anticlastic surfaces as shown in Figure 5.7 with the curvature $1/\rho$ convex in the yz plane. The moment of inertia per unit length is $I = h^3/12$ and therefore by the simple bending theory of the

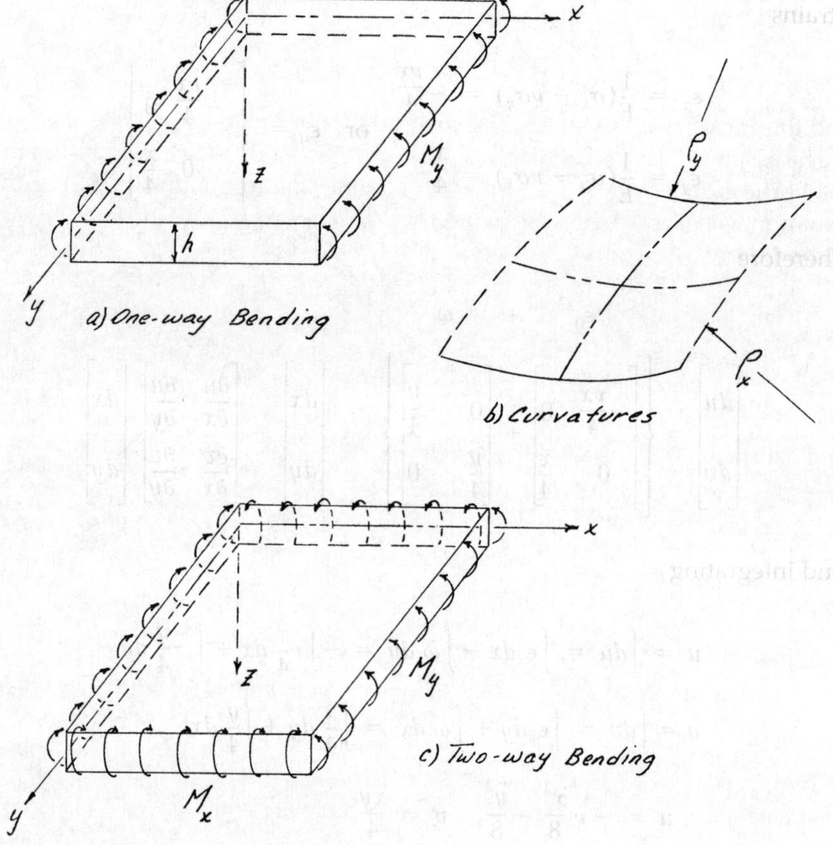

FIGURE 5.7
Pure bending of plates.

previous section:

$$\frac{1}{\rho_y} = \frac{M_y}{EI} = \frac{12M_y}{Eh^3} \quad \text{and} \quad \frac{1}{\rho_x} = -\frac{\nu}{\rho_y} = -\frac{12\nu M_y}{Eh^3} \tag{5.25}$$

With distributed moments along both perpendicular edges (Figure 5.7c), the curvatures, stresses, and deflections can be found by superposition. With a bending moment per unit length around the x axis, M_x, as well as M_y

$$\frac{1}{\rho_y} = \frac{12}{Eh^3}(M_y - \nu M_x), \qquad \frac{1}{\rho_x} = \frac{12}{Eh^3}(M_x - \nu M_y) \tag{5.26}$$

or, solving for moments:

$$M_x = \frac{Eh^3}{12(1-\nu^2)}\left(\frac{1}{\rho_x} + \nu\frac{1}{\rho_y}\right), \qquad M_y = \frac{Eh^3}{12(1-\nu^2)}\left(\frac{1}{\rho_y} + \nu\frac{1}{\rho_x}\right) \tag{5.27}$$

and the corresponding stresses are $\sigma_i = M_i z/I$.

The assumption of an anticlastic surface corresponds to the assumption that the vertical deflection is very small in comparison to the thickness. For this situation, we can use the approximation that the curvatures are given by the second derivatives of the deflection:

$$\frac{1}{\rho_x} = -\frac{\partial^2 w}{\partial y^2}, \qquad \frac{1}{\rho_y} = -\frac{\partial^2 w}{\partial x^2}$$

and calling the flexural rigidity

$$D = \frac{Eh^3}{12(1-\nu^2)} \tag{5.28}$$

Then

$$M_x = -D\left(\frac{\partial^2 w}{\partial y^2} + \nu\frac{\partial^2 w}{\partial x^2}\right), \qquad M_y = -D\left(\frac{\partial^2 w}{\partial x^2} + \nu\frac{\partial^2 w}{\partial y^2}\right) \tag{5.29}$$

Two special cases are of particular note. To bend the plate into a cylindrical surface, say one with the longitudinal direction parallel to the y axis, $\frac{\partial^2 w}{\partial y^2} = 0$ and:

$$M_x = -\nu D\frac{\partial^2 w}{\partial x^2}; \qquad M_y = -D\frac{\partial^2 w}{\partial x^2} \quad \therefore \quad \frac{M_x}{M_y} = \nu$$

For the plate to be bent into a spherical surface:

$$\frac{1}{\rho_x} = \frac{1}{\rho_y} = \frac{1}{\rho} \quad \text{and} \quad M_x = M_y = M = \frac{D(1+\nu)}{\rho}$$

The maximum stresses top and bottom on both faces will equal:

$$\sigma_{\max} = \pm\frac{6M}{h^2} = \frac{Eh}{2(1-\nu)}\cdot\frac{1}{\rho}$$

Usually the moments in plates are not constant, even away from the boundaries due to surface loads that introduce shears. However, just as for beams, these formulas for pure bending can be used with confidence away from stress-concentration points if the thickness is small compared to the width and length of the plate.

5.7 Problems and Questions

P5.1 Consider a layer of the half-space of depth H loaded on the surface $z = 0$ by a uniform shear force per unit area, s, in the x direction. Determine the stress, strain, and displacement fields for:

a. The full three-dimensional case

b. Plane Stress ($\sigma_y = \tau_{yx} = \tau_{yz} = 0$)

c. Plane Strain ($\epsilon_y = \gamma_{yx} = \gamma_{yz} = 0$)

P5.2 Make up a chart on one page showing the fields associated with a uniform load on the half-space ($\sigma_z = -q$) for: (a) the general 3D case); (b) 2D plane stress and plane strain cases; and (c) the uniaxial case ($\sigma_x = \sigma_y = 0$). Include the stress, strain, rotation, and displacement fields for each.

P5.3 Why, for plane stress, does the stress field for the half-space with either uniform constant surface loading or self weight include Poisson's ratio, ν, when it was shown in Chapter 4 that 2D stress fields are independent of material properties?

P5.4 Analogous to the solution for a block of unit weight γ supported at its base, derive the comparable solution for displacements if the block is stretched by its own weight being supported at the top. Plot the deformed shape and determine the shape of the supporting surface.

P5.5 Estimate (sketch with approximate values) the distribution of contact stresses at the base of a tall block under gravity loading (Figure 5.4) if the smooth support surface were flat. Justify your result by a superposition argument.

P5.6 Refer to Figure P5.6. Compare the ratio of the max shear stress, $\tau_{max} = (\sigma_1 - \sigma_3)/2$, at points 1 and 2 (at the base and center) for two thin strips in plane stress compressed between rigid plates to which they are bonded at the interface. Assume $(\sigma_y)_1 = (\sigma_y)_2$. Plot the ratio as a function ν. Check the experimental result for the long plastic strip between steel plates of 0.75 [Exp. Mech., Proc. SEAS, Vol. XXIV, No. 2, pp. 381–383].

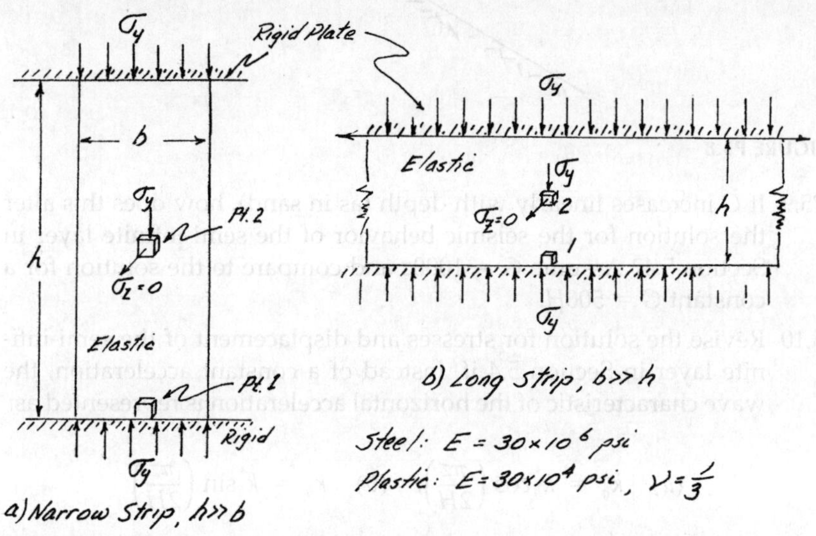

FIGURE P5.6

P5.7 Rock (say granite) is very much like strong concrete with nominal elastic properties: $E = 4 \times 10^6$ psi, $\nu = 1/4$ and unconfined compressive strength $Y_c = f'_c = 8{,}000$ psi. If it is considered an ideally elastic-plastic material, at what height will an unconfined block yield and become plastic (will flow rather than accept further shear)? Does this make sense (Mt. Everest is 29,000 ft.)?

P5.8 Refer to Figure P5.8. Determine the elastic fields in a semi-infinite slope of thickness, H, of a material with unit weight γ at an angle of α to the horizontal. (Find the stresses and displacement.) Compare this to the solution for a semi-infinite horizontal layer of this material subjected to seismic acceleration components $k_h g$ and $k_v g$ as presented in Section 5.4.

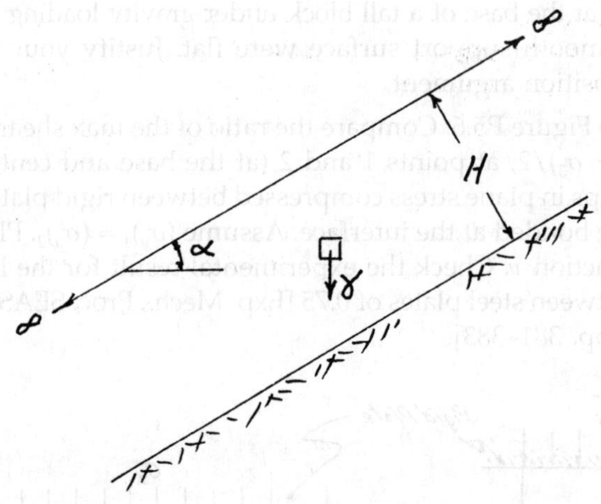

FIGURE P5.8

P5.9 If G increases linearly with depth (as in sand), how does this alter the solution for the seismic behavior of the semi-infinite layer in Section 5.4? Assume $G = 1000z$ and compare to the solution for a constant $G = 500H$.

P5.10 Revise the solution for stresses and displacement of the semi-infinite layer in Section 5.4 if, instead of a constant acceleration, the wave characteristic of the horizontal acceleration is represented as:

$$(a)\quad k_h = \overset{*}{k}_h \cos\left(\frac{\pi z}{2H}\right), \quad (b)\quad k_h = \overset{*}{k}_h \sin\left(\frac{\pi z}{2H}\right)$$

Assume $k_v = 0$, G = const. and compare your solution for τ_{xz} and u to that in Section 5.4 with k_h = const. = $\overset{*}{k}_h$. (See Figure P5.10.)

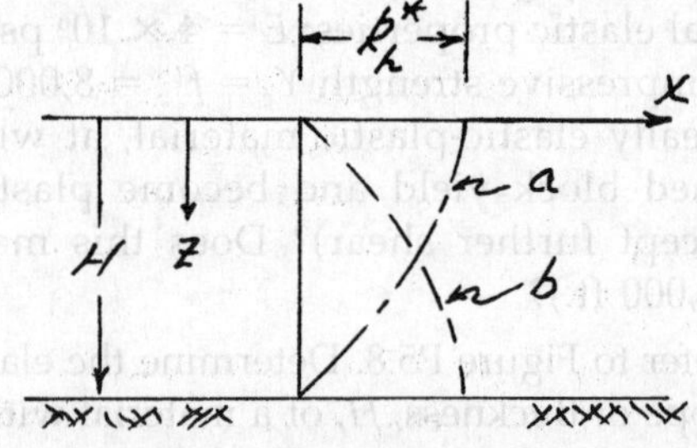

FIGURE P5.10

6

Two-Dimensional Solutions for Straight and Circular Beams

6.1 The Classic Stress-Function Approach

Most solutions in two dimensions, for other than linear stress fields where direct integration is possible, can be obtained most easily by the semi-inverse method using stress functions. Usually Airy's stress function is employed which, in two dimensions, reduces the three field equations to one fourth-order partial differential equation.* This "biharmonic equation" is exact for plane strain and nearly so for plane stress if the plate is reasonably thin.**

For simple geometry and loadings, a wide range of practical problems in various coordinate systems have been solved since elasticity was formalized. These results are used extensively for design since many structures can be idealized to plane stress or plane strain without serious error. Moreover, such two-dimensional solutions are essential benchmarks to, on the one hand, establish the range where the simpler strength-of-materials type analysis is adequate and, on the other, to validate the more complicated numerical or experimental models necessary to determine fields in specific structures that

---

* G.B. Airy was one of the first to introduce applied mathematics at Cambridge. After a brilliant undergraduate record at Trinity College, he was appointed Professor of Astronomy in 1826 and Astronomer Royal in 1835.

** Many elasticity texts have a discussion of "pseudo" or "equivalent" plane stress showing that the single compatibility equation in the plane [Equation (2.7)] is insufficient. The additional compatibility equations [Equation (2.6)] which, in two dimensions, reduce to:

$$\frac{\partial^2 \epsilon_z}{\partial x^2} = 0, \qquad \frac{\partial^2 \epsilon_z}{\partial y^2} = 0, \qquad \frac{\partial^2 \epsilon_z}{\partial x \partial y} = 0$$

must also be satisfied. There is obviously no problem for plane strain where $\epsilon_z = 0$, but for plane stress where $\epsilon_z = -\frac{\nu}{E}(\sigma_x + \sigma_y)$, difficulties may arise near high stress gradients as, for example, near concentrated loads or near the tip of a crack in fracture mechanics. Such problems are too complicated for this text and we shall assume a plate thin enough to avoid them.

cannot be safely idealized to a shape or loading pattern where a closed-form analytic solution is feasible.

Hundreds of solutions are available and there is space in this and the next two chapters to present only a limited number* selected on the basis of one or more of the following criteria:

a. to best demonstrate the stress-function method,
b. to give insight as to the fundamental behavior of isotropic and deviatoric stress fields,
c. to present questions, perhaps unanswered, concerning the theory,
d. to suggest a new problem not yet solved that might excite a student's interest, or
e. to display particular utility for design in themselves or by superposition.

The presentation will be in a somewhat diagrammatic form with frequent reference to figures augmented by sketches and notes. This approach is intended to emphasize the rational aspects of the semi-inverse method based on physical reasoning.**

6.2 Airy's Stress Function in Cartesian Coordinates

To review once again, the basic field equations in terms of stress are:

Equilibrium:

$$\frac{\partial \sigma_x}{\partial x} + \frac{\partial \tau_{xy}}{\partial y} + B_x = 0$$
$$\frac{\partial \tau_{xy}}{\partial x} + \frac{\partial \sigma_y}{\partial y} + B_y = 0 \tag{5.7}$$

* Entire areas of interest such as thermal stress, contact problems, bending of nonprismatic or thin-walled sections, plates and shells, vibrations, stress waves, etc., are omitted, not because they are unimportant (they are often crucial in practice), but because a clear understanding of basic solution procedures best develops ability in rational mechanics necessary to modify the fundamental theory and go on to more advanced topics. The mechanics of deformable solids is an incredibly rich subject and students who enjoy thinking about its intricacies must build up a personal library to include specialized texts more advanced and exhaustive in their coverage.

** The problems at the end of these chapters ask the reader to fill in the details of the solutions. This may involve (a) drawing isotropic and deviatoric contours, perhaps to compare to experimental results; (b) to complete the solution by attempting to integrate for displacements and thereby required geometric boundary conditions, or (c) to deal with ambiguity or uncertainty in some of the supposedly exact solutions. Others even ask you to attempt new problems. A thorough investigation of the Chapter Problems and Questions (some of the answers are ill-defined) will be as rewarding and certainly more fun than the prerequisite reading of the chapter itself.

and:

$$\nabla^2(\sigma_x + \sigma_y) = -(1 - \nu)\left(\frac{\partial B_x}{\partial x} + \frac{\partial B_y}{\partial y}\right) \tag{5.6}$$

Once the stress field is determined, the elastic strains and rotations are found from the stress–strain and conjugate relationships giving the components of the relative deformation vector *du*, *dv*, *dw* which can then be integrated for the displacement field.

In 1862, G.B. Airy introduced his potential function defined as:

$$\sigma_x = \frac{\partial^2 F}{\partial y^2} + \Omega$$
$$\sigma_y = \frac{\partial^2 F}{\partial x^2} + \Omega \tag{6.1a}$$
$$\tau_{xy} = -\frac{\partial^2 F}{\partial x \partial y}$$

where $B_x = -\frac{\partial \Omega}{\partial x}$ and $B_y = -\frac{\partial \Omega}{\partial y}$. On substitution, the equilibrium equations are satisfied "by definition" and the compatibility equation for plane stress becomes

$$\frac{\partial^4 F}{\partial x^4} + 2\frac{\partial^4 F}{\partial x^2 \partial y^2} + \frac{\partial^4 F}{\partial y^4} = -(1 - \nu)\left(\frac{\partial B_x}{\partial x} + \frac{\partial B_y}{\partial y}\right) \tag{6.1b}$$

or

$$\nabla^2(\nabla^2 F) = \nabla^4 F = -(1 - \nu)\nabla^2 \Omega \tag{6.2}$$

For plane strain, the fictitious $\nu' = \frac{\nu}{1 - \nu}$ is substituted for ν. If the body forces are given by a harmonic function (e.g., constant, linear, magnetic fields, seepage, etc.), Equation (6.2) reduces to the homogeneous elliptic form known as the biharmonic equation:

$$\nabla^4 F = 0 \tag{6.3}$$

With this ingenious approach, Airy reduces the two-dimensional problem to solving one equation.* Generally, if body forces are significant, their

* Also, as discussed in Chapter 4, the conditions for existence and/or uniqueness regarding proper boundary conditions and self-equilibrating traction on internal boundaries of multiconnected structures must be observed.

effect is solved for separately as in Chapter 5 and introduced by superposition. Therefore, unless specifically stated to be otherwise, body forces will be neglected and Equation (6.3) used in this and succeeding chapters.

6.3 Polynomial Solutions and Straight Beams

A number of stress fields in rectangular coordinates will be presented or asked for in problems in later chapters. In most cases, these results are obtained most easily by solving in polar coordinates and transforming them to Cartesian form by Mohr's Circle. However for beams (long rectangular strips) polynomial stress functions in Cartesian coordinates are preferable because boundary conditions are easily expressed. It is also possible to compare the more exact elasticity solutions thus generated directly to the standard strength-of-materials type solutions,* which are expressed in powers of x and y. To get ahead of ourselves, we will show that, away from concentrated loads or support reactions (which are a separate problem), the elasticity modifications of elementary formulae are significant for design only for very short beams (where the span and depth are comparable).

A polynomial of nth degree with unknown coefficients can be written:

$$\begin{aligned} F = {} & a_0 + a_1x + b_1y + a_2x^2 + b_2xy + c_2y^2 \\ & + a_3x^3 + b_3x^2y + c_3xy^2 + d_3y^3 \\ & + a_4x^4 + b_4x^3y + c_4x^2y^2 + d_4xy^3 + e_4y^4 \\ & + a_5x^5 + b_5x^4y + c_5x^3y^2 + d_5x^2y^3 + e_5xy^4 + f_5y^5 \\ & + a_6x^6 + \cdots \; etc \end{aligned} \tag{6.4}$$

The constant a_0 and first-order terms give no stresses and can be omitted. The second-order terms give constant stress components and the third-order terms, linear stress components corresponding to the free fields of Chapter 5. Since the polynomial terms through the third order automatically satisfy the biharmonic equation $\nabla^4 F = 0$, the unknown coefficients can be adjusted at

* Elementary solutions neglect σ_y and are based on the assumption that plane-sections remain plane. This is only true for pure bending. The elementary formulae for stresses in prismatic beams are:

$\sigma_x = -\frac{My}{I}$ where M, as a function of x, plots as the moment diagram.

σ_y is of unknown distribution but known on top and bottom boundaries.

$\tau_{xy} = \frac{-VQ}{Ib}$ where $V = \frac{dM}{dx}$ is the total shear force on the cross-section and Q/b (the shear flow) is parabolic in y. The slope, which is the rotation of the cross-section and the vertical deflection both as functions of x, are given by the first and second integral of the $\frac{M}{EI}$ diagram, respectively.

will to satisfy the boundary conditions.* However, for polynomials of higher than third degree, there must be relationships between the coefficients if geometric compatibility, expressed by Equation (6.3), is to be maintained. For example, if all the fourth-order terms were included, then Equation (6.3) would be satisfied only if:

$$a_4 + 2c_4 + e_4 = 0$$

Consider, for example, the case of a cantilever beam with a uniform load as shown in Figure 6.1. Using the strength-of-materials approach based on plane sections, gives the elementary formulas:

$$\sigma_x = -\frac{My}{I} = -\frac{qy}{I}\left(-\frac{L^2}{2} + Lx - \frac{x^2}{2}\right)$$

$$\sigma_y = \text{unspecified} \quad \left[= -q\left(\frac{1}{2} + \frac{y}{h}\right)\text{if linear}\right]$$

$$\tau_{xy} = -\frac{VQ}{Ib} = -q\frac{(L-x)}{2I}\left(\frac{h^2}{4} - y^2\right)$$

Let us now develop an elasticity solution using a polynomial stress function and the semi-inverse approach of rational mechanics in choosing appropriate terms. There are two stress boundary conditions on each of the three exposed faces and two overall equilibrium conditions at $x = 0$. We will also get an equation relating coefficients from Equation (6.3). Thus we should apparently assume a polynomial with at least nine terms. From our physical understanding, we know we should not include terms with even powers of y greater than two since σ_y changes sign top and bottom. Also, since σ_y should not vary with x, we should eliminate all $x^n > x^2$. Therefore, starting with the second-order terms, assume the stress function.**

$$F = a_2x^2 + b_2xy + b_3x^2y + c_3xy^2 + d_3y^3 + c_4x^2y^2 + d_4xy^3 + d_5x^2y^3 + f_5y^5 \tag{6.5}$$

* Stephenson's original design for his famous Britannia Bridge included suspension "chains" to help support the immense wrought iron box girders of 460-ft span through which the trains would run. Experimental and an analytical analysis of flexure by Airy helped convince him they were unnecessary (saving a good deal of money). In his paper on beams in which he introduced his stress function, Airy at this time did not consider compatibility (or body forces) so he did not explicitly derive the biharmonic equation. However, since he only considered third-order terms, he essentially derived the elementary formulas for flexure, which are perfectly adequate for beam design (the shear stress taken as its average valve V/A).

** Students will find, on working out some of the chapter problems, that the reasoning behind choosing terms for a trial stress function is not this simple. Only in retrospect, in presenting a known result, does it appear straightforward. Actually what you do is make an educated guess and see where it leads. If there is an extra boundary condition, then another term will be needed. Remember that the closed-form result can be checked for a few particular cases by finite elements or an experiment.

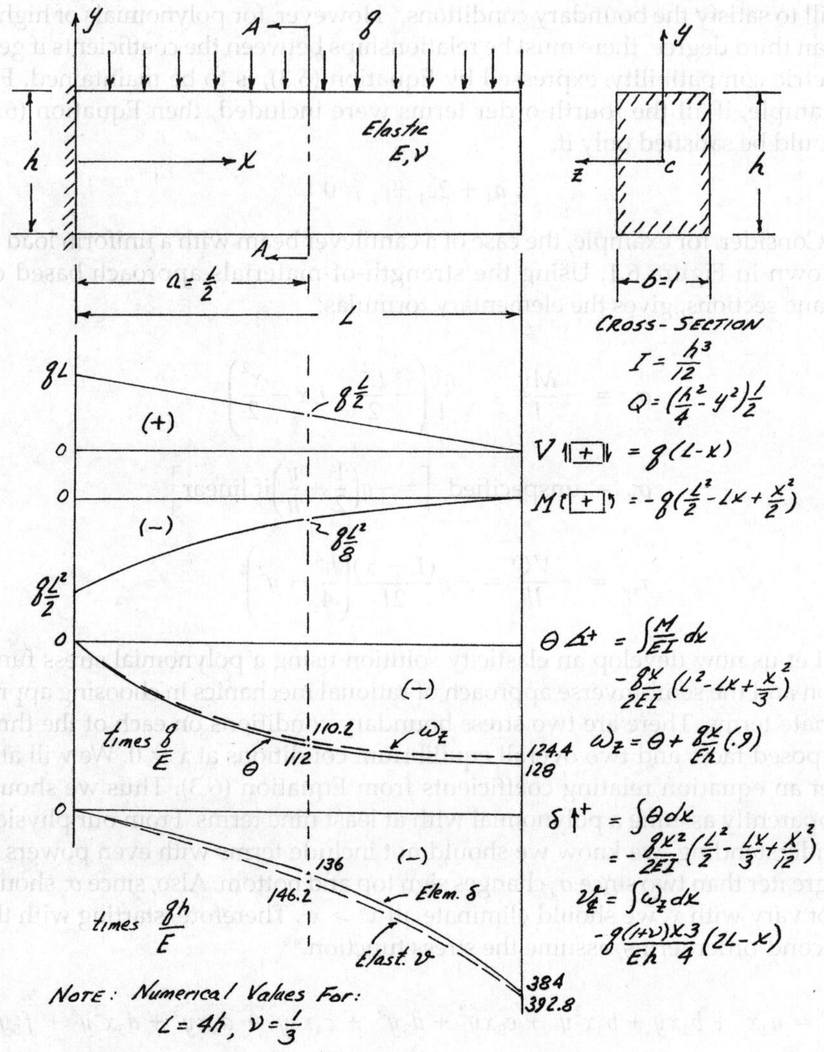

FIGURE 6.1
Elementary θ & δ____vs. elasticity ω_z & $v_{ȼ}$_ _ _ for a cantilever beam.

From the field equation:

$$\nabla^4 F = \frac{\partial^4 F}{\partial x^4} + 2\frac{\partial^4 F}{\partial x^2 \partial y^2} + \frac{\partial^4 F}{\partial y^4} = 8c_4 + 24d_5 y + 120 f_5 y = 0$$

we learn that:

$$c_4 = 0 \qquad \text{and} \qquad f_5 = -\frac{1}{5}d_5$$

The stresses, then, by definition are:

$$\sigma_x = \frac{\partial^2 F}{\partial y^2} = 2c_3x + 6d_3y + 6d_4xy + 6d_5x^2y + 4d_5y^3$$

$$\sigma_y = \frac{\partial^2 F}{\partial x^2} = 2a_2 + 2b_3y + 2d_5y^3$$

$$\tau_{xy} = -\frac{\partial^2 F}{\partial x\,\partial y} = -b_2 - 2b_3x - 2c_3y - 3d_4y^2 - 6d_5xy^2$$

Looking at σ_x and τ_{xy}, it is clear that the c_3 term is unrealistic. There should not be a large normal stress at the neutral axis ($y = 0$) or a linear shear term that changes sign. Therefore let $c_3 = 0$

Applying the boundary conditions at the top and bottom:

i. at $y = +\frac{h}{2}$, $x =$ any: $\sigma_y = -q = 2a_2 + 2b_3\frac{h}{2} + 2d_5\frac{h^3}{8}$ (a)

$$\tau_{xy} = 0 = -b_2 - 2b_3x - 3d_4\frac{h^2}{4} - 6d_5x\frac{h^2}{4} \quad \text{(b)}$$

ii. at $y = -\frac{h}{2}$, $x =$ any: $\sigma_y = 0 = 2a_2 - 2b_3\frac{h}{2} - 2d_5\frac{h^3}{8}$ (c)

$$\tau_{xy} = 0 = \text{same as (b)}$$

While it looks like we are losing a boundary condition, the $x =$ any requirement on (b) gives two requirements:

$$-b_2 = -3d_4\frac{h^2}{4}\text{..............(b}'\text{)} \quad \text{and} \quad -2b_3 = -6d_5\frac{h^2}{4}\text{..............(b}''\text{)}$$

Solving (a), (c), and (b″):

$$a_2 = -\frac{q}{4}; \quad b_3 = -\frac{3q}{4h}; \quad d_5 = \frac{q}{h^3}$$

For equilibrium at the built-in end, $x = 0$:

$$-\int_{-h/2}^{h/2} \tau_{xy}\,dy = -V = -qL = -b_2h - 2d_4\frac{h^3}{8} \quad \text{(d)}$$

which with (b′) results in: $b_2 = \frac{3}{2}\frac{qL}{h}$ and $d_4 = -2\frac{qL}{h^3}$

Finally,

$$-\int_{-h/2}^{h/2} \sigma_x y\,dy = M = \frac{qL^2}{2} = \left[2d^3 y^3 - 4d_5 \frac{y^5}{5}\right]_{-\frac{h}{2}}^{\frac{h}{2}} \tag{e}$$

$$= d_3 \frac{h^3}{2} - \frac{qh^2}{20}$$

giving

$$d_3 = \frac{qL^2}{h^3} + \frac{q}{10h}$$

Thus all the coefficients are determined and the stress field is:

	Elementary Strength of Materials Solution	+	Elasticity Modification
σ_x	$= \frac{qy}{I}\left(\frac{L^2}{2} - Lx + \frac{x^2}{2}\right)$	+	$\frac{qy}{h}\left(.6 - \frac{4y^2}{h^2}\right)$
σ_y	$= -q\left(\frac{1}{2} + \frac{y}{h}\right)$	−	$\frac{qy}{h}\left(\frac{1}{2} - \frac{2y^2}{h^2}\right)$
τ_{xy}	$= -q\frac{(L-x)}{2I}\left(\frac{h^2}{4} - y^2\right)$	+	0

(6.6)

The shear stress distribution is exactly that given by the elementary formula. The "elasticity correction" to σ_x is plotted in Figure 6.2a, as is the distribution of σ_y, which is not considered at all in the elementary theory. Both are independent of x so that at $x = L$ the boundary condition, that the end of the beam be free of normal stresses, is not satisfied.* However, these stresses are very small and the net resultant $= \int \sigma_x\,dA$ is in fact zero so that only very close to the end of the beam will this elasticity correction be inexact.

Note that σ_y and the elasticity correction to σ_x are plotted in Figure 6.2a at greatly exaggerated scale so they are recognizable compared to the basic elementary portion of σ_x. In practice they are of no engineering significance. At $x = 0$, where σ_x will be maximum when $y = \pm h/2$

$$\sigma_x = \pm 3q\frac{L^2}{h^2} \mp 0.2q$$

Even for a very short beam, say $L = 2h$, the elasticity contribution is less than 2%. Other uncertainties introduced by the geometric conditions of a real built-in support at $x = 0$ (not rigid and not plane stress) would be more important.

* In fact this boundary condition was not even used. Actually, each order of polynomial terms used with $n > 3$ gives a separate equation in order that $\nabla^4 F = 0$. In retrospect, we were "lucky" since, by using fourth and fifth order terms we generated an extra condition from $\nabla^4 F = 0$, so that our "guess" to use nine terms in our trial stress function still "worked out" even after dropping the c_3 term.

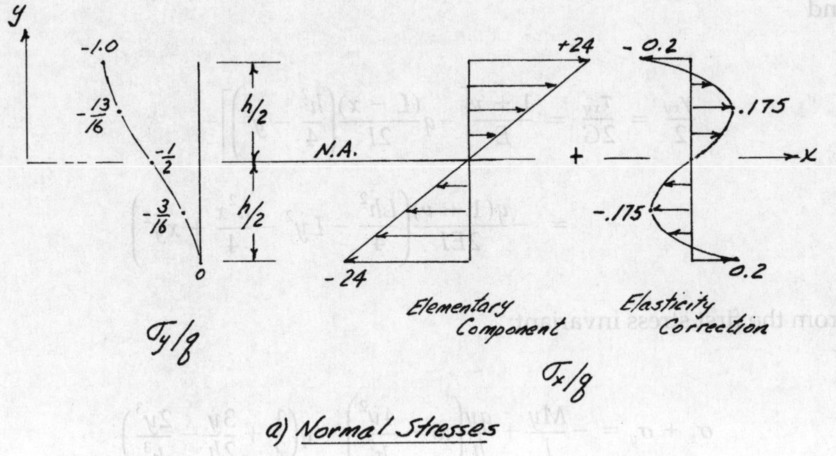

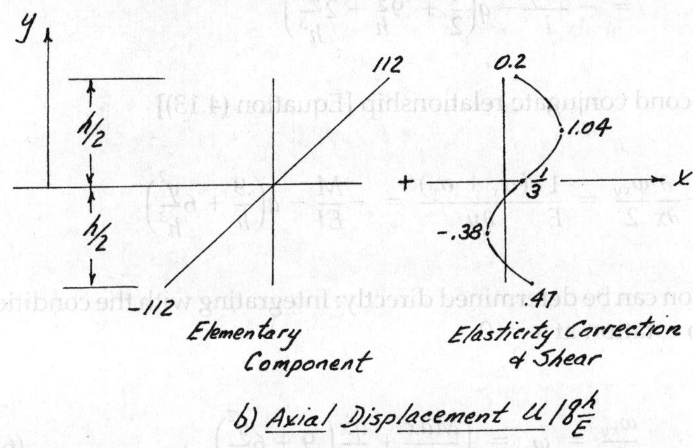

FIGURE 6.2
Longitudinal stresses and displacement for the cantilever beam of Figure 6.1 at $x = 2h = L/2$ and $\nu = 1/3$.

The deflection of the beam can be determined by integrating the relative displacements given by Equation (4.18). From the stress–strain relationships:

$$\epsilon_x = \frac{1}{E}(\sigma_x - \nu\sigma_y) = \frac{qy}{EI}\left(\frac{L^2}{2} - Lx + \frac{x^2}{2}\right) + \frac{qy}{Eh}\left(.6 - \frac{4y^2}{h^2}\right) + \frac{\nu}{E}q\left(-\frac{3y}{h}\frac{y^3}{h^3} + \frac{1}{2}\right)$$

$$\epsilon_y = \frac{1}{E}(\sigma_y - \nu\sigma_x) = \frac{q}{E}\left(\frac{1}{2} + \frac{3y}{2h} - \frac{2y^3}{h^3}\right) - \frac{\nu qy}{EI}\left(\frac{L^2}{2} - Lx - \frac{x^2}{2}\right) - \frac{\nu qy}{Eh}\left(.6 - \frac{4y^2}{h^2}\right)$$

and

$$\frac{\gamma_{xy}}{2} = \frac{\tau_{xy}}{2G} = \frac{1+\nu}{E}\left[-q\frac{(L-x)}{2I}\left(\frac{h^2}{4} - y^2\right)\right]$$

$$= -\frac{q(1+\nu)}{2EI}\left(\frac{Lh^2}{4} - Ly^2 - \frac{h^2x}{4} + xy^2\right)$$

From the first stress invariant:

$$\sigma_x + \sigma_y = -\frac{My}{I} + \frac{qy}{h}\left(.6 - \frac{4y^2}{h^2}\right) - q\left(\frac{1}{2} + \frac{3y}{2h} - \frac{2y^3}{h^3}\right)$$

$$= -\frac{My}{I} - q\left(\frac{1}{2} + .9\frac{y}{h} + 2\frac{y^3}{h^3}\right) \tag{6.7}$$

and from the second conjugate relationship [Equation (4.13)]

$$-\frac{\partial}{\partial x}\frac{\omega_{xy}}{2} = \frac{1}{E}\frac{\partial(\sigma_x + \sigma_y)}{\partial y} = -\frac{M}{EI} - q\left(\frac{.9}{h} + 6\frac{y^2}{h^3}\right)$$

the elastic rotation can be determined directly: Integrating with the condition that there be no rotation at $x = 0$.

$$\frac{\omega_{xy}}{2} = \omega_z = \int\frac{Mdx}{EI} + \frac{qx}{Eh}\left(.9 + 6\frac{y^2}{h^2}\right) \tag{6.8}$$

The first term, being the integral of the moment diagram, is the same as the slope, θ, given by elementary theory and the second, the elasticity contribution, is due primarily to σ_y now being included. Substituting the expression for the Moment Diagram:

$$\frac{\omega_{xy}}{2} = -\frac{qx}{2EI}\left(L^2 - Lx + \frac{x^2}{3}\right) + \frac{qx}{Eh}\left(.9 + \frac{6y^2}{h^2}\right) \tag{6.9}$$

Since this is plane stress, there will be elastic rotation ω_x and anticlastic curvature, but we will concern ourselves with the displacement of the centroidal plane of symmetry $z = 0$.

The displacements are found by integrating Equation (4.17) directly with the support condition that at $x = 0$ and $y =$ any, $u = v = 0$.

$$u = \int du = \int \epsilon_x dx + \int \frac{\gamma_{xy}}{2} dy - \int \frac{\omega_{xy}}{2} dy$$

$$= \frac{qxy}{2EI}\left(L^2 - Lx + \frac{x^2}{3}\right) + \frac{qxy}{Eh}\left(.6 - \frac{4y^2}{h^2}\right) + \frac{\nu qx}{E}\left(\frac{1}{2} + \frac{3y}{2h} - 2\frac{y^3}{h^3}\right)$$

$$+ \frac{q(1+\nu)xy}{2EI}\left(\frac{h^2}{4} - \frac{y^2}{3}\right) + \frac{qxy}{2EI}\left(L^2 - Lx + \frac{x^2}{3}\right) - \frac{qxy}{Eh}\left(.9 + \frac{2y^2}{h^2}\right)$$

collecting in ascending powers of y

$$u = \frac{\nu qx}{2E} + \frac{qxy}{EI}\left(L^2 - Lx + \frac{x^2}{3} + 0.1h^2 + \frac{\nu}{4}h^2\right) - \frac{(2+\nu)}{3EI}qxy^3 \quad (6.10)$$

The first term not in the elementary theory is the Poisson-ratio effect due to $(\sigma_y)_{avg} = -q/2$ giving $(\epsilon_x)_{avg} = +\upsilon q/2E$ which integrates to a linear function of x. The second term, linear in y, conforms with the assumption of plane sections with the last two h^2 components being a contribution from elasticity. The final y^3 term warps the cross-section and is a result of ϵ_y, the elasticity correction to ϵ_x and the shear strain γ_{xy}, all of which are not included in the elementary solution for displacements. The warping of an arbitrary cross-section at $x = a = 2h$ for a short beam $L = 4h$ is shown in Figure 6.2b. It is clear that from an engineering standpoint the elementary formula is perfectly adequate.

Following the same procedure to find the vertical displacement

$$v = \int dv = \int \frac{\omega_{xy}}{2} dx + \int \frac{\gamma_{xy}}{2} dx + \int \epsilon_y dy$$

where the third term does not contribute directly since v must equal zero at $x = 0$. Therefore:

$$v = -\frac{q}{2EI}\left(\frac{L^2x^2}{2} - \frac{Lx^3}{3} + \frac{x^4}{12}\right) + \frac{qx^2}{2EI}\left(.9 + \frac{6y^2}{h^2}\right)$$

$$- \frac{q(1+\nu)}{2EI}\left(\frac{Lh^2x}{4} - Ly^2x - \frac{h^2x^2}{8} + \frac{x^2y^2}{2}\right) \quad (6.11)$$

The first term is the integral of the slope diagram as in the elementary theory, and the second and third terms are the contributions from the elasticity correction and the shear, respectively. The deflection of the centroidal axis at $y = 0$ is therefore:

$$v = -\frac{qx^2}{2EI}\left(\frac{L^2}{2} - \frac{Lx}{3} + \frac{x^2}{12}\right) + \frac{qx^2}{2Eh}(.9) - \frac{3q(1+\nu)x}{2Eh}\left(L - \frac{x}{2}\right) \tag{6.12}$$

The rotation ω_z and vertical deflection v of the centroidal axis from elasticity are compared to the values for Θ and δ from elementary theory in Figure 6.1. As was true for stresses and longitudinal displacements, the difference between them must be exaggerated in plotting to be recognizable.

This conclusion that elementary analysis for straight beams is very accurate will be true for all other cases of distributed loading. Therefore, while most texts include a few more examples, here they are left to the chapter problems. Some further remarks concerning straight beams are included in the final section.

Example 6.1

A thin elastic square plate of $2c$ on a side is free of body forces and is subjected to a loading around its four edges such that the stress function is found to be $F = \frac{cxy^3}{3}$.

a. Determine the stress components σ_x, σ_y, τ_{xy}.
b. Sketch the boundary tractions and verify that equilibrium for the plate as a whole is satisfied.
c. Plot the flow net for σ_m and ω_z for the plate.
d. Determine the displacement, u, v, of the point $x = 2c$, $y = 0$ relative to $x = 0$.

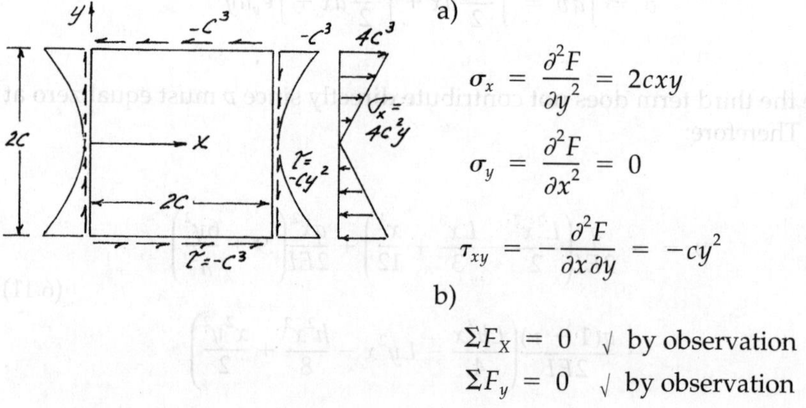

a)

$$\sigma_x = \frac{\partial^2 F}{\partial y^2} = 2cxy$$

$$\sigma_y = \frac{\partial^2 F}{\partial x^2} = 0$$

$$\tau_{xy} = -\frac{\partial^2 F}{\partial x\,\partial y} = -cy^2$$

b)

$\Sigma F_X = 0$ √ by observation

$\Sigma F_y = 0$ √ by observation

$$\Sigma M_o = 2(2c)(c^3)(c) + \int_{-c}^{c} -cy^2 dy(2c) - \int_{-c}^{c} 4c^2y^2 dy = c^5\left(4 - \frac{4}{3} - \frac{8}{3}\right) = 0$$

c) Flow Net ——— $2\sigma_m/c^3$

--------- $\omega_z E/c^3$

$$\frac{(\sigma_x + \sigma_y)}{2} = \sigma_m = cxy$$

$$\therefore \frac{\partial \sigma_m}{\partial x} = cy = \frac{E}{2}\frac{\partial \omega_z}{\partial y} \qquad \therefore \omega_z = \frac{\omega_{xy}}{2}$$

$$\frac{\partial \sigma_m}{\partial y} = cx = -\frac{E}{2}\frac{\partial \omega_z}{\partial x} \qquad = \frac{C}{E}(y^2 - x^2)$$

d) $\epsilon_x = \frac{1}{E}(\sigma_x - \nu \sigma_y^{\,0})\frac{2cxy}{E}$

$$\epsilon_y = -\nu\epsilon_x = -\nu\frac{2cxy}{E}, \quad \frac{\gamma_{xy}}{2} = \frac{cy^2}{2G}$$

$\therefore$ Assuming at $x = 0,\ y = 0;\ u = v = 0$

$$\text{at } x = 2c,\ y = 0 \quad u = \int_0^{2c} \epsilon_x dx + \int_0^{2c} \frac{\gamma_{xy}}{2} dy - \int_0^{2c} \frac{\omega_{xy}}{2} dy = 0$$

$$v = \int_0^{2c} \epsilon_y dy + \int_0^{2c} \frac{\gamma_{xy}}{2} dx + \int_0^{2c} \frac{\omega_{xy}}{2} dx = -\frac{c}{E}\frac{x^3}{3}\bigg]_0^{2c} = -\frac{8c^4}{3E}$$

6.4 Polar Coordinates and Airy's Stress Function

The fundamental relationships for plane polar coordinates as given in Section 2.11 were obtained as a special case of general curvilinear coordinates. Since polar coordinates are so useful, let us re-derive them from basic principles. First consider equilibrium of a differential element as shown in Figure 6.3a.

$$\Sigma F_r = 0 = \frac{\partial \sigma_r}{\partial r} dr(rd\theta) + \frac{\partial \tau_{r\theta}}{\partial \theta} d\theta dr + 2\sigma_\theta dr\frac{d\theta}{2} + \sigma_r dr d\theta + B_r dr(rd\theta)$$

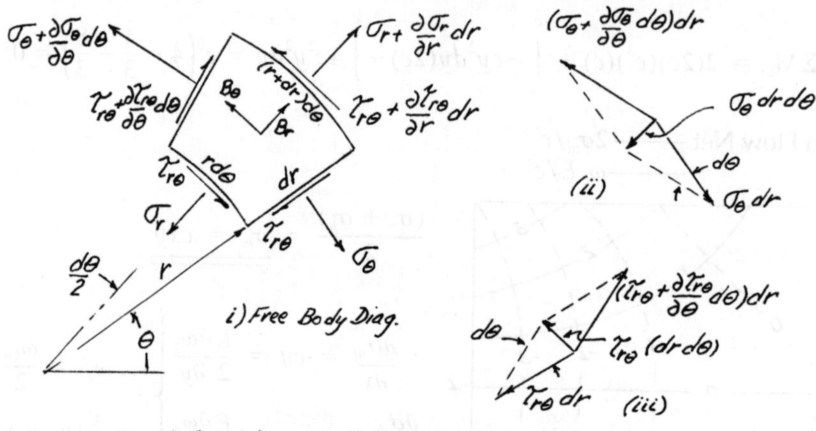

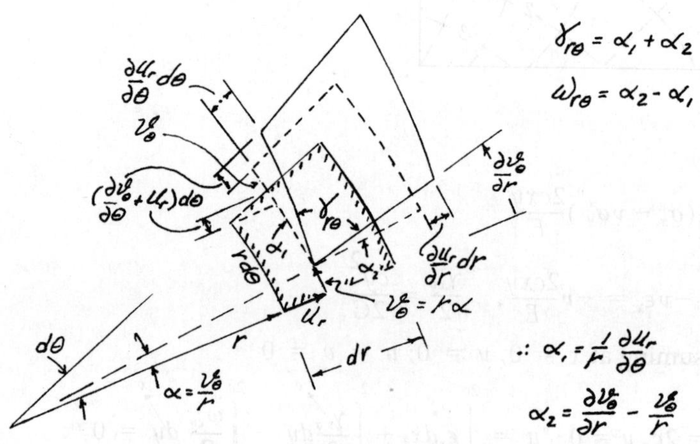

FIGURE 6.3
Stresses and geometry of deformation in polar coordinates.

Therefore:

$$\frac{\partial \sigma_r}{\partial r} + \frac{1}{r}\frac{\partial \tau_{r\theta}}{\partial \theta} + \frac{(\sigma_r - \sigma_\theta)}{r} + B_r = 0 \tag{6.13}$$

Similarly:

$$\frac{1}{r}\frac{\partial \sigma_\theta}{\partial \theta} + \frac{\partial \tau_{r\theta}}{\partial r} + 2\frac{\tau_{r\theta}}{r} + B_\theta = 0$$

Next consider the geometry of deformation from Figure 6.3b.

$$\epsilon_r = \frac{\partial u_r}{\partial r}; \qquad \epsilon_\theta = \frac{u_r}{r} + \frac{1}{r}\frac{\partial v_\theta}{\partial \theta}$$

$$\gamma_{r\theta} = \frac{1}{r}\frac{\partial u_r}{\partial \theta} + \frac{\partial v_\theta}{\partial r} - \frac{v_\theta}{r} \tag{6.14}$$

$$\omega_{r\theta} = \omega_{xy} = 2\omega_z = \frac{\partial v_\theta}{\partial r} - \frac{v_\theta}{r} - \frac{1}{r}\frac{\partial u_r}{\partial \theta}$$

Since the in-plane compatibility relationship in terms of stress, involves only the first invariant ($\sigma_r + \sigma_\theta = \sigma_x + \sigma_y$) and the gradient body force,*

$$\nabla^2(\sigma_r + \sigma_\theta) = \left(\frac{\partial^2}{\partial r^2} + \frac{1}{r}\frac{\partial}{\partial r} + \frac{1}{r^2}\frac{\partial^2}{\partial \theta^2}\right)(\sigma_r + \sigma_\theta) = -(1+\nu)\left(\frac{B_r}{r} + \frac{\partial B_r}{\partial r} + \frac{1}{r}\frac{\partial B_\theta}{\partial \theta}\right) \tag{6.15}$$

To formulate the solution of Equations (6.13) and (6.15) in terms of the stress function $F(r, \theta)$, define for the case where $B_\theta = 0$:

$$\sigma_r = \frac{1}{r}\frac{\partial F}{\partial r} + \frac{1}{r^2}\frac{\partial^2 F}{\partial \theta^2}$$

$$\sigma_\theta = \frac{\partial^2 F}{\partial r^2} + r\frac{d\Omega}{dr} \tag{6.16}$$

$$\tau_{r\theta} = -\frac{\partial}{\partial r}\left(\frac{1}{r}\frac{\partial F}{\partial \theta}\right)$$

where B_r is derived from the potential function $\Omega = \Omega(r)$ by the relationship $B_r = \partial\Omega/\partial r$. The equilibrium [Equations (6.13)] are, thereby, satisfied "by definition" and the compatibility equation becomes:

$$\nabla^2(\nabla^2 F) = \nabla^4 F = -(1-\nu)\nabla^2\Omega \tag{6.17}$$

---

* In plane polar coordinates, the out-of-plane compatibility requirements on ϵ_z become:

$$r\frac{\partial \epsilon_z}{\partial r} + \frac{\partial^2 \epsilon_z}{\partial \theta^2} = 0; \qquad \frac{\partial}{\partial r}\left(\frac{1}{r}\frac{\partial \epsilon_z}{\partial \theta}\right) = 0; \qquad \frac{\partial^2 \epsilon_z}{\partial r^2} = 0$$

Thus, as with rectangular coordinates, there is the question of these equations not being satisfied for plane stress. Again we will assume loads are symmetric with z and we consider the plate to be thin enough to avoid any serious difficulties.

where in polar coordinates:

$$\nabla^2 = \left(\frac{\partial^2}{\partial r^2} + \frac{1}{r}\frac{\partial}{\partial r} + \frac{1}{r^2}\frac{\partial^2}{\partial \theta^2}\right) \tag{6.18}$$

For plane strain, $\nu' = \nu/(1-\nu)$ is substituted for ν in Equations (6.15) and (6.17). Thus again, the two-dimensional problem is reduced to finding a solution to one, fourth-order, partial differential equation which, if body forces are neglected or can be derived from a potential function, reduces to $\nabla^4 F = 0$.

There are an infinite number of terms in a general expression for F in polar coordinates that will satisfy this biharmonic equation. In this and the succeeding two chapters, we will concentrate on three basic types:

a. Type 1: $F = F(r)$ = a function of r only and $\therefore \tau_{r\theta} = \omega_{r\theta} = 0$

therefore $$\nabla^4 F = \frac{d^4F}{\partial r^4} + \frac{2}{r}\frac{d^3F}{dr^3} - \frac{1}{r^2}\frac{d^2F}{dr^2} + \frac{1}{r^3}\frac{dF}{dr} \underset{\text{must}}{=} 0$$

with the solution

$$\underline{F = A\ln r + Br^2\ln r + Cr^2 + D\ldots}\begin{cases} \sigma_r = 2C + \dfrac{A}{r^2} + B(1+2\ln r) \\ \tau_{r\theta} = 0 \\ \sigma_\theta = 2C - \dfrac{A}{r^2} + B(3+2\ln r) \end{cases} \tag{I}$$

b. Type 2: $F = R(r)\cos m\theta$

Then

$$\nabla^2 F = \left(\frac{d^2R}{dr^2} + \frac{1}{r}\frac{dR}{dr} - \frac{m^2}{r^2}R\right)\cos m\theta = 0$$

Assume $R = r^n$ and then solving this differential equation

$$F = (Ar^{2+m} + Br^m + Cr^{2-m} + Dr^{-m})(\sin m\theta \text{ or } \cos m\theta) \tag{II}$$

note that if $m = 1$, B and C merge. Therefore, write

$$F = (Ar^3 + Br + Cr\ln r + Dr^{-1})(\sin\theta \text{ or } \cos\theta) \tag{II'}$$

c. Type 3: $F = r^m T(\theta) \quad \therefore \nabla^2 F = r^{m-2}\left[m^2 T(\theta) + \dfrac{d^2 T(\theta)}{d\theta^2}\right]$

For $\nabla^2 F = 0$, then $T = -C_1 \sin m\theta + C_2 \cos m\theta$ if $m^2 T + \dfrac{d^2 T}{d\theta^2} = 0$. In general for $\nabla^4 F = 0$

$$F = r^m[A\cos m\theta + B\sin m\theta + C\cos(m-2)\theta + D\sin(m-2)\theta] \qquad \text{(III)}$$

note that if $m = 0, 1,$ or 2, terms vanish or constants merge, and therefore:

For $m = 0$, write $F = A + B\theta + c\cos 2\theta + D\sin 2\theta$
For $m = 1$, write $F = r[A\cos\theta + B\theta\sin\theta + C\sin\theta + D\theta\cos\theta]$
For $m = 2$, write $F = r^2[A\cos 2\theta + B\sin 2\theta + C + D\theta]$

(III′)

In every case there are four terms with unknown coefficients, A, B, C, and D to determine. The choice of which type of stress function and what terms to use will become apparent from considering the anticipated structural action, the boundary conditions available, and comparison to special cases approximated by elementary analysis.

Example 6.2

The casement for a particle accelerator is a thick steel ring where the magnetic field introduces a body force in the radial direction, which dies off linearly with the radius as shown below.
Determine the elasticity solution. Assume *plane strain* and determine both the stress field and the displacements.

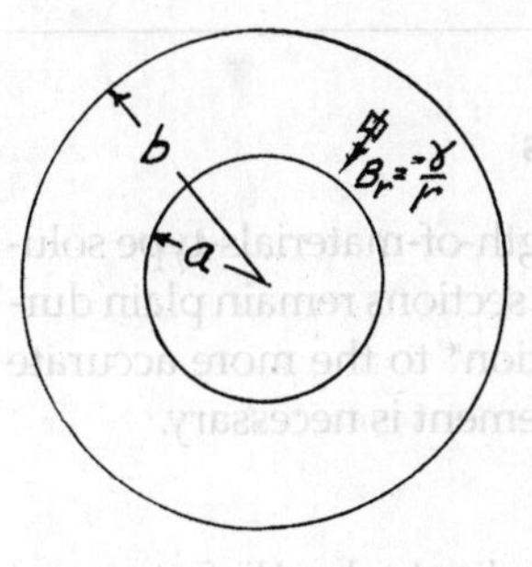

From symmetry (nothing a function of θ)
$\therefore \tau_{r\theta} = 0 \ \& \ \omega_{r\theta} = 0$
one equilibrium equation: [Equation (6.13)]

$$\frac{d\sigma_r}{dr} + \frac{1}{r}(\sigma_r - \sigma_\theta) - \frac{\gamma}{r} = 0 \ldots\ldots(1)$$

and compatibility [Equation (6.15)]

$$\nabla^2(\sigma_r + \sigma_\theta) = -(1+\nu)\left(\frac{\partial Br}{\partial r} + \frac{Br}{r}\right)$$

$$= -(1+\nu)\left(\frac{\gamma}{r^2} - \frac{\gamma}{r^2}\right)$$

$\therefore (\sigma_r + \sigma_\theta)$ is harmonic since $\nabla^2(\sigma_r + \sigma_\theta) = 0 \ldots\ldots(2)$

Since $\omega_z = 0$, from the conjungate relations $\quad (\sigma_r + \sigma_\theta) = \text{const.} \quad (3)$

For rings where $F = F(r)$ use $F = A\ln r + Br^2\ln r + Cr^2 + D$

Giving stresses

$$\sigma_r = \frac{1}{r}\frac{\partial F}{\partial r} = \frac{A}{r^2} + B(1 + 2\ln r) + 2C$$

$$\text{and} \quad \sigma_\theta = \frac{\partial^2 F}{\partial r^2} + r\left(-\frac{\gamma}{r}\right) = -\frac{A}{r^2} + B(3 + 2\ln r) + 2C - \gamma$$

B.C. $@r = a,b;\ \sigma_r = 0$ But from (3) $\sigma_r + \sigma_\theta = \text{const}$

$$\therefore B = 0$$

$$\left.\begin{aligned} \therefore A/a^2 + 2C &= 0 \\ A/b^2 + 2C &= 0 \end{aligned}\right\} \quad \therefore \frac{A}{a^2} - \frac{A}{b^2} = 0 \quad \therefore A = 0 \quad \therefore C = 0$$

$$\therefore \sigma_r = 0 \quad \text{and} \quad \sigma_\theta = -\gamma$$

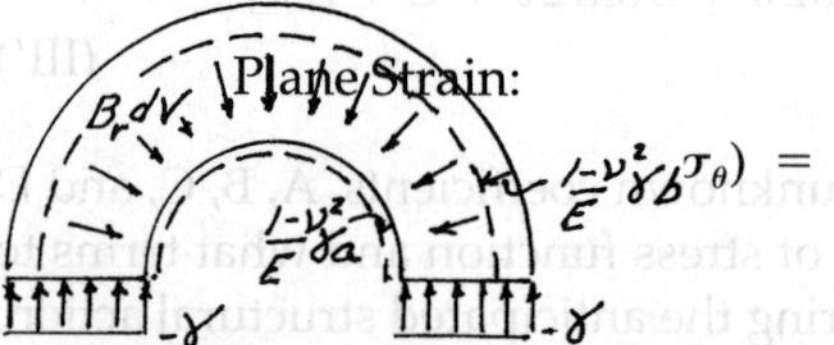

Plane Strain:

$$\sigma_\theta) = 0 \qquad \therefore \sigma_z = \nu\sigma_\theta = -\nu\gamma$$

Solution (A ring of optimum const. strength)

$$\epsilon_\theta = \frac{1}{E}(\sigma_\theta - \nu\sigma_r - \nu\sigma_z)$$

$$= \frac{1-\nu^2}{E}\sigma_\theta = -\frac{1-\nu^2}{E}\gamma$$

$$\therefore u = \epsilon_\theta r = -\frac{1-\nu^2}{E}\gamma r, \quad v = w = 0$$

displacement field

6.5 Simplified Analysis of Curved Beams

As for straight beams, it is possible to develop a strength-of-materials-type solution for curved beams from the assumption that plane sections remain plain during flexure. We can then compare this Winkler solution* to the more accurate results from elasticity theory to see when added refinement is necessary.

* E. Winkler (1835–1888) was a theoretical engineer and an outstanding teacher. His first paper at the age of 23 on curved beams and rotating discs was written two years before receiving his Ph.D. in 1860 from Leipzig University. His dissertation was on retaining walls followed shortly by one on continuous beams. Study of the deformation of railroad rails led to the publication of his most famous paper solving the beam on an elastic foundation. Modeling soil by closely spaced springs without shear stiffness is known as a "Winkler foundation." After appointment

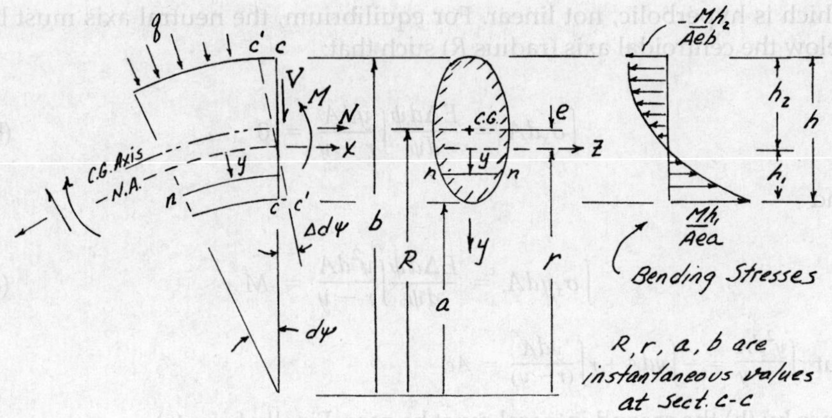

FIGURE 6.4
Elementary theory of curved beams (Winkler).

Consider a differential slice from a curved beam and call the tangent to the neutral axis at any cross-section x, the radial coordinate, y, positive inward as shown in Figure 6.4. Assume that:

a. The transverse loading as well as the cross-section is symmetric and the bending moment is in the plane of symmetry (the xy plane).
b. The radial stress is neglected.
c. If there is shear and/or normal force on the cross-section they induce shear and/or normal stresses as in the elementary straight beam formulas.
d. Transverse sections remain plane (i.e., plane c–c rotates around the z axis an angle $\Delta d\psi$).

While these are the same assumptions used for the elementary analysis of straight beams, for a curved beam the longitudinal fibers are not all the same original base length. Therefore, linear displacements produce nonlinear strains. For a layer of longitudinal fibers n–n, a distance y from the neutral axis at radius r, the rotation of the cross-section $\Delta d\psi$ due to M gives:

$$\epsilon_x = \frac{\Delta L}{L_o} = \frac{y \Delta d\psi}{(r - y) d\psi} \quad \text{and} \quad \sigma_x = E\epsilon_x = \frac{Ey\Delta d\psi}{(r - y)d\psi} \qquad \text{(a)}$$

as Professor at the Vienna Polytechnicium, he did pioneering experimental work and wrote the most complete book on Strength of Materials in the 19th century. His early death was a blow to German structural design. Although skilled in mathematics, Winkler appreciated approximate and experimental analysis and would have opposed the developing German fixation on not building civil engineering structures that could not be rigorously analyzed mathematically. The Swiss and French who believed in graphic analysis and visual methods, became the leaders in design with the new, uncertain, inexact material of reinforced concrete.

which is hyperbolic, not linear. For equilibrium, the neutral axis must be below the centroidal axis (radius R) such that:

$$\int \sigma_x dA = \frac{E\Delta d\psi}{d\psi}\int \frac{ydA}{r-y} = 0 \tag{b}$$

and

$$\int \sigma_x y dA = \frac{E\Delta d\psi}{d\psi}\int \frac{y^2 dA}{r-y} = M \tag{c}$$

But $\int \frac{y^2 dA}{r-y} = -\int ydA + r\int \frac{ydA}{(r-y)} = Ae$

since by (b) the second integral must be zero. Finally from (c)

$$-\int ydA = \int \frac{y^2 dA}{r-y} = Ae = \frac{Md\psi}{E\Delta d\psi}$$

So due to bending:

$$\sigma_x = \frac{My}{Ae(r-y)}$$

Thus for a curved beam, the formulas for flexure with N and V by this Winkler analysis are:

$$\sigma_x = \frac{My}{Ae(r-y)} + \frac{N}{A'}; \qquad \sigma_y \text{ is unknown}$$

$$\tau_{xy} = \frac{VQ}{Ib} \tag{6.19}$$

$$\Delta d\psi = \frac{Mds}{AeRE} - \frac{Nds}{AER} \qquad \text{where} \quad ds = Rd\psi$$

where e, the distance from the centroidal axis to the neutral axis, is a property of the cross-section such that (b) is satisfied. Formulas for various useful shapes are given in many books. For the rectangular cross-section of unit thickness necessary for comparison to an elasticity solution:

$$e \cong \frac{h^2}{12R}\left[1 + \frac{4}{15}\left(\frac{h}{2R}\right)^2\right] \quad \text{approximately} \tag{6.20}$$

Therefore: $AeR \cong \frac{bh^3}{12}\left(1 + \frac{1}{15}\frac{h^2}{R^2}\right) = I_c + \frac{1}{15}\frac{h^2}{R^2}$ (d)

showing that only for very deep beams is there an appreciable increase in the flexural stiffness.

6.6 Pure bending of a Beam of Circular Arc

Consider a circular beam of constant rectangular cross-section bent by end moments in the plane of curvature (Figure 6.5). Since the moment diagram is constant, stresses are only a function of r and a stress function of type I is called for:

$$F = F(r) = A\ln r + Br^2\ln r + Cr^2 + D \tag{6.21}$$

giving the stress field:

$$\begin{aligned}
\sigma_r &= \frac{1}{r}\frac{\partial F}{\partial r} + \frac{1}{r^2}\frac{\partial^2 F}{\partial \theta^2} = \frac{1}{r^2}A + B(1 + 2\ln r) + 2C \\
\sigma_\theta &= \frac{\partial^2 F}{\partial r^2} = -\frac{1}{r^2}A + B(3 + 2\ln r) + 2C \\
\tau_{r\theta} &= -\frac{\partial}{\partial r}\left(\frac{1}{r}\frac{\partial F}{\partial \theta}\right) = 0
\end{aligned} \tag{6.22}$$

Since shear stresses, $\tau_{r\theta}$, are zero, only two boundary conditions remain.

$$(\sigma_r)_{r=a} = 0 = \frac{1}{a^2}A + (1 + 2\ln a)B + 2C \tag{a}$$

$$(\sigma_r)_{r=b} = 0 = \frac{1}{b^2}A + (1 + 2\ln b)B + 2C \tag{b}$$

Also, equilibrium requires that $\Sigma M = 0$ and therefore:

$$\int \sigma_\theta(dr \cdot 1)r = M = -\int_a^b \frac{A}{r}dr + \int_a^b B(3r + 2r\ln r)dr + \int_a^b 2Cr\,dr$$

giving

$$-A\ln\frac{b}{a} + B(b^2 - a^2) + B(b^2\ln b - a^2\ln a) + C(b^2 - a^2) = M \tag{c}$$

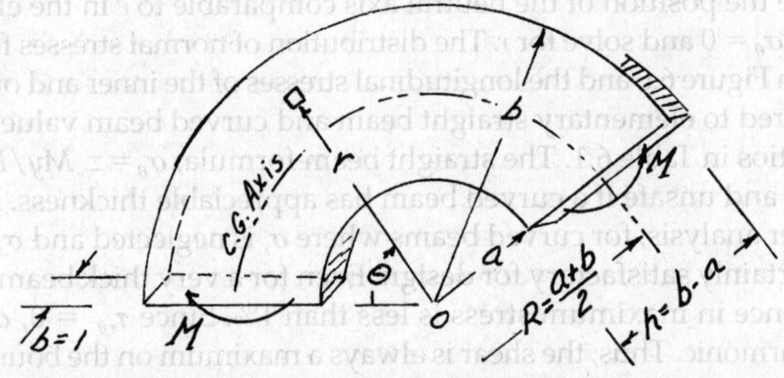

FIGURE 6.5
Beam of circular arc.

The force equilibrium equation

$$\Sigma F_\theta = 0 = \int \sigma_\theta(dr \cdot 1) = \int_a^b \left[-\frac{A}{r^2} + B(3 + 2\ln r) + 2C \right] dr$$

evaluates to $\underbrace{\left[\frac{A}{b} + B(b + 2b\ln b) + 2Cb\right]}_{\text{same as (b)}} - \underbrace{\left[\frac{A}{a} + B(a + 2a\ln a) + 2Ca\right]}_{\text{same as (a)}}$

giving no new information. Thus it turns out that we have just three equations to evaluate exactly three coefficients. Letting:

$$N = (b^2 - a^2)^2 - 4a^2b^2\left(\ln\frac{b}{a}\right)^2$$

then:

$$A = -\frac{4M}{N}a^2b^2\ln\frac{b}{a}; \qquad B = -\frac{2M}{N}(b^2 - a^2)$$

and

$$C = \frac{M}{N}[b^2 - a^2 + 2(b^2\ln b - a^2\ln a)]$$

giving the stress field:

$$\begin{aligned}
\sigma_r &= -\frac{4M}{N}\left[\frac{a^2b^2}{r^2}\ln\frac{b}{a} + b^2\ln\frac{r}{b} + a^2\ln\frac{a}{r}\right] \\
\sigma_\theta &= -\frac{4M}{N}\left[-\frac{a^2b^2}{r^2}\ln\frac{b}{a} + b^2\ln\frac{r}{b} + a^2\ln\frac{a}{r} + b^2 - a^2\right] \qquad (6.23) \\
\tau_{r\theta} &= 0
\end{aligned}$$

To locate the position of the neutral axis comparable to e in the elementary theory, set $\sigma_\theta = 0$ and solve for r. The distribution of normal stresses for $b/a = 3$ is shown in Figure 6.6 and the longitudinal stresses of the inner and outer fibers are compared to elementary straight beam and curved beam values for various b/a ratios in Table 6.1. The straight beam formula, $\sigma_\theta = \pm\, My/I$ is clearly inaccurate and unsafe if a curved beam has appreciable thickness. However, the Winkler analysis, for curved beams where σ_r is neglected and σ_θ is hyperbolic, is certainly satisfactory for design. Even for a very thick beam ($b/a = 3$) the difference in maximum stress is less than 1%. Since $\tau_{r\theta} = 0$, $\sigma_r + \sigma_\theta = 2\,\tau_{max}$, is harmonic. Thus, the shear is always a maximum on the boundary and the radial stress from elasticity does not affect yield for an EPS material.

TABLE 6.1
Coefficients, m, for $\sigma_\theta = mM/a^2$

b/a	Straight Beam Anal.	Winkler Analysis Inner	Winkler Analysis Outer	Elastic Analysis Inner	Elastic Analysis Outer
1.3	±66.67	+72.98	−61.27	+73.05	−61.35
2	±6.00	+7.725	−4.863	+7.755	−4.917
3	±1.50	+2.285	−1.095	+2.292	−1.130

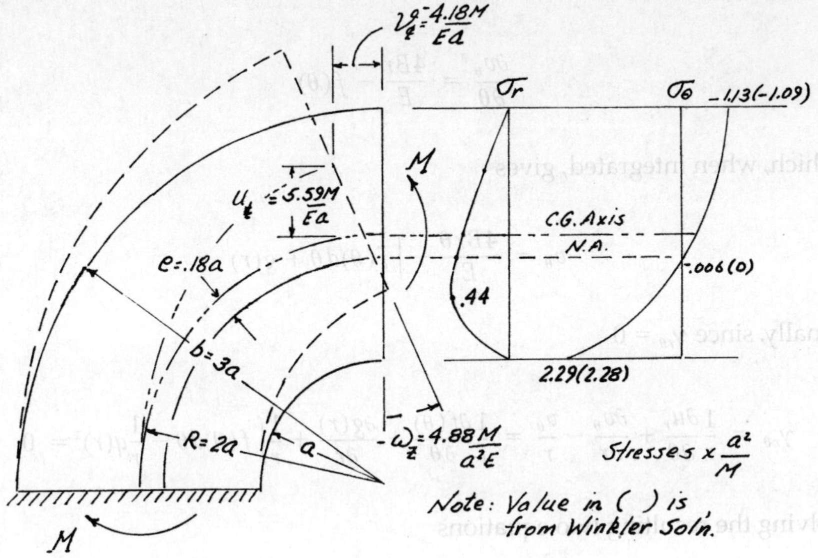

FIGURE 6.6
Example—pure bending of a cantilever beam.

Displacements are not difficult to determine for symmetrical strain distributions since the strains are not a function of θ. From the stress–strain relation for plane stress,

$$\begin{aligned}\epsilon_r &= \frac{1}{E}(\sigma_r - \nu\sigma_\theta) = \frac{\partial u}{\partial r} \\ &= \frac{1}{E}\left[\frac{(1+\nu)A}{r^2} + 2(1-\nu)B\ln r + (1-3\nu)B + 2(1-\nu)C\right]\end{aligned} \tag{d}$$

which can be integrated to give:

$$u_r = \frac{1}{E}\left[-\frac{(1+\nu)}{r}A + 2(1-\nu)Br\ln r - B(1+\nu)r + 2C(1-\nu)r\right] + f(\theta) \tag{e}$$

Similarly:

$$\epsilon_\theta = \frac{1}{E}(\sigma_\theta - \nu\sigma_r) = \frac{u_r}{r} + \frac{1}{r}\frac{\partial v_\theta}{\partial \theta}$$
$$= \frac{1}{E}\left[\frac{-(1+\nu)A}{r^2} + 2(1-\nu)B\ln r + (3-\nu)B + 2(1+\nu)C\right] \qquad \text{(f)}$$

and therefore with (e)

$$\frac{\partial v_\theta}{\partial \theta} = \frac{4Br}{E} - f(\theta) \qquad \text{(g)}$$

which, when integrated, gives

$$v_\theta = \frac{4Br\theta}{E} - \int f(\theta)d\theta + g(r) \qquad \text{(h)}$$

Finally, since $\gamma_{r\theta} = 0$

$$\gamma_{r\theta} = \frac{1}{r}\frac{\partial u_r}{\partial \theta} + \frac{\partial v_\theta}{\partial r} - \frac{v_\theta}{r} = \frac{1}{r}\frac{\partial f(\theta)}{\partial \theta} + \frac{\partial g(r)}{\partial r} + \frac{1}{r}\int f(\theta)d\theta - \frac{1}{r}g(r) = 0 \qquad \text{(i)}$$

Solving the resulting two equations

$$g(r) = Fr \qquad \text{and} \qquad f(\theta) = H\sin\theta + K\cos\theta \qquad \text{(j)}$$

where F, H, and K are constants to be determined by the geometric conditions associated with the constraints holding the structure in equilibrium. Substituting (j) into (e) and (g), the general displacement equations are:

$$u_r = \frac{1}{E}\left[-\frac{(1+\nu)A}{r} + 2(1-\nu)Br\ln r - B(1+\nu)r + 2C(1-\nu)r\right]$$
$$+H\sin\theta + K\cos\theta \qquad (6.24)$$

$$v_\theta = \frac{4\,Br\theta}{E} + F(r) + H\cos\theta - K\sin\theta \qquad (6.25)$$

It is interesting to look at the invariants. From Equation (6.22)

$$\sigma_r + \sigma_\theta = 4B(1 + \ln r) + 4C \qquad (6.26)$$

In polar coordinates, the Cauchy-Riemann conditions are

$$\frac{\partial\phi}{\partial r} + \frac{1}{r}\frac{\partial\psi}{\partial\theta} \quad \text{and} \quad \frac{1}{r}\frac{\partial\phi}{\partial\theta} = -\frac{\partial\psi}{\partial r} \tag{k}$$

or from Equation (4.13) with no body forces:

$$\frac{1}{E}\frac{\partial(\sigma_r + \sigma_\theta)}{\partial r} = \frac{4B}{Er} = \frac{1}{r}\frac{\partial\omega_z}{\partial\theta} \tag{6.27}$$

Therefore

$$\omega_z = \frac{4B\theta}{E} \tag{6.28}$$

and we see that the assumption of plane sections for the Winkler analysis is, in fact, correct. A particular cross-section rotates an angle $4B\theta/E$ and moves circumferentially a constant amount $-K\sin\theta$. The maximum shear stress

$$\tau_{\max} = \tau_q = \frac{\sigma_r - \sigma_\theta}{2} = \frac{2A}{r^2} - 2B \tag{6.29}$$

is always at 45° to the r–θ orientation since $\tau_{r\theta} = 0$.

Coming back now to the pure bending of a beam of circular arc, assume the cross-section at $\theta = 0$, in Figure 6.5, although allowed to expand or contract in the radial direction, is fixed against rotation and pinned at the centerline. Thus:

$$F = H = 0 \quad \text{since} \quad u_r = v_\theta + \frac{\partial v}{\partial r} = 0 \text{ at } \theta = 0$$

and from Equation (6.24)

$$K = \frac{1}{E}\left[\frac{(1+\nu)A}{R} - 2(1-\nu)BR\ln R + B(1+\nu)R - 2(1-\nu)CR\right] \tag{l}$$

Equations (6.24) and (6.25) can now be evaluated for the displacements since all the constants A, B, C, D, F, H, and K are known.

The rotation and the displacements for a particular beam can be computed from the Winkler and elasticity solutions for comparison. Results for a deep beam (Figure 6.6) are presented in Table 6.2. The Winkler or elementary values of the displacements at the free end, $\theta = \pi/2$, are computed by integration

i) For Elasticity Calculation: $\frac{b}{a} = 3$ $\quad N = +20.55a^4$

if $\nu = \frac{1}{4}$, $K = -\frac{5.59M}{Ea}$ $\quad A = -1.9246\,M$

$B = -0.7786\,M/a^2$

$C = +1.8912\,M/a^2$

ii) For Winkler Calculation:

$$x = -R\cos\theta,\ y = -R(1-\cos\theta) \qquad e = 0.18a$$

$$\Delta\psi = \int \frac{Mds}{AeRE}; \quad u_x = \int \frac{Myds}{AeRE}; \quad v_y = \int \frac{Mxds}{AeRE}$$

TABLE 6.2

Coefficients α and β for u_r or $v_\theta = \alpha \frac{M}{Ea}$ and $\omega_z = \beta \frac{M}{a^2E}$.

	Solution	$r = b$	℄	N.A.	$r = a$
u_r	Winkler			5.56	
	Elastic	5.95	5.59	5.55	5.41
v_θ	Winkler	−8.32	−3.98	−3.17	+0.39
	Elastic	−9.06	−4.18	−3.30	+0.70
ω_z	Winkler			−4.36	
	Elastic			−4.88	

of the slope (rotation) diagram. Thus, at the neutral axis,

$$\Delta\psi = \int_0^{\pi/2} \frac{Mds}{AeRE} = 4.36\frac{M}{a^2E}$$

$$u_x = v_\theta = \int_0^{\pi/2} \frac{Myds}{AeRE} = \int_0^{\pi/2} \frac{-MR^2(1-\sin\theta)d\theta}{AeRE} = 3.17\frac{M}{aE}$$

$$v_y = -u_r = \int_0^{\pi/2} \frac{Mxds}{EI} = \int_0^{\pi/2} \frac{-MR^2\cos\theta d\theta}{AeRE} = -5.56\frac{M}{aE}$$

Comparing the two solutions, we see that again, agreement is excellent. At the neutral axis, u_r is the same. The elasticity values are computed assuming $v = 1/4$ and therefore $K = -5.585\,M/aE$. If, instead $\nu = 0$, then $K = -5.78\,M/aE$ and the agreement in v_θ would essentially be exact. The discrepancy in v_θ at the inner and outer fibers, which is still small, is due primarily to the difference in the rotation that accumulates when integrated over the arc length and is then multiplied by the distance from the neutral axis in computing the longitudinal displacement.

Although the radial displacement at the centerline depends slightly on Poisson's ratio due to the K term, the small radial change in thickness:

$$(u_r)_b - (u_r)_a = \frac{0.54M}{aE}$$

is constant for any Poisson's ratio since the net axial force is zero.

Thus, considering both stresses and displacements, the close agreement of values computed from the simple Winkler analysis indicates that the added refinement from elasticity theory, while very satisfying, is of little practical significance.

6.7 Circular Beams with End Loads

Consider the semicircular beam of rectangular cross-section bent by radial loads P applied at either end as shown in Figure 6.7. The cross-section is rectangular of unit width and assume plane stress. The bending moment at any radial cut is $PR\sin\theta$ and from the Winkler solution we would expect that $\sigma_\theta = f(r)\sin\theta$. Therefore from Section 6.4 we need a stress function of Type II with $m = 1$.

$$F = \left(Ar^3 + \frac{B}{r} + Cr + Dr\ln r\right)\sin\theta \tag{6.30}$$

in which case the stresses are:

$$\begin{aligned}
\sigma_r &= \frac{1}{r}\frac{\partial F}{\partial r} + \frac{1}{r^2}\frac{\partial^2 F}{\partial \theta^2} &&= \left(2Ar - \frac{2B}{r^3} + \frac{D}{r}\right)\sin\theta \\
\sigma_\theta &= \frac{\partial^2 F}{\partial r^2} &&= \left(6Ar + \frac{2B}{r^3} + \frac{D}{r}\right)\sin\theta \\
\tau_{r\theta} &= -\frac{\partial}{\partial r}\left(\frac{1}{r}\frac{\partial F}{\partial \theta}\right) &&= -\left(2Ar - \frac{2B}{r^3} + \frac{D}{r}\right)\cos\theta
\end{aligned} \tag{6.31}$$

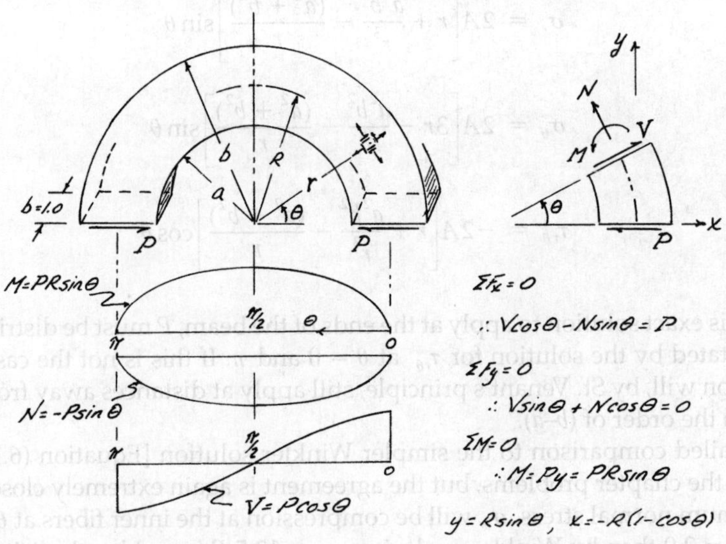

FIGURE 6.7
Circular beam with radial end loads.

The boundary conditions (that at $r = a$ and $r = b$, both the shear and normal stress are zero) give only two independent conditions on the coefficients:

$$2Aa - \frac{2B}{a^3} + \frac{D}{a} = 0 \tag{a}$$

$$2Ab - \frac{2B}{b^3} + \frac{D}{b} = 0 \tag{b}$$

The shear at any cross-section is $P\cos\theta$ so, $\int_a^b \tau_{r\theta} dr = V$, and

$$-A(b^2 - a^2) + \frac{B(b^2 - a^2)}{a^2b^2} - D\ln\frac{b}{a} = P \tag{c}$$

Finally solving (a), (b) and (c)

and $$2A = \frac{P}{a^2 - b^2 + (a^2 + b^2)\ln(b/a)}$$

$$B = -A(a^2b^2)\,; \qquad C = -2A(a^2 + b^2)$$

which gives the final field:

$$\begin{aligned}
\sigma_r &= 2A\left[r + \frac{a^2b^2}{r^3} - \frac{(a^2 + b^2)}{r}\right]\sin\theta \\
\sigma_\theta &= 2A\left[3r - \frac{a^2b^2}{r^3} - \frac{(a^2 + b^2)}{r}\right]\sin\theta \\
\tau_{r\theta} &= -2A\left[r + \frac{a^2b^2}{r^3} - \frac{(a^2 + b^2)}{r}\right]\cos\theta
\end{aligned} \tag{6.32}$$

For this exact solution to apply at the ends of the beam, P must be distributed as dictated by the solution for $\tau_{r\theta}$ at $\theta = 0$ and π. If this is not the case, the solution will, by St. Venant's principle, still apply at distances away from the end in the order of $(b\text{–}a)$.

Detailed comparison to the simpler Winkler solution [Equation (6.19)] is left to the chapter problems, but the agreement is again extremely close. The maximum normal stress, σ_θ, will be compression at the inner fibers at $\theta = \frac{\pi}{2}$. If $b/a = 2.0$ then by Winkler analysis, $\sigma_\theta = -12.5\ P/a$ and by elasticity $\sigma_\theta = -12.9\ P/a$ (a difference of 3%). The maximum $\tau_{r\theta}$ is 1.566 P/h vs. 1.5 P/h for the parabolic distribution (a difference of about 4%).

The elastic rotation, ω_z, satisfying Laplace's equation can be found directly from the Cauchy-Riemann conditions. The first stress invariant is:

$$\sigma_r + \sigma_\theta = 2A\left[4r - \frac{2(a^2 + b^2)}{r}\right]\sin\theta \tag{6.33}$$

Therefore:

$$\frac{1}{E}\frac{\partial(\sigma_r + \sigma_\theta)}{\partial r} = \frac{1}{r}\frac{\partial \omega_z}{\partial \theta} = \frac{2A}{E}\left[4 + 2\frac{(a^2 + b^2)}{r^2}\right]\sin\theta \tag{d}$$

and

$$\frac{1}{E}\frac{\partial(\sigma_r + \sigma_\theta)}{r\partial\theta} = -\frac{\partial \omega_z}{\partial r} = \frac{2A}{E}\left[4 - 2\frac{(a^2 + b^2)}{r^2}\right]\cos\theta \tag{e}$$

giving

$$\omega_z = -\frac{2A}{E}\left[4r + \frac{2(a^2 + b^2)}{r}\right]\cos\theta \tag{6.34}$$

Thus plane sections do not remain plane. For example if $a/b = 3$, then at the centerline $r = R = 2a$, $2A = .3349\, P/a^2$ and

$$\omega_z = -6.02\frac{P}{Ea}\cos\theta$$

In contrast for $a/b = 3$, the eccentricity for the Winkler solution is $0.18a$ and the rotation of the cross-section of the neutral axis is:

$$\Delta\psi = \int \frac{Mds}{AeRE} = \int \frac{PR\sin\theta(Rd\theta)}{R(.18a)RE} = -5.56\frac{P}{Ea}\cos\theta$$

This is roughly the same relative difference as it was for pure bending.

The elastic displacement field can be determined by integrating the strains in the same way as was done for pure bending. Following this procedure:

$$u_r = -\frac{2D}{E}\theta\cos\theta + \frac{\sin\theta}{E}\left[D(1-\nu)\ln r + A(1-3\nu)r^2 + B\frac{(1+\nu)}{r^2}\right] + K\sin\theta + L\cos\theta \tag{6.35}$$

$$v_\theta = \frac{2D}{E}\theta\sin\theta - \frac{\cos\theta}{E}\left[A(5+\nu)r^2 + \frac{B(1+\nu)}{r^2} - D(1-\nu)\ln r\right] + \frac{D(1+\nu)}{E}\cos\theta + K\cos\theta - L\sin\theta + Hr \tag{6.36}$$

where the constants H, K, and L are calculated from the geometric conditions associated with the supports. If we assume the centerline is pinned at $\theta = \pi/2$, then u, v, and $\partial v/\partial r = 0$ there, so:

$$H = 0, \qquad L = \frac{D\pi}{E},$$

and

$$K = -\frac{1}{E}\left[D(1-\nu)\ln R + A(1-3\nu)R^2 + B\frac{(1+\nu)}{R^2}\right] \tag{f}$$

and the displacements can be evaluated everywhere. For example at the free end, $\theta = 0$, from Equation (6.35).

$$u_r = L = \frac{D\pi}{E} = \frac{P\pi(a^2+b^2)}{E\left[(a^2-b^2)+(a^2+b^2)\ln\frac{b}{a}\right]} \tag{g}$$

For a thin beam where $b - a = h$ is small compared to a, then

$$\ln\frac{b}{a} = \ln\left(1+\frac{h}{a}\right) = \frac{h}{a} - \frac{1}{2}\frac{h^2}{a^2} + \frac{1}{3}\frac{h^3}{a^3} \tag{h}$$

and

$$(u_r)_{\theta=0} = -\frac{3\pi a^3 P}{Eh^3} = \int_0^{\pi/2}\frac{Myds}{EI}$$

corresponding to elementary analysis. Further comparison of the deflections of thick rings is left to the chapter problems, but again the results from elasticity and Winkler analysis are in close agreement.

6.8 Concluding Remarks

Other texts such as, for example, Timoshenko's *Theory of Elasticity*, give or refer to many other two-dimensional solutions for other cases of straight and curved beams. These are primarily for distributed loads which can, no matter how complicated, be represented by Fourier series to any accuracy.

However, one purpose in developing a complete field theory such as elasticity is to determine when it is not necessary. Throughout the chapter it is emphasized, perhaps excessively, that even for short deep beams, approximate analysis based on the assumption of plane sections turns out to be perfectly adequate for engineering purposes. Although elasticity includes transverse stresses and gives shear exactly, these refinements are not significant. Thus, for simple or complicated loading, one can proceed with the methods of determinate or indeterminate structural analysis based on shear and bending moment diagrams computing stresses, rotations, and displacements from elementary straight beam or the Winkler formulae.

Moreover, throughout the chapter we have referred to elasticity as more advanced, more sophisticated while strength-of-materials-type analysis is called elementary. A designer would argue the reverse. Elasticity is a complete field theory and mathematically sophisticated, but for beams it is limited. The two-dimensional solutions apply only to beams of constant curvature and rectangular cross-section.

Thus elastic solutions are physically elementary when it comes to beams. Straight-beam or Winkler analysis in comparison is much more sophisticated. Any symmetric cross-section is fine, e.g., I-Beams, T-Beams, circular, or elliptic sections, solid or hollow. They can have any variation in curvature or, for that matter, changing depth and thickness. To designers, so-called elementary analysis is the much more powerful alternative.

Consequently, books on structural mechanics here diverge from elasticity and go on to develop this highly versatile approach in all its complexity. For straight and curved beams, the idea of shear flow and the concept of shear center are developed for nonsymmetric and for thin-walled sections. Buckling and stability and the interaction effects of combined bending and axial forces become all important in design and the "elementary" approach becomes anything but simple. However, rather than follow that route, we will go on to analysis and design problems where elasticity is not only preferable, but often essential.

6.9 Problems and Questions

P6.1 Determine the stress function $F(x,z)$ for the linear geostatic field with horizontal acceleration in plane strain (Section 5.5) giving the stresses:

$$\sigma_z = -\gamma z, \qquad \sigma_x = \frac{\nu}{1-\nu}\sigma_z = -K_o \gamma z, \qquad \tau_{xz} = k_h \gamma z$$

P6.2 What is the stress function $F(x,y)$ for pure bending (Section 5.6)?

P6.3 What are the displacements u, v, w of the prismatic beam in pure bending (Section 5.6) in plane strain compared to those given for plane stress?

P6.4 What is the stress function $F(x,z)$ for the constant stress field due to a uniform normal load, q, and uniform shear, s, on the half-space in plane strain (Section 6.3)?

P6.5 Plot to scale the flow net for $2\,\sigma_q = \sigma_y + \sigma_y$ and its conjugate $E\omega_z$ for the cantilever beam with a uniform load [i.e., Equations (6.7) and (6.9)]. Assume $L = 3h = 6$ *in.* (150 mm). Show at least 6 contours for each function and show the value for each contour so that the conjugate relationship is demonstrated.

P6.6 A cantilever beam of rectangular cross-section, $b = 1.0$, has an end load P distributed parabolically (Figure P6.6). Assume plane stress and a stress function of the form

$$F = b_2xy + c_3y^3 + d_4xy^3$$

Determine the coefficients from the BCs, equilibrium at $x = 0$, and $\nabla^4F = 0$. Compare the elasticity stresses to other elementary formulas.

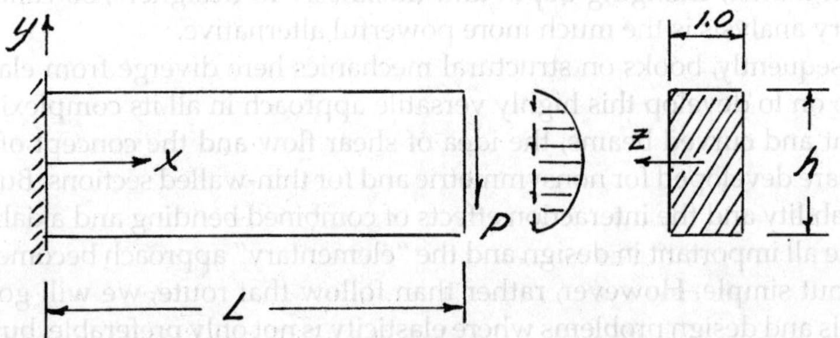

FIGURE P6.6

P6.7 From your solution for stresses in the cantilever beam of P6.6, determine ω_z, u, and v and compare to θ and δ obtained from integrating the moment diagram (elementary analysis). From your result for v, deduce an expression for the shear stiffness of a differential beam element (of length dx) under a unit shear force V (comparable to EI for M and AE for N).

P6.8 From your solution for the cantilever beam of P6.6, plot the flow net for this case to the same specification as required for P6.5. Also plot the isochromatics [contours of equal $(\sigma_1 - \sigma_2)/2$] and the isoclinics (contours for equal orientation of principal stress).

P6.9 Determine what problem in plane stress is solved by the stress function:

$$F = \frac{3Q}{2h}\left(xy - \frac{4xy^3}{3h^2}\right) + \frac{N}{2}y^2$$

P6.10 Determine what problem in plane stress is solved by the stress function:

$$F = -\frac{8Q}{h^3}xy^2\left(\frac{3}{2}h - 2y\right)$$

P6.11 Plot the flow net $\sigma_x + \sigma_y$ vs. $E\omega_z$ for the stress field of P6.10.

P6.12 A simple beam where the support reactions are idealized as shear forces distributed "parabolically" is loaded uniformly as shown in Figure P6.12. Solve for the stress and displacement field by the stress function method. Note: The result can be obtained by superposition of the cantilever beams of Figure P6.6. Compare σ_2 and v at the centerline to their strength-of-materials counterparts.

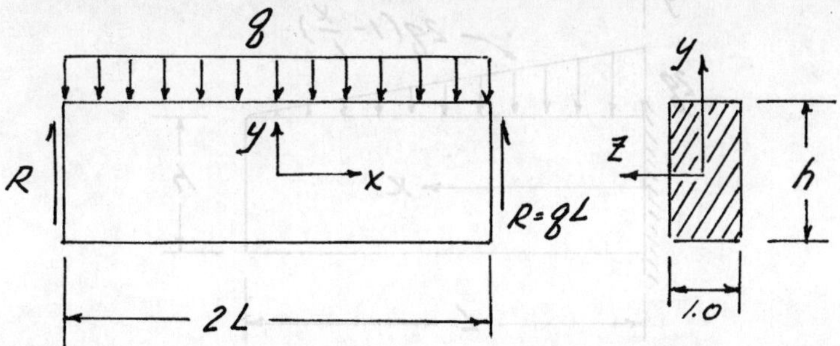

FIGURE P6.12

Note: For P6.13–6.16, assume a rectangular cross-section b = 1.0 *in plane stress and solve for stresses.*

P6.13 Derive an elasticity solution for a straight cantilever beam loaded by a inform shear on its top surface as shown in Figure P6.13. Where is the solution not perfect?

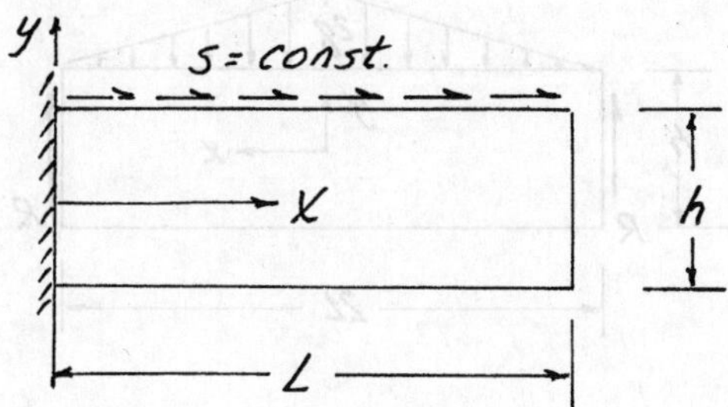

FIGURE P6.13

P6.14 Derive a solution for a straight cantilever beam with a linearly decreasing shear on its top surface as shown in Figure P6.14.

P6.15 Derive a solution for a straight cantilever beam with a linearly decreasing normal load on its top surface as shown in Figure P6.15.

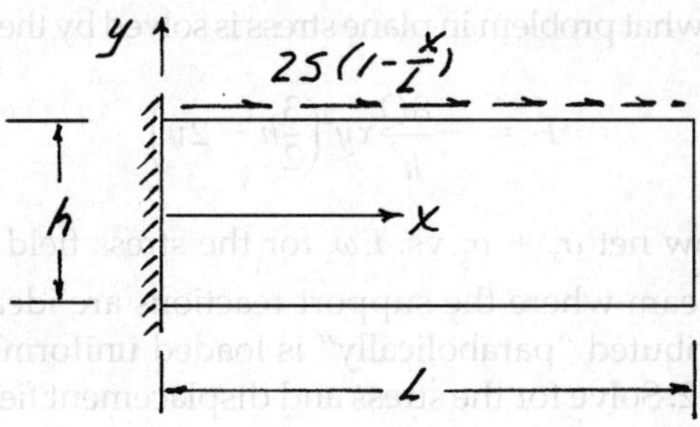

FIGURE P6.14

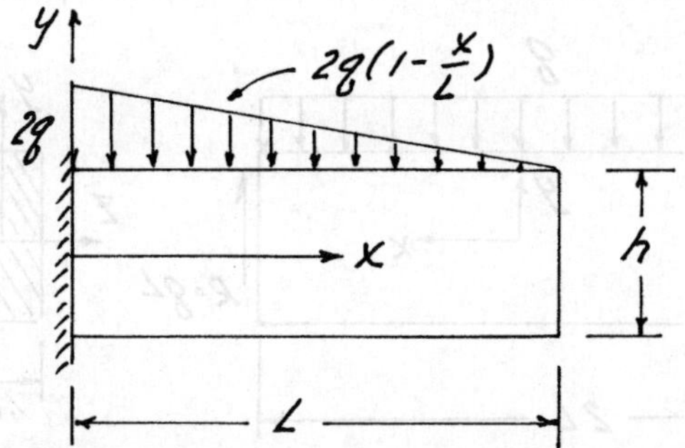

FIGURE P6.15

P6.16 A simple beam where the support reactions are idealized as shear forces distributed parabolically is loaded as shown in Figure P6.16. Derive the elasticity solution. Note: This is P6.6 and P6.15 combined.

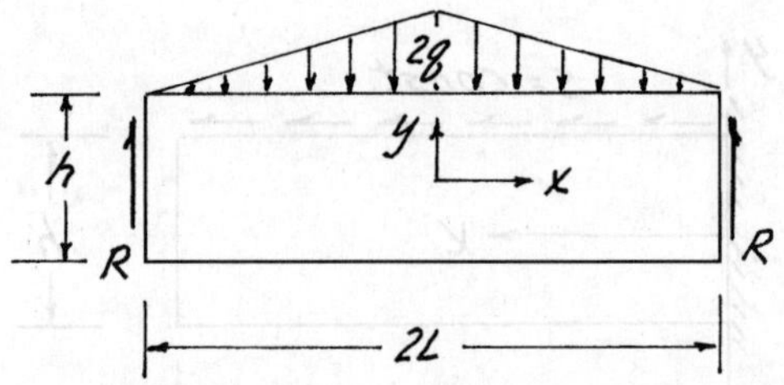

FIGURE P6.16

P6.17 Show that the in-plane compatability equation in polar coordinates is

$$r\frac{\partial \epsilon_r}{\partial r} - \frac{\partial^2 \epsilon_r}{\partial \theta^2} - \frac{\partial}{\partial r}\left[r\left(\frac{\partial \epsilon_\theta}{\partial r} - \frac{\partial \gamma_{r\theta}}{\partial \theta}\right)\right] = 0$$

Combining this equation with the equilibrium equations, derive Equation (6.15).

P6.18 Using the relationships $r^2 = x^2 + y^2$ and $\tan\theta = \frac{x}{y}$ obtain Equations (6.15), (6.16), and (6.17) by transformation of their counterparts in Cartesian coordinates. Finally transform Equation (6.17) to express du as du_r and dv as dv_θ in polar coordinates.

P6.19 Plot the flow net for $\sigma_r + \sigma_\theta$ and $E\omega_z$ for the cantilever beam of Figure 6.6 with $b/a = 3$. Plot the isochromatics ($\tau_q = \tau_{12}$) and the isoclinics (orientation of the principal stresses to the horizontal and vertical). Use the scale $a = 2$ in. (50 mm) and have at least 6 contour lines for each of the four functions.

P6.20 Compare for pure bending of a circular beam, the circumferential stress σ_θ at the inner and outer fibers by (i) straight beam equation; (ii) Winkler analysis, and (iii) elasticity as done in Table 6.1 for

a) $\frac{b}{a} = 1.5$, b) $\frac{b}{a} = 4$, c) $\frac{b}{a} = 5$

Also plot the distribution of the radial stress σ_r for each case.

P6.21 Compare the rotation and deflection by Winkler and elasticity analysis at the free end of a cantilever beam in pure bending such as shown in Figure 6.6 but for

a) $\frac{b}{a} = 2$, b) $\frac{b}{a} = 4$

P6.22 Plot the flow net for the conjugate harmonic functions given by Equations (6.33) and (6.34) for the semicircular beam of Figure 6.7 for $\frac{b}{a} = 3.0$ and $0 \leqslant \theta \leqslant \frac{\pi}{2}$. Use the scale $a = 2$ in. (50 mm) and have at least 6 contour lines for each function.

P6.23 Compare deflections by the elasticity and Winkler solutions at $\theta = 0$ for the semicircular beam of Figure 6.7 for $\frac{b}{a} = 3.0$.

P6.24 Compare the distribution of the shear stress given by the elasticity solution for the semicircular beam of Figure 6.7 to the parabolic distribution $\tau_{r\theta} = \frac{VQ}{Ib}$. Use $\frac{b}{a} = 1.5, 2$ and 3 and plot $\tau h/P$ vs. $h = b - a$ at $\theta = 0$.

P6.25 A thin slice of angle α is cut from a thick ring between two adjacent cross-sections as shown in Figure P6.25. If the ring is now joined and welded, determine the bending moment in the ring. Why do we know there is no resultant axial force or shear and there is, therefore, only pure bending?

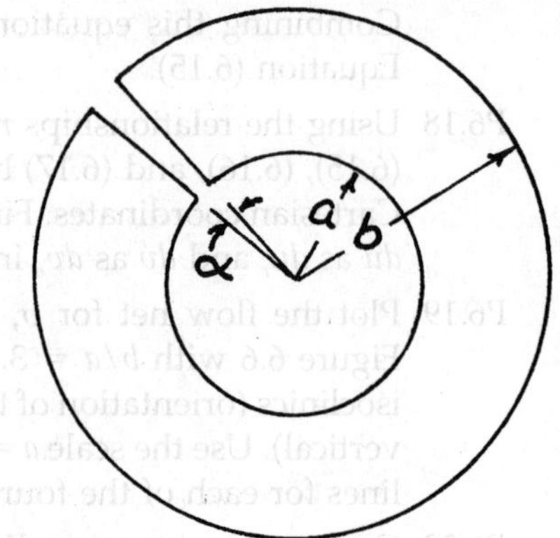

FIGURE P6.25

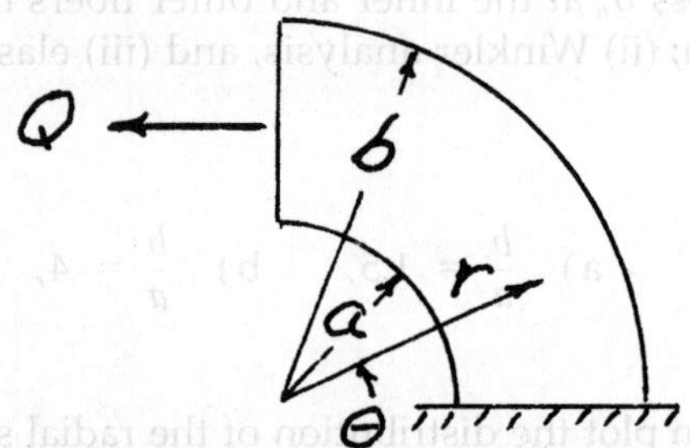

FIGURE P6.26

P6.26 Determine the elastic stress field for a circular cantilever beam loaded with a circumferential normal force Q at $\theta = \frac{\pi}{2}$ as shown in Figure P6.26. Assume a stress function of the form $F = f(r)\cos\theta$. Could this solution have been obtained by superposition? Compare σ_θ at $\theta = 0$ to the Winkler solution for $\frac{b}{a} = 2$ and 3.

7

Rings, Holes, and Inverse Problems

7.1 Lamé's Solution for Rings under Pressure

A pressurized cylindrical ring might have been discussed in Chapter 6 since the field equations can be integrated directly. However, since it and derivative solutions are so elegant, practical, and intriguing because of their optimal properties, this simple case has been reserved as an introduction to this chapter devoted to this general class of structures.

Consider a thick ring as shown in Figure 7.1a of outer radius, b, and inner radius, a, subjected to uniform internal and external pressures p_a and p_b. This might be a transverse section in generalized plane strain cut from a cylinder away from the ends. From symmetry, each radial line (longitudinal section) must simply move radially along its original position the same as any adjacent radial line (Figure 7.1b). Therefore, nothing can be a function of θ and there is no shear strain (or stress) or elastic rotation.

Thus the equilibrium equations reduce to a single requirement:

$$\frac{d\sigma_r}{dr} + \frac{\sigma_r - \sigma_\theta}{r} = 0 \tag{a}$$

and from the conjugate relationship between the first invariant and elastic rotation [Equation (5.13)]

$$\sigma_r + \sigma_\theta = \text{const.} = c_1 \tag{7.1}$$

since

$$\frac{\omega_{r\theta}}{2} = \omega_z = 0 \tag{7.2}$$

Substituting, $\sigma_\theta = c_1 - \sigma_r$ and integrating the equilibrium equation (a):

$$\sigma_r = \frac{c_1}{2} + \frac{c_2}{r^2}; \qquad \sigma_\theta = \frac{c_1}{2} - \frac{c_2}{r_2} \tag{b}$$

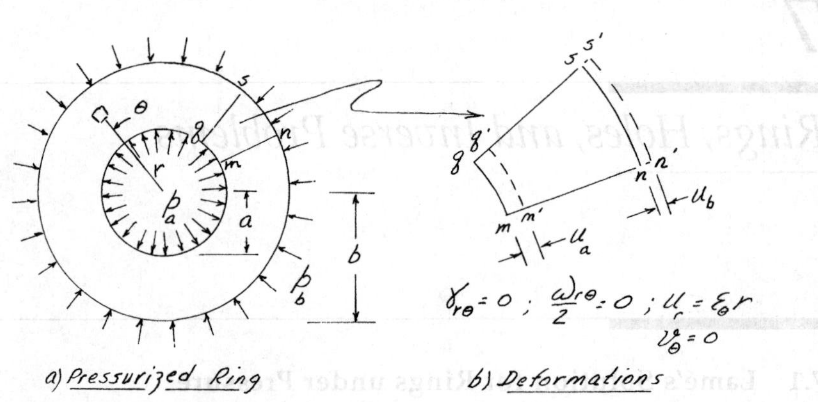

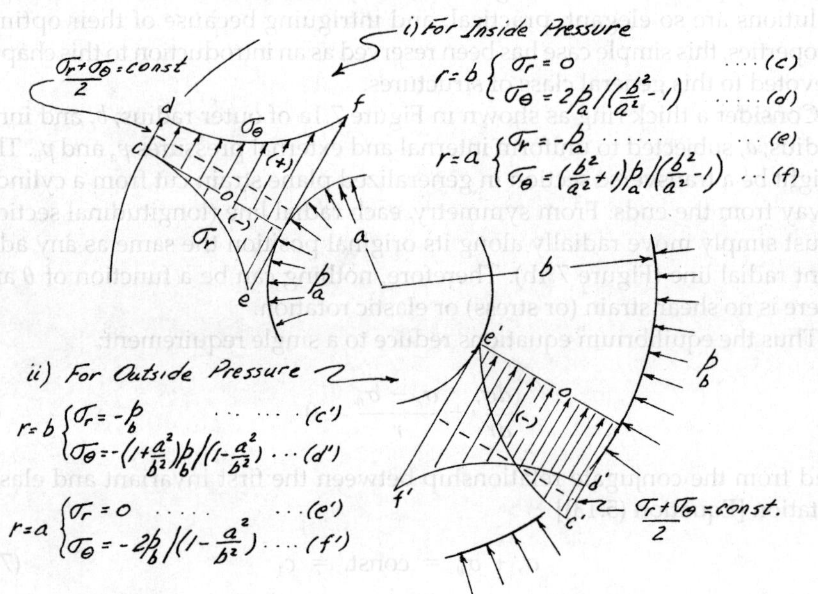

FIGURE 7.1
Lamé's solution.

Finally, by imposing the boundary conditions:

$$
\begin{aligned}
(\sigma_r)_{r=a} &= -p_a = \frac{c_1}{2} + \frac{c_2}{a_2}; \\
(\sigma_r)_{r=b} &= -p_b = \frac{c_1}{2} + \frac{c_2}{b_2}
\end{aligned}
\qquad \text{(c)}
$$

gives:

$$c_1 = \frac{p_a a^2 - p_b b^2}{b^2 - a^2}; \qquad c_2 = \frac{a^2 b^2 (p_b - p_a)}{b^2 - a^2} \tag{d}$$

and therefore the stress field is:

$$\sigma_r = -\frac{\left(\frac{b^2}{r^2} - 1\right)}{\left(\frac{b^2}{a^2} - 1\right)} p_a - \frac{\left(1 - \frac{a^2}{r^2}\right)}{\left(1 - \frac{a^2}{b^2}\right)} p_b$$

$$\sigma_\theta = \frac{\left(\frac{b^2}{r^2} + 1\right)}{\left(\frac{b^2}{a^2} - 1\right)} p_a - \frac{\left(1 + \frac{a^2}{r^2}\right)}{\left(1 - \frac{a^2}{b^2}\right)} p_b \tag{7.3}$$

This stress distribution for internal and external pressure considered separately is plotted in Figure 7.1c.

The solution by the stress function approach is equally straightforward. Since nothing can be a function of θ, the stress function must be of type I.

$$F = A \ln r + B r^2 \ln r + C r^2 + D \tag{a}$$

We have seen, when considering the pure bending of a ring, that the B term contributes to the displacements by introducing [Equation (6.28)] an elastic rotation $\omega_z = 4B\theta/E$, which is not single-valued and therefore not possible for complete rings. Therefore, taking B as zero, the stresses are:

$$\sigma_r = \frac{A}{r^2} + 2C$$

$$\sigma_\theta = -\frac{A}{r^2} + 2C \tag{7.4}$$

$$\tau_{r\theta} = 0$$

From the boundary conditions

$$A = \frac{a^2 b^2 (p_b - p_a)}{b^2 - a^2}; \qquad C = \frac{p_a a^2 - p_b b^2}{2(b^2 - a^2)} \tag{7.5}$$

giving the same solution for the stress field [Equation (7.3)].

Since there are no shears or rotation, all points move radially outward and

$$v_\theta = 0; \qquad u_r = \epsilon_\theta r \tag{7.6}$$

where ϵ_θ is obtained directly from the stress–strain relationships. Moreover, since $(\sigma_\theta + \sigma_r)$ is constant, so is ϵ_z and the assumption of generalized plane-strain for a pressurized cylinder is valid.

Clearly this Lamé solution* leads to many practical applications. For thin rings $b - a = t << r$, the strength-of-materials approximation: $\sigma_r = 0$ (assumed) $\sigma_\theta \cong (p_a - p_b)r/t$ is adequate for design. For thick-walled rings, the advantages of prestressing to apply p_b to reduce the inner tensile hoop stress produced by internal pressure p_a is obvious. A number of examples are given as chapter problems.

Example 7.1

A thick, oval, cylindrical pressure vessel has the cross-sectional dimensions shown below. Assume plane stress. By superpoisition of elasticity solutions, determine the maximum stress and the yield pressure.

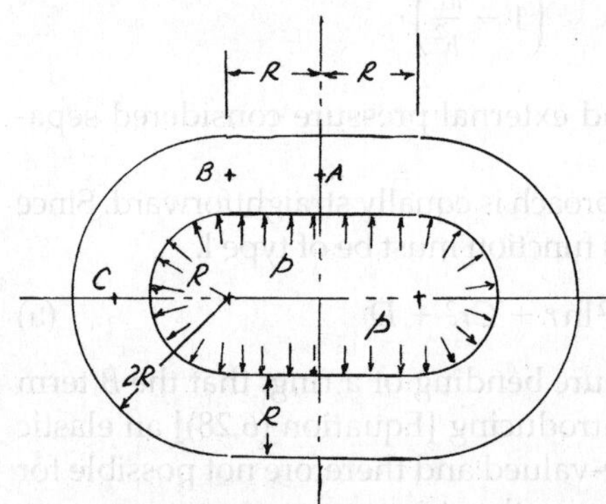

Assume $\sigma_z = 0$
Outside Radius = $2R$
Inside Radius = R
$\therefore \frac{b}{a} = 2$, e = $0.057R$
Since $V_A = V_C = 0$
I degree indeterminate
$\therefore$ Use flex. approach with pin at A.

By Superposition

ring, beam, B, A, M_A, pR, $r_c = 1.5R$, C, p, M_c, $2pR$ = θ_1^{ring}, θ_1, pR, p, 1., M_c', $2pR$ + M_A, θ_2^{ring}, 2. θ_2, M_c^2

$$\text{where:} \quad \theta = \theta^{\text{ ring}} + \theta^{\text{ beam}} \tag{1}$$

* Gabriel Lamé (1795–1870) did much of his early work with Clapeyron when they were on the faculty together at the Institute of Engineers in St. Petersburg. His solution for cylinders subjected to internal and external pressures was published in 1833. Thereafter, explosion of boilers for steam engines was largely due to faulty materials or captains circumventing relief valves while racing or attempting speed records. Lamé in 1852 published the first book on the theory of elasticity, which was notable for using Cauchy's formulation in terms of two elastic constants and for its wealth of examples with a "philosophical discussion" of them in relation to theory.

and θ_1 = Error in geometry introduced at A by allowing rotation
θ_2 = Correction from applying M_A to restore the:

$$\text{Compatibility Condition} \quad \theta_A = \theta_1 + \theta_2 = 0 \tag{2}$$

For 1:

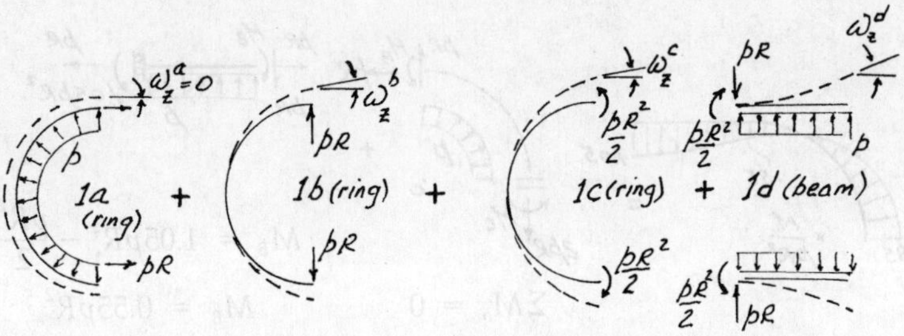
ω_z^a = 0
p
1a (ring)
+
pR
ω_z^b
1b (ring)
+
ω_z^c
pR²/2
1c (ring)
+
1d (beam)
ω_z^d
p

For 1b: From Equation (6.34): $\omega_z^b = -\frac{2A}{E}[4r + 2(a^2 + b^2)]\cos\theta$

where $2A = \dfrac{V}{a^2 - 4a^2 + 5a^2\ln\frac{b}{a}} = V/.456a^2$ at $r = r_C = 1.5a, \cos\theta = 1$

$\therefore \omega_z^b = -27.25V/Ea$ with $V = -pR, \quad a = R$: $\quad \omega_z^b = \dfrac{27.25p}{E}$

For 1c: From Equation (6.28) $\omega_z^c = \dfrac{4B\theta}{E}$

where $B = \dfrac{-2M}{N}(b^2 - a^2)$; $\quad N = (b^2 - a^2)^2 - 4a^2b^2\left(\ln\frac{b}{a}\right)^2 = 1.313a^4$

$\therefore \omega_z^b = -4.57\dfrac{M}{a^2}(4)\left(\dfrac{\pi}{2E}\right)$ $\therefore$ with $M = \dfrac{-pR^2}{2}$, $a = R$ $\quad \omega_z^c = \dfrac{14.36p}{E}$

For 1d: From Equation (6.9) $\omega_z^d = -\dfrac{qx}{2EI}\left[L^2 - Lx + \dfrac{x^2}{3}\right] + \dfrac{qx}{Eh}\left(.9 + \dfrac{6y^2}{h^2}\right)$

with: $q = -p, I = \frac{R^3}{12}, x = L = R, y = 0, h = R \qquad \therefore \omega_z^d = \dfrac{1.1p}{E}$

Combining: $\theta_1 = \omega_z^a + \omega_z^b + \omega_z^c + \omega_z^d = \dfrac{42.7p}{E}$

For 2:

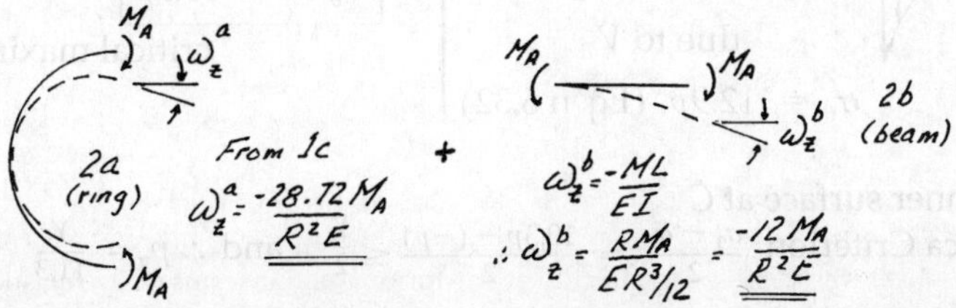
M_A
ω_z^a
2a (ring)
From 1c
ω_z^a = -28.72 M_A / R²E
+
M_A
2b (beam)
ω_z^b
ω_z^b = -ML / EI
∴ ω_z^b = RM_A / (ER³/12) = -12 M_A / R²E

Combining: $\theta_2 = \omega_z^a + \omega_z^b = -40.7\dfrac{M_A}{ER^2}$

$\therefore$ From (2): $\theta_A = \theta_1 + \theta_2 = \dfrac{42.7p}{E} - \dfrac{40.7M_A}{ER^2} = 0$ $\quad \boxed{M_A = 1.05pR^2}$

and the indeterminate analysis is complete

$\therefore$ Mom. Diag. (comp. side)

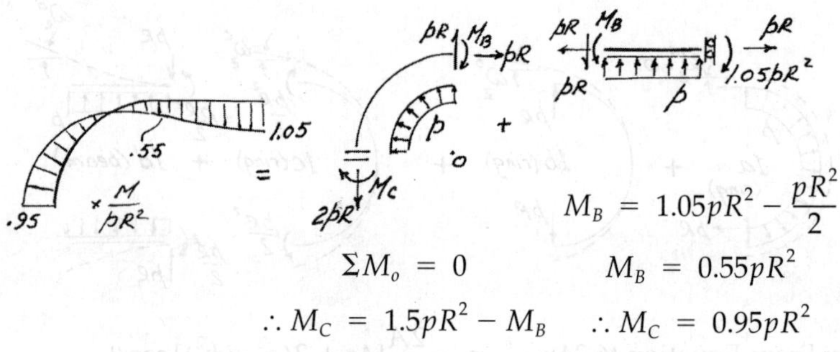

$$M_B = 1.05pR^2 - \frac{pR^2}{2}$$

$$\Sigma M_o = 0 \qquad M_B = 0.55pR^2$$

$$\therefore M_C = 1.5pR^2 - M_B \qquad \therefore M_C = 0.95pR^2$$

Stresses
Clearly either Section A or C is critical

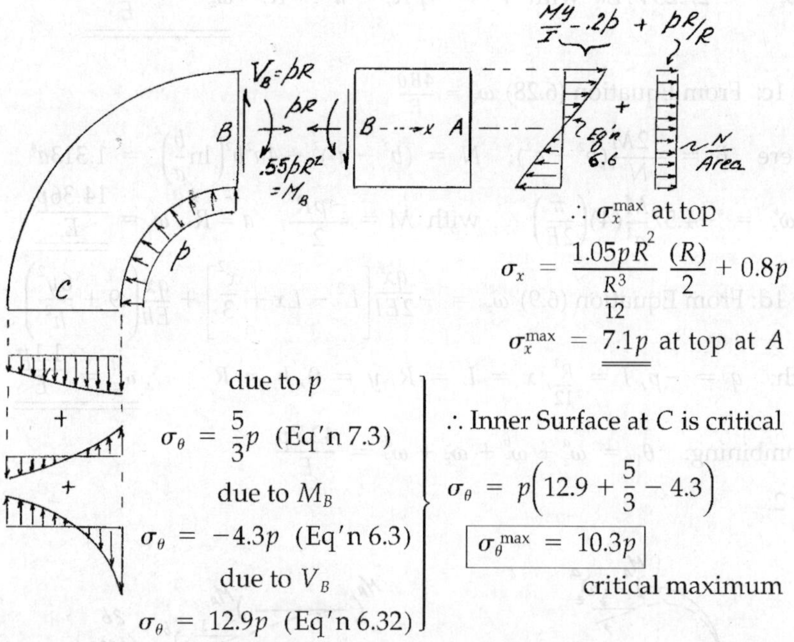

$\therefore \sigma_x^{max}$ at top

$$\sigma_x = \frac{1.05pR^2}{\frac{R^3}{12}}\frac{(R)}{2} + 0.8p$$

$$\sigma_x^{max} = \underline{\underline{7.1p}} \text{ at top at } A$$

due to p

$$\sigma_\theta = \frac{5}{3}p \quad \text{(Eq'n 7.3)}$$

due to M_B

$$\sigma_\theta = -4.3p \quad \text{(Eq'n 6.3)}$$

due to V_B

$$\sigma_\theta = 12.9p \quad \text{(Eq'n 6.32)}$$

$\therefore$ Inner Surface at C is critical

$$\sigma_\theta = p\left(12.9 + \frac{5}{3} - 4.3\right)$$

$$\boxed{\sigma_\theta^{max} = 10.3p}$$

critical maximum

Yield at inner surface at C

By Tresca Criterion $\dfrac{\sigma_1 - \sigma_3}{2} = \dfrac{10.3p - (-p)}{2} \le \dfrac{Y}{2}$ and $\therefore p_y = \dfrac{Y}{11.3}$

$$\text{By von Mises Criterion } \tau_{oct} = \frac{1}{3}\sqrt{(\sigma_1-\sigma_2)^2+(\sigma_2-\sigma_3)^2+(\sigma_3-\sigma_1)^2}$$

$$\tau_{oct} = \frac{1}{3}\sqrt{10.3^2+1^2+11.3^2} = 5.10p \leq \frac{\sqrt{2}}{3}Y$$

and $p_Y = \frac{Y}{10.8}$ ANS.

Note: Slight discrepancy at B where the neutral axis of the beam is at the centroid and of the ring at $-e = -.057R$. However, this section is not critical and resulting error at C is, by St. Venant's principle, negligible.

7.2 Small Circular Holes in Plates, Tunnels, and Inclusions

Small holes are often cast or drilled in plates for everything from conduits to windows. Corresponding situations in plane strain are tunnels or shafts. A related question is the inclusion problem where the hole is filled with a material of different stiffness than the plate. Rather than the normal and shear stresses disappearing on the hole boundary, displacement conditions can be specified. The practical design situation of lined holes falls between the two extremes of an unreinforced hole and a rigid inclusion.

Solutions for circular holes and inclusions are relatively straightforward in polar coordinates if the external boundary is distant enough so that the perturbation from the hole essentially damps out by St. Venant's principle into the free field and the boundaries do not interact. This will be the assumption in this section. For noncircular holes or when the hole and external boundary interact, the problem usually requires either conformal mapping with complex calculus, experimental analysis, or numerical solution. We shall see, however, that, in certain optimum design cases such as harmonic or neutral holes, some or all of these restraints on geometry disappear.

7.2.1 Isotropic Field

Lamé's solution for external pressure can be used directly for a small circular hole in an isotropic free field (i.e., Mohr's Circle is everywhere a point). If, as in Figure 7.2 we assume $b \gg a$, then for uniform tension, σ_m, on the external boundary, the stress function is:

$$F = -\sigma_m a^2 \ln r + \frac{\sigma_m}{2} r^2 \tag{7.7}$$

giving the stress field

$$\sigma_r = \left(1 - \frac{a^2}{r^2}\right)\sigma_m ; \quad \sigma_\theta = \left(1 + \frac{a^2}{r^2}\right)\sigma_m ; \quad \tau_{r\theta} = 0 \tag{7.8}$$

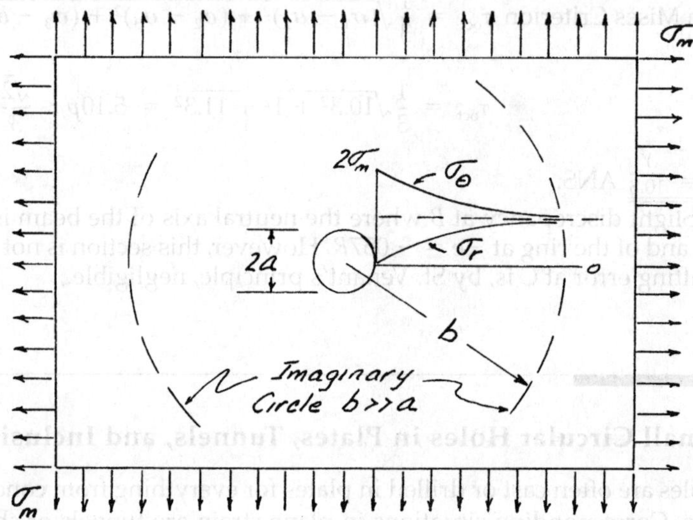

FIGURE 7.2
Small hole in an isotropic field.

Thus the effect of the hole is to introduce purely deviatoric stress and there is no perturbation of the original isotropic field. In other words, everywhere:

$$\frac{\sigma_r + \sigma_\theta}{2} = \sigma_m \quad \text{and} \quad \frac{\sigma_\theta - \sigma_r}{2} = \sigma_m \frac{a^2}{r^2} \tag{a}$$

The hole's effect is entirely captured in the first term of the stress function which is, itself, harmonic. This remarkable result, as plotted in Figure 7.2, leads directly to the design criterion for harmonic holes. That is, a harmonic hole shape is that which does not perturb (anywhere) the isotropic component of the stress field into which it is placed and, therefore, introduces only deviatoric stresses everywhere.

The stresses damp out parabolically as r^2. For example, at $r = 5a$ the difference from the original field (i.e., zero) is 4%. At the hole boundary, $r = a$:

$$\sigma_r = 0; \quad \sigma_\theta = 2\sigma_m \tag{b}$$

giving a stress concentration factor for normal stress

$$(SCF)_\sigma = \frac{\sigma_{\max}}{\sigma_{\text{nominal}}} = 2.0 \tag{c}$$

The solution for a rigid circular inclusion follows directly. At the boundary, $r = a$, the radial displacement, u_r, must be zero. Therefore, the radial contact stress, σ_c, must be such that in plane stress:

$$\epsilon_\theta = 0 = \frac{1}{E}(\sigma_\theta - \nu\sigma_c) \quad \text{or} \quad \sigma_\theta = \nu\sigma_c \tag{d}$$

Since $\sigma_r + \sigma_\theta = 2\sigma_m$ everywhere, at the interface $\nu\sigma_c + \sigma_c = 2\sigma_m$ and therefore at $r = a$

$$\sigma_c = \sigma_r = \frac{2\sigma_m}{1+\nu}; \qquad \sigma_\theta = \frac{2\nu\sigma_m}{1+\nu} \tag{7.9}$$

The deviatoric perturbation, which must damp out as r^2, must then be:

$$\frac{\sigma_r - \sigma_\theta}{2} = \frac{\sigma_m(1-\nu)}{1+\nu}\frac{a^2}{r^2} \tag{7.10}$$

Note that in this case $\sigma_r > \sigma_\theta$ and the stress concentration factor:

$$(SCF)_\sigma = \frac{\sigma_c}{\sigma_m} = \frac{2}{1+\nu} \tag{e}$$

is less than for a hole.

Example 7.2

A thick brass ring ($b = 2a$) with internal pressure $p = 6000$ psi is enclosed by a rigid boundary as shown below. Assume $E = 12 \times 10^6$ psi, $\nu = 1/3$, $Y =$ 12,000 psi.

(a) Determine the stress field and plot.
(b) What is the factor of safety against yield?

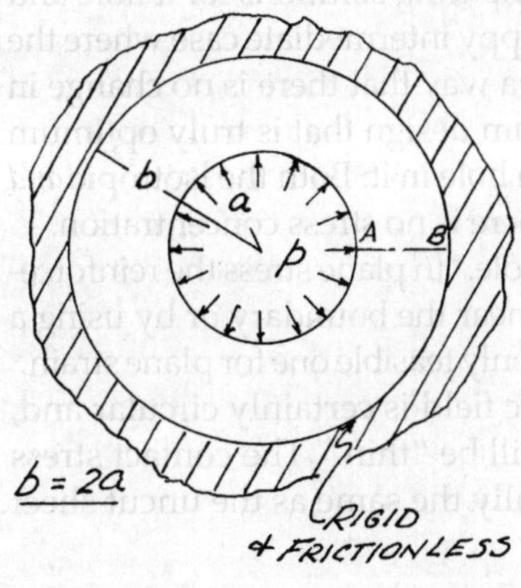

a) Statically Indeterminate
To determine contact pressure at
$r = b = 2a$

$$(u_r)_b = \epsilon_\theta r = \epsilon_\theta b = 0 \tag{1}$$

From Equation (7.3) with $r = b = 2a$

$$\sigma_\theta = \left(\frac{1+1}{4-1}\right)p_a - \left(\frac{1+1/4}{1-1/4}\right)p_b$$

$$\therefore \text{at } r = b \quad \sigma_r = -p_b$$

Assume $\quad (\sigma_z = 0) \quad \sigma_\theta = \frac{2}{3}p_a - \frac{5}{3}p_b$

$$\therefore (\epsilon_\theta)_b = \frac{1}{E}(\sigma_\theta - \nu\sigma_r) = \frac{2}{3}p_a - \frac{4}{3}p_b = 0$$

$$\therefore p_b = (\sigma_r)_b = \frac{1}{2}p_a = -3000 \text{ psi}$$

at $r = a$ [From Equation (7.3)]

$$\sigma_\theta = \frac{4+1}{4-1}p_a - \frac{1+1}{1-1/4}p_b = \frac{5}{3}p_a - \frac{8}{3}p_b = \frac{1}{3}p_a \qquad \therefore (\sigma_\theta)_a = 2000 \text{ psi}$$

$$(\sigma_r)_a = -6000 \text{ psi}$$

b) 1st yield at inner surface
$\sigma_\theta = \sigma_1 = 2000$ psi
$\sigma_z = 0$ (plane stress)
$\sigma_r = \sigma_3 = -6000$ psi

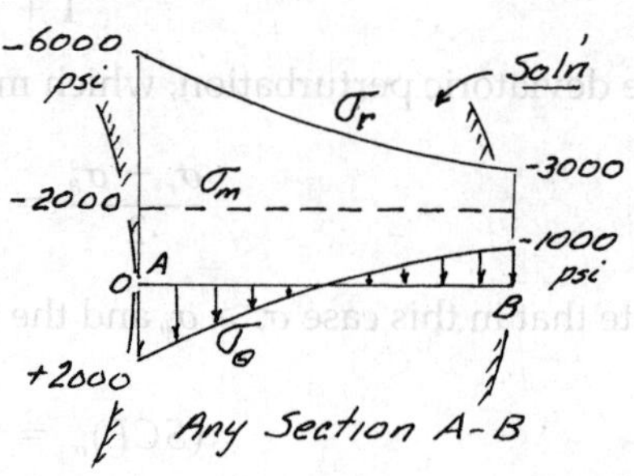

i) Tresca Criterion, $\tau_{max} = 4000$

$$\text{F.S.} = \frac{Y/2}{\tau_{max}} = \frac{6}{4} = \underline{\underline{1.5}}$$

ii) von Mises Criterion

$$\tau_{oct} = \frac{1}{3}\sqrt{(\sigma_1-\sigma_2)^2 + (\sigma_2-\sigma_3)^2 + (\sigma_3-\sigma_1)^2} = \frac{1000}{3}\sqrt{2^2+6^2+8^2}$$

$$\sqrt{104}$$

$$F.S. = \frac{\sqrt{2}}{3}Y/\tau_{oct} = \frac{6}{\sqrt{13}} = 1.66 \sim 11\% \text{ Higher}$$

To get ahead of ourselves, it is clear from comparing solutions for a hole and for a rigid inclusion, that there must be some happy intermediate case where the hole boundary is reinforced (stiffened) in such a way that there is no change in the field stresses whatsoever. This is an optimum design that is truly optimum in that the plate will never know there is even a hole in it. Both the isotropic *and* deviatoric components are unperturbed and there is no stress concentration.

We will call this absolute optimum a neutral hole.* In plane stress the reinforcement can be provided by thickening the sheet near the boundary or by using a liner of stiffer material. The second option is the only feasible one for plane strain.

The shape of the neutral hole in an isotropic field is certainly circular and, if we assume a liner with $E_\ell >> E_s$, then it will be "thin". The contact stress must be σ_m and the interface must move radially the same as the uncut sheet at $r = a$. That is:

$$u_r \text{ at } r = a = \epsilon_\theta a = \frac{a}{E_s}(\sigma_\theta - \nu\sigma_r) = \frac{\sigma_m a}{E_s}(1-\nu_s) \qquad \text{(f)}$$

Therefore the thin liner with a uniform boundary traction, σ_m, must be of constant area such that

* This is the name given them by E.G. Mansfield who first presented the concept in 1953 in a paper with the descriptive title Neutral Holes in Plane Sheet-Reinforced Holes which Are Elastically Equivalent to the Uncut Sheet, *Quart. Jour. of Mechanics and Applied Math.*, Vol. VI, Pt. 3, pp. 370–378. His shapes, as we shall see, are only for membrane liners, but the idea is applicable to the design of any reinforced cutout.

$$\epsilon_\theta a = \frac{\sigma_\theta a}{E_l} = \frac{\sigma_m a^2}{E_l A} \tag{g}$$

Equating (f) and (g)

$$\frac{\sigma_m a^2}{E_l A} = \frac{\sigma_m a}{E_s}(1 - \nu_s) \tag{h}$$

giving the area required (or thickness for a liner of the same thickness as the sheet):

$$A_{\text{req}} = \frac{E_s a}{E_l(1 - \nu_s)} \tag{7.11}$$

The three cases of a hole, a rigid inclusion, and a hole reinformed with a neutral liner are shown together in Figure 7.3.

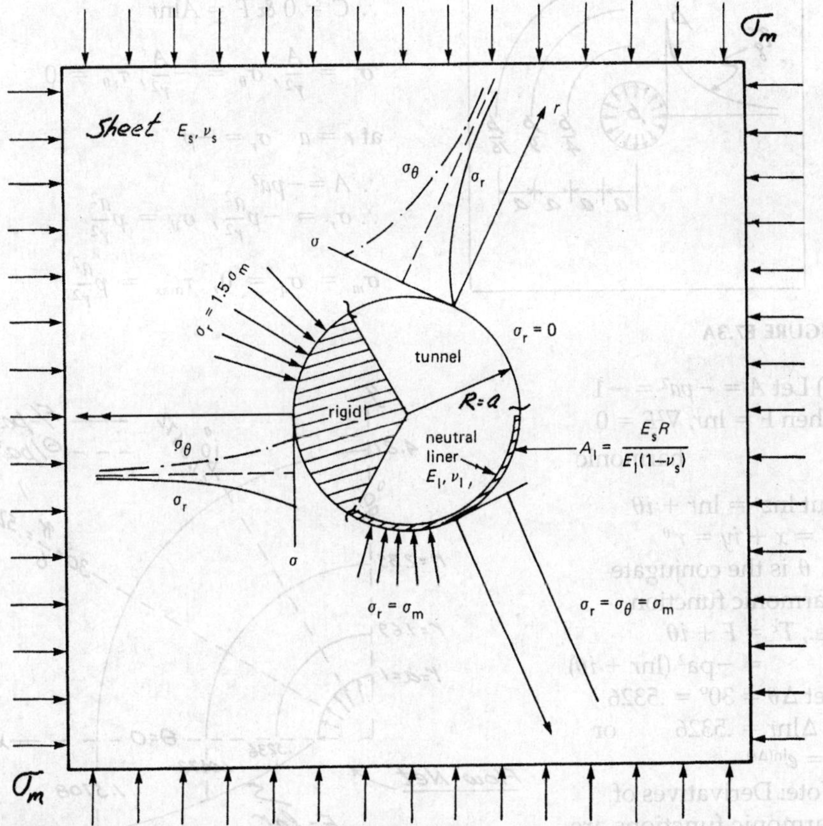

FIGURE 7.3
Neutral liner for an isotropic field.

Example 7.3

Derive the solution (stress field) for uniform pressure, p, applied to a small hole (radius $= a$) in a large plate.

a. Plot contours of σ_q, τ_q.
b. Is the stress function harmonic? If so determine its conjugate and plot the flow net.
c. Determine the displacement field in Cartesian coordinates and plot contours of u and v. Discuss.
d. Derive at least one more set of conjugate harmonic functions and plot. (Hint: Consider the equilibrium equations in Cartesian coordinates.)

a) For Lamé's solution [Equation(8.4)]

$F = A\ln r + Cr^2$

but as $r \to \infty$ $\quad \sigma_r = \sigma_\theta = 0$

$\therefore C = 0$ & $F = A\ln r$

$$\sigma_r = \frac{A}{r^2},\ \sigma_\theta = -\frac{A}{r^2},\ \tau_{r\theta} = 0$$

at $r = a \quad \sigma_r = -p$

$\therefore A = -pa^2$

$$\therefore \sigma_r = -p\frac{a^2}{r^2},\ \sigma_\theta = p\frac{a^2}{r^2}$$

$$\sigma_m = \sigma_q = 0,\ \tau_{\max} = p\frac{a^2}{r^2}$$

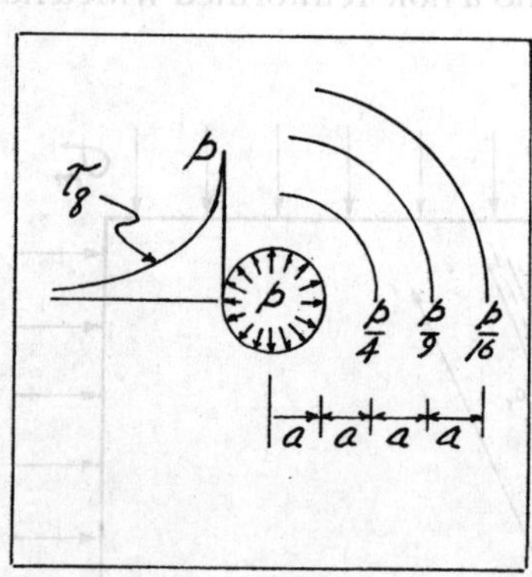

FIGURE E7.3A

b) Let $A = -pa^2 = -1$
Then $F = \ln r$, $\nabla^2 F = 0$
harmonic

but $\ln z^c = \ln r + i\theta$
$z^c = x + iy = r^{i\theta}$
$\therefore \theta$ is the conjugate harmonic function
i.e., $T^c = F + i\theta$
$= -pa^2 (\ln r + i\theta)$
Let $\Delta\theta = 30° = .5326$
$\therefore \Delta \ln r = .5326$ or
$r = e^{\ln(\Delta r)}$
Note: Derivatives of harmonic functions are harmonic.

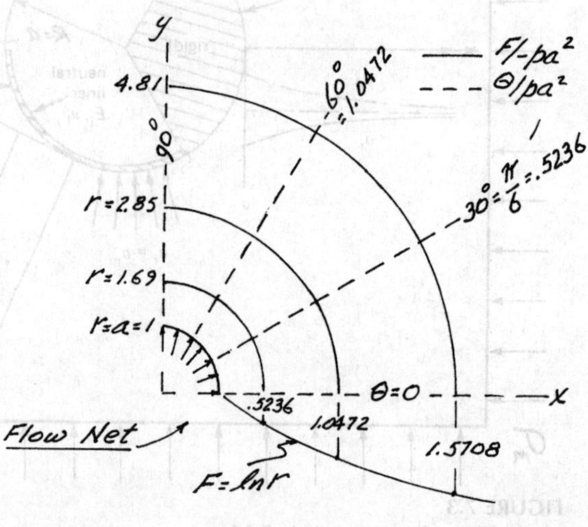

FIGURE E7.3B

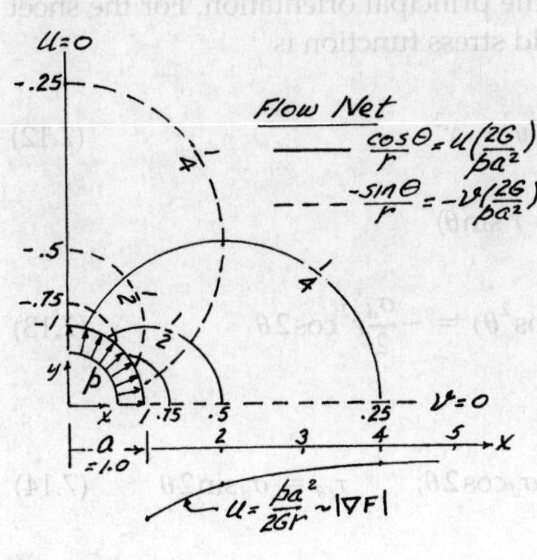

FIGURE E7.3C

c) 1st Derivative

$$\frac{dT^c}{dz} = \frac{-d}{dz}pa^2(\ln r + i\theta)$$

$$= -pa^2\frac{d}{dz}(\ln z^c)$$

$$= -\frac{pa^2}{z^c} = -pa^2r^{-1}e^{i(-\theta)}$$

$$= -pa^2\left(\frac{\cos\theta}{r} + i\frac{-\sin\theta}{r}\right)$$

conj. functions

but $u_r = \epsilon_\theta r = \frac{r}{E}(\sigma_\theta - \nu\sigma_r) = \frac{pa^2}{2Gr}$

and

and $u = u_r\cos\theta$

$v = u_r\sin\theta$

$$\therefore \frac{dT^c}{dz} = 2G(u - iv)$$

and u & $-v$ are the conj. harmonic functions which plot as circles dying off as $1/r$

d) 2nd Derivative

$$\frac{d^2}{dz^2} = pa^2r^{-2}e^{i(2\theta)} =$$

$$pa^2\left[\frac{\cos 2\theta}{r^2} + \frac{i(-\sin 2\theta)}{r^2}\right]$$

conj. functions

Mohr's Circle!

$$\sigma_y = -\sigma_x$$

$$= \frac{pa^2}{r^2}\cos(2\theta)$$

$$\tau_{xy} = \frac{-pa^2}{r^2}\sin 2\theta$$

Note: since $\sigma_y = -\sigma_x$ the equilibrium equations [Equation (2.45)] become the conj. Cauchy-Riemann conditions

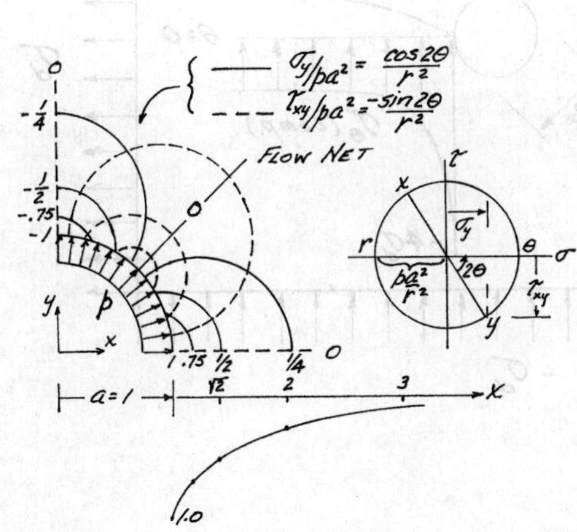

$$\sigma_y = -\sigma_x = \tau_{max} = \frac{pa^2}{r^2} = \nabla(|\nabla F|) = \frac{d^2F}{dn^2}$$

FIGURE E7.3D

7.2.2 Deviatoric Field

A purely deviatoric field, where there is no isotropic component, can be thought of as pure shear at 45° from the principal orientation. For the sheet without a hole, the deviatoric, free-field stress function is

$$F = \frac{\sigma_d}{2}(y^2 - x^2) \tag{7.12}$$

or in polar coordinates ($y = r\sin\theta$, $x = r\sin\theta$)

$$F = \frac{\sigma_d r^2}{2}(\sin^2\theta - \cos^2\theta) = -\frac{\sigma_d}{2}r^2\cos 2\theta \tag{7.13}$$

giving the free-field stress distribution

$$\sigma_\theta = -\sigma_d\cos 2\theta; \qquad \sigma_r = \sigma_d\cos 2\theta; \qquad \tau_{r\theta} = \sigma_d\sin 2\theta \tag{7.14}$$

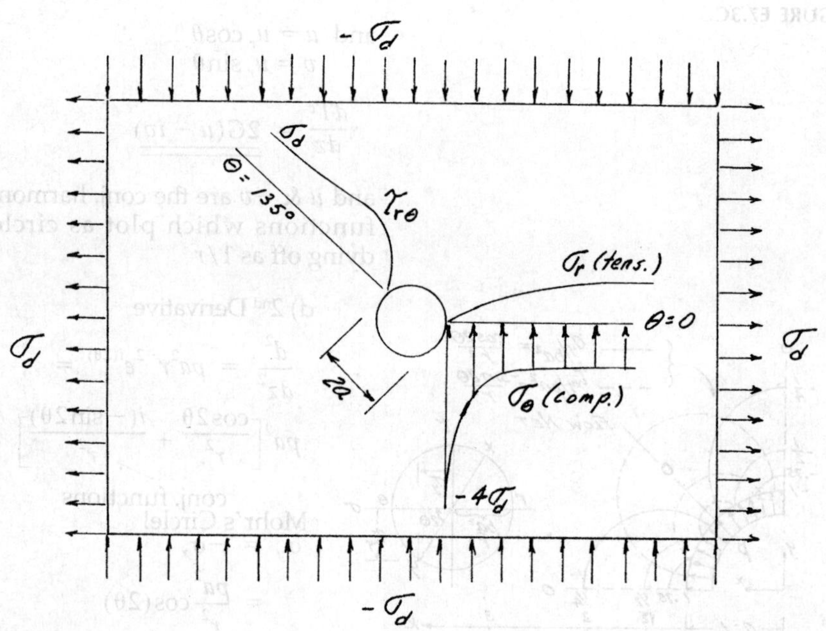

FIGURE 7.4
Small hole in a deviatoric field.

If we now drill a small circular hole as in Figure 7.4, the perturbation must rapidly damp to this free field and thus a stress function of type II is required with $m = 2$. However, the first term in the general form:

$$F = (A^{2+m} + B^{m} + C^{2-m} + D^{-m})\cos 2\theta$$

can be eliminated since stresses would increase as r^2. Therefore, assume

$$F = (Br^2 + C + Dr^{-2})\cos 2\theta \tag{7.15}$$

which gives

(1)

$$\begin{aligned}
\sigma_r &= -\left(2B + \frac{4C}{r^2} + \frac{6D}{r^4}\right)\cos 2\theta \\
\sigma_\theta &= +\left(2B + \frac{6D}{r^4}\right)\cos 2\theta \\
\tau_{r\theta} &= \left(2B - \frac{2C}{r^2} - \frac{6D}{r^4}\right)\sin 2\theta
\end{aligned} \tag{7.16}$$

We know that at $r \gg a$ to capture the free field

(2)

$$B = -\frac{\sigma_d}{2} \tag{a}$$

and at the hole boundary $r = a$, $\sigma_t = \tau_{r\theta} = 0$ for any θ. Therefore:

$$\begin{aligned}
2B + \frac{4C}{a^2} + \frac{6D}{a^4} &= 0 \\
2B - \frac{2C}{a^2} - \frac{6D}{a^4} &= 0
\end{aligned} \tag{b}$$

which gives: $$C = \sigma_d a^2, \qquad D = -\frac{\sigma_d}{2}a^4 \tag{c}$$

so that

$$\sigma_r = \sigma_d\left(1 - \frac{4a^2}{r^2} + \frac{3a^4}{r^4}\right)\cos 2\theta$$

$$\sigma_\theta = -\sigma_d\left(1 + \frac{3a^4}{r^4}\right)\cos 2\theta \tag{7.17}$$

$$\tau_{r\theta} = -\sigma_d\left(1 + \frac{2a^2}{r^2} - \frac{3a^4}{r^4}\right)\sin 2\theta$$

in this case the harmonic isotropic component of the stress field

$$\nabla^2 F = \sigma_r + \sigma_\theta = \frac{\partial^2 F}{\partial r^2} + \frac{1}{r}\frac{\partial F}{\partial r} + \frac{1}{r^2}\frac{\partial^2 F}{\partial \theta} = \frac{-4\sigma_d a^2}{r^2}\cos 2\theta \tag{d}$$

damping out parabolically to the free-field value of zero as the deviatoric component damps out to σ_d. Thus on the hole boundary where $\sigma_r = \tau_{r\theta} = 0$

$$\sigma_r + \sigma_\theta = \sigma_\theta = -4\sigma_d \cos 2\theta \tag{e}$$

which, compared to the free field stress, gives a stress concentration factor

$$SCF = \frac{(\sigma_\theta)}{(\sigma_\theta)_{f.f.}} = \frac{-4\sigma_d \cos 2\theta}{-\sigma_d \cos 2\theta} = 4.0 \tag{f}$$

Although the stress concentration is twice that for the isotropic case it is, as shown in Figure 7.4, much more localized.

7.2.3 General Biaxial Field

By superimposing isotropic and deviatoric components in differing ratios, the stress distribution introduced by a small hole in any biaxial field can be obtained. The particular case of uniaxial tension, w, is shown in Figure 7.5. Setting $\sigma_m = \sigma_d = w/2$ the Kirsch solution for the stress field is

$$\sigma_r = \frac{w}{2}\left(1 - \frac{a^2}{r^2}\right) + \frac{w}{2}\left(1 + \frac{3a^4}{r^4} - \frac{4a^2}{r^2}\right)\cos 2\theta$$

$$\sigma_\theta = \frac{w}{2}\left(1 + \frac{a^2}{r^2}\right) - \frac{w}{2}\left(1 + \frac{3a^4}{r^4}\right)\cos 2\theta \tag{7.18}$$

$$\tau_{r\theta} = -\frac{w}{2}\left(1 - \frac{3a^4}{r^4} + \frac{2a^2}{r^2}\right)\sin 2\theta$$

It is evident that the hoop stress, σ_θ, is greatest at $\theta = \frac{\pi}{2}$ and $\frac{3}{2}\pi$ when $\cos 2\theta = -1$, in which case

$$\tau_{r\theta} = 0, \qquad \sigma_\theta = \frac{w}{2}\left(2 + \frac{a^2}{r^2} + \frac{3a^4}{r^4}\right) \tag{g}$$

again dying away rapidly from the hole boundary, although not quite as quickly as for the deviatoric free field.

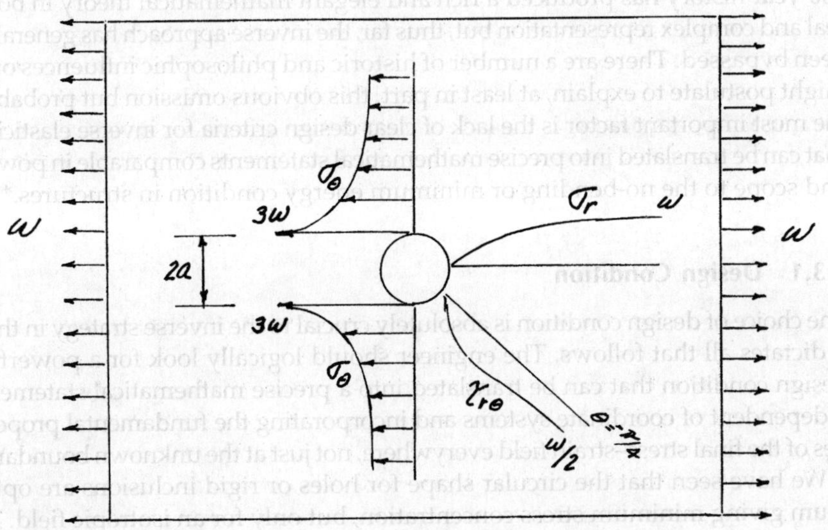

FIGURE 7.5
Small hole in a uniaxial field.

7.3 Harmonic Holes and the Inverse Problem

Large stress concentrations at the edges of holes produce serious structural problems. Overstressed areas, no matter how local, often lead to fatigue cracks greatly reducing the life of a structure or even triggering collapse if they are not detected or if they propagate rapidly.

However, two design modifications may be beneficial. The shape of the hole might be changed and/or the hole boundary reinforced by thickening the plate locally or providing a liner of stiffer material. Such an inverse approach, where the geometry of the boundary is left as an unknown, might be called "shape mechanics." It involves a fundamental reorientation of solid mechanics to thinking about design conditions and optimum shapes to control fields rather than accept unsatisfactory fields dictated by prescribed geometry.

The search for optimum structural shapes has traditionally been conducted by engineers using a strategy distinctly different from that of standard mathematical analysis where the governing equations are solved for a given set of boundary conditions on a body of specified geometry. Instead, the engineer leaves at least some geometric parameters unspecified but imposes extra conditions on the stress–strain field sufficient to then derive a configuration "optimized" to that preselected design criterion.

Linear and shell structures provide a classic example of this alternative design strategy in which the search for optimum shapes to eliminate bending and the resulting effects on building forms is a major historical theme. In startling contrast, there is essentially no counterpart to the design tradition in elasticity. A 200-year history has produced a rich and elegant mathematical theory in both real and complex representation but, thus far, the inverse approach has generally been bypassed. There are a number of historic and philosophic influences one might postulate to explain, at least in part, this obvious omission but probably the most important factor is the lack of clear design criteria for inverse elasticity that can be translated into precise mathematical statements comparable in power and scope to the no-bending or minimum energy condition in structures.*

7.3.1 Design Condition

The choice of design condition is absolutely crucial to the inverse strategy in that it dictates all that follows. The engineer should logically look for a powerful design condition that can be translated into a precise mathematical statement independent of coordinate systems and incorporating the fundamental properties of the final stress–strain field everywhere, not just at the unknown boundary.

We have seen that the circular shape for holes or rigid inclusions are optimum giving minimum stress concentration, but only for an isotropic field. In this case, not only are the boundary stresses constant and their sum equal to

* Hooke's experimental solution in 1671 to "the riddle of the arch" is simplicity itself: "As hangs a (loaded) flexible cable, so inverted, stand the touching pieces of a (similarly loaded) arch." This concept is now used for example to design shell roofs by loading a draped-cloth model, spraying it with epoxy, and inverting the resulting shape. Analytic analysis to find optimum shapes based on Hooke's insight has been "a", if not "the", central theme of structural design for 300 years. Current research, as in the past, leads to creative insights of surprising simplicity, elegance, and beauty. We shall see that, as equations degenerate, the real structure becomes stronger until, for optimum shapes, the mathematically trivial becomes physically profound.

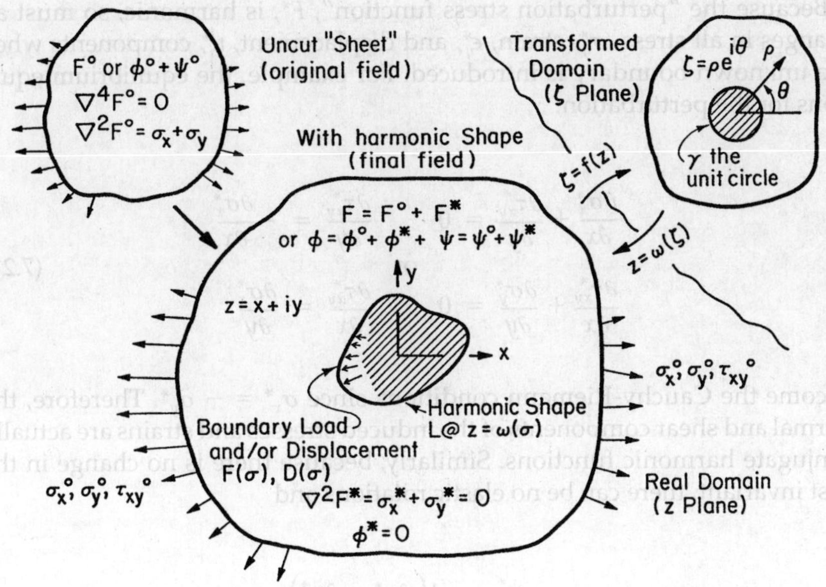

FIGURE 7.6
Inverse strategy for harmonic shapes.

twice the isotropic stress (i.e., the value of the first invariant) but, quite remarkably, this is also true everywhere in the resulting field as the perturbations of the individual stress components damp out. With the circular shape as a model, let us therefore choose for the design criterion the "harmonic field condition" that, when the hole or inclusion is introduced, the first invariant of the stress (or strain) remain unperturbed everywhere in the field. The shape of the new boundary that achieves this is, in turn, called a "harmonic shape."

This name is used for obvious reasons. If, in any portion of an uncut two-dimensional domain in plane stress or plane strain defined by a known "original" stress function F^o, a new boundary is introduced, such as the hole in Figure 7.6, then the final field, F, is given by

$$F = F^o + F^* \tag{7.19}$$

in which
$$\nabla^2 F^* = \sigma_{ii}^* + \sigma_{jj}^* = 0 \tag{7.20}$$

since by the design condition

$$\nabla^2 F = \sigma_{ii} + \sigma_{jj} = \nabla^2 F^o = \sigma_{ii}^o + \sigma_{jj}^o$$

The superscript ^{o} indicates an original quantity before a harmonic shape is introduced, and the superscript * indicates the change or perturbation of this quantity due to the new harmonic boundary. The final state, being always the sum of these two components, is given no superscript.

Because the "perturbation stress function", F^*, is harmonic, so must all changes in all stress, σ_{ij}^*, strain, ϵ_{ij}^*, and displacement, u_i^*, components when the unknown boundary is introduced. For example, the equilibrium equations for the perturbation:

$$\begin{aligned} \frac{\partial \sigma_x^*}{\partial x} + \frac{\partial \tau_{xy}^*}{\partial y} = 0 \qquad & \frac{\partial \tau_{xy}^*}{\partial y} = -\frac{\partial \sigma_x^*}{\partial x} \\ \text{or} \qquad & \\ \frac{\partial \tau_{xy}^*}{\partial x} + \frac{\partial \sigma_y^*}{\partial y} = 0 \qquad & \frac{\partial \tau_{xy}^*}{\partial x} = \frac{\partial \sigma_x^*}{\partial y} \end{aligned} \tag{7.21}$$

become the Cauchy-Riemann conditions since $\sigma_x^* = -\sigma_y^*$. Therefore, the normal and shear components of the induced stresses and strains are actually conjugate harmonic functions. Similarly, because there is no change in the first invariant, there can be no elastic rotations and

$$\omega_z^* = \frac{\omega_{xy}^*}{2} = \frac{1}{2}\left(\frac{\partial v^*}{\partial x} - \frac{\partial u^*}{\partial y}\right) = 0 \tag{7.22}$$

Since $\epsilon_x^* = -\epsilon_y^*$, u^* and $-v^*$ must be conjugate harmonic function as well.

Thus, the analogy between the harmonic design condition and the membrane shapes for arches and shells is complete. For membrane structures the "no bending" condition is actually, of course, a requirement that there be no elastic rotations of any cross-section and the harmonic condition gives a full two-dimensional counterpart imposing "no bending" anywhere in the field when the harmonic shape is introduced.* Thus, as engineers, we might expect, with the benefit of historical perspective, optimum behavior from designs fulfilling the harmonic condition.

Harmonic shapes can be derived in a number of ways. It is, after all, only the perturbation, F^*,which is sought and, by imposing the harmonic condi-

* The properties of harmonic functions discussed briefly in Chapter 5 lead to many significant conclusions not requiring much in the way of formal proof. For example, G.A. Cherepanov in his paper "Inverse Problems in the Plane Theory of Elasticity" (*Prikladraya Mathematika Mechanika*, 38, 6, 1974) uses the design condition that "the tangential stress at the hole boundary be constant" in deriving his "holes of constant strength." However, if the tangential stress on a uniformly loaded or unloaded boundary is constant, so is the first invariant, which is harmonic. But harmonic functions that are constant on a boundary must be constant throughout the field. Immediately then, the theorem that the maximum and minimum value of harmonic functions must occur on the boundary offers direct proof that, within a constant field, a hole (or several holes for that matter) of equal or uniform strength does, in fact, produce minimum stress concentrations. The converse of the first observation is also clearly true in that for the case of a field that is nonconstant, a hole of uniform strength cannot exist except perhaps with very special internal boundary loading. Thus the hole of constant strength is simply a subset of harmonic holes only possible for constant fields.

tion, the biharmonic field has been reduced to the simplest elliptic form. It can also be shown that both F^* and its normal derivative can be completely specified on the unknown boundary in terms of F°. Essentially then, the problem becomes directly analogous to the classic question of determining the shape of a free-surface in Laplacian fields and all the experimental, numerical, and mathematical techniques employed in the various disciplines where this problem arises might conceivably be used.*

Since we seek the geometry of an unknown boundary, a complex variable formulation such as presented by Muschelishvili** offers a particularly powerful analytic approach. As shown in Figure 7.6 , we can describe an opening of essentially any shape as a function of the complex variable $z = x + iy$, and it becomes convenient to map from a region containing a circular opening using the inverse transformation $z = \omega(\zeta)$. Stresses, strains, and displacements for the plane problem without body forces can be expressed in terms of two complex functions $\phi(\zeta)$ and $\psi(\zeta)$ which, in the standard problems of elasticity, are determined from the boundary condition:

$$\phi(\sigma) + \frac{\omega(\sigma)\overline{\phi'(\sigma)}}{\overline{\omega'(\sigma)}} + \overline{\psi(\sigma)} = F(\sigma) \qquad \text{(a)}$$

or

$$\kappa\phi(\sigma) - \omega(\sigma)\frac{\overline{\phi'(\sigma)}}{\overline{\omega'(\sigma)}} - \overline{\psi(\sigma)} = 2\mu(g_1 + ig_2) = D(\sigma) \qquad \text{(b)}$$

in which σ (without a subscript) equals the value of ζ on the boundary of the unit circle γ. The first of these equations is for the first fundamental problem of elasticity in which $F(\sigma)$ represents any self-equilibrating boundary loading, while the second is for the second fundamental problem in which g_1 and g_2 are

* For example, N.J. Hardiman in his paper "Elliptic Elastic Inclusion in an Infinite Elastic Plate", *Quart. Journal of Mech. and Applied Math.*, VII, 2, 1954, 226–230, notes that:

> constant stresses at infinity in an elastic plate containing an elliptic inclusion of different material induce constant stresses in the elliptic inclusion, a result similar to the known fact that a uniform two-dimensional electric field will induce in an elliptic dielectric cylinder placed in it a uniform field in the dielectric. It is interesting that similar results should hold in the electric and elastic problem, since the first is essentially a harmonic problem, whilst the second involves biharmonic functions.

Not realizing that by fortuitously choosing to analyze an optimum elliptic shape the harmonic condition was made operable, Hardiman was unable to explain this correlation or extend it, for example, to nonconstant fields.

** Many two-dimensional problems in elasticity can be solved with complex variables. The method was developed in Russia by G.V. Kolosoff and his pupil N.I. Muschelishvili beginning in 1914. Muschelishvili presents the method in his book *Some Basic Problems in the Mathematical Theory of Elasticity*, 2nd Ed., P. Noordhoff Ltd., (1953). G.N. Savin's book *Stress Distribution around Holes*, Naukova Domka Press, Kiev, 1968 translated by NASA is 997 pages of solutions for holes with complex variables. Like his predecessors, Savin never refers to the question of inverting the analysis procedure to the design perspective of the inverse problem.

the known boundary values of any displacement components u and v imposed on γ. The material properties κ and 2μ may also be written:

$$\text{for plane stress:} \qquad 2\mu = 2G = \frac{E}{(1+\nu)}; \qquad \kappa = \frac{3-\nu}{1+\nu} \tag{c}$$

$$\text{for plane strain:} \qquad 2\mu = 2G = \frac{E}{(1+\nu)}; \qquad \kappa = 3 - 4\nu \tag{d}$$

in terms of Young's modulus, E, and Poisson's ratio, ν.

For standard analysis where geometry is known, the problem becomes the determination of the complex potentials ϕ and ψ. For the inverse problem on the other hand, the desired shape will be given directly by the unknown mapping function $\omega(\sigma)$, which can be found from the applicable boundary condition only if the unknown perturbation components ϕ^* and ψ^* can be eliminated. The harmonic condition allows us to do this since it requires that:

$$\phi^*(\zeta) = 0 \tag{e}$$

and that the Cauchy-type singular integral

$$\int_\gamma \frac{\psi^*(\sigma)d\sigma}{\sigma - \zeta} = 0 \tag{f}$$

Thus, multiplying each term in Equations (a) and (b) by $d\sigma/(\sigma - \zeta)$ and integrating over the boundary, the basic equations for the form of the harmonic shape (opening or inclusion) are found to be:

$$\int_\gamma \frac{\phi^o(\sigma)d\sigma}{\sigma - \zeta} + \int_\gamma \frac{\omega(\sigma)\overline{\phi^{o'}(\sigma)}d\sigma}{\overline{\omega'(\sigma)}(\sigma - \zeta)} + \int_\gamma \frac{\overline{\psi^o(\sigma)}d\sigma}{\sigma - \zeta} = \int_\gamma \frac{F(\sigma)d\sigma}{\sigma - \zeta} \tag{7.23a}$$

or

$$\int_\gamma \frac{\kappa\phi^o(\sigma)d\sigma}{\sigma - \zeta} - \int_\gamma \frac{\omega(\sigma)\overline{\phi^{o'}(\sigma)}d\sigma}{\overline{\omega'(\sigma)}(\sigma - \zeta)} - \int_\gamma \frac{\overline{\psi^o(\sigma)}d\sigma}{\sigma - \zeta} = \int_\gamma \frac{D(\sigma)d\sigma}{\sigma - \zeta} \tag{7.23b}$$

7.4 Harmonic Holes for Free Fields

Solutions for harmonic holes from the basic equations presented in the previous section are possible, in theory, for any elasticity field where the "opening" (unloaded, loaded, or an inclusion) is far enough from any other boundary so as to avoid significant interaction effects. Here we will concentrate on the results* for self-sustaining biaxial and gradient fields because of their fundamental interest and immediate applicability to design.

The basic equations are derived with general boundary loading and displacement functions $F(\sigma)$ and $D(\sigma)$, and the significance of this capability to control the harmonic shape for particular applications should be emphasized. Practically any combination of internal tractions and displacements expressed as a function of an angular coordinate, θ, could be examined as, for example, prestress with expandable liners or "shrink fitting." However, to emphasize the basic idea and because so little work has been done on the inverse problem, harmonic shapes will be derived for holes with uniform loading and for rigid inclusions.

7.4.1 Harmonic Holes for Biaxial Fields

Since the biaxial field is so commonly encountered in structural mechanics, it is selected first to illustrate the application of the general functional equation. In addition, a uniformly distributed normal traction of magnitude P (positive if tension) applied to the unknown hole boundary is also included. Thus the boundary loading function is

$$F_1(z) = Pz \tag{g}$$

where P is zero for the unloaded hole.

For the plane problem, consider the biaxial free field specified by the principal stresses N_1, N_2, which have constant magnitude and direction at every point in the plane and where it will always be assumed that the direction of N_1 coincides with the x-axis and that $|N_1| \geq |N_2|$. For these conditions

$$\begin{aligned} \phi_1(z) &= \Gamma z \\ \psi_1(z) &= \Gamma' z \end{aligned} \tag{h}$$

* Only a very brief outline of the solutions will be given. More details can be found in papers by Bjorkman, G., Jr. and Richards, R. Jr:

a. Harmonic Holes—An Inverse Problem in Elasticity, *Jour. of Applied Mech., ASME*, 43, Series E, 3, Sept. 1976, 414–418.
b. Harmonic Holes for Nonconstant Fields, *Jour. of Applied Mech., ASME*, 46, 3, Sept. 1979, 573–576.
c. The Rigid Harmonic Inclusion, *Proceedings, 3rd ASCE Eng. Mech. Div. Specialty Conf.*, Austin, TX, Sept. 17–19, 1979.

where

$$\Gamma = \frac{1}{4}(N_1 + N_2)$$
$$\Gamma' = -\frac{1}{2}(N_1 - N_2) \qquad \text{(i)}$$

Transforming to the ζ plane:

$$F(\zeta) = P\omega(\zeta)$$
$$\phi_0(\zeta) = \Gamma\omega(\zeta) \qquad \text{(j)}$$
$$\psi_0(\zeta) = \Gamma'\omega(\zeta)$$

Finally, substituting these expressions into the general integral equation and combining similar terms, one obtains

$$(2\Gamma - P)\int_\gamma \frac{\omega(\sigma)d\sigma}{\omega - \zeta} + \Gamma'\int_\gamma \frac{\overline{\omega(\sigma)}d\sigma}{\sigma - \zeta} = 0 \qquad (7.24)$$

as the functional equation to be solved for $\omega(\zeta)$.

Letting $\omega(\zeta)$ have the form

$$\omega(\zeta) = R\left[\zeta + \frac{a_1}{\zeta} + \frac{a^2}{\zeta^2} + ...\right] \qquad \text{(k)}$$

where the a's are constants and R is a real scaling constant and using the fact that $\omega(\sigma)$ and $\overline{\omega(\sigma)}$ are, respectively, the boundary values of $\omega(\zeta)$ and $\bar{\omega}$ $(1/\zeta)$, the integral equation becomes

$$-(2\Gamma - P)R\left[\frac{a_1}{\zeta} + \frac{a^2}{\zeta^2} + ...\right] - \Gamma' R\left[\frac{1}{\zeta}\right] = 0 \qquad \text{(l)}$$

and by equating similar powers of ζ one obtains

$$a_2 = a_3 = \ldots = 0$$
$$a_1 = \frac{-\Gamma'}{2\Gamma - P} \qquad \text{(m)}$$

Substituting, the coefficient a_1 can be expressed in terms of the principal stresses as

$$a_1 = \frac{N_1 - N_2}{N_1 + N_2 - 2P}$$

Thus the mapping function whose boundary value describes the geometry of the harmonic hole for a biaxial field with uniform boundary loading is

$$\omega(\zeta) = R\left[\zeta + \frac{a_1}{\zeta}\right]$$

In the z-plane the boundary value of this mapping function describes an elliptic hole with major axis, a, oriented parallel to the direction of the major principal stress N_1, and where the ratio of the major to minor axes of the ellipse is

$$\frac{a}{b} = \frac{N_1 - P}{N_2 - P} \tag{7.25}$$

Harmonic holes with stresses along the x-axis for a particular biaxial field are shown in Figure 7.7a.

Conditions to ensure the existence of a harmonic hole shape in a biaxial field correspond to a breakdown in the conformality of $\omega(\zeta)$ which, in turn, coincide with physical conditions on the original stress field. Considering the first derivative of the mapping function on the boundary,

$$\omega'(\sigma) = R\left[1 - \frac{a_1}{\sigma^2}\right]$$

the conformality of $\omega(\sigma)$ breaks down when $a_1 = \pm 1$. However, if the direction of N_1 is to be parallel with the direction of the major axis of the ellipse, as has been specified, a_1 can only vary between 0 and +1. Obviously, if $a_1 = 0$, a circular hole is the harmonic shape; and this occurs when $N_1 = N_2$ regardless of the value of P. For $a_1 = +1$, the harmonic hole is a slit parallel to the direction of N_1. In the case of an unloaded hole (i.e., $P = 0$), the limitations on the values for the coefficient a_1, establish the following limitation on the field:

$$0 \preceq \frac{N_1 - N_2}{N_1 + N_2} \preceq 1.$$

This relationship is satisfied only if N_1 and N_2 have the same sign (i.e., both tension or both compression). In the case of a uniformly loaded hole

$$0 \preceq \frac{N_1 - N_2}{N_1 + N_2 - 2P} \preceq 1 \tag{n}$$

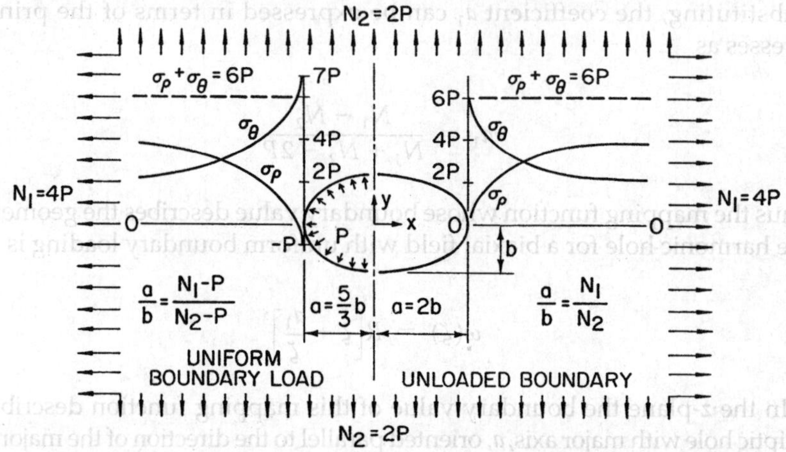

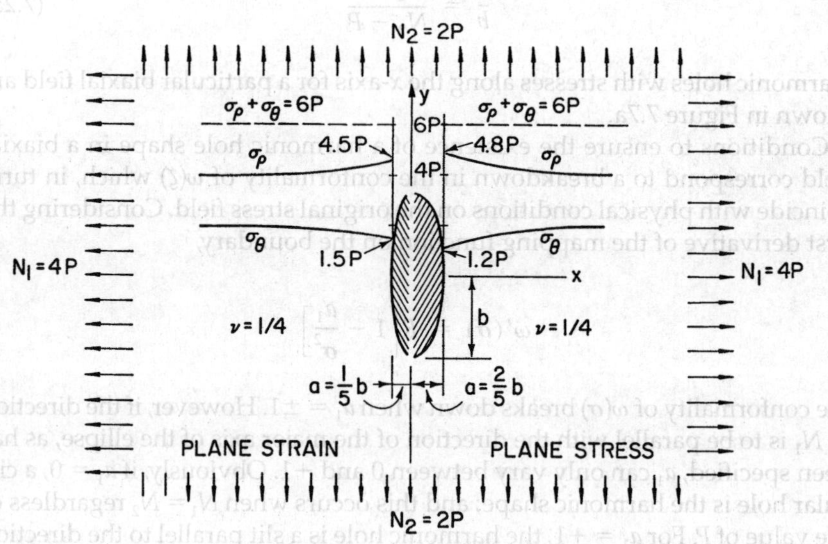

FIGURE 7.7
Harmonic openings for a biaxial field.

This relationship is satisfied when N_1 is positive (i.e., tension) only if $N_2 \geq P$ and when N_1 is negative (i.e., compression) only if $N_2 \leq P$. It is helpful to note that these limitations on the type of biaxial field for which a harmonic hole exists may be more easily visualized by realizing that the ratio a/b in Equation (7.25) must always be greater than or equal to one.

At the other extreme of a rigid inclusion, the harmonic shape can be found from Equation (7.23b) following an analogous procedure. Substituting, Equation (7.24) becomes

$$\Gamma(\kappa - 1)\int_{\gamma} \frac{\omega(\sigma)d\sigma}{(\sigma - \zeta)} = 0$$

as the functional equation to be solved for $\omega(\sigma)$.

Again letting $\omega(\sigma)$ have the form

$$\omega(\zeta) = R\left[\zeta + \frac{a_1}{\zeta} + \frac{a^2}{\zeta_2} + \dots\right]$$

the integral can be evaluated to give

$$-\Gamma(\kappa - 1)R\left[\frac{a_1}{\zeta} + \frac{a_2}{\zeta^2} + \dots\right] + \Gamma' R\left[\frac{1}{\zeta}\right] = 0$$

in which the a_i quantities are constants and R is a real scaling constant. Equating similar powers of ζ, one obtains:

$$a_2 = a_3 = \dots = 0; \qquad a_1 = \frac{\Gamma'}{\Gamma(\kappa - 1)}$$

Thus, the rigid harmonic inclusion is an ellipse of arbitrary size with the ratio of major to minor axes:

$$\frac{a}{b} = \frac{(\kappa - 3)N_1 + (\kappa + 1)N_2}{(\kappa + 1)N_1 + (\kappa - 3)N_2} \tag{7.26a}$$

or, in terms of principal normal strains, simply

$$\frac{a}{b} = \frac{\epsilon_2}{\epsilon_1} \tag{7.26b}$$

the variable a coinciding with the x-axis. In an isotropic field where $N_1 = N_2$ the shape, as expected, is a circle.

The stresses on the boundary of any rigid inclusion are

$$\sigma_\rho = (1+\kappa)Re\left[\frac{\phi'(\sigma)}{\omega'(\sigma)}\right]; \qquad \sigma_\theta = (3-\kappa)Re\left[\frac{\phi'(\sigma)}{\omega'(\sigma)}\right];$$

$$\tau_{\rho\theta} = (1+\kappa)Im\left[\frac{\phi'(\sigma)}{\omega'(\sigma)}\right]$$

However, because of the harmonic condition $\phi^* = 0$

$$\sigma_\rho = \frac{(\kappa+1)}{4}(N_1+N_2); \qquad \sigma_\theta = \frac{3-\kappa}{4}(N_1+N_2); \qquad \tau_{\rho\theta} = 0 \tag{7.27}$$

As can be seen, harmonic inclusions not only give the required condition $\sigma_\rho + \sigma_\theta = N_1 + N_2$ at the interface, but also no shears so that, in fact, the contact stresses are principal with no tendency for slip to occur at the interface.

To demonstrate these results, the same test case $N_1 = 2N_2 = 4P$ with $\nu = 1/4$ is shown in Figure 7.7b for both plane stress and plane strain. Both harmonic holes and harmonic inclusions are ellipses oriented with the principal stresses but in opposite directions. The unloaded harmonic hole has a ratio of the axes proportional to the ratio of the original principal stresses while this ratio for the elliptic harmonic inclusion is inversely proportional to the original principal strains (irrespective of the planar condition).

For the rigid inclusion, the condition for existence is similar to that for holes in that $a/b > 0$ and therefore it can exist only when

$$0 \preceq \frac{\epsilon_2}{\epsilon_1} \preceq 1$$

It is clear that holes that are harmonic must produce minimum stress concentration in constant fields since maximums and minimums in Laplacian fields must occur on the boundary. It can now be shown that the same is true for a rigid harmonic inclusion. For example, in plane stress the in-plane normal strains are given by:

$$\epsilon_\theta = \frac{1}{E}(\sigma_\theta - \nu\sigma_\rho); \qquad \epsilon_\rho = \frac{1}{E}(\sigma_\rho - \nu\sigma_\rho)$$

Thus, at the boundary of any rigid inclusion:

$$\epsilon_\theta = 0; \qquad \sigma_\theta = \nu\sigma_\rho$$

and therefore

$$\sigma_\rho + \sigma_\theta = (1+\nu)\sigma_\rho = \left(\frac{1+\nu}{\nu}\right)\sigma_\theta.$$

But the minimum value possible for the first invariant in a constant field is simply $N_1 + N_2$ so the minimum stresses possible on the boundary are:

$$\sigma_\rho = \frac{N_1 + N_2}{1+\nu}; \qquad \sigma_\theta + \frac{\nu}{1+\nu}(N_1 + N_2)$$

which are in fact, those for the harmonic inclusion [Equation (7.27)]. The proof for plane strain is similar. Although the standard uniqueness theorems of elasticity do not apply directly to the "inverse problem," it is hard to conceive of other shapes that might fulfill the harmonic design requirement for a given original stress field since this would be comparable to finding two different shapes that have precisely the same value of stress or displacement at every point on the boundary. Again comparing the two "extreme" alternatives of holes and rigid inclusions, it seems, from Figure 7.7 that a rigid inclusion is somewhat closer than an unloaded hole to the absolute optimum of a neutral hole with no stress concentration whatsoever. Comparing each harmonic shape for the test case $N_1 = 2N_2$, $\nu = 1/4$, to the corresponding circular shape, the stress concentration factor is reduced from 2.5 to 1.5 for an unloaded hole and from 1.56 to 1.2 for a rigid inclusion in plane stress. In either case, the practical benefits that accrue from using harmonic shapes are important.

7.4.2 Harmonic Holes for Gradient Fields

Using the complex-variable representation of the plane theory of elasticity, it has been possible to invert the standard problems of classical elasticity to solve for the geometry of a hole that does not anywhere perturb the first invariant of the original free field. This design criterion and the resulting shapes are termed "harmonic" not only because the first invariant, which is harmonic, remains everywhere unchanged, but also because all perturbations of the original stress–strain field are harmonic as well. For a constant field the harmonic ellipse does produce constant boundary stresses and a minimum stress concentration and it is postulated that such harmonic shapes are, in fact, "optimum" in that they produce minimum stress concentrations in any field for which they exist.

In theory at least, the harmonic field condition, Equation (7.20), and the resulting boundary integral Equations (7.23) apply to any field. The only unknown in these equations is the harmonic hole geometry $\omega(\sigma)$, which is the boundary value of the conformal mapping $z = \omega(\zeta)$ from the exterior $|\zeta| \geq 1$ of the unit circle in the ζ- plane. The complex potentials $\phi_0(\sigma)$ and $\psi_o(\sigma)$, which define the original free field, and $F(\sigma)$ and $D(\sigma)$, which describe the boundary loading or deflection are always known functions of $\omega(\sigma)$. In these terms, σ (no subscript) is the value of $\zeta = \rho e^{i\theta}$ on the boundary of the unit circular hole γ with center at the origin (i.e., $\sigma = e^{i\theta}$).

A solution of this equation for $\omega(\sigma)$ therefore determines the harmonic hole shape. In addition, it should be reemphasized that the equation admits great generality in that one may obtain solutions for either constant or nonconstant fields with either constant or nonconstant boundary loads.

Consider the simple case of a linear gradient field (e.g., bending) acting within a region large in comparison with the size of a hole to be cut. Superimposing on this field an isotropic field, the resulting components of the combined field in the absence of a hole are

$$\sigma_y = N, \qquad \sigma_x = -Ay + N, \tag{o}$$

where A is a real constant that determines the slope of the linear field and N is the magnitude of the isotropic field component. The fact that this stress field is unbounded at infinity causes no particular difficulty since one need only consider a sufficiently large finite region in which all stresses are bounded.

For this combined field with a uniform pressure "P" (positive if tension) on the hole boundary:

$$F_1(z) = Pz, \qquad \phi_1(z) = \frac{Aiz^2}{8} + \Gamma z, \qquad \psi_1(z) = -\frac{Aiz^2}{8} \tag{p}$$

where $\Gamma = (1/4)(N_1 + N_2)$ and N_1, N_2 are the principal stresses. Transforming these equations to the ζ-plane they become

$$F(\zeta) = P\omega(\zeta), \qquad \phi_0(\zeta) = \frac{Ai[\omega(\zeta)]^2}{8} + \Gamma\omega(\zeta), \qquad \psi_0(\zeta) = \frac{-Ai[\omega(\zeta)]^2}{8}$$

Finally, substituting the foregoing expressions into the harmonic hole equation and combining terms one obtains as the functional equation to be solved for $\omega(\sigma)$:

$$(2\Gamma - P)\int_\gamma \frac{\omega(\sigma)d\sigma}{\sigma - \zeta} + \frac{Ai}{8}\int_\gamma \frac{[\omega(\sigma) - \overline{\omega(\sigma)}]^2 d\sigma}{\sigma - \zeta} = 0 \tag{q}$$

It is important to note that unlike integral equations encountered in the solution of the standard elasticity problem where the mapping function is known, Equation (q) expresses the condition that must be satisfied by the *boundary value* of the *unknown* mapping function. Since it is the boundary value of the final mapping function that must satisfy Equation (q), it is only necessary that functional representation of $\omega(\sigma)$ be shown as a general rational polynomial containing both negative and positive powers of

σ, i.e.,

$$\omega(\sigma) = R\left(\sigma + \frac{a_1}{\sigma} + \frac{a_2}{\sigma^2} + \ldots + b_1\sigma + b_2\sigma^2 + \ldots\right) \tag{r}$$

In this equation R is a real scaling constant and the a's and b's are complex constants. As can be seen, such an expression allows a far more concise representation of the hole geometry than could be obtained if the positive powers of $\sigma \geq 2$ were omitted and, in fact, if these positive powers of σ are not included, a closed-form polynomial solution cannot be obtained.

Substituting Equation (r) into Equation (q) and evaluating, one can show that a solution exists when

$$\begin{aligned} a_1 &= a_3 = a_4 \ldots = 0, \qquad & b_1 &= b_3 = b_4 = \ldots = 0 \\ a_2 &= -Gi/16, & b_2 &= Gi/16 \end{aligned} \tag{s}$$

Here, G is a nondimensional parameter called the "gradient field strength," which relates the relative strength of the original field components (A, Γ), the half depth of the hole (R), and the boundary loading

$$G = \frac{2AR}{2\Gamma - P} \tag{7.28}$$

In the present case of a gradient superimposed on an isotropic field, $G = 2AR/(N - P)$. Thus the boundary value of the mapping function that describes the shape of the harmonic hole for an isotropic field with superimposed linear gradient and uniform boundary loading is

$$\omega(\sigma) = R\left[\sigma + \frac{Gi}{16}\left(\sigma^2 - \frac{1}{\sigma^2}\right)\right] \tag{t}$$

This harmonic hole shape is called the "deloid." When the point σ in Equation (t) describes the unit circle γ, the point (x, y) describes the deloid in the z-plane the parametric representation of which is

$$x = R\left(\cos\theta - \frac{G}{8}\sin 2\theta\right), \qquad y = R(\sin\theta) \tag{7.29}$$

The geometry of a deloid with an unloaded boundary for an original stress field of $G = 2.0$ and a deloid in the same original stress field, but with a uniform boundary loading that results in a $G = 3.0$ is shown in Figure 7.8a.

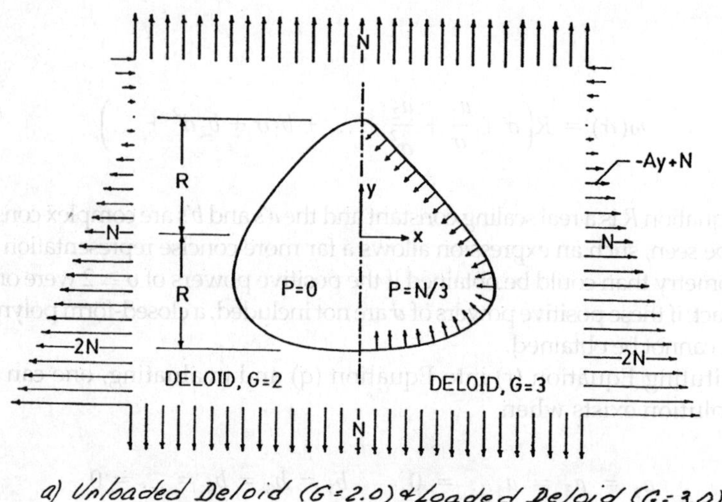

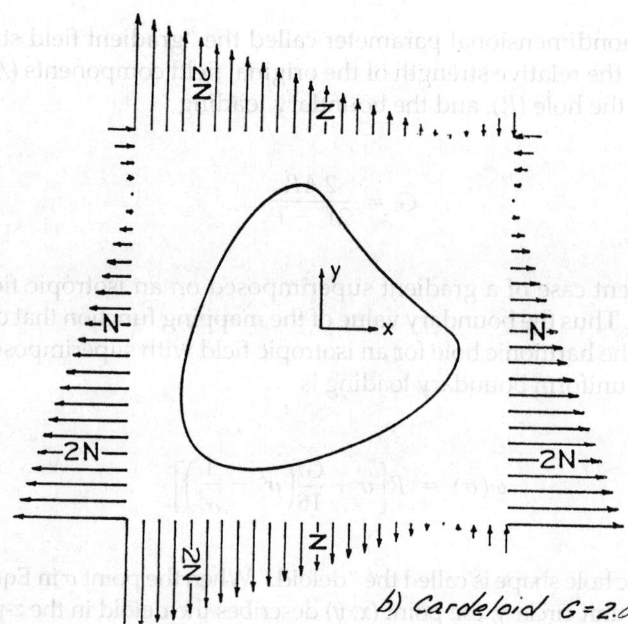

FIGURE 7.8
Harmonic hole shapes for linear gradient fields.

Following the same procedure for a stress field where both principal stresses vary linearly, the "cardeloid" can be obtained with the parametric representation:

$$x = [R\cos\theta - (G/8)\sin 2\theta], \qquad y = [R\sin\theta - (G/8)\sin 2\theta] \tag{7.30}$$

Using Equation (t) the range of values of G for which $\omega(\sigma)$ presents a conformal transformation can be obtained. This range is the same for both the deloid and cardeloid and is simply $|G| < 4.0$. In the case of the unloaded hole (i.e., $P = 0$) this condition establishes the following limitation on the stress field and hole size:

$$-4 < \frac{2AR}{N} < 4$$

which is physically satisfied only if the entire hole boundary occupies a region in the original stress field where the first invariant (or mean stress) has the same sign.

While it has been postulated that harmonic holes are also optimum in that they produce minimum values of stress concentration for any hole shape in any free field for which they exist, this postulate has only been proven for the constant free-field. Nonetheless one can obtain an idea of the extent to which harmonic holes can reduce stress concentrations in nonconstant fields by comparing the stresses on the boundary of an unloaded harmonic hole to those for a circular hole most commonly encountered in practice.

For example, Figure 7.9 shows the boundary values of the tangential normal stress σ_θ nondimensionalized by the isotropic field stress N, for the deloid $G = 3.0$ and a circular hole of the same size in the same position. It is clear that the

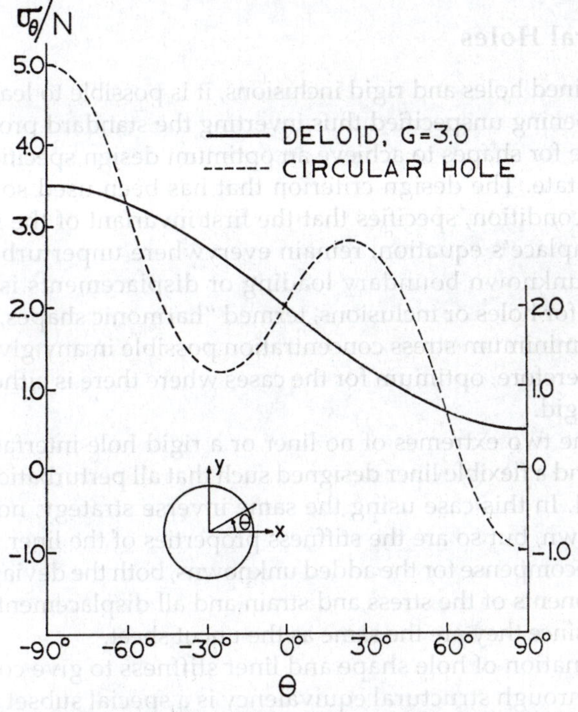

FIGURE 7.9
Comparison of hoop stresses around an unloaded hole in a gradient field.

deloid has not only significantly reduced the maximum stress, but it has also significantly reduced the total variation of stress around the hole boundary.

In the next two sections we will go on to the question of the neutral hole where both the shape and the properties of reinforcement to eliminate all stress concentration are presented for the same simple free fields discussed here for unlined harmonic holes. However, the essence of inverting elastic analysis to a design orientation should not be lost in the details of a particular solution. Since the inverse problem in elasticity has received so little attention and less "theoretical" treatment, time has been spent here on formulating it properly, deriving some basic equations, and emphasizing the overriding importance and effects of the design condition rather than generating solutions for many specific design situations. Essentially, any initial field for which the stress function is known can now be investigated and shapes derived of practical interest and importance. There are also a myriad of possibilities and directions for further basic study. The exterior problem, questions of existence and uniqueness, nonhomogeneous or nonisotropic materials, energy formulations, other design conditions, and related problems in other areas of applied science are a few of the obvious possibilities.

7.5 Neutral Holes

For both unlined holes and rigid inclusions, it is possible to leave the geometry of an opening unspecified thus inverting the standard problem in elasticity to solve for shapes to achieve an optimum design specification on the final stress state. The design criterion that has been used so far, the harmonic field condition, specifies that the first invariant of the stress tensor, satisfying Laplace's equation, remain everywhere unperturbed when the shape with unknown boundary loading or displacements is introduced. Such shapes for holes or inclusions, termed "harmonic shapes," apparently produce the minimum stress concentration possible in any given free field. These are, therefore, optimum for the cases where there is either no liner or the liner is rigid.

Between the two extremes of no liner or a rigid hole interface, it may be possible to find a flexible liner designed such that all perturbation of the field is eliminated. In this case using the same inverse strategy, not only is the shape unknown, but so are the stiffness properties of the liner. However, in this case as recompense for the added unknowns, both the deviatoric and isotropic components of the stress and strain and all displacements are known everywhere since they are the same as the uncut sheet.

The combination of hole shape and liner stiffness to give complete field restoration through structural equivalency is a special subset of harmonic

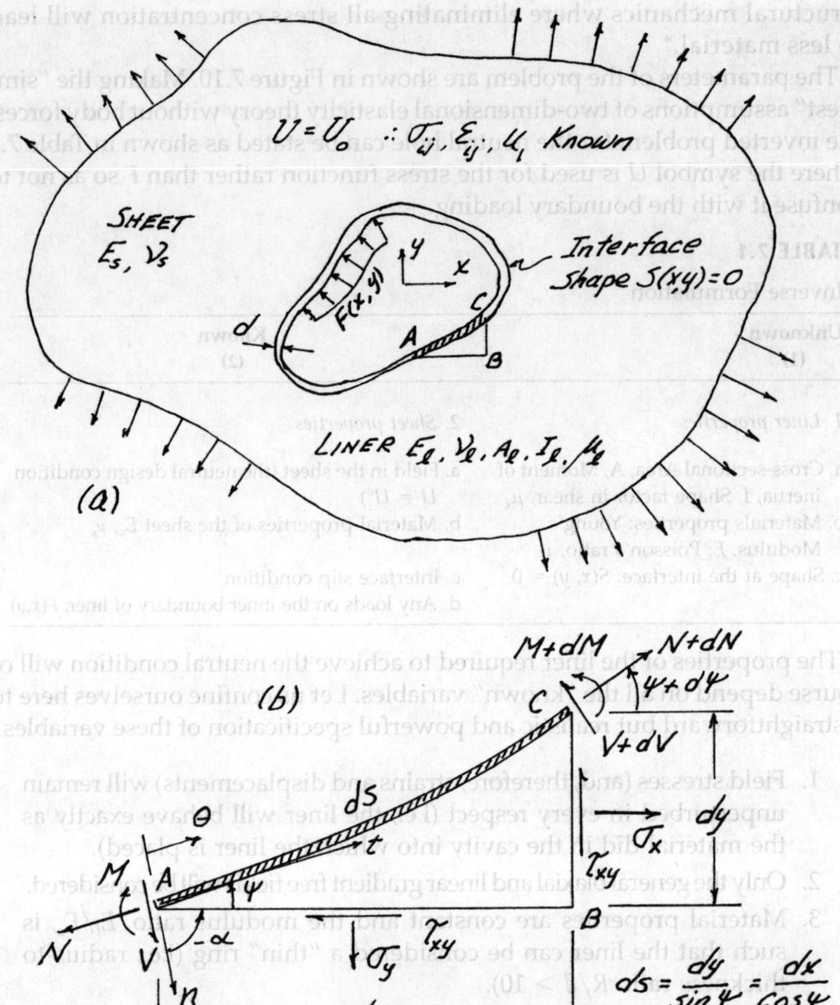

FIGURE 7.10
Lined hole in an elastic sheet.

shape called a "neutral hole" by Mansfield. He was the first to clearly define this inverse problem and develop a theory for the special case of membrane reinforcement. A more general theory, with examples where flexural as well as axial stiffness of the liner is included, should be of immediate practical consideration for new design in almost every area of structural mechanics where eliminating all stress concentration will lead to less material.*

The parameters of the problem are shown in Figure 7.10. Making the "simplest" assumptions of two-dimensional elasticity theory without body forces, the inverted problem for the neutral hole can be stated as shown in Table 7.1 where the symbol U is used for the stress function rather than F so as not to confuse it with the boundary loading.

TABLE 7.1

Inverse Formulation

Unknown (1)	**Known (2)**
1. Liner properties	*2. Sheet properties*
a. Cross-sectional Area, A, Moment of inertia, I, Shape factor in shear, μ_ℓ	a. Field in the sheet (the neutral design condition $U = U^o$)
b. Materials properties: Young's Modulus, E; Poisson's ratio, ν	b. Material properties of the sheet E_s, ν_s
c. Shape at the interface: $S(x, y) = 0$	c. Interface slip condition
	d. Any loads on the inner boundary of liner, $F(x,y)$

The properties of the liner required to achieve the neutral condition will of course depend on all the "known" variables. Let us confine ourselves here to a straightforward but realistic and powerful specification of these variables.

1. Field stresses (and, therefore, strains and displacements) will remain unperturbed in every respect (i.e., the liner will behave exactly as the material did in the cavity into which the liner is placed).
2. Only the general biaxial and linear gradient free fields will be considered.
3. Material properties are constant and the modulus ratio, E_l/E_s, is such that the liner can be considered a "thin" ring (i.e., radius to thickness ratio $R/d > 10$).
4. A no slip condition exists at the interface.
5. The inner surface of the liner is not loaded [i.e., $F(x,y) = 0$].

Within this framework, the problem may be "underspecified" for a unique solution emphasizing the "design" nature of this approach.

* The development of the theory of neutral holes presented here follows that given by: Richards, R., Jr. and Bjorkman, G.S. Jr, in "Neutral Holes: Theory and Design", *Jour. of Eng. Mech. Div., Proc. ASCE*, 108, EMS, Oct. 1982. Applications to pipe design and lined tunnels is given in Chapter 42, Buried Structures, by R. Richards, Jr., *Ground Engineers Reference Book*, Edited by F.G. Bell, Butterworths, Boston, 1987. An alternative derivation using the imaginary part of the complex formulation is given by Bjorkman, G.S. Jr. and Richards, R., Jr., "On the Deviation of Neutral Holes Using Complex Variable Methods", *Proc. ASCE Eng. Mech. Div. Specialty Conf.*, Perdue University, May 23–25, 1983. The original derivation given here, while less elegant, is more physically descriptive and introduces the force resultants in the liner as related to the stress function for the uncut sheet.

Generalizing Mansfield's derivation to include flexural stiffness, consider first the equilibrium of the diferential element of the liner shown in Figure 7.10b:

$$\Sigma F_z = 0 \qquad \therefore d(N\cos\psi) - d(V\sin\psi) = (\tau_{xy}dx - \sigma_x dy) \qquad \text{(a)}$$

$$\Sigma F_y = 0 \qquad \therefore d(N\sin\psi) + d(V\cos\psi) = (\sigma_y dx - \tau_{xy}dy) \qquad \text{(b)}$$

However, since the right hand side of Equations (a) and (b) are total differentials in terms of the stress function, U, they can be integrated to give:

$$N\cos\psi - V\sin\psi = -\left(\frac{\partial U}{\partial y} + b\right) \qquad \text{(c)}$$

$$N\sin\psi - V\cos\psi = \left(\frac{\partial U}{\partial y} + a\right) \qquad \text{(d)}$$

in which (a) and (b) are arbitrary constants. Solving for the normal and shear force resultants separately:

$$N = \left(\frac{\partial U}{\partial x} + a\right)\frac{dy}{ds} - \left(\frac{\partial U}{\partial y} + b\right)\frac{dx}{ds} \qquad \text{(7.31a)}$$

$$V = \left(\frac{\partial U}{\partial y} + b\right)\frac{dy}{ds} + \left(\frac{\partial U}{\partial x} + a\right)\frac{dx}{ds} \qquad \text{(7.31b)}$$

For moment equilibrium:

$$dM + N\sin\psi dx - N\cos\psi dy + V\cos 0\psi dx + V\sin\psi dy$$

$$-\left(\frac{\sigma_x dy^2}{2} + \frac{\sigma_y dx^2}{2} - \tau_{xy}dxdy\right) = 0 \qquad \text{(e)}$$

Neglecting the second-order stress terms and substituting Equations (c) and (d)

$$dM + \left(\frac{\partial U}{\partial y} + b\right)dy + \left(\frac{\partial U}{\partial x} + a\right)dx = 0 \qquad \text{(f)}$$

Thus from Equation (7.31) we obtain the classic relationship:

$$dM = -V ds \quad \text{or} \quad \frac{dM}{ds} = -V \qquad \text{(g)}$$

and integrating

$$M = -U - ax - by - c \qquad \text{(7.31c)}$$

in which c is again an arbitrary constant.

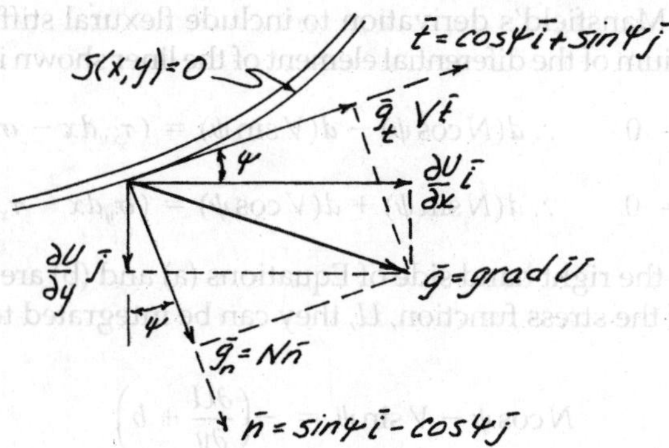

FIGURE 7.11
Force resultants as vectors.

Thus, with Equations (7.31), the internal force resultants and moment can be determined for the liner for a prescribed shape in any given field, U, independent of material properties. Since terms such as ax, by, and c can be added to U without affecting stresses in the sheet, they can be used to adjust the V, N, and M diagrams or the position and size of the hole. For the case of a neutral membrane liner without flexural stiffness, Mansfield's equation for the neutral shape is obtained directly from Equation (7.31c) as

$$U = 0 \tag{7.32}$$

if the arbitrary constants are neglected.

These basic equilibrium equations can also be considered in terms of vector components as shown in Figure 7.11 where t and n are unit tangent and normal vectors. If we neglect the arbitrary constants a, b, and c, and define the vector $\bar{g} = Pi + Qj = \text{grad } U$, the internal force equilibrium conditions are simply that N and V are the magnitudes of the normal and tangential components of the directional derivative of U. That is

$$N = g_n = n \cdot \text{grad } U = \nabla_n g = \frac{\partial U}{\partial n} \tag{7.33a}$$

$$V = g_t = t \cdot \text{grad } U = \nabla_t g = \frac{\partial U}{\partial t} \tag{7.33b}$$

If for membrane reinforcement the moment is zero, then the shear must be zero and the neutral shape is a level line of U.

It should be emphasized that equilibrium Equations (7.31) apply in general to any liner or, for that matter, all problems involving a flexible member interacting with an elastic medium. As such they are quite remarkable in that they give a physical interpretation of the stress function and its gradient at the interface. However, only for the inverse problem where U is known

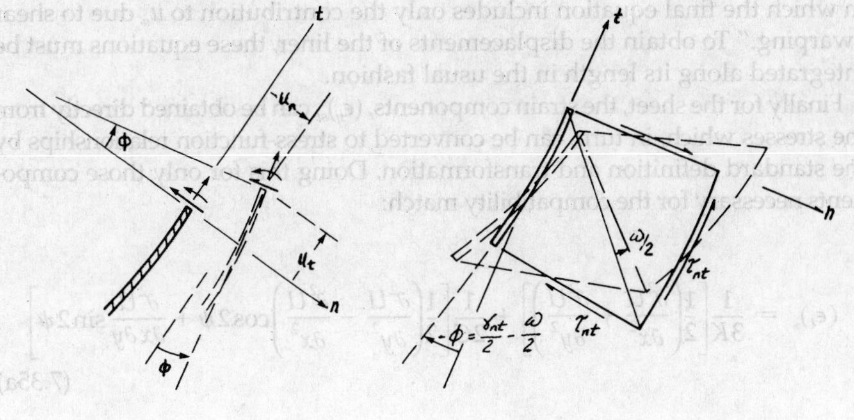

FIGURE 7.12
Geometry of liner deformation.

(or perhaps for the standard problem if the moment and force resultants were measured experimentally) is it possible to use them directly.*

Displacements of the liner and strains at its neutral axis due to the internal moment and force resultants can now be found by the so-called "generalized equations of Bresse."** The cross-section will be displaced and will rotate as shown positively in Figure 7.12a. Considering a differential length, assuming plane sections, and neglecting any contribution from the normal expansion or contraction of the liner, then in curvalinear coordinates:

$$(\epsilon_t)_\ell = \frac{N}{A_\ell E_\ell} + \frac{M}{A_\ell E_\ell r} \tag{7.34a}$$

$$\left(\frac{d\phi}{ds}\right)_\ell = \frac{M}{E_\ell I_\ell} + \frac{N}{A_\ell E_\ell r} + \frac{M}{A_\ell E_\ell r^2} \tag{7.34b}$$

$$-\gamma_{nt} = \frac{\partial u_n}{\partial s} = \gamma_\ell = \frac{V}{\mu A_\ell E_\ell} \tag{7.34c}$$

---

* What about Harmonic Holes where $U = U^o + U^*$ is harmonic? Is it possible to derive harmonic hole shapes from the stress boundary conditions expressed in terms of the stress function and avoid complex variables? This is an interesting topic for research.

** Jacques Antoine Charles Bresse was another of the accomplished engineers who graduated from L'École Polytechnique and taught at L'École des Ponts et Chaussees helping to lead France to preeminence in applied elasticity and solid mechanics prior to the Panama Canal disaster. Navier in his theory of the deflection of curved bars included only the change in curvature due to the bending moment. Bresse, by including the normal force, derives Equations (8.34a, b). The effect of shear warping Equation (8.34c) was recognized later. One of the best presentations of the "generalized equations of Bresse" and their use is in the book: *Mechanics of Engineering Structures*, by G.L. Rogers and M.L. Causey, John Wiley & Sons, New York, 1962, 246–252.

in which the final equation includes only the contribution to u_n due to shear "warping." To obtain the displacements of the liner, these equations must be integrated along its length in the usual fashion.

Finally for the sheet, the strain components, $(\epsilon_{ij})_s$, can be obtained directly from the stresses which, in turn, can be converted to stress-function relationships by the standard definition and transformation. Doing this for only those components necessary for the compatibility match:

$$(\epsilon_t)_s = \frac{1}{3K}\left[\frac{1}{2}\left(\frac{\partial^2 U}{\partial x^2}+\frac{\partial^2 U}{\partial y^2}\right)\right]+\frac{1}{2G}\left[\frac{1}{2}\left(\frac{\partial^2 U}{\partial y^2}-\frac{\partial^2 U}{\partial x^2}\right)\cos 2\psi+\frac{\partial^2 U}{\partial x\,\partial y}\sin 2\psi\right] \tag{7.35a}$$

$$\left(\frac{\gamma_{nt}}{2}\right)_s = \frac{1}{2G}\left[\frac{1}{2}\left(\frac{\partial^2 U}{\partial x^2}-\frac{\partial^2 U}{\partial y^2}\right)\sin 2\psi+\frac{\partial^2 U}{\partial x\,\partial y}\cos 2\psi\right] \tag{7.35b}$$

The rotation of the sheet at the interface, as shown in Figure 7.12b, is given directly by

$$\left(\frac{d\phi}{ds}\right)_s = \frac{d}{ds}\left(\frac{\omega_{nt}}{2}-\frac{\gamma_{nt}}{2}\right) \tag{7.36}$$

Since the elastic rotation, $\omega_{nt} = \omega_{xy}$, is the invariant conjugate harmonic function of the mean stress,

$$\frac{\partial\omega_{xy}}{\partial y} = \frac{1}{E}\frac{\partial}{\partial x}(\nabla^2 U); \qquad \frac{\partial\omega_{xy}}{\partial x} = -\frac{1}{E}\frac{\partial}{\partial y}(\nabla^2 U) \tag{7.37}$$

the relative rotation of the interface [by Equation (7.19)] can also be expressed in terms of the stress function. Displacements of the sheet, $(u_i)_s$, can be found by integrating the strains directly.

7.6 Solution Tactics for Neutral Holes—Examples

For the inverse problem, a solution matching displacements at the interface using the integral equations, will necessarily be one of trial and error even for an assumed shape since the unknown geometric properties of the cross-section will, in general, be functions of position and must remain inside the integral. A better tactic is to enforce compatibility on a differential scale by using Equations (7.34) with Equations (7.31) and (7.32). The displacements will then necessarily conform aside from initial conditions u_0, v_0, and ϕ_0. These along with any unknown integration constants in the equilibrium equations can then be found in the standard way once the geometry of the hole and liner is known.

The theory as presented essentially reduces to a "strength-of-materials" solution for a ring where the geometry of its deformation must match that of the sheet at the interface. It is, of course, approximate to the extent that all strength-of-material solutions are approximate. Thick-ring effects could easily be incorporated if necessary by including the eccentricity of the neutral axis and adding the bending contribution at the interface.

However, as is usually done, the equations will be simplified further by neglecting the curvature contribution and reducing the liner to the "elastica" at the neutral axis. The significance of this "thin ring" assumption can be estimated by considering a circular liner of radius R in an isotropic field.

For this situation the ring simply expands uniformly and, since there is no rotation of the cross-section, $(d\phi/ds)_\ell = 0 = M/E_\ell I_\ell + (\epsilon_\theta)_\ell/R$; $(\epsilon_\theta)_\ell = N/A_\ell E_\ell + M/A_\ell E_\ell R$. Solving simultaneously: $N = E_\ell \epsilon_\theta (A_\ell + I_\ell/R^2)$. For a liner of uniform depth $d = r/10$ (an upper limit for a thin ring), $I/R^2 = A/1{,}200$, and the contribution to ϵ_θ of the coupling term M/AER is 1/1,200 that of the axial force N/AE. Thus, the final term in the first two equations of Bresse will be negligible and will not be included hereafter.

Two special cases of obvious importance will be considered in the examples that follow. First, membrane reinforcement may be possible in certain fields by adjusting the hole shape to "eliminate" bending. This is the Mansfield solution and, as shown previously, can occur only for that shape which is a level line of U given by Equation (7.32). This "short-circuits" the equations of Bresse in that if $V = 0$, both μ and I are indeterminate. The area required can be found directly from Equations (7.34a) with (7.35a). However, an actual liner will have some bending stiffness inducing secondary moments (i.e., a gradient of normal stress through the thickness) violating the original membrane assumption. This dilemma, not considered by Mansfield, is a direct result of the strength-of-materials approach. Such secondary effects will be shown to be very small for thin liners.

Circular liners are a second important class of potential solutions. In this case the shape is prescribed and I_ℓ, A_ℓ, and μ are to be determined to satisfy the neutral condition. The strains in the sheet in polar coordinates r, and θ become:

$$\epsilon_n = \epsilon_r = \frac{\partial u_r}{\partial r} \tag{7.38a}$$

$$\epsilon_t = \epsilon_\theta = \frac{u_r}{r} + \frac{1}{r}\frac{\partial u_\theta}{\partial \theta} \tag{7.38b}$$

$$\frac{\gamma_{nt}}{2} = \frac{\gamma_r \theta}{2} = \frac{1}{2}\left(\frac{1}{r}\frac{\partial u_r}{\partial \theta} + \frac{\partial u_\theta}{\partial r} - \frac{u_\theta}{\partial r}\right) \tag{7.38c}$$

$$\frac{\omega_{xy}}{2} = \frac{\omega_{r\theta}}{2} = \frac{1}{r}\left[\frac{1}{2}\frac{\partial}{\partial r}(ru_\theta) - \frac{1}{r}\frac{\partial u_r}{\partial \theta}\right] \tag{7.38d}$$

Therefore, the relative rotation by Equation (7.36) is:

$$\left(\frac{d\phi}{ds}\right)_s = \frac{1}{r}\frac{\partial}{\partial \theta}\left(-\frac{\partial u_r}{r\partial \theta} + \frac{u_\theta}{r}\right) = \frac{1}{r^2}\left(-\frac{\partial^2 u_r}{\partial \theta^2} + \frac{\partial^2 u_\theta}{\partial \theta}\right) \tag{7.39}$$

Similarly, since the radius is a constant, the differential equations of Bresse are easily solved for the required liner properties.

7.6.1 Isotropic Field

With equal normal stress in orthogonal directions, σ_m, the stress function is:

$$U = \frac{\sigma_m}{2}(x^2 + y^2) = \frac{\sigma_m}{2}r^2 \tag{7.40}$$

and the obvious shape is circular. Thus, by the equilibrium equations

$$M = -\frac{\sigma_m}{3}R^2 - c \tag{a}$$

$$V = R\sin\theta\cos\theta(\sigma_x - \sigma_y) = 0 \tag{b}$$

$$N = \sigma_m R(\cos^2\theta + \sin^2\theta) = \sigma_m R \tag{c}$$

From the equations of Bresse:

$$\epsilon_\theta = \frac{(1-\nu_s)}{E_s}\sigma_m = \frac{u_r}{r} = \frac{N}{A_\ell E_\ell} \tag{d}$$

which is not possible if $M = 0$. However, we know from Lamé's exact solution that there is, in fact, a constant moment in the ring since the hoop stress is not uniform through the thickness. Combining (c) and (d):

$$M = \frac{NE_\ell I_\ell}{A_\ell E_\ell R} = -\sigma_m\frac{I_\ell}{R}$$

Thus, in Equation (a)

$$c = \sigma_m \left(\frac{I_\ell}{R} - \frac{R^2}{2} \right)$$

The area required to match the hoop strain, therefore, is:

$$A_\ell^i = \frac{N}{E_\ell \epsilon_\theta} = \frac{E_s R}{E_\ell (1 - \nu_s)}$$

per unit thickness which we derived earlier.

There will be no requirement on the moment of inertia. Thus, a "neutral" liner will have a uniform cross-section and be "thin" as defined by $d/r < 10$ if $E_s/E_\ell(1 - \nu_s) < 10$. This is nearly true for a steel liner in concrete or rock and true for a concrete or steel liner in soil.

7.6.2 Deviatoric Field

For the pure shear or deviatoric field with equal but opposite principal stresses, $\pm\sigma_d$

$$U = \frac{\sigma_d}{2}(y^2 - x^2) \tag{7.41}$$

Therefore, the membrane shape is a self-equilibriating system of parabolic "arch" and "suspension" segments which, while conceivable, is not likely in practice.

On the other hand, a circular, neutral liner may be feasible. Using cylindrical coordinates where $\phi = \psi - 90°$:

$$M = \frac{\sigma_d}{2} R^2 \cos 2\theta - ax - by - c$$

$$V = \sigma_d R \sin 2\theta - a \sin\theta + b \cos\theta$$

$$N = -\sigma_d R \cos 2\theta + a \cos\theta + b \sin\theta$$

However, from symmetry at $\theta = 0$, and $\pi/2$, the shear $V = 0$ and both a and b must be zero. Through symmetry the moment at the springing line must be equal and opposite that at the crown. Thus, c must be zero. Therefore, the stress resultants in the liner are

$$M = \frac{\sigma_d R^2}{2} \cos 2\theta, \quad V = \sigma_d R \sin 2\theta, \quad N = -\sigma_d R \cos 2\theta \tag{7.42}$$

Displacements of the sheet at the interface are:

$$u_r = u_n = \frac{\sigma_d R}{2G_s}\cos 2\theta \tag{e}$$

$$u_\theta = u_t = -\frac{\sigma_d R}{2G_s}\sin 2\theta \tag{f}$$

and therefore, for compatibility:

$$\epsilon_\theta = \frac{u_r}{r} + \frac{1}{r}\frac{\partial u_\theta}{\partial \theta} = \frac{\sigma_d \cos 2\theta}{2G_s} = \frac{\sigma_d R \cos 2\theta}{A_\ell E_\ell} \tag{g}$$

$$\frac{d\phi}{ds} = \frac{1}{R^2}\left(\frac{\partial u_\theta}{\partial \theta} - \frac{\partial^2 u_r}{\partial \theta^2}\right) = \frac{\sigma_d \cos 2\theta}{G_s R} = \frac{\sigma_d R \cos 2\theta}{2E_\ell I_\ell} + \frac{\epsilon_\theta}{R} \tag{h}$$

Finally, solving for the required geometric properties of the liner:

$$A_\ell^d = \frac{2G_s R}{E_\ell} = \frac{E_s R}{E_\ell(1+\nu_s)} \quad \text{per unit thickness} \tag{7.42a}$$

which is the same as an isotropic liner and

$$I_\ell^d = \frac{G_s R^3}{3E_\ell} = \frac{E_s R^3}{6E_\ell(1+\nu_s)} \quad \text{per unit thickness} \tag{7.42b}$$

These results are startling. It is possible to design a neutral circular liner completely restoring the original field, which has constant A and I, and therefore, is easy to build. Checking shear compatibility from Equation (7.34c)

$$-\gamma_{r\theta} - \frac{\sigma_d}{G_s}\sin 2\theta = \frac{V}{\mu A_\ell G_\ell} = \sigma_d R \frac{\sin 2\theta}{\mu A_\ell G_\ell}$$

and the required area is

$$A_{\text{shear}}^d = \frac{RG_s}{\mu G_\ell} = \frac{RE_s(1+\nu_\ell)}{\mu E_\ell(1+\nu_s)} \quad \text{per unit thickness} \tag{7.42c}$$

This is close to that required by Equation (7.42a) in order to match the hoop strain since $1 + \nu_\ell \cong \mu$ (e.g., for a steel H section $\mu \cong 1.3$).

It is, however, difficult to design A and I simultaneously since $(I/A)_{req} = R^2/6$ is so large. As an example, assume a steel liner $E_\ell = 30 \times 10^6$ psi; $\nu = 1/3$ in clay $E_s = 3 \times 10^3$ psi; $\nu = 1/2$ and a pipe diameter of 20 ft. Then $I^d_{req} = 120^3/6(10{,}000)(1.5) = 19.2$ in.^{4}; $A^d_{req} = 120/10{,}000(1.5) = 0.0008$ in.^{2} A 10 WF 17 would give the required moment of inertia, but over 100 times too much area, and slip joints to reduce the effective A would be necessary. Clearly, as one might expect for a circular shape in pure shear, bending dominates entirely and we might term this remarkable neutral liner with constant cross-section the flexural counterpart to the circular, axial-force liner for the isotropic field.

7.6.3 General Biaxial Field

A general biaxial field will be a combination of isotropic and deviatoric components. Assuming that the x axis is in the direction of the maximum principal stress:

$$\begin{aligned} U &= \frac{\sigma_1}{2}y^2 + \frac{\sigma_2}{2}x^2 + ax + by + c \\ &\equiv \frac{\sigma_m}{2}(x^2 + y^2) + \frac{\sigma_d}{2}(y^2 - x^2) + ax + by + c \end{aligned} \tag{7.43}$$

Thus, the membrane shape will be a combination of a circle and parabola. The hole boundary can only be closed if $\sigma_m > \sigma_d$, in which case an ellipse results with the principal axes in the ratio $a/b = \sqrt{\sigma_1/\sigma_2}$. This important and practical case is given by Mansfield, but was actually found nearly a century before by Rankine using parallel projection.

The neutral shape and liner for no stress concentration is compared to the unlined harmonic shape (stress concentration of 1.5) in Figure 7.13a for a 2 to 1 biaxial field. If ν_s is 1/4, then the area of the membrane liner at the "springing line" (i.e., $y = 0$) would be about 2.5 times that required at the "crown" (i.e., $x = 0$). If $E_\ell/E_s = 1{,}000$ and R is 10 ft, then the A_ℓ, at the springing line, should be 0.34 in.^{2}/in. (219 mm^2/mm). This is roughly that of the average corrugated plate section now used in practice for large-span elliptic culverts and arches in soil (Figure 7.13b). Therefore, if the axial stiffness were gradually reduced by a factor of 2.5 from the springing line to the crown in such structures by slip joints or by using lighter plates, the moments could be "eliminated" and the design improved to the neutral case.*

* This conclusion also correlates with a similar but nonquantifiable recommendation based on previous parametric variation studies of particular shapes of large span elliptic pipe using finite elements and photoelastic models [Abel, J.F., Mark, R. and Richards, R., Jr., Stresses Around Flexible Elliptic Pipes, *Journal of the Soil Mechanics and Foundation Division, ASCE*, 99, SM7, July 1973, 509–526.]

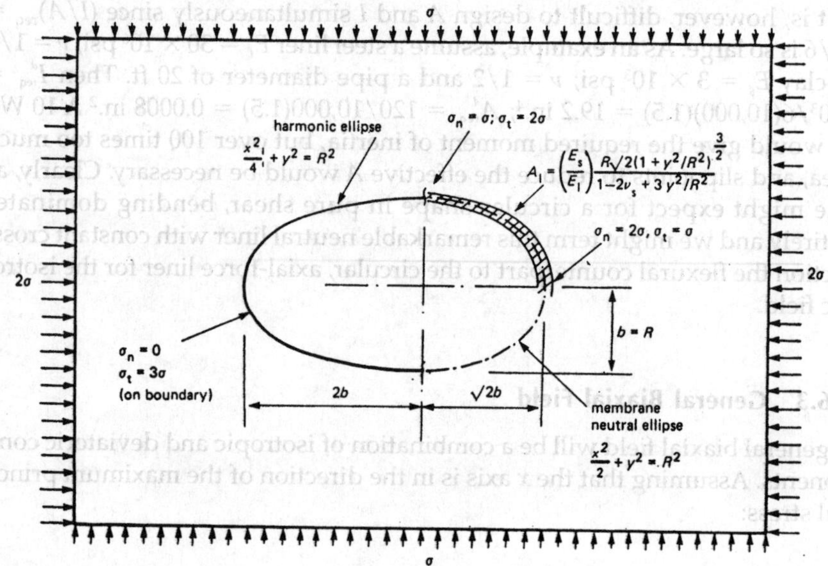

FIGURE 7.13
Design for holes in biaxial fields. Optimum elliptic shapes.

For a shape other than the membrane ellipse, moments will be introduced and the liner must also have the correct flexural stiffness to achieve neutrality. Equations based on the general theory and conditions for their existence can be derived and studied. However, from a practical design standpoint, it is doubtful if such intermediate cases are as useful as the four special cases:

1. Harmonic hole ellipse (7.25)
2. Rigid harmonic inclusion ellipse (7.26)
3. Membrane neutral ellipse ($a/b = \sqrt{\sigma_1 / \sigma_2}$)
4. Flexurally neutral circle (7.42)

If a hole is to be placed in a loaded body (e.g., a tunnel), the harmonic shape may be preferable to minimize stress concentration in the sheet and then reinforcement provided to take active loads. For a sheet with a reinforced hole that is then loaded (e.g., buried pipe or most structural applications in plates and shells), the designer would normally choose the membrane configuration or, for ease in construction, a circle. Only if forced to use a rigid liner, would the designer choose the harmonic inclusion shape.

7.6.4 Gradient Fields with an Isotropic Component

In many engineering applications, there is an isotropic component in the stress field in addition to the gradient induced by bending or geostatic conditions. This is the general case in pressure vessels, shell structures or masonry walls, tunnels, pipe, or in other buried structures. For this case the stress function becomes

$$U = -\frac{Ay^3}{6} + \frac{\sigma_m}{2}(x^2 + y^2) + ax + by + c \tag{7.44}$$

in which A is the slope of the linear stress distribution of σ_x.

Although Mansfield does not consider this case, a simple closed membrane shape is possible if the stress function can be put in the form:

$$U = kf^2(x^2 + y^2 - R^2) \tag{7.45}$$

which, when equal to zero, gives the neutral hole shape. To do this let: $x = fx_o$ and $y = y_o$, where x_o, and y_o refer to points on the circle of radius R and f is the "mapping function" which transforms this circle to the membrane shape. The stress function [Equation (7.44)] can then be rewritten:

$$U = \frac{\sigma_m}{2}\left(1 - \frac{Ay_o}{3\sigma_m}\right)y_o^2 + \frac{\sigma_m}{2}\left(1 - \frac{Ay_o}{3\sigma_m}\right)x_o^2 - \frac{\sigma_m R^2}{2}\left(1 - \frac{Ay_o}{3\sigma_m}\right) \tag{7.46}$$

which is the desired form if

$$k = \frac{\sigma_m}{2}; \qquad f^2 = 1 - \frac{Ay}{3\sigma_m} \tag{i}$$

Moreover, the constants in the stress function are:

$$a = 0, \qquad b = \frac{AR^2}{6\sigma_m}, \qquad c = -\frac{\sigma_m R^2}{2} \tag{j}$$

which are not arbitrary and depend on the size and position of the hole. To nondimensionalize the gradient, let $G = 2AR/\sigma_m$, in which case:

$$f = \left(1 - \frac{Gy_o}{6R}\right)^{1/2} \tag{k}$$

limiting the possible gradient field for a membrane shape to $G < 6$.

A more stringent limitation is imposed by the area required of the liner for compatibility. The normal force in the liner is by Equation (7.31a)

$$N = \sigma_m(\sin\psi) - \left(-\frac{Ay^2}{2} + \sigma_m y + \frac{AR^2}{6}\right)\cos\psi \tag{l}$$

and the tangential strain must be

$$(\epsilon_t)_s = \frac{1}{2E_s}[(2\sigma_m - Ay)(1 - 2\nu_s) - Ay(1 + \nu_s)\cos 2\psi] \tag{m}$$

Thus, the area required for the membrane liner will be:

$$A_l = \frac{N}{\epsilon E_\ell} = \frac{2E_s}{E_\ell}\left[\frac{\sigma_m(\sin\psi) + \left(\frac{Ay^2}{2} - \sigma_m y - \frac{AR^2}{6}\right)\cos\psi}{(1 - 2\nu_s)(2\sigma_m - Ay) - (1 + \nu_s)Ay\cos 2\psi}\right] \tag{7.47}$$

Since N in the numerator is always positive, ϵ_t must also be positive. This requirement is critical when $y = R$ in which case, for $\epsilon_t > 0$:

$$A < \frac{\sigma_m}{R}(1 - \nu) \quad \text{or} \quad G < 2(1 - \nu)$$

giving a much more severe restriction on the size of a neutral membrane hole that can be put in any given gradient field.

The neutral "deloid" shape giving no stress concentration with a membrane liner, is compared in Figure 7.14a to the harmonic deloid for an unlined hole for the extreme case, $G = 2$. They are both of the same generic type, but the neutral shape is much closer to a circle. The nondimensionalized area required of the membrane liner for $G = 1.0$, $\nu = 1/3$ is shown in Figure 7.14b. Comparing the result to the membrane liner for a circular hole in an isotropic field where $E_\ell A/E_s R = 1.5$, this neutral deloid design is, for a gradient field, just as feasible. Even for a concrete liner in stiff soil where E_ℓ/E_s might be 100, the depth of a prismatic liner for a 20 ft diameter opening would be only 1.2 in. at the bottom, 1.8 in. at mid-height, and 6 in. at the top.

To achieve a circular neutral liner, rotation compatibility is critical in that

$$M = \frac{Ay^2}{6} - \frac{\sigma_m}{2}(x^2 + y^2) \tag{7.48}$$

with a, b, and c all zero by symmetry arguments. The rotation of the sheet at the interface by Equation (7.36) is

$$\left(\frac{d\phi}{ds}\right)_s = -\frac{A}{E_s}[(1 + \nu_s)\cos 2\theta - \sin\theta] \tag{n}$$

so to match the liner rotations:

$$I_l^b = \frac{E_s R^3}{6E_\ell}\sin^3\theta[(1 + \nu_s)(-1 + 2\sin^2\theta + \sin^3\theta) + (1 - \nu_s)\sin\theta]^{-1} \tag{7.49}$$

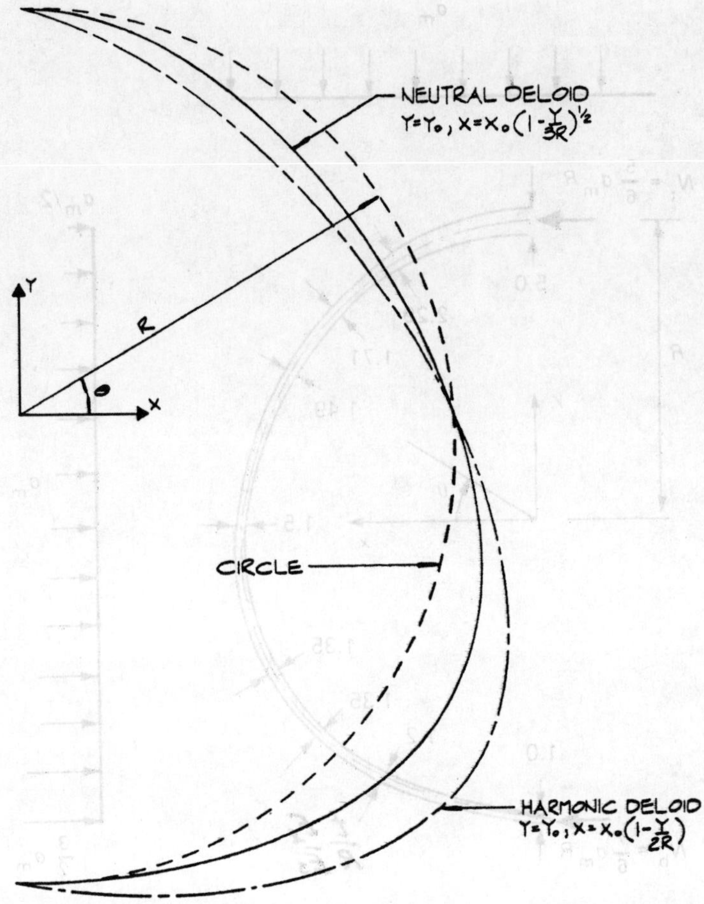

FIGURE 7.14a
Optimum holes for a one-way gradient field. a) Neutral and Harmonic Shapes ($G = 2.0$, $A = \sigma_m/R$)

Therefore, designing a liner for a circular neutral hole in a gradient field is not practical since the moment of inertia required is slightly negative near the neutral axis (unless $\nu_s = 0$). However, this might be feasible in concrete or steel where the sheet near the neutral axis could be made thinner.

7.6.5 Summary

It has been possible to extend the Mansfield theory for membrane neutral liners to the general case of flexural reinforcement. The resulting expressions for equilibrium and compatibility in terms of the stress function U for the sheet are, in fact, not limited to the neutral condition, but are valid for any thin reinforcement and thus, are fundamental to the general interaction problem.

In the classical interaction problem, the total, $U = U^o + U^*$, is unknown since the perturbation in the field U^* due to the hole in unspecified. By inverting

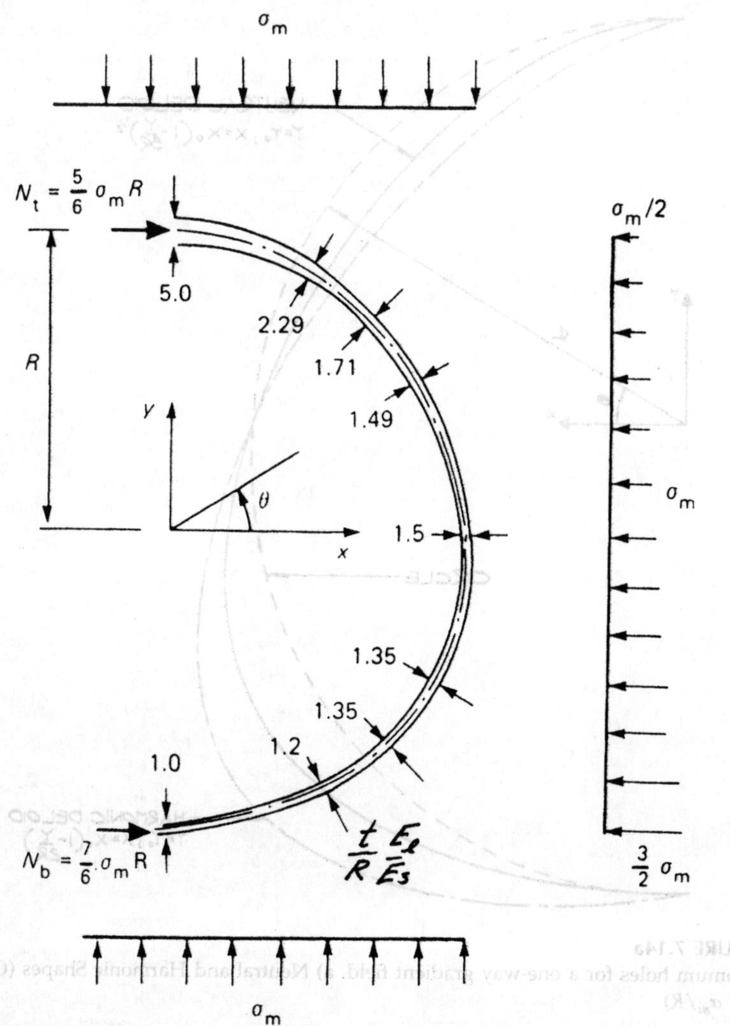

FIGURE 7.14b
b) Neutral Membrane Liner, $G = 1.0$, $\nu_s = \frac{1}{3}$ (values of $t_\ell E_\ell / E_s R$).

the problem to a design mode where U^* is specified (to be zero for the neutral condition), the stresses, strains, and displacements in the field are entirely known, as are the stress resultants in the liner. Thus, the tangential strain and the rotation of the liner can be found from the equations of Bresse and matched at the interface to the same known quantities in the field to give expressions for the required liner area and bending stiffness.

Closed-form solutions are given here for both the circular and the membrane shape for a variety of free fields. A review of these results reveals two interesting limitations to achieving, at least in a simple way, the neutral conditions with a thin continuous liner: Most important is the "Poisson ratio effect." For a positive area of the liner: $(\epsilon_t)_s$ must be exactly in phase with N, i.e., have the same sign at every point on the interface. This might seem obvious, but apparently was not recognized by Mansfield in his work on membrane reinforcement. Similarly, if there is bending, the rotation change $(d\phi/ds)_s$ must be exactly in phase with M to have a positive moment of inertia. Thus, only in special cases such as the purely deviatoric field or biaxial bending can a neutral liner be found for a "simple" closed hole in a region of a field where either U or its gradient change sign or even approach zero.

Within this constraint, neutral liner designs are found for two basic situations important in practice: (1) Circular holes in deviatoric and biaxial fields (flexural liner); and (2) membrane "deloid shape" and liner for the gradient field with an isotropic component. The second case is particularly useful in that it allows a design within a bending field and closely resembles the geostatic field condition for shallow pipes and tunnels. Moreover, since the "deloid" neutral shape is of the same generic type as the harmonic hole for minimum stress concentration with no liner, its use may serve a double purpose both in construction to reduce stresses around the hole and then with reinforcement to best withstand gradient service loads.

Example 7.4

A circular shaft (radius = 20 ft) is to be excavated in a clay layer ($\gamma = 120$ lb/ft³, $E = 3{,}000$ psi, $\nu = 2/5$, $c = Y/2 = 10$ psi. Assuming initial static free-field stresses determine, as a function of depth, the minimum thickness of a *thin* circumferential wooden liner ($E = 1 \times 10^6$ psi, $\nu = 1/3$, $Y = 1000$ psi) to prevent *yield* of the clay (and wood).

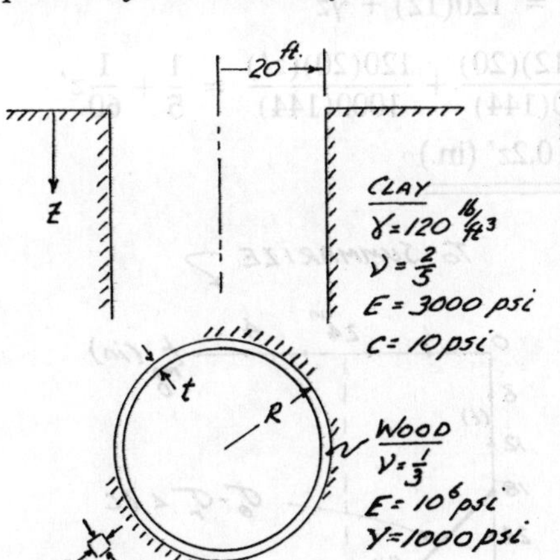

Use soil-mech. sign convention

Assume instantaneous gravity loading

$$K_o = \frac{\nu}{1-\nu} = \frac{2}{3}$$

$\therefore$ in the free field ($r \gg R$)

$$\sigma_z = \gamma z, \sigma_r = \sigma_\theta = 2\gamma z/3$$

$$\therefore \text{since } \sigma_r + \sigma_\theta = \text{const.} = \frac{4}{3}\gamma z$$

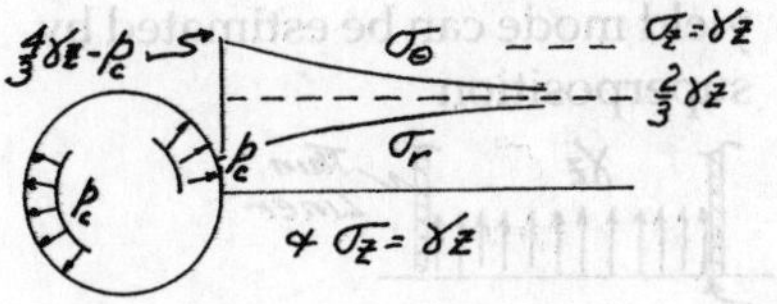

i) Assume no liner $\therefore p_c = 0 \quad \therefore \tau_{\max} = \dfrac{\sigma_\theta - \sigma_r}{2} = \dfrac{2}{3}\gamma z = C$

$$\therefore z_o = \frac{3C}{2\gamma} = \frac{3(10)(144)}{2(120)} = \underline{\underline{18 \text{ ft.}}} \text{ (no liner needed)}$$

ii) z > 18 ft

$\sigma_z = \gamma z, \quad \sigma_r = p_c, \quad \sigma_\theta = \frac{4}{3}\gamma z - p_c, \quad \tau_{rz}$ negl.

let $z^* = z - z_o = z - 18$

$$\therefore \tau_{max} = \frac{\sigma_\theta - \sigma_r}{2} = \frac{2}{3}\gamma(z_o + z^*) - p_c \quad \text{if } \sigma_z < \sigma_\theta$$

$$\therefore -p_c = c - \frac{2}{3}\gamma z^o - \frac{2}{3}\gamma z^* = 10(144) - 12(120) - \frac{2}{3}\gamma z^*$$

and $\therefore p_c = \frac{2}{3}\gamma z^*$

but for a thin ring $\sigma_\theta \cong \frac{pr}{t}, \quad \sigma_r \cong 0 \quad \therefore t \geq \frac{pr}{Y}$

$$t_{req} = \frac{2(120)(20)z^*}{3(1000)(144)} = \frac{z^*}{90}(\text{ft}) = .1333z^*(\text{in.})$$

But at some z(say z′) $\sigma_z = \sigma_\theta$ beyond which $\tau_{max} = \frac{\sigma_z - \sigma_r}{2}$

at $z' = z_o + z^* \quad \sigma_z = \gamma z' = \frac{4}{3}\gamma z' - p_c = \sigma_\theta$

or $\gamma(z_o + z^*) = \frac{4}{3}\gamma(z_o + z^*) - \frac{2}{3}\gamma z^*$

$$\therefore z^*\left(\frac{4}{3}\gamma - \frac{2}{3}\gamma - \gamma\right) = z_o\left(\gamma - \frac{4}{3}\gamma\right) \quad \therefore z^* = z_o = 18' \quad \text{where } \sigma_\theta = \sigma_z$$

$$t_{reg} \text{ at } z^* = 18 \ (z = 36 \text{ ft}) = \frac{18}{90} = \frac{1}{5} = \underline{2.4 \text{ in.}}$$

iii) Below z = 36 ft. σ_θ becomes the intermediate princ. stress

$$\tau_{max} = \frac{\sigma_2 - \sigma_r}{2} = \frac{\gamma z - p_c}{2} \leq C \quad \text{or} \quad \gamma(36) + \gamma z' - p_c = 2C$$

$$\therefore p_c = 120(36) - 20(144) + \gamma z = 120(12) + \gamma z'$$

$$\text{again } (\sigma_\theta)_{wood} \leq \frac{pr}{t} \therefore t \geq \frac{120(12)(20)}{1000(144)} + \frac{120(20)(z')}{1000(144)} = \frac{1}{5} + \frac{1}{60}z'$$

or for $z' = z - 36 \quad t_{req} = \underline{\underline{2.4 + 0.2z' \text{ (in.)}}}$

iv) Note: At some depth z > 36 ft. (say z″) the clay will yield at the free surface z = z″, r = 20 − t leading to base heave. This yield mode can be estimated by superposition

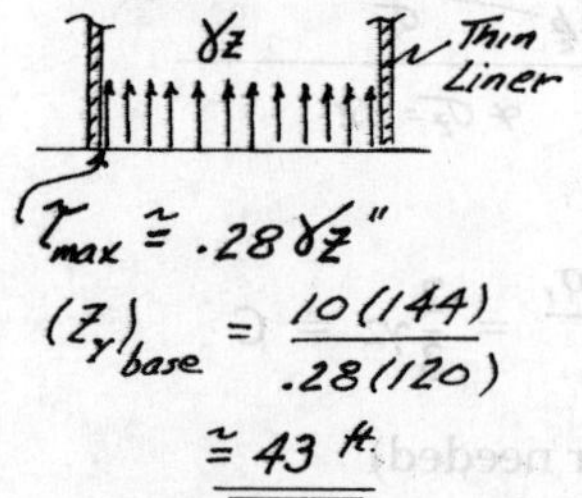

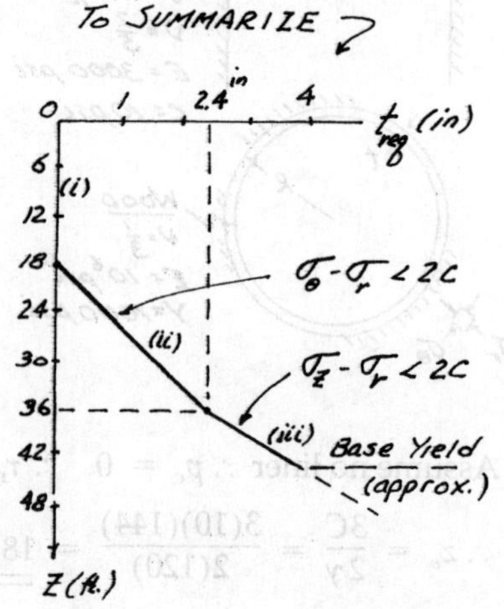

7.7 Rotating Disks and Rings

In the first section of this chapter we discussed the situation where, in polar coordinates, nothing is a function of θ and therefore all radical lines must simply move outward or inward the same amount as any adjacent radial line. We considered pressurized rings to derive Lamé's solution. Let us now consider radial body force due to angular rotation of velocity ω. The effect of spin may be of critical importance in turbines or motors or for flywheels where achieving high angular velocities is, of course, fundamental to its purpose for storing energy. If both pressure and high rates of rotation occur as, for example, where a spinning disk is attached to a drive shaft by a shrink fit, the two effects can be combined by superposition.

7.7.1 Disk of Constant Thickness

For a thin circular disk (Figure 7.15a) the stresses induced by the rotational body force are plane and only a function of r. Therefore, there is only one equilibrium equation:

$$\frac{d\sigma_r}{dr} + \frac{(\sigma_r - \sigma_\theta)}{r} + \rho\omega^2 r = 0 \tag{7.50}$$

where ρ is the mass per unit volume and ω the angular velocity. The strain components, because of symmetry, reduce to

$$\epsilon_r = \frac{du}{dr}, \qquad \epsilon_\theta = \frac{u}{r} \tag{7.51}$$

and the strain-compatibility relation becomes

$$\frac{d\epsilon_\theta}{dr} + \frac{\epsilon_\theta - \epsilon_r}{r} = 0 \tag{7.52}$$

Using the stress–strain relationships, Equations (7.50) through (7.52) can be combined into either a stress equation

$$r^2\frac{d^2\sigma_r}{dr^2} + 3r\frac{d\sigma_r}{dr} = \frac{1}{r}\frac{d}{dr}\left(r^3\frac{d\sigma_r}{dr}\right) = -(3+\nu)\rho\omega^2 r^2 \tag{7.53}$$

or the displacement relationship

$$r^2\frac{d^2u}{dr^2} + r\frac{du}{dr} - u = -\frac{1-\nu^2}{E}\rho\omega^2 r^2 \tag{7.54}$$

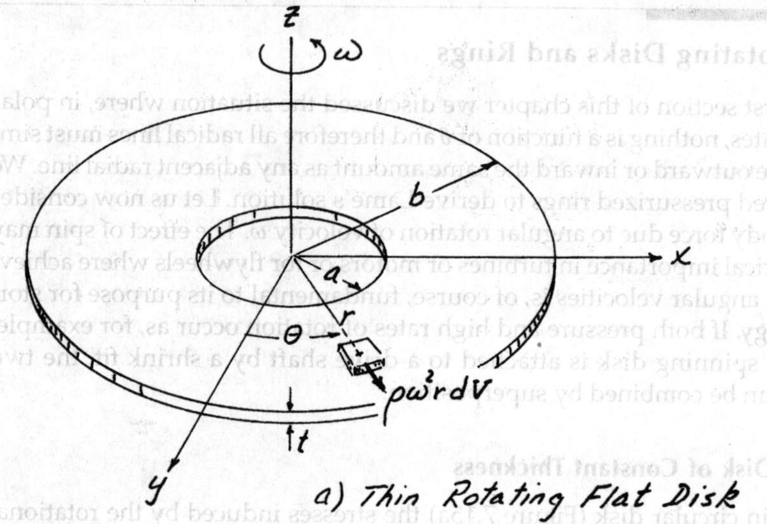

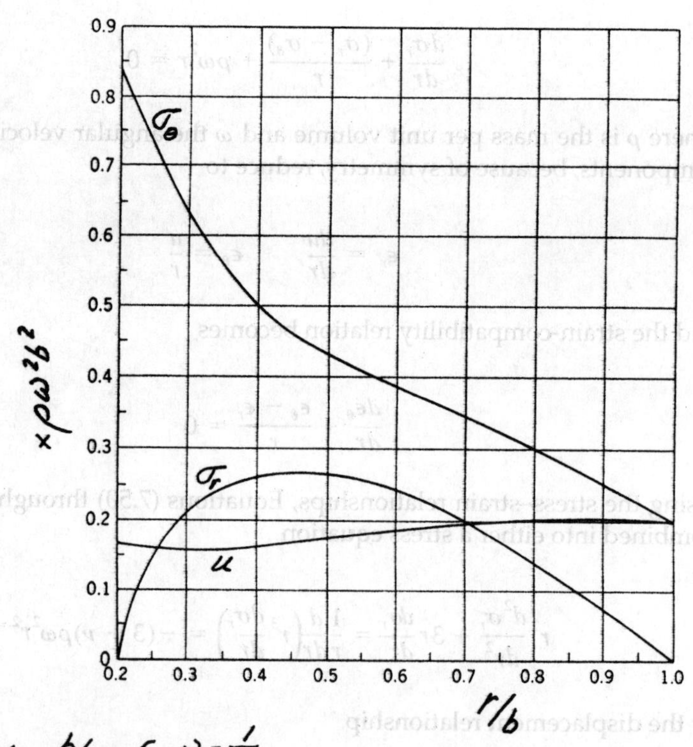

FIGURE 7.15
Annular disk analysis.

Successive integration gives the solution

$$\sigma_r = C_1 + C_2\frac{1}{r^2} - \frac{3+\nu}{8}\rho\omega^2 r^2 \tag{7.55a}$$

$$\sigma_\theta = C_1 - C_2\frac{1}{r^2} - \frac{1+3\nu}{8}\rho\omega^2 r^2 \tag{7.55b}$$

$$u = \frac{1}{E}\left[(1-\nu)C_1 r - (1+\nu)C_1\frac{1}{r} - \frac{1-\nu^2}{8}\rho\omega^2 r^2\right] \tag{7.56}$$

where the integration constants are to be determined from the boundary conditions.

For a *solid* disk ($a = 0$) the stress at $r = 0$ must be finite so $C_2 = 0$ and at $r = b$, $\sigma_r = 0$ so

$$C_1 = \frac{3+\nu}{8}\rho\omega^2 b^2$$

Thus the stresses and displacement are:

$$\begin{aligned}
\sigma_r &= \frac{3+\nu}{8}\rho\omega^2(b^2 - r^2) \\
\sigma_\theta &= \frac{3+\nu}{8}\rho\omega^2 b^2 - \frac{1+3\nu}{8}\rho\omega^2 r^2 \\
u &= \frac{1-\nu}{8E}[(3+\nu)b^2 - (1+\nu)r^2]\rho\omega^2 r
\end{aligned} \tag{7.57}$$

The stress components are a maximum at the center where they both equal

$$\sigma_r = \sigma_\theta = C_1 = \frac{3+\nu}{8}\rho\omega^2 b^2 \tag{7.58}$$

For a disk with a *circular hole* at radius a (an annular disk), the boundary conditions at the center require that $\sigma_r = 0$ at $r = a$ and b. Thus

$$C_1 = \frac{3+\nu}{8}\rho\omega^2(b^2 + a^2), \qquad C_2 = -\frac{3+\nu}{8}\rho\omega^2 a^2 b^2$$

and therefore:

$$\begin{aligned}
\sigma_r &= \frac{3+\nu}{8}\rho\omega^2\left(b^2 + a^2 - \frac{a^2b^2}{r^2} - r^2\right) \\
\sigma_\theta &= \frac{3+\nu}{8}\rho\omega^2\left(b^2 + a^2 + \frac{a^2b^2}{r^2} - \frac{1+3\nu}{3+\nu}r^2\right) \\
u &= \frac{(3+\nu)(1-\nu)}{8E}\left(a^2 + b^2 - \frac{1+\nu}{3+\nu}r^2 + \frac{1+\nu}{1-\nu}\frac{a^2b^2}{r^2}\right)\rho\omega^2 r
\end{aligned} \tag{7.59}$$

The maximum radial stress is at $r = \sqrt{ab}$ where

$$(\sigma_r)_{\max} = \frac{3+\nu}{8}\rho\omega^2(b-a)^2$$

and the maximum hoop stress is on the inner boundary where

$$(\sigma_\theta)_{\max} = \frac{3+\nu}{4}\rho\omega^2\left(b^2 + \frac{1-\nu}{3+\nu}a^2\right)$$

Results for a particular annular disk are shown in Figure 7.14b. Since the stress is isotropic at the center of a solid rotating disk we would expect that a small hole would introduce a stress concentration factor of 2.0. By letting, a, become small in Equation (7.59), we see that this is the case. Thus, it is better to attach a drive shaft to either side of a solid disk rather than pass it through a central hole.

7.7.2 Variable Thickness and the Inverse Problem

From our results for uniform thickness we see why most turbine disks are tapered, being thicker at the center where the stresses are high. Such disks save material and reduce rotational inertia, both of which are beneficial in turbine design (but not for flywheels).

To analyze tapered discs, the fundamental equilibrium equation must be modified to incorporate a change in thickness $h + \Delta h$ at a radius $r + dr$. Thus Equation (7.50) becomes:

$$\frac{1}{rh}\frac{d}{dr}(\sigma_r rh) - \frac{\sigma_\theta}{r} + \rho\omega^2 r = 0 \tag{7.60}$$

If the taper is not severe, the equations of deformation do not change, and therefore Equation (7.54) becomes:

$$\frac{du^2}{dr^2} + \frac{1}{r}\frac{du}{dr} + \frac{u}{r^2} + \frac{1}{h}\left(\frac{du}{dr} + \nu\frac{u}{r}\right)\frac{dh}{dr} = -\frac{1-\nu^2}{E}\rho\omega^2 r \tag{7.61}$$

While Equation (7.61) is easy to derive, it is practically impossible to solve in closed form. The only useful solution is for a hyperbolic profile such as shown in Figure 7.16a where $h \sim H/r$. But a hyperbolic shape is not optimum, just better than a flat disk. Moreover, any profile can be analyzed by numerical methods quickly and with less effort.

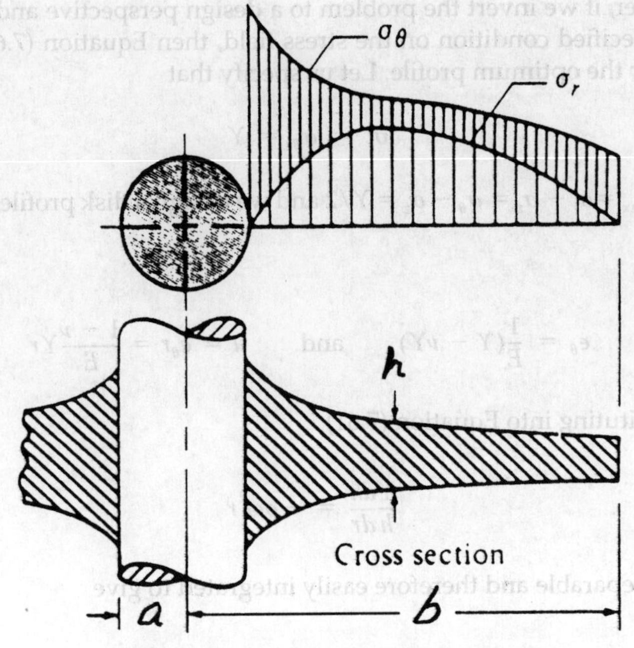

a) Stresses in a Spinning Hyperbolic Disk

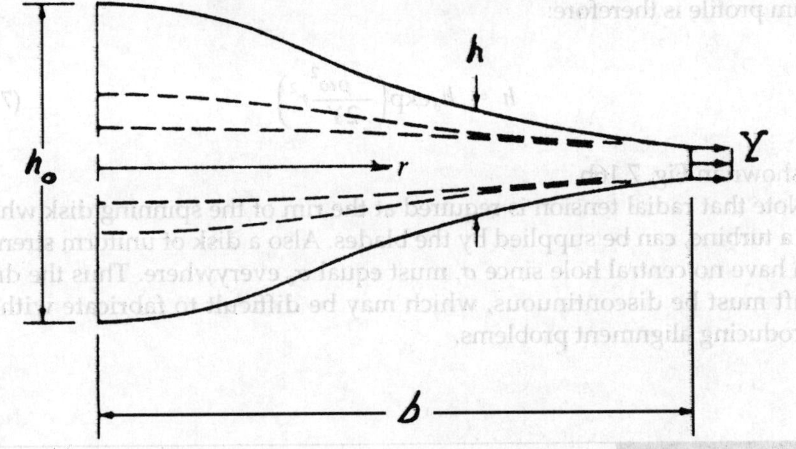

b) Profiles of Spinning Disks of "Uniform Strength"

FIGURE 7.16
Analysis of tapered disks.

However, if we invert the problem to a design perspective and determine h for a specified condition on the stress field, then Equation (7.61) is easily solved for the optimum profile. Let us specify that

$$\sigma_r = \sigma_\theta = Y \tag{7.62}$$

so that $\tau_{max} = \sigma_r - \sigma_z = \sigma_\theta - \sigma_z = Y/2$ and we design a disk profile of uniform *strength*.

Therefore

$$\epsilon_\theta = \frac{1}{E}(Y - \nu Y) \quad \text{and} \quad u = \epsilon_\theta r = \frac{1-\nu}{E} Y r \tag{7.63}$$

and substituting into Equation (7.61)

$$\frac{Y}{h}\frac{dh}{dr} = -\rho\omega^2 r$$

which is separable and therefore easily integrated to give

$$\ln h = -\frac{\rho\omega^2}{2Y}r^2 + \ln C$$

The integration constant C must equal h_o (the thickness at $r = 0$) and the optimum profile is therefore:

$$h = h_o \exp\left(\frac{-\rho\omega^2}{2Y}r^2\right) \tag{7.64}$$

as shown in Fig. 7.16b.

Note that radial tension is required at the rim of the spinning disk which, for a turbine, can be supplied by the blades. Also a disk of uniform strength can have no central hole since σ_r must equal σ_θ everywhere. Thus the drive shaft must be discontinuous, which may be difficult to fabricate without introducing alignment problems.

7.8 Problems and Questions

P7.1 Rather than use either approach presented in the text for a pressurized ring, directly integrate the field equations in polar coordinates [Equations (6.13) and (6.15)] to derive Equation (7.1) and thus the solution. Remember nothing can be function of θ.

P7.2 A short steel rod of 2 in. diameter is subjected to an axial compressive load of 60,000 lb. It is surrounded by a steel sleeve 1/2 in. thick, slightly shorter than the rod so that the load is carried only by the rod. Assuming a close fit before the load is applied and, neglecting friction, find the pressure between the sleeve and the rod and the maximum tensile stress in the sleeve ($\nu = 0.3$).

P7.3 A brass rod is fitted firmly inside a steel tube whose inner and outer diameters are 1 in. and 2 in., respectively, when the materials are at a temperature of 60 °F. If the rod and the tube are both heated to a temperature of 300 °F, determine the maximum stress in the brass and in the steel. The coefficients of expansion for steel and brass are 6×10^{-6} and 10×10^{-6} per degree Fahrenheit, respectively. Young's modulus is 30×10^6 psi for steel and 12.5×10^6 psi for brass. Poisson's ratio is 0.3 for steel and 0.34 for brass.

P7.4 A 100 ft diameter water tank 20 ft high is to be built of concrete ($f_c' = Y_c = 4000$ psi). To eliminate tension, it is prestressed with external ½ diameter circumferential steel tendons at a stress of 160,000 psi. Determine the tendon spacing and the thickness for the bottom foot of the tank. Assume a factor of safety of 2 against crushing.

P7.5 Rather than using a single steel cylinder $b = 3a$, a compound cylinder is proposed consisting of an outer steel ring $b_o = 3a$, $a_o = 2a - \Delta$, and an inner steel ring $b_i = 2a + \Delta$ shrink-fitted together as shown in Figure P7.5. This is done by heating the outer ring, which

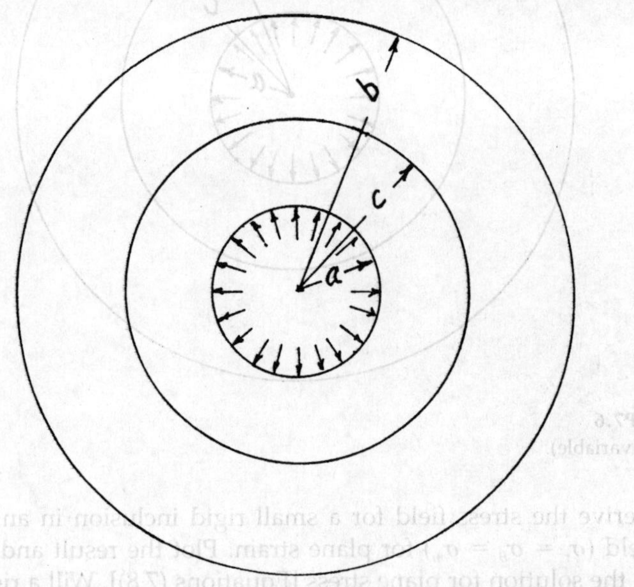

FIGURE P7.5
($b = 3a$, $c = 2a$).

is machined such that $b_i - a_o = 2\Delta$, and slipping it over the inner ring. Thus, when the assembly cools to ambient temperature, a contact pressure, p_c is introduced between them prestressing the rings. Using the Tresca yield criterion, determine the optimum mismatch dimension 2Δ and therefore the temperature change ΔT required for the shrink-fit. Compare the added capacity of the compound ring to a single ring by comparing internal pressure which would cause yield in each. Assume $E = 30 \times 10^6$ psi, $\nu = 0.3$, $\alpha = 6 \times 10^{-6}/°\text{F}$.

P7.6 Refer to Figure P7.6. Develop a general theory of compound rings following the procedure used in P7.5 but leaving the radii variable with the radius between the inner at outer rings, c, to be optimized in addition to 2Δ. Compare to a single ring and to the ring of problem 7.5 where

$$c = \frac{a + b}{2} = 2a$$

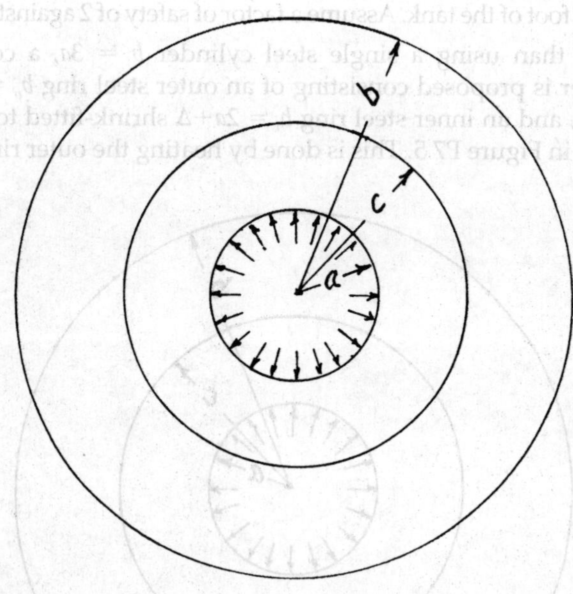

FIGURE P7.6
(a, b, & c variable).

P7.7 Derive the stress field for a small rigid inclusion in an isotropic field ($\sigma_r = \sigma_\theta = \sigma_m$) for plane strain. Plot the result and compare to the solution for plane stress [Equations (7.8)]. Will a rigid inclusion perturb the field if $\nu = 1/2$? Why?

P7.8 Solve the general problem of a circular elastic inclusion of variable E_i, ν_i in an isotropic field. Plot the result to compare with the hole and the rigid inclusion as the limiting cases.

P7.9 Design a neutral liner for the isotropic case where the liner has only twice the Young's modulus of the sheet and the Poisson's ratios are the same (i.e., $E_\ell = 2\ E_s$ and $\nu_\ell = \nu_s$).

P7.10 Determine and plot the deformation of the *boundary* of the hole in the deviatoric field corresponding to the stress field given by Equation (7.17).

P7.11 For the circular hole in a deviatoric field, what is the maximum value of $\tau_{r\theta}$ and where does it occur?

P7.12 For the circular hole in a deviatoric field, plot for multiples of $\sigma_d / 2$ for the first quadrant for $r \leq 3a$:

a. The isochromatics τ_{max}
b. The flow net for $\sigma_1 + \sigma_2$ and $E\omega_z$
c. The isoclinics (the angle at which the principal stresses are oriented to the horizontal and vertical).

P7.13 Make the field plots called for in P7.12 for the case of uniaxial tension.

P7.14 For the cases covered in Section 7.2, it would seem that there is a relationship for the stress concentration factor due to circular holes as a function of the far free field. Derive and plot such a relationship and check it out for $\sigma_1 = 2\sigma_2$, $\sigma_1 = -\sigma_2/2$.

P7.15 Compare the error in using the solutions of Section 7.2 for a plate of finite width $w = nd$ ($2 \leq n \leq 6$) if d is the diameter of the circular hole.

P7.16 Determine the Harmonic Shape (elliptic) for a rigid inclusion in a plate with $\nu = 0$ and $\nu = 1/2$ and compare to that for the $N_1 = 2N_2$ field with $\nu = 1/4$.

P7.17 Compare the stress concentration for mean stress, $(\sigma_r + \sigma_\theta)/2$, maximum shear stress, and hoop stress, σ_θ, for the unloaded harmonic hole in a biaxial field to that for a circular hole as a function of the N_1/N_2 ratio.

P7.18 Can the shape for a harmonic rigid inclusion in a linear gradient field be inferred from that for an unloaded or uniformly loaded hole? How might you approach this problem.

P7.19 What shape to minimize stress concentration would you propose for a small hole, $R = a/10$, centered at $r = (a + b)/2$ in a thick ring $b = 2a$ loaded by external pressure?

P7.20 Design the membrane neutral liner shape for an isotropic field with biaxial gradients $G = 1.5$. Compare the shape to the harmonic cardeloid [Equation (7.30)].

P7.21 Design a *circular* neutral liner for the 2 to 1 biaxial field of Figure 7.13a. Is it feasible?

P7.22 Express the general stress boundary conditions in two dimensions [Equation (4.1)] in terms of the Stress Function and compare to those for a liner [Equation (7.31)]. What special form do they take for a harmonic hole boundary (with and without a constant boundary traction)?

P7.23 What effect does a geostatic body force have on the derivation for the force resultants in a thin liner [Equation (7.31)]? Is a membrane or circular neutral liner possible in a geostatic field (Figure 5.3)?

P7.24 As shown in Section 5.4, a horizontal acceleration of the half-space introduces simple shear $\tau_{zx} = k_h \gamma z$. Is a membrane or circular neutral liner possible?

P7.25 Compare the stresses and displacement for a solid rotating disk of the same thickness, outer radius, and material properties given for the annular disk in Figure 7.15b. Make a comparative plot.

P7.26 A flat steel disk (a = 4 in., b = 16 in.) which is 4 inches thick is shrunk onto a steel shaft. If the assembly is to run safely at 5400 rpm, determine:

a. the shrink allowance

b. the maximum stress when not rotating

c. the maximum stress when rotating

The material properties are = 30 × 10^6 psi, $\nu = \frac{1}{3}$, and $\rho = 0.00072$ lb-sec^2/in^4. Assume the shaft can be treated as a solid disk in plane stress. The shrinkage allowance is the original mismatch in the shrink-fit process necessary such that the disk and shaft just separate at the design speed. Plot the variation in stresses in the disk as a function of ω.

8

Wedges and the Half-Space

Structures that can be idealized to wedges are found at every scale from massive gravity dams to gear teeth. Moreover, for wedge angles greater than 90°, no strength-of-materials type analysis is possible and only for very small angles should we expect such simplified analysis to give even a decent approximation. Therefore, as for thick rings and plates with small holes, two-dimensional solutions from elastic field theory are mandatory for the design of such structures.

8.1 Concentrated Loadings at the Apex

Consider first a wedge of angle 2ϕ in plane stress such as shown in Figure 8.1 with an axial load, P, at the apex. If the wedge angle were very small, then we might expect $\sigma_x \cong \sigma_r \cong P/A = P/(2r\phi)$. Let us assume, then, a stress function of type III:

$$F = r^m[A\cos m\theta + B\sin m\theta + C\cos(m-2)\theta + D\sin(m-2)\theta]$$

to satisfy $\nabla^4 F = 0$. In this case we expect that $m = 1$ since we anticipate that the radical stress varies as $1/r$. Therefore, as we have seen, assume

$$F = r[\underbrace{A'\cos\theta + B'\theta\sin\theta}_{\text{symmetric}} + \underbrace{C'\sin\theta + D'\theta\cos\theta}_{\text{asymmetric}}] \tag{a}$$

But we know the stresses must be symmetric with θ and the first term gives nothing so, through this physical reasoning process, we can deduce that:

$$F = Br\theta\sin\theta \tag{8.1}$$

Therefore, the stresses are:

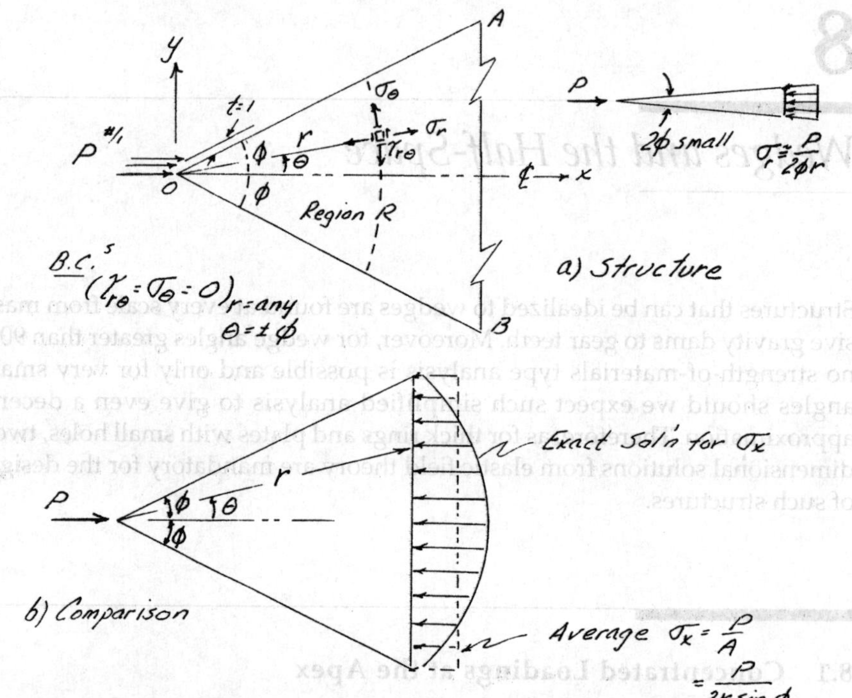

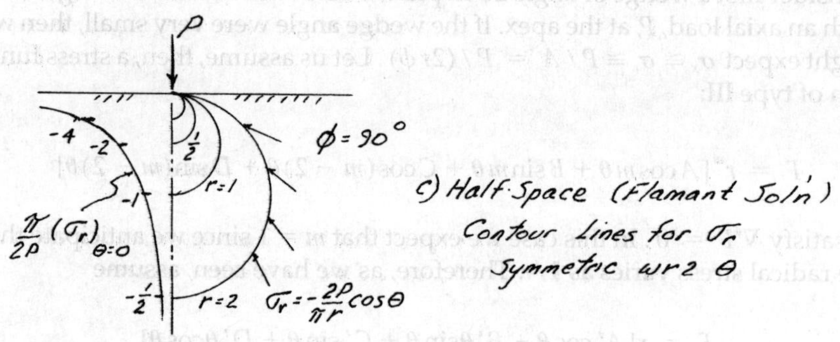

FIGURE 8.1
Axial loading of a plane wedge.

$$\sigma_r = \frac{1}{r}\frac{\partial F}{\partial r} + \frac{1}{r^2}\frac{\partial^2 F}{\partial \theta^2} = \frac{2B\cos\theta}{r}$$

$$\sigma_\theta = \frac{\partial^2 F}{\partial r^2} = 0 \qquad \text{(b)}$$

$$\tau_{r\theta} = -\frac{\partial}{\partial r}\left(\frac{1}{r}\frac{\partial F}{\partial \theta}\right) = 0$$

with the boundary conditions on the faces $\theta = \pm\,\phi$ automatically satisfied.

In addition, overall equilibrium of a region R requires that:

$$\Sigma F_x = \int_{-\phi}^{\phi} \sigma_r \, dA \cos\theta = \int_{-\phi}^{\phi} \sigma_r (r d\theta \cdot 1) \cos\theta = -P$$

Therefore:

$$-2\int_{0}^{\phi} 2B\frac{\cos\theta r}{r}\cos\theta \, d\theta = P$$

so that

$$B = \frac{-P}{2\phi + \sin 2\phi}$$

and the stress field is:

$$\sigma_r = -\frac{2P\cos\theta}{(2\phi + \sin 2\phi)r}; \qquad \sigma_\theta = \tau_{r\theta} = 0 \tag{8.2}$$

A comparison of σ_x from the elasticity solution to the average value $\sigma_x = P/A$ is shown in Figure 8.1b. We see that only for small wedge angles is the strength-of-materials approximation adequate.

This solution when applied to the half-space is particularly important as it corresponds to the punch or bearing capacity situation. It also serves as one of the fundamental building blocks for 2D boundary-element modeling. The result for $2\phi = \pi$, often called the Flamant Solution,

$$\sigma_r = \frac{-2P}{\pi}\frac{\cos\theta}{r}, \qquad \sigma_\theta = \tau_{r\theta} = 0 \tag{8.3}$$

is shown in Figure 8.1c. The stresses die off as $1/r$ and the isopachics and isochroniatics (contour lines of equal $\sigma_1 + \sigma_2$ and $\sigma_1 - \sigma_2$) are identical circles:

The load, P, must actually be distributed over a small "circular" area near the apex in which plastic flow takes place redistributing the stress to its yield value at $r = r_{\text{yield}}$. Outside this plasticized zone the strains can be computed from the stresses using the elastic relationships and integrated to determine displacements.*

* See Timoshenko and Goodier, *Theory of Elasticity*, 3rd Ed., McGraw-Hill, 1951, pp. 102–104 for a discussion. Since the harmonic invariant $\omega_z = \omega_{r\theta}$, can be found directly from its conjugate $\sigma_x + \sigma_y = \sigma_r + \sigma_\theta$, the procedure is straightforward.

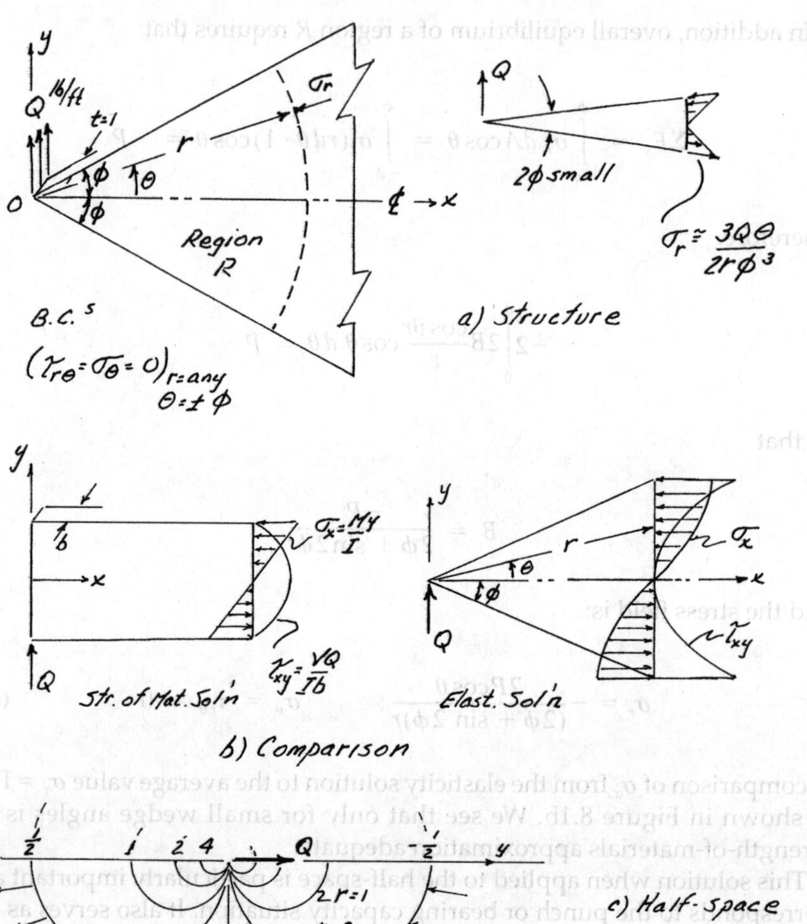

FIGURE 8.2
Transverse loading of a plane wedge.

For a vertical transverse load Q at the apex, the reasoning is similar. For a very small wedge angle, we would expect that $\sigma_r \approx \sigma_x \cong My/I$ where, as shown in Figure 8.2, $M = Qr$, $y = r\theta$, and $I = (2r\phi)^3/12$. Therefore $\sigma_r \cong 3Q\theta/2r\phi^3$, so again the radial stress dies off as $1/r$ and we should assume $m = 1$ in the stress function (a). In this case, however, since there is bending and no axial force, only the asymmetric terms apply. The term $Cr\sin\theta$ gives no stress and therefore:

$$F = Dr\theta\cos\theta \tag{8.4}$$

Again $\sigma_\theta = \tau_{r\theta} = 0$, so the only stress in radial

$$\sigma_r = \frac{1}{r}\frac{\partial F}{\partial r} + \frac{1}{r^2}\frac{\partial^2 F}{\partial \theta^2} = -2D\frac{\sin\theta}{r} \tag{c}$$

The boundary conditions like the overall moment and axial-force equilibrium requirements are already satisfied in Region R but to satisfy vertical equilibrium:

$$Q = -\int_{-\phi}^{\phi} \sigma_r \; dA \sin\theta = -2\int_0^{\phi} -2D\frac{\sin\theta}{r}\sin\theta(rd\theta)$$

and

$$D = \frac{Q}{2\phi - \sin 2\phi}$$

giving the stress field:

$$\sigma_\theta = \tau_{r\theta} = 0, \qquad \sigma_r = -\frac{2Q}{2\phi - \sin 2\phi}\frac{\sin\theta}{r} \tag{8.5}$$

For any significant wedge angle, a strength-of-material approximation is unsatisfactory. For the half-space, Q becomes a concentrated shear producing radial stress bulbs,

$$\sigma_r = \frac{-2Q}{\pi}\frac{\sin\theta}{r}, \qquad \sigma_\theta = \tau_{r\theta} = 0 \tag{8.6}$$

as shown in Figure 8.2c. They are identical to those for P but 90° out of phase.

Clearly the solutions for concentrated line loads P and Q at the apex of a wedge can be superimposed to give the field for a line load at any inclination. Moreover, a line load within an "infinite" elastic body ($\phi = \pi$), sometimes called the Kelvin Problem, can also be solved.*

Finally let us solve for the stresses in a wedge with a concentrated line moment at the apex (Figure 8.3). Again, considering a narrow wedge with ϕ

* See Timoshenko and Goodier, *Theory of Elasticity*, 3rd Ed., McGraw-Hill, 1951, pp. 127–132 for a presentation of the solution to the Kelvin Problem and other cases related to it. Most interesting is that the stress field at $\phi = \pi$ "suddenly" introduces Poisson's ratio indicating a three-dimensional aspect to the field introduced by displacement boundary conditions.

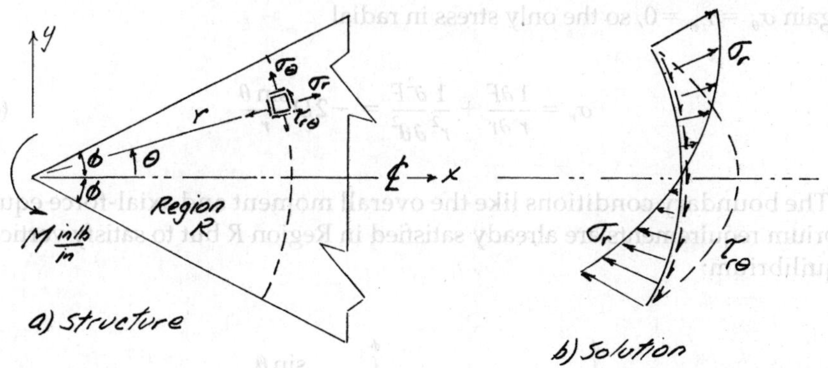

FIGURE 8.3
Line moment at the apex.

small, $\sigma_r \cong \sigma_x = My/I$ and therefore since $\sigma_r \sim 1/r^2$ this time we should assume $m = 0$. Eliminating the symmetric terms, assume

$$F = B\theta + D\sin 2\theta \tag{8.7}$$

The boundary condition on σ_θ is satisfied, but the requirement that the surfaces at $\theta = \pm\phi$ have no shear gives:

$$\tau_{r\theta} = -\frac{\partial}{\partial r}\left(\frac{1}{r}\frac{\partial F}{\partial \theta}\right) = -\frac{B}{r^2} - \frac{2D}{r^2}\cos 2\phi = 0$$

and therefore

$$B = -2D\cos 2\phi \tag{d}$$

so

$$\tau_{r\theta} = \frac{-2D}{r^2}(\cos 2\theta - \cos 2\phi) \tag{e}$$

Moment equilibrium of any Region R tells us that

$$\int_{-\phi}^{\phi} \tau_{r\theta}(r\,d\theta \cdot 1)r = M = -\int_{-\phi}^{\phi}(B + 2D\cos 2\theta)\,d\theta \tag{f}$$

and therefore:

$$M = -2B\phi - 2D\sin 2\phi \tag{g}$$

Solving (d) and (g) simultaneously

$$D = -\frac{M}{2}\left(\frac{1}{\sin 2\phi - 2\phi\cos 2\phi}\right) \quad \text{and} \quad B = \frac{M\cos 2\phi}{\sin 2\phi - 2\phi\cos 2\phi}$$

Thus the solution for the stress field is:

$$\begin{aligned}
\sigma_r &= \frac{1}{r}\frac{\partial F}{\partial r} + \frac{1}{r^2}\frac{\partial^2 F}{\partial \theta^2} = \frac{2M}{\sin 2\phi - 2\phi\cos 2\phi}\frac{\sin 2\theta}{r^2} \\
\sigma_\theta &= \frac{\partial^2 F}{\partial r^2} = 0 \\
\tau_{r\theta} &= -\frac{\partial}{\partial r}\left(\frac{1}{r}\frac{\partial F}{\partial \theta}\right) = \frac{M}{\sin 2\phi - 2\phi\cos 2\phi}\cdot\frac{\cos 2\phi - \cos 2\theta}{r^2}
\end{aligned} \tag{8.8}$$

On the boundaries $\theta = \pm\phi$, $\sigma_x = \sigma_r\cos^2\phi$ so that the strength-of-materials approximation breaks down quickly for significant wedge angles. The comparison of normal stress is similar to that for a transverse load in Figure 8.2b, but the shears are very different. More significant, perhaps, is that the field is highly localized near the point of application since the stresses induced by the moment die off as $1/r^2$ compared to $1/r$ for a concentrated load.

For the half-space,* with $\phi = \frac{\pi}{2}$:

$$\sigma_r = \frac{2M}{\pi}\frac{\sin 2\theta}{r^2}, \qquad \sigma_\theta = 0, \qquad \tau_{r\theta} = \frac{-2M}{\pi}\frac{\cos^2\theta}{r^2} \tag{8.9}$$

For larger wedge angles the solution is suspect. At $2\phi = 257.4°$ the denominator $\sin 2\phi - 2\phi\cos 2\phi$ goes to zero and the solution blows up. Sternberg and Koiter term this ambiguity in the two-dimensional solution for a line moment on a wedge for $2\phi > \pi$ a paradox,** which they do not entirely resolve. We shall see that such ambiguity is present in a number of wedge solutions and that all questions in "simple" two-dimensional elasticity theory are not yet answered.

Example 8.1

A tapered Epoxy beam (wedge) in plane stress is loaded by shear tractions of $+s$ on the upper face and $-s$ on the lower face as shown. The stress function $F = Ax^2 - By^2$ is proposed. Evaluate A and B and show that this is the solution. Are the support conditions appropriate? For $\nu = 1/3$, E = 200,000 psi determine the deformed shape of a 60° wedge (2ϕ=60°) if ℓ=10 ft. for s =10,000 lb/ft per inch thickness. Sketch the deformed shape.

---

* Analysis of this solution for the half-space is covered in the chapter problems.

** Sternberg, E., and Koiter, W.T., The Wedge Under a Concentrated Couple: A Paradox in Two-Dimensional Theory of Elasticity, *Transactions*, Am. Soc of Mech. Eng'rs., Paper n58-A-15, *J. of Appl. Mech.*, pp. 575–581, Dec. 1958.

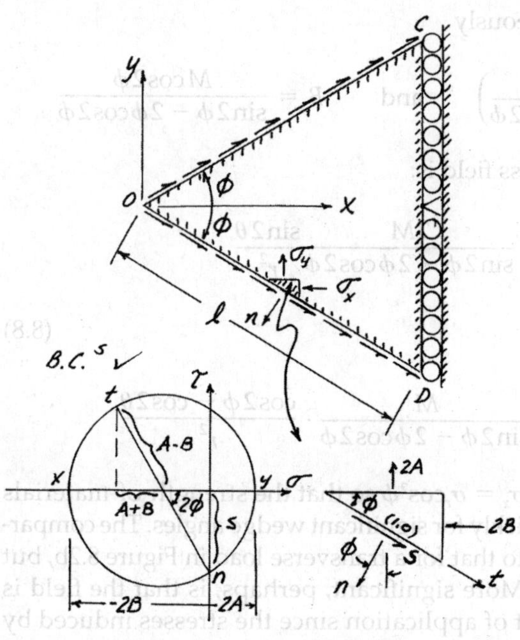

$$F = Ax^2 - By^2$$

Assuming no body forces

$$\sigma_x = \frac{\partial^2 F}{\partial y^2} = -2B$$

$$\sigma_y = \frac{\partial^2 F}{\partial x^2} = -2A$$

$$\tau_{xy} = -\frac{\partial^2 F}{\partial x \partial y} = 0$$

$\therefore$ x & y are principal dir. & stresses are const. as are the invariants.

$$\therefore \gamma_{xy} = \omega_z = 0$$

$$\Sigma F_x = 0$$

$$\therefore 2B\sin\phi = s(1)(\cos\phi)$$

$$B = s/2\tan\phi$$

$$\Sigma F_y = 0 \;\therefore 2A\cos\phi = s(1)\sin\phi \qquad \therefore A = \frac{s}{2}\tan\phi$$

$\therefore$ This is the solution and support conditions are perfect. For $\phi = 30°$, $\tan\phi = \frac{1}{\sqrt{3}}$

$$\therefore \sigma_x = -2B = -s\sqrt{3};\; \sigma_y = 2A = s\sqrt{3}$$

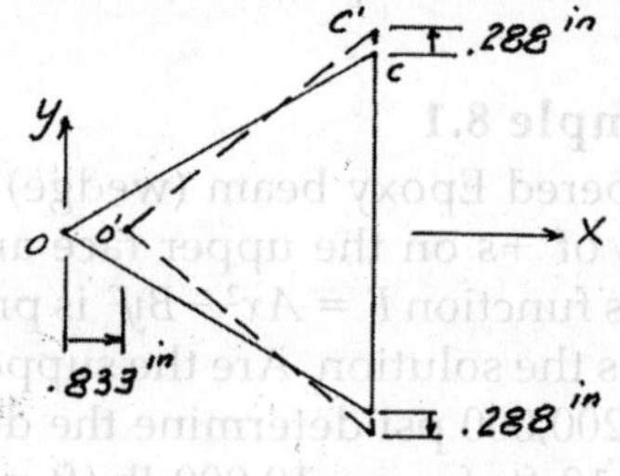

$$\therefore \epsilon_x = \frac{1}{E}(-s\sqrt{3} - \nu s/\sqrt{3}) = -\frac{10s}{3\sqrt{3}E}$$

$$\epsilon_y = \frac{1}{E}(s/\sqrt{3} - \nu s\sqrt{3}) = \frac{2s}{\sqrt{3}E}$$

$$\therefore u_0 = \int_{5\sqrt{3}}^{0} \epsilon_x\, dx = \frac{50s}{3E} = .833\rightarrow$$

$$v_c = \int_0^5 \epsilon_y\, dy = \frac{10s}{\sqrt{3}E} = .288\uparrow$$

8.2 Uniform Loading Cases

A similar paradox and a clue to its source and possible resolution can be seen clearly in the solution for a uniform (isotropic) loading on each face (Figure 8.4a) of a wedge. From the boundary condition that at any radius when, $\theta = \pm\phi$

$$\sigma_\theta = -p = \frac{\partial^2 F}{\partial r^2}$$

m must equal 2 and using only the symmetric terms

$$F = Ar^2 \cos 2\theta + Cr^2 \tag{8.10}$$

Therefore

$$\sigma_r = \frac{1}{r}\frac{\partial F}{\partial r} + \frac{1}{r^2}\frac{\partial^2 F}{\partial \theta^2} = -2A\cos 2\theta + 2C$$

$$\sigma_\theta = \frac{\partial^2 F}{\partial r^2} = 2A\cos 2\theta + 2C$$

$$\tau_{r\theta} = -\frac{\partial}{\partial r}\left(\frac{1}{r}\frac{\partial F}{\partial \theta}\right) = 2A\sin 2\theta$$

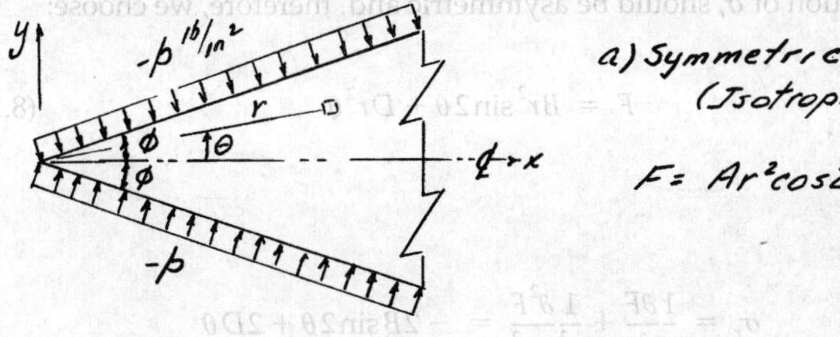

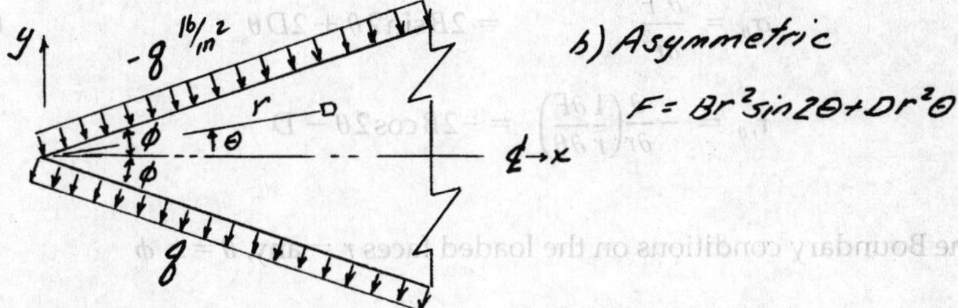

FIGURE 8.4
Wedges with uniform surface loads.

but when $\theta = \pm\phi$ for any r, $\tau_{r\theta} = 0$ and A must be zero. Similarly, at $\theta\pm\phi$, $\sigma_\theta = -p$ so $2C = -p$ and therefore the field is simply a constant isotropic value.

$$\sigma_r = \sigma_\theta = -p; \qquad \tau_{r\theta} = 0 \tag{8.11}$$

Everywhere Mohr's Circle is a point and the rotation is zero as is the shear strain and $\epsilon_x = \epsilon_y = \epsilon_r = \epsilon_\theta = -p(1-\nu)/E$.

But we know this solution is wrong for the half-space where, since $\epsilon_y = 0$, we deduced in Chapter 5 that the horizontal stress is $\sigma_y = -\nu p$ or in plane strain:

$$\sigma_y = \sigma_z = -\frac{\nu}{1-\nu}p \tag{a}$$

Clearly, since there is no singularity in the solution, its nonexistence for $2\phi = 180°$ introduces a more fundamental ambiguity. For wedge angles less than 90° where there is no constraint on ϵ_θ no matter what the radius, the isotropic field seems reasonable, but for $2\phi > 90°$, it seems suspect. Certainly equilibrium, compatibility, and the stress boundary conditions are satisfied for any wedge angle, but there seems to be some added condition on deformations that precludes a simple isotropic field.

The same is true for the accepted solution for a wedge with uniform tension and compression on the faces (Figure 8.4b). Again, by similar reasoning, $m = 2$, the distribution of σ_r should be asymmetric and, therefore, we choose:

$$F = Br^2\sin 2\theta + Dr^2\theta \tag{8.12}$$

Therefore:

$$\begin{aligned}
\sigma_r &= \frac{1}{r}\frac{\partial F}{\partial r} + \frac{1}{r^2}\frac{\partial^2 F}{\partial\theta^2} &&= -2B\sin 2\theta + 2D\theta \\
\sigma_\theta &= \frac{\partial^2 F}{\partial r^2} &&= 2B\sin 2\theta + 2D\theta \\
\tau_{r\theta} &= -\frac{\partial}{\partial r}\left(\frac{1}{r}\frac{\partial F}{\partial\theta}\right) &&= -2B\cos 2\theta - D
\end{aligned} \tag{b}$$

The Boundary conditions on the loaded faces $r =$ any, $\theta = \pm\,\phi$

$$\begin{aligned}
\tau_{r\theta} &= 0 = -2B\cos 2\phi - D \\
\sigma_\theta &= -q = 2B\sin 2\phi + 2D\phi
\end{aligned} \tag{c}$$

give

$$B = -\frac{q}{2}\left(\frac{1}{\sin 2\phi - 2\phi\cos 2\phi}\right), \qquad D = q\left(\frac{1}{\sin 2\phi - 2\phi\cos 2\phi}\right) \tag{d}$$

so:

$$\begin{aligned}
\sigma_r &= q\frac{(\sin 2\theta + 2\theta\cos 2\phi)}{(\sin 2\phi - 2\phi\cos 2\phi)}\\
\sigma_\theta &= q\frac{(-\sin 2\theta + 2\theta\cos 2\phi)}{(\sin 2\phi - 2\phi\cos 2\phi)}\\
\tau_{r\theta} &= q\frac{(\cos 2\theta - \cos 2\phi)}{(\sin 2\phi - 2\phi\cos 2\phi)}
\end{aligned} \tag{8.13}$$

Again this solution cannot be correct in the far field for the half-space where the horizontal stress σ_y is a function of ν. Moreover, the solution has the same singularity at $2\phi = 257.4°$ as in the case of a line moment at the apex [Equation (8.8)]. Again the denominator $\sin 2\phi - 2\phi\cos 2\phi$ goes to zero and the solution blows up.

Finally, if we add the two previous cases together as in Figure 8.4 where $w/2 = p = q$, this gives the solution for a wedge with a uniform load along the top surface (Figure 8.5).

$$\begin{aligned}
\sigma_r &= -\frac{w}{2} + \frac{w}{2}\frac{(\sin 2\theta + 2\theta\cos 2\phi)}{(\sin 2\phi - 2\phi\cos 2\phi)}\\
\sigma_\theta &= -\frac{w}{2} - \frac{w}{2}\frac{(\sin 2\theta - 2\theta\cos 2\phi)}{(\sin 2\phi - 2\phi\cos 2\phi)}\\
\tau_{r\theta} &= \;\;0 + \frac{w}{2}\frac{(\cos 2\theta - \cos 2\phi)}{(\sin 2\phi - 2\phi\cos 2\phi)}
\end{aligned} \tag{8.14}$$

satisfying the boundary conditions that at r = any $\sigma_\theta = \tau_{r\theta} = 0$ at $\theta = -\phi$, and $\sigma_\theta = -w$ and $\tau_{r\theta} = 0$ at $\theta = +\phi$.

The special cases of a vertical cut, $\phi = 45°$, and the half-space are shown in Figure 8.5 b and c. Again in the far field under the load, the solution is suspect, but near the origin it gives sensible results. An example of how wedge solutions can be combined to give a solution for a haunched beam with uniform loading is shown in Figure 8.5d. Other examples are given in the chapter problems.*

* We could also continue this process and derive solutions for parabolic, cubic, and higher-order distributed loadings on one or both faces. A few are given as chapter problems and many others are found in various texts.

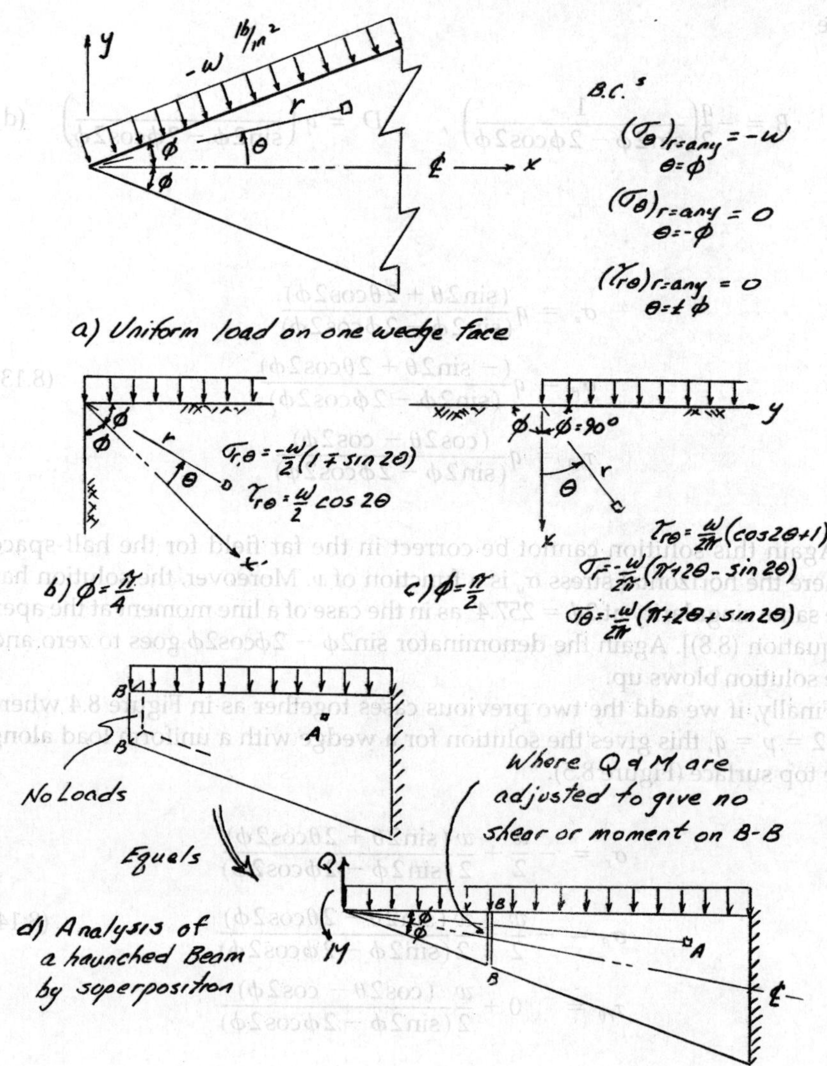

FIGURE 8.5
Wedges with uniform loading on one face.

Example 8.2

A wedge is loaded by a uniform shearing traction, *s*, on its "upper" face as shown below. Determine the general stress function in terms of unknown coefficients *A*, *B*, *C*, and *D*. *For* $\phi = 45°$, evaluate *A*, *B*, *C*, and *D* and plot stresses at the cross-section $r = 1.0$.

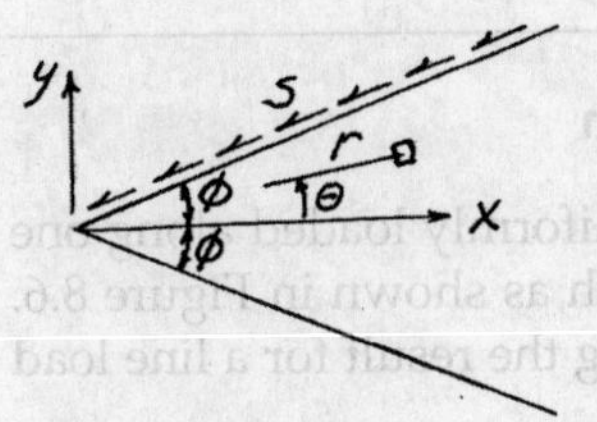

For φ small

$$\sigma_r \cong sr/2r\phi \cong s/2\phi$$

$$\therefore Try\ m = 2 \text{ and}$$

$$\therefore F = r^2(A\theta - B\sin 2\theta - C\cos 2\theta - D) \quad (1)$$

Giving stresses

$$\sigma_r = \frac{1}{r}\frac{\partial F}{\partial r} + \frac{1}{r^2}\frac{\partial^2 F}{\partial \theta} = 2A\theta - 2B\sin 2\theta - 2C\cos 2\theta - 2D + 4B\sin 2\theta + 4C\cos 2\theta$$

$$\therefore \sigma_r = 2A\theta - 2B\sin 2\theta - 2C\cos 2\theta - 2D \quad (2)$$

$$\sigma_\theta = \frac{\partial^2 F}{\partial r^2} = 2A\theta - 2B\sin 2\theta - 2C\cos 2\theta - 2D \quad (3)$$

$$\tau_{r\theta} = -\frac{\partial}{\partial r}\left(\frac{1}{r}\frac{\partial F}{\partial \theta}\right) = -A + 2B\cos 2\theta - 2C\sin 2\theta \quad (4)$$

B.C.s for $\phi = 45°$, r = any: $\sigma_\theta = \tau_{r\theta} = -s$

$$\therefore 2A\frac{\pi}{4} - 2B - 2D = 0 \quad \text{(a)} \qquad \text{and} \quad -A - 2C = -s \quad \text{(b)}$$

for $\phi = -45°$, r = any: $\sigma_\theta = \tau_{r\theta} = 0$

$$\therefore -2A\frac{\pi}{4} + 2B - 2D = 0 \quad \text{(c)} \qquad \text{and} \quad -A - 2C = 0 \quad \text{(d)}$$

& solving (a), (b), (c), (d) simultaneously :

$$A = \frac{s}{2},\ D = 0,\ C = \frac{s}{4},\ B = \frac{s\pi}{8}$$

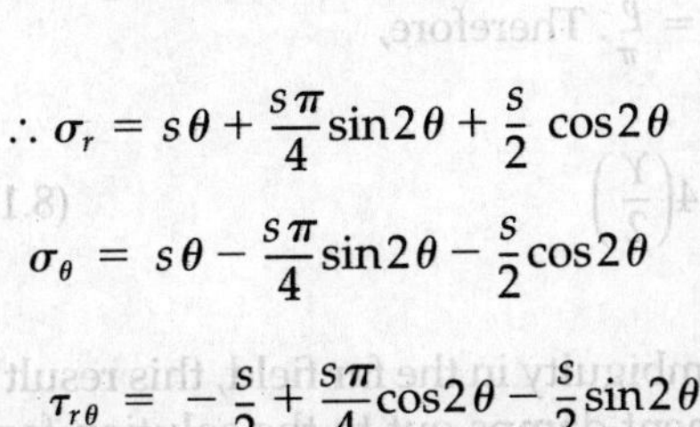

$$\therefore \sigma_r = s\theta + \frac{s\pi}{4}\sin 2\theta + \frac{s}{2}\cos 2\theta$$

$$\sigma_\theta = s\theta - \frac{s\pi}{4}\sin 2\theta - \frac{s}{2}\cos 2\theta$$

$$\tau_{r\theta} = -\frac{s}{2} + \frac{s\pi}{4}\cos 2\theta - \frac{s}{2}\sin 2\theta$$

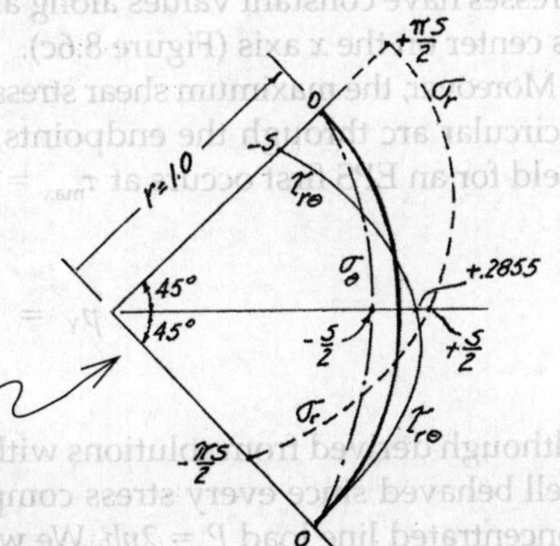

8.3 Uniform Loading over a Finite Width

Superposition of the solution for the half-space uniformly loaded along one axis gives us the solution for a strip of finite width as shown in Figure 8.6. The same result can also be obtained by integrating the result for a line load (Section 8.1) over the interval $-b \leq y \leq b$.

There are various ways to express the solution. In Cartesian coordinates:

$$\begin{aligned} \sigma_y &= -\frac{p}{\pi}\left[\alpha_2 - \alpha_1 + \frac{1}{2}(\sin 2\alpha_2 - \sin 2\alpha_1)\right] \\ \sigma_x &= -\frac{p}{\pi}\left[\alpha_2 - \alpha_1 - \frac{1}{2}(\sin 2\alpha_2 - \sin 2\alpha_1)\right] \qquad (8.15) \\ \tau_{xy} &= +\frac{p}{2\pi}(\cos 2\alpha_2 - \cos 2\alpha_1) \end{aligned}$$

If we consider principal stresses, the stress field is particularly simple:

$$\sigma_{1,3} = -\frac{p}{\pi}(\beta \mp \sin\beta) \qquad \text{(a)}$$

and therefore the invariants are:

$$\frac{\sigma_1 - \sigma_3}{2} = \tau_{max} = -\frac{p}{\pi}\sin\beta, \qquad \frac{\sigma_1 + \sigma_3}{2} = \sigma_m = -\frac{p}{\pi}\beta \qquad (8.16)$$

where $\beta = \alpha_2 - \alpha_1$. Since $\beta > \sin\beta$ there is no tension anywhere.

The stress trajectories are confocal ellipses and hyperbolas having the end points of the strip, F_1, and F_2, as foci as shown in Figure 8.6b. Both principal stresses have constant values along any circle passing through F_1 and F_2 with its center on the x axis (Figure 8.6c).

Moreover, the maximum shear stress has a maximum where $\sin\beta = 1$. This is a circular arc through the endpoints, as shown in Figure 8.6d, along which yield for an EPS first occurs at $\tau_{max} = \frac{Y}{2} = \frac{p}{\pi}$. Therefore,

$$p_Y = 3.14\left(\frac{Y}{2}\right) \qquad (8.17)$$

Although derived from solutions with ambiguity in the far field, this result is well behaved since every stress component damps out to the solution for a concentrated line load $P = 2pb$. We will see later that these results are useful

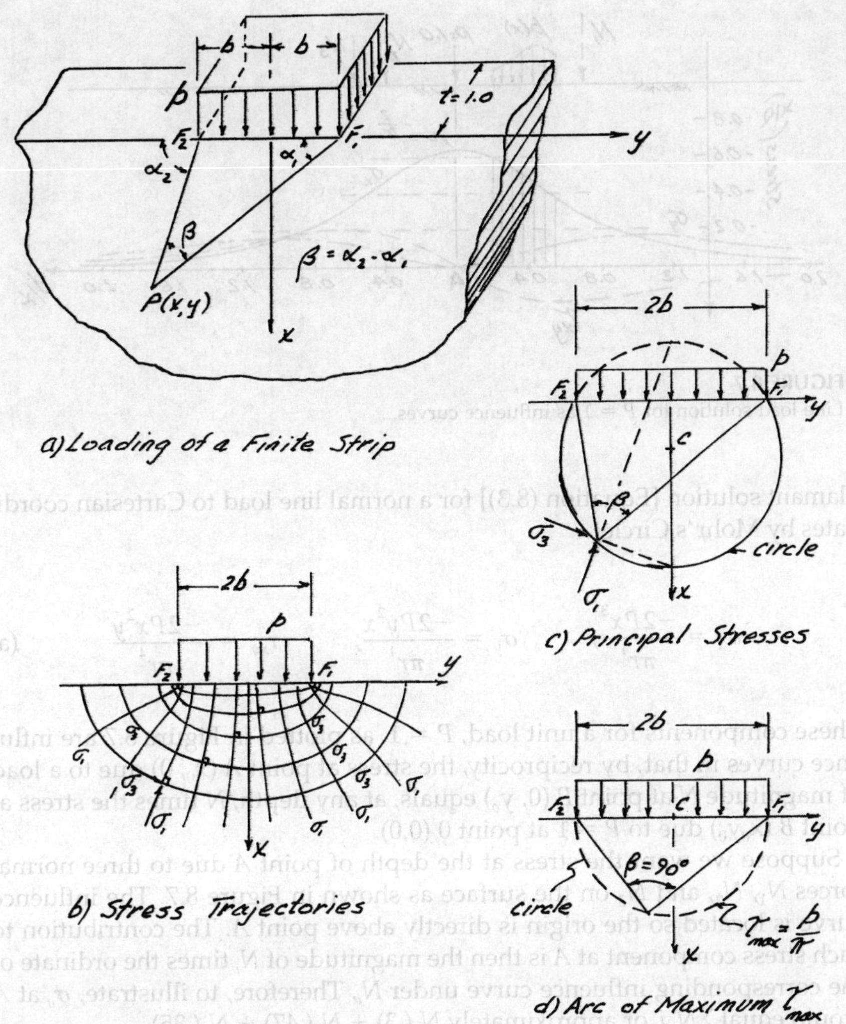

FIGURE 8.6
Uniform loading of a strip on the half-space.

when we consider slip-line theory to calculate the collapse load for a uniformly loaded strip.

8.4 Nonuniform Loadings on the Half-Space

It has been mentioned that this result for a uniform load on a finite width of the half-space could as easily be obtained by integrating the point-load solution. This is most easily seen in terms of *influence curves*. Transforming the

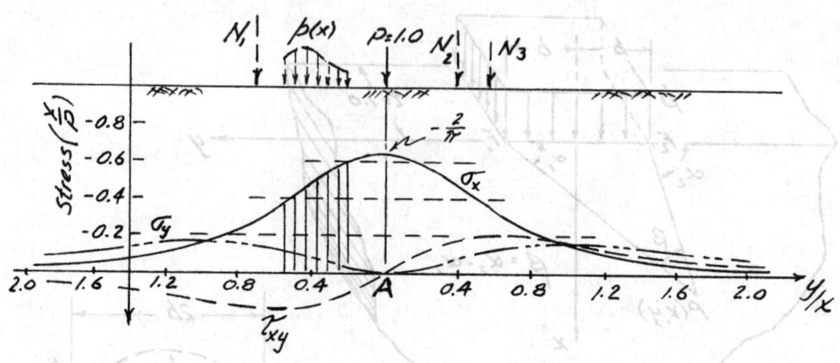

FIGURE 8.7
Line load solution for $P = 1$ as influence curves.

Flamant solution [Equation (8.3)] for a normal line load to Cartesian coordinates by Mohr's Circle:

$$\sigma_x = \frac{-2Px^3}{\pi r^4}, \qquad \sigma_y = \frac{-2Py^2x}{\pi r^4}, \qquad \tau_{xy} = \frac{-2Px^2y}{\pi r^4} \tag{a}$$

These components for a unit load, $P = 1$, as plotted in Figure 8.7 are influence curves in that, by reciprocity, the stress at point A $(x_0, 0)$ due to a load of magnitude N at point P $(0, y_o)$ equals, at any depth, N times the stress at point B $(x_0 y_0)$ due to $P = 1$ at point 0 (0,0).

Suppose we want the stress at the depth of point A due to three normal forces N_1, N_2, and N_3 on the surface as shown in Figure 8.7. The influence curve is located so the origin is directly above point A. The contribution to each stress component at A is then the magnitude of N_i times the ordinate of the corresponding influence curve under N_i. Therefore, to illustrate, σ_y at A would equal $\Sigma N_i\, i_i$ or approximately $N_1(.3) + N_2(.47) + N_3(.35)$.

Similarly, by superposition, the stresses under any nonuniform normal loading can be found to any degree of accuracy desired by numerical integration. The load $p(x)$ in Figure 8.7 is simply divided into increments, Δx, each of which is treated as a concentrated load $p_L \Delta x_i$ to be multiplied by the average ordinate of the influence curve under it. If $p(x)$ were constant as in the previous section, then σ_y at A would equal p times the shaded area or approximately 0.5 pb.

Closed-form analytic solutions are available for a number of nonuniform strip loadings (e.g., linear, parabolic, etc.). From an engineering standpoint however, computation of stress from influence lines given by point load solutions is perfectly adequate for design and inherently more versatile. In fact, such influence lines, often called Green's Functions, are the fundamental building blocks of the boundary element method for numerical analysis.

Another approach for nonconstant loading is to use either a Fourier or polynomial series solution. Similarly, the solutions for constant loading can be superimposed as step functions to approximate any nonconstant distribution. Some of these solutions for nonuniform loading of the half-space are referred to in the chapter problems as are the influence curves for tangential loading.

8.5 Line Loads within the Half-Space

To complete our repertoire of 2D solutions for the half-space, we need a solution for a vertical and horizontal line load acting at some depth, d, below the surface of a semi-infinite mass. Referring to Figure 8.8, the solution for plane strain*

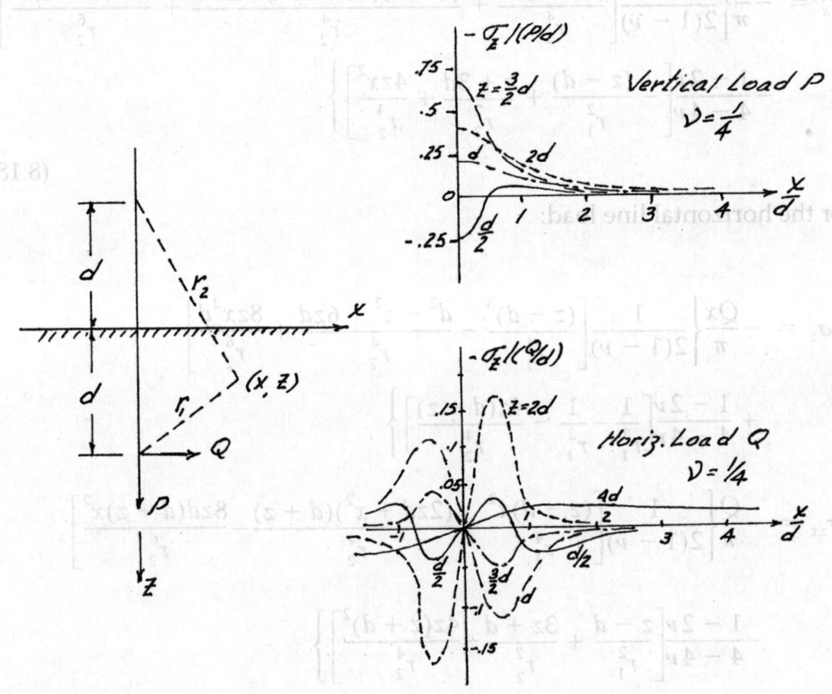

FIGURE 8.8
Line load within the half-space.

* Harr, M.E., *Foundations of Theoretical Soil Mechanics*, McGraw-Hill, New York, 1966, 66–69.

for a vertical line load is:

$$\sigma_z = -\frac{P}{\pi}\left\{\frac{1}{2(1-\nu)}\left[\frac{(z-d)^3}{r_1^4} + \frac{(z+d)(z^2+4zd+d^2)}{r_2^4} - \frac{8zd(d+z)x^2}{r_2^6}\right] - \frac{1-2\nu}{4-4\nu}\left(\frac{z-d}{r_1^2} + \frac{3z+d}{r_2^2} - \frac{4zx^2}{r_2^4}\right)\right\}$$

$$\tau_{zx} = -\frac{Px}{\pi}\left\{\frac{1}{2(1-v)}\left[\frac{(z-d)^2}{r_1^4} + \frac{z^2-2zd-d^2}{r_2^4} + \frac{8zd(d+z)^2}{r_2^6}\right] - \frac{1-2v}{4-4v}\left[\frac{1}{r_1^2} - \frac{1}{r_2^2} + \frac{4z(d+z)}{r_2^4}\right]\right\}$$

$$\sigma_x = -\frac{P}{\pi}\left\{\frac{1}{2(1-\nu)}\left[\frac{(z-d)\,x^2}{r_1^4} + \frac{(z+d)(x^2+2d^2)-2x^2d}{r_2^4} + \frac{8zd(d+z)x^2}{r_2^6}\right] - \frac{1-2\nu}{4-4\nu}\left[-\frac{(z-d)}{r_1^2} + \frac{z+3d}{r_2^2} + \frac{4zx^2}{r_2^4}\right]\right\} \tag{8.18}$$

For the horizontal line load:

$$\sigma_z = -\frac{Qx}{\pi}\left\{\frac{1}{2(1-\nu)}\left[\frac{(z-d)^3}{r_1^4} - \frac{d^2-z^2+6zd}{r_2^4} + \frac{8zx^2d}{r_2^6}\right] + \frac{1-2\nu}{4-4\nu}\left[\frac{1}{r_1^2} - \frac{1}{r_1^2} - \frac{4z(d+z)}{r_2^4}\right]\right\}$$

$$\tau_{zx} = -\frac{Q}{\pi}\left\{\frac{1}{2(1-\nu)}\left[\frac{(z-d)x^2}{r_1^4} + \frac{(2zd+x^2)(d+z)}{r_2^4} - \frac{8zd(d+z)x^2}{r_2^6}\right] - \frac{1-2\nu}{4-4\nu}\left[\frac{z-d}{r_1^2} + \frac{3z+d}{r_2^2} - \frac{4z(z+d)^2}{r_2^4}\right]\right\}$$

$$\sigma_x = -\frac{Qx}{\pi}\left\{\frac{1}{2(1-\nu)}\left[\frac{x^2}{r_1^4} + \frac{x^2+8zd+6d^2}{r_2^4} + \frac{8zd(d+z)^2}{r_2^6}\right] - \frac{1-2\nu}{4-4\nu}\left[\frac{1}{r_1^2} + \frac{3}{r_2^2} - \frac{4z(d+z)}{r_2^4}\right]\right\} \tag{8.19}$$

This solution is again very useful for numerical analysis. It is important to recognize that Poisson's ratio now appears in what would seem to be a two-dimensional problem. This is due to the displacement conditions in the far field which we hypothesize is probably the cause of difficulties with some of the previous wedge solutions. We will return to this question of existence and uniqueness in the concluding section of this chapter after the solution for constant body force loading of wedges is presented.

8.6 Diametric Loading of a Circular Disk

It is somewhat surprising that wedge solutions can be applied to determine stresses in disks under self-equilibrating loads. The simplest example is a disk with diametric, concentrated loads as shown in Figure 8.9.*

From the solution for the half-space with a concentrated load, $F = -\frac{P}{\pi} r\theta \sin\theta$ giving a constant radial stress $\sigma_r = \frac{-P}{\pi R}$ on the outline of a circle of radius R as shown in Figure 8.9a. Therefore, (Figure 8.9b), a disk of radius R with loads top and bottom would be in equilibrium with a constant boundary pressure $\frac{P}{\pi R}$. We can then remove this pressure to obtain the final result.

More specifically, if we associate the stress function F_1 with the top load and F_2 with the bottom load then:

$$\sigma_{r_1} = \frac{1}{r}\frac{\partial F_1}{\partial r_1} + \frac{1}{r_1^2}\frac{\partial^2 F_1}{\partial \theta_1^2} = -\frac{2P}{\pi r_1}\cos\theta_1 = -\frac{P}{\pi R} \quad \text{on the boundary} \tag{a}$$

$$\sigma_{\theta_1} = \tau_{r_1\theta_1} = 0$$

and

$$\sigma_{r_2} = -\frac{2P}{\pi r_2}\cos\theta_2 = -\frac{P}{\pi R} \quad \text{on the boundary} \tag{b}$$

$$\sigma_{\theta_2} = \tau_{r_2\theta_2} = 0$$

Finally, applying a uniform boundary tension from the Lamé solution with the origin at the center of the disk

$$F_3 = \frac{P}{2\pi R} r_3^2 \qquad \text{and} \qquad \sigma_{r_3} = \sigma_{\theta_3} = \frac{P}{\pi R} \tag{c}$$

* Many other cases are discussed with references in *Theory of Elasticity* by Timoshenko and Goodier. The solution for a heavy disk is given by Love in his classic book: *A Treatise on the Mathematical Theory of Elasticity*. Some of these solutions are referred to in the chapter problems.

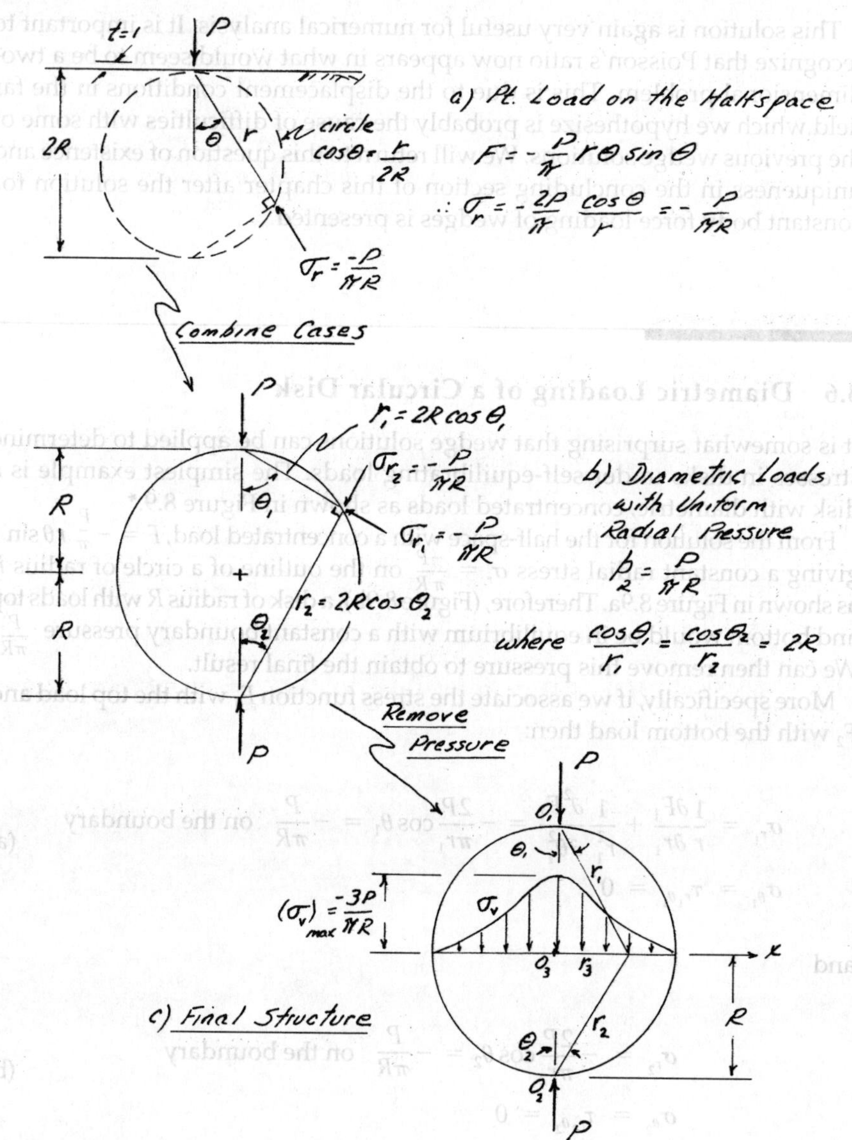

FIGURE 8.9
Solution procedure for a thin disk with diametric loads.

Superimposing the three cases

$$F = F_1 + F_2 + F_3 = -\frac{P}{\pi}\left(r_1\theta_1 \sin\theta_1 - r_2\theta_2 \sin\theta_2 + \frac{r_3^2}{2R} \right)$$

giving the stresses at any point as the sum of three states, which must be transformed to a common orientation. For example, at the center of the disk where $r_1 = R$, $\theta_1 = 0$, $r_2 = R$, $\theta_2 = 0$, $r_3 = 0$

$$\begin{aligned}\sigma_{vert} &= -\frac{2P}{\pi}\frac{1}{R} - \frac{2P}{\pi}\frac{1}{R} + \frac{P}{\pi R} = -\frac{3P}{\pi R} \\ \sigma_{horiz} &= 0 + 0 + \frac{P}{\pi R} = \frac{P}{\pi R} \\ \tau_{hv} &= 0\end{aligned} \tag{d}$$

The distribution of vertical stress along the horizontal diameter given by the expression

$$\sigma_v = \sigma_y = \frac{P}{\pi R}\left[1 - \frac{4R^4}{(R^2 + x^2)^2}\right] \tag{8.20}$$

is shown in Figure 8.9d.

8.7 Wedges with Constant Body Forces

Gravity loading of massive concrete, stone, and soil structures often produce the largest components of stress for which the designer must provide. For slopes, retaining walls, and geologic formations, body force may, in fact, be the only load. Seismic loading is often approximated by the assumption of a constant acceleration applied to the structure and even when the acceleration is not constant, the assumption of a constant peak value gives the engineer a worst-case estimate.

A solution can be obtained in polar coordinates, but is more clearly presented in the Cartesian reference frame* shown in Figure 8.10. Letting γ_x and γ_y be the body-force components due to gravity in the subscripted directions and $K = \tan\beta$, the stress function:

$$F_\gamma = \frac{1}{6K^2}[\gamma_x K^2 x^2 + (\gamma_x K + \gamma_y K^2 - 2\gamma_y)y^3 - 3\gamma_y K + y^2] \tag{8.21}$$

---

* A general presentation of various solution procedures for a variety of wedges and comparison to experimental results is given in *Body-Force Stresses in Wedge-Shaped Gravity Structures* by R. Richards, Jr., Ph.D. Thesis, Princeton University, 1964. This section is largely a synopsis of this work and subsequent analysis over the last 30 years. We will see that this remains an open problem.

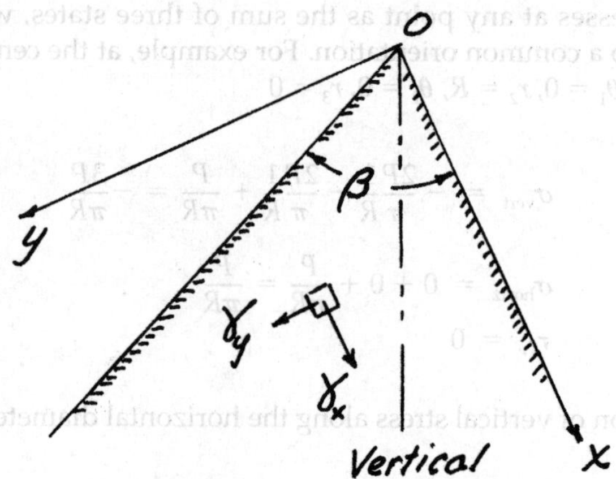

FIGURE 8.10
Semi-infinite wedge.

leads to the stress field:

$$\sigma_x = \frac{\partial^2 F}{\partial y^2} - \gamma_x x - \gamma_y y = \frac{1}{K}\left[(\gamma_y - \gamma_x K)x + \left(\gamma_x - \frac{2\gamma_y}{K}\right)y\right]$$

$$\sigma_y = \frac{\partial^2 F}{\partial x^2} - \gamma_x x - \gamma_y y = -\gamma_y y \tag{8.22}$$

$$\tau_{xy} = -\frac{\partial^2 F}{\partial x dy} = -\frac{\gamma_y y}{K}$$

This is a wonderfully simple solution since it predicts a linear stress field. Therefore $\nabla^2(\sigma_x + \sigma_y) = 0$ in a "trivial" fashion. We see, however, that the so-called "paradox of elasticity" we discovered in Section 8.2 again appears since for the half-space $\beta = 180°$, $K = 0$, and the solution blows up. In fact, for wedges where $\beta > 90°$, this solution probably does not exist. Shown in Figure 8.11 are the analytic results for a 45° slope, which are certainly peculiar. The normal stress becomes larger and larger along the upper horizontal face as the distance from the crest, x, increases. This is nonsense. Yet as shown in Figure 8.11b, the seemingly unrealistic result certainly does give equilibrium and satisfy the boundary conditions on the exposed faces of the wedge.*

* An ingenious elasticity solution for slopes built up in layers of small thickness, Δ, (fill rather than cut construction) is developed by L.E. Goodman and C.B. Brown in their paper Dead Load Stresses and the Instability of Slopes, [*Journ. of Soil Mech. & Found. Div., ASCE*, 81, 3, May, 1963].

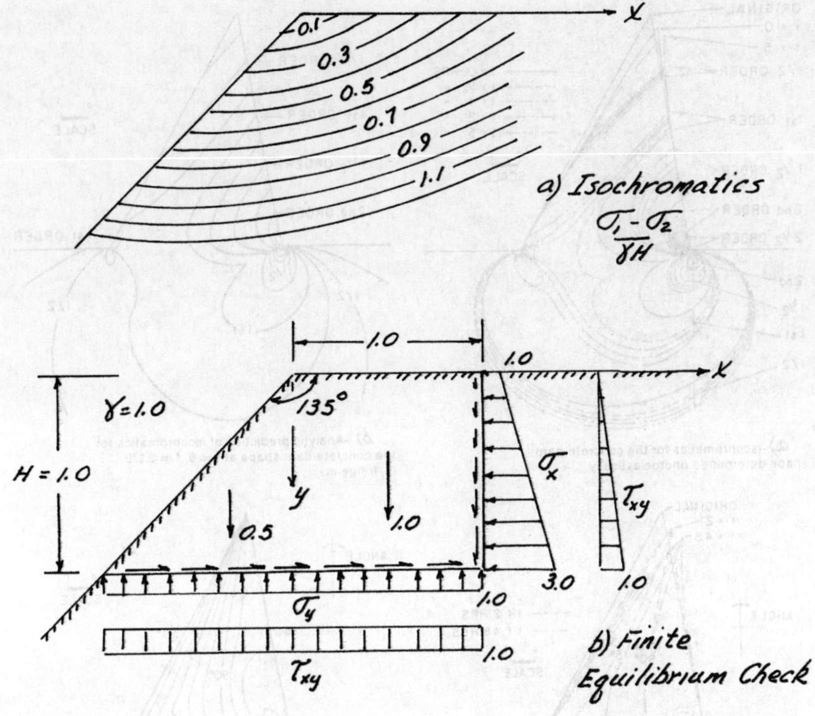

FIGURE 8.11
Semi-infinite results for a 45° slope (analytic).

On the other hand, at $\beta \leq 90°$ the analytic solution is perfectly reasonable and is used in the design of concrete gravity dams. Figure 8.12 shows a comparison of the analytic solution and experimental results for a gelatin model for a concrete dam shape in terms of isochromatics and isoclinics. The agreement is excellent. The same is true for a vertical cut ($\beta = 90°$) shown in Figure 8.13.

It would appear then that this "paradox" is a problem with uniqueness, which, in retrospect, is not surprising. From our discussion in Chapter 4 of the Laplace equation $\nabla^2(\sigma_x + \sigma_y) = 0$, we know that we can only guarantee a unique solution if we specify proper conditions over a *closed* boundary. Since no conditions are specified at r = large in any of these wedge solutions, we must be

---

The solution is obtained by using the solution for a uniform load $w = -\rho g \Delta$ on the horizontal surface [Equation (9.14)] repeatedly. Each layer must satisfy equilibrium and compatibility, but the stress field in the final slope will not satisfy the compatibility equation $\nabla^2(\sigma_x + \sigma_y) = 0$ since each layer "relaxes" before the next is placed. This incremental solution is reasonable in that it is independent of x and remains finite for finite y. The field is therefore well behaved and seems to be reasonably close to that observed photoelasticity for a 45 degree slope.

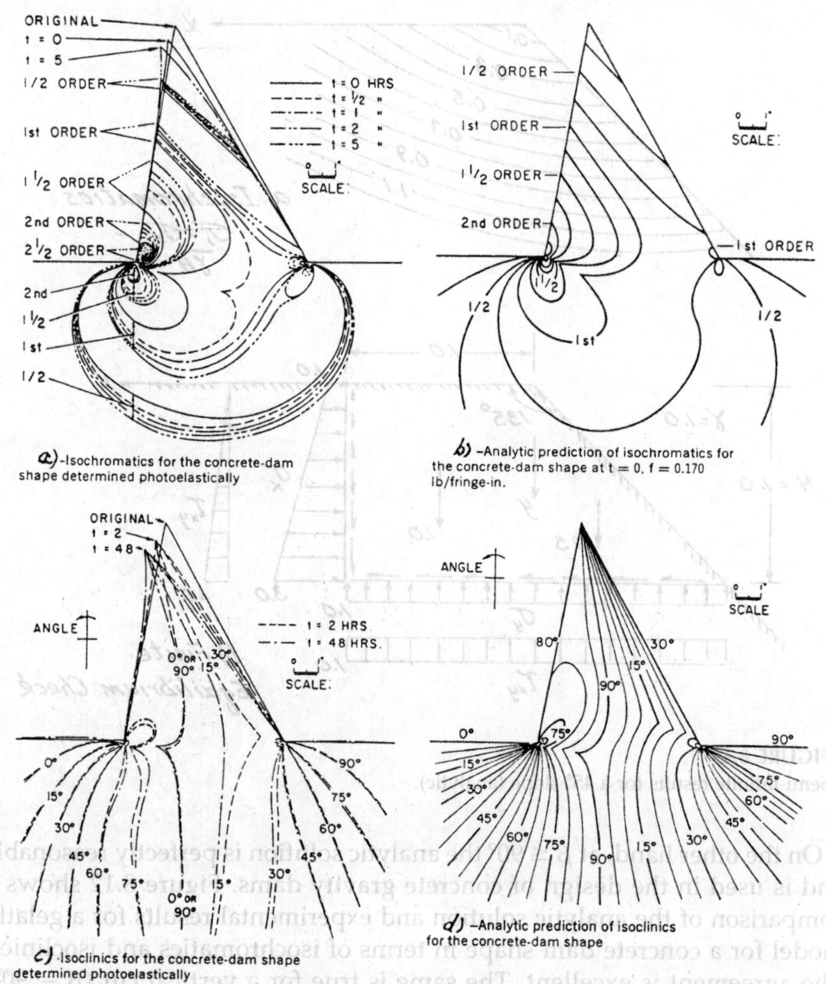

a) -Isochromatics for the concrete-dam shape determined photoelastically

b) -Analytic prediction of isochromatics for the concrete-dam shape at t = 0, f = 0.170 lb/fringe-in.

c) -Isoclinics for the concrete-dam shape determined photoelastically

d) -Analytic prediction of isoclinics for the concrete-dam shape

FIGURE 8.12
Comparison of analytic and experimental results for a concrete gravity dam shape.

skeptical of results where the stress field does not either damp out as $r \to \infty$ or approaches the free field.*

---

* We are left then with certain wedge solutions involving surface or body force loading, which give nonexistent results for wedge angles $2\phi = \beta > 90°$ even though they satisfy equilibrium, compatibility, and boundary conditions on the faces. This so-called paradox is not a paradox, but a result of not specifying conditions over a *closed* boundary (namely at large r). The only closed-form solutions available for these cases are by conformal mapping where wedges with $\beta > 90°$ are mapped to wedges with $\beta < 90°$. For example this is done by Savage for gravity loading in his 1994 paper Gravity Induced Stresses in Finite Slopes, *Int. Jour. of Rock Mech and Min. Sci*, Geomech. Abstracts, 31(5), 471–483. All such solutions indicate that Poisson's ratio becomes significant implying that displacement conditions must be introduced.

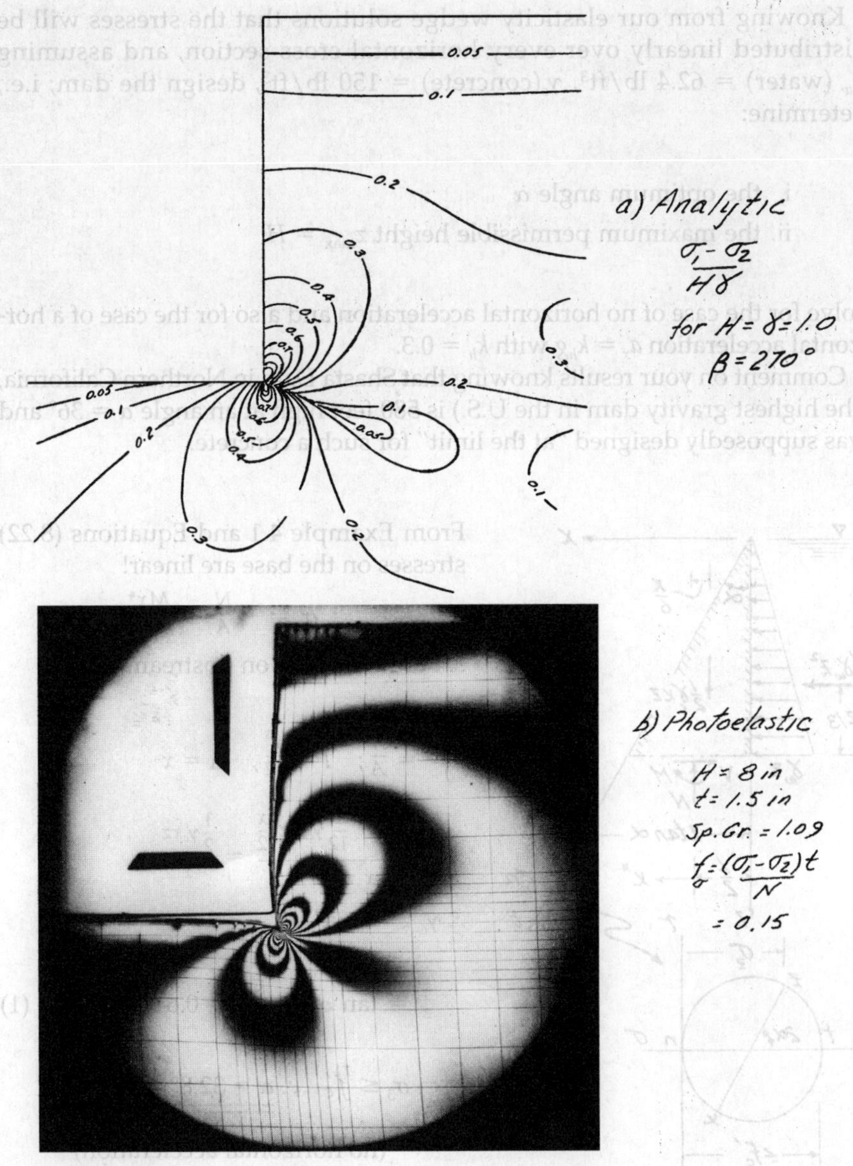

FIGURE 8.13
Isochromatics for a vertical slope.

Example 8.3

A concrete gravity dam with a vertical upstream face is to be built so that, with a full reservoir, there will be no tension and an adequate factor of safety against crushing ($f_c' = 1125$ psi = allowable compressive stress).

Knowing from our elasticity wedge solutions that the stresses will be distributed linearly over every horizontal cross-section, and assuming γ_w (water) = 62.4 lb/ft³, γ_c(concrete) = 150 lb/ft³, design the dam; i.e., determine:

i. the optimum angle α
ii. the maximum permissible height $z_{max} = H$

Solve for the case of no horizontal acceleration and also for the case of a horizontal acceleration $a_x = k_h g$ with $k_h = 0.3$.

Comment on your results knowing that Shasta Dam in Northern California, (the highest gravity dam in the U.S.) is 580 feet high at an angle $\alpha = 36°$ and was supposedly designed "at the limit" for such a concrete.

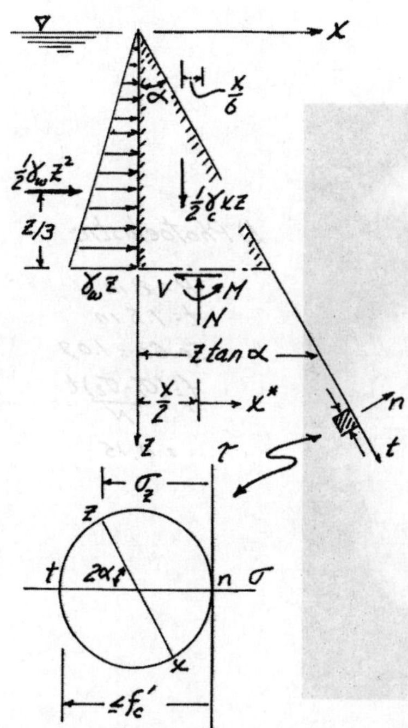

From Example 4.1 and Equations (8.22) stresses on the base are linear!

$$\therefore \sigma_z = -\frac{N}{A} \pm \frac{Mx^*}{I}$$

$\therefore$ For no tension on upstream face

$$\frac{M\frac{x}{2}}{I} = \frac{N}{A}; \quad I = \frac{x^3}{12}, \quad A = x$$

$$\therefore \frac{\left(\frac{1}{6}\gamma_w z^3 - \frac{1}{12}\gamma_c x^2 z\right)\frac{x}{2}}{x^3/12} = \frac{\frac{1}{2}\gamma_c xz}{x}$$

$$\therefore \gamma_\omega z^2 = \gamma_c x^2$$

$$\frac{x}{z} = \tan\alpha = \sqrt{\frac{\gamma_\omega}{\gamma_c}} = 0.645\ldots \qquad (1)$$

$$\sigma_t = \sigma_3 \leq f_c' \quad \therefore \alpha = 32.8°$$

(no horizontal acceleration)

Crushing on downstream face will limit the height

$$\sigma_z = -\frac{N}{A} - \frac{Mx^*}{I} = -2\frac{N}{A} = -2\frac{1}{2}\frac{\gamma_c xz}{x} = \gamma_c z\,(comp.)$$

$$\text{but from Mohr's Circle } \sigma_z \le \frac{f_c'}{2} + \frac{f_c'}{2}\cos 2\alpha = \frac{f_c'}{2}(1+\cos 2\alpha) \le \gamma_c z$$

$$z_{max} = \frac{f_c'}{2\gamma_c}(1+\cos 2\alpha)....(2) \qquad 795 = \frac{150}{12^2} z_{max}$$

$\underline{\underline{z_{max} = 765\text{ft.}}}$ (no horizontal acceleration).

b) with horizontal acceleration $a_h = -k_h g$ (towards water)

i) Neglect inertial effect of the water

ii) Assume $k_h = \text{const.}$

Then added horizontal force $= k_h\,\gamma_c\,\frac{xz}{2}$ at $\frac{z}{3}$

$$\therefore M = \frac{1}{2}\gamma_w z^2 \frac{z}{3} + \frac{1}{2}\gamma_c x z k_h \frac{z}{3} - \frac{1}{2}\gamma_c x z\left(\frac{x}{6}\right)$$

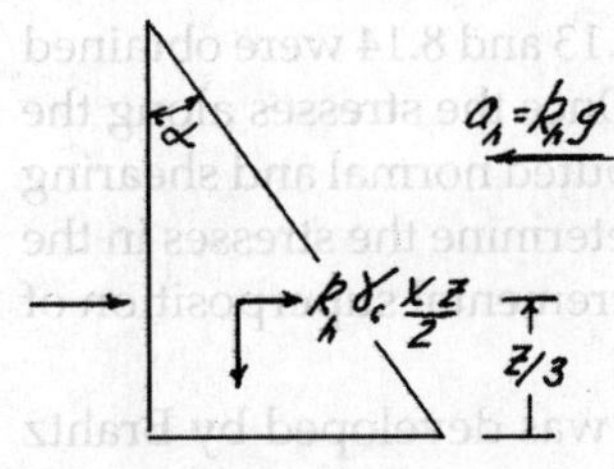

For no tension

$$\frac{M\frac{x}{2}}{\frac{x^3}{12}} = \frac{1}{2}\gamma_c \frac{xz}{x} \quad \text{as before}$$

$$\therefore \gamma_w z^3 + \gamma_c x z^2 k_h - \frac{1}{2}\gamma_c x^2 z = \frac{1}{2}\gamma_c x^2 z$$

$$\text{but } x = z\tan\alpha \;\therefore\; \tan^2\alpha - k_h \tan^2\alpha - \frac{\gamma_w}{\gamma_c} = 0 \qquad (1)$$

$$\text{which gives eqn (1) if } k_h = 0 \quad \therefore \tan\alpha = \frac{k_h + \sqrt{k_h^2 + 4\gamma_w/\gamma_c}}{2}$$

for

$$k_h = .3 \qquad \alpha = 39°$$

$$k_h = .5 \qquad \alpha = 43.3°$$

$$\text{from (2)} \qquad z_{max} = \frac{f_c'}{2\gamma_c}(1+\cos 2\alpha)$$

$\therefore$ for $k_h = .3$ $\quad z_{max} = \underline{\underline{651\text{ ft}}}$

for $k_h = .5$ $\quad z_{max} = \underline{\underline{572\text{ ft}}}$

∴ Shasta Dam is a good design. Actually Shasta Dam looks more like the illustration at right. Concrete will take some tension and downward water force, F_w, will reduce or eliminate seismic tension at the heel. By bringing the downstream slope toward the vertical $\sigma_t \rightarrow \sigma_z$ reducing crushing. Highest gravity dam is in Switzerland (no earthquakes) at $z_{max} > 900$ ft. Reasonable since if $\sigma_z = \sigma_t$, $z_{max} = f_c' / \gamma_c = 1080$ ft with stronger concrete even greater heights are possible in nonseismic zones.

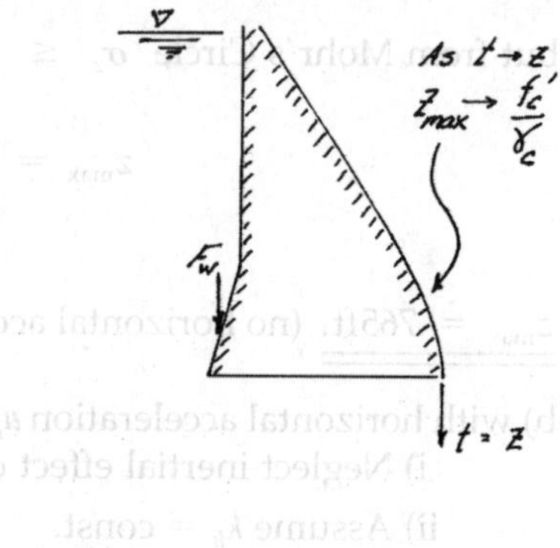

8.8 Corner Effects—Eigenfunction Strategy

The full-field analytic solutions shown in Figures 8.13 and 8.14 were obtained by including both foundation and corner effects. Once the stresses along the base are determined, the known solution for distributed normal and shearing strip loads on the half-space can be combined to determine the stresses in the foundation to any degree of accuracy using the incremental superposition of different load levels as discussed in Section 8.4.

A powerful method to determine corner effects was developed by Brahtz and is presented in a somewhat simpler form by Silverman. Later, Silverman, following the initial work of Jacobson, obtained closely similar results by using a variational method. However, the eigenfunction strategy introduced by Brahtz has the advantage of yielding a stress function that can be directly superimposed with any of those already derived for a semi-infinite wedge.*

Let us apply this to the gravity dam shape shown in Figure 8.12. Considering either the upstream, *U*, or downstream, *D*, corner of the finite wedge redrawn in Figure 8.14a, a stress function can be found that will, while excluding body and surface forces, only affect the internal stress distribution. Assuming a general stress function of the type

$$F_k = r^{n+1} f(\theta) \tag{8.23}$$

* Brahtz, J.H.A., et al., Mathematical Analysis of Stresses in Grand Coulee Dam (Study No. 1), *Technical Memorandum No. 1*, U.S. Department of the Interior, Bureau of Reclamation, Denver, Colo., 1934 and "Stress Distribution in a Reentrant Corner", *Transactions*, American Society of Mechanical Engineers, 55, 1933, 31–37. Also "The Stress Function and Photo-Elasticity Applied to Dams", *Proceedings*, ASCE. 61, 7, September 1935, 983–1020. Silverman, I.K., discussion of "The Stress Function and Photoelasticity Applied to Dams", by John H.A. Brahtz, *Proceedings*, ASCE, 61, 9, November 1935, 1409–1412 and "Approximate Stress Functions for Triangular Wedges", *Journal of Applied Mechanics*, American Society of Mechanical Engineers, 77, March 1955, 123–128. Finally, Jacobsen, B.F., "Stresses in Gravity Dams by the Principle of Least Work", *Transactions*, ASCE, 96, 1932, 489–513.

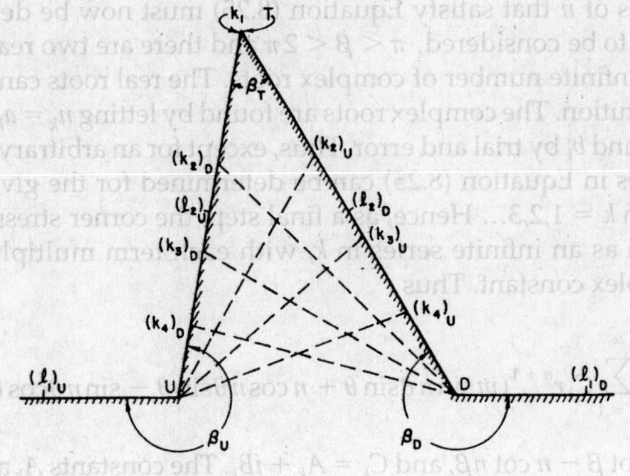

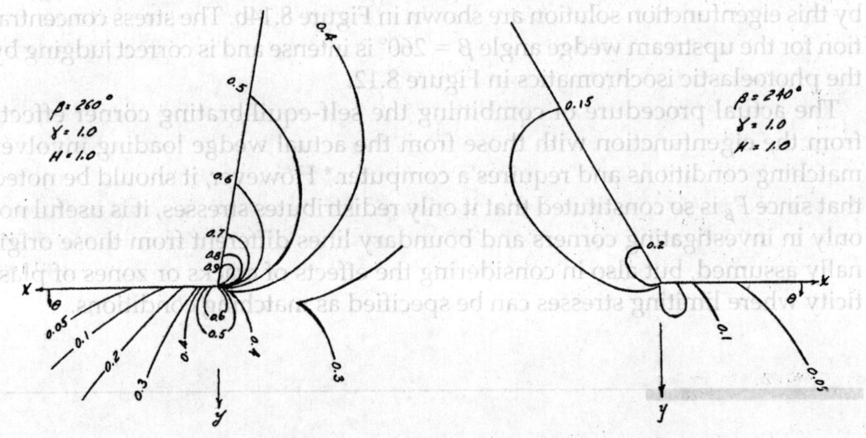

FIGURE 8.14
Corner effects for concrete dam shape.

and substituting into the compatibility requirement $\nabla^4 F = 0$

$$F_k = r^{n+1}(C_1 \cos n\theta \cos\theta + C_2 \sin n\theta \cos\theta + C_3 \cos n\theta \sin\theta + C_4 \sin n\theta \sin\theta) \tag{8.24}$$

which, with the boundary conditions, reduces to

$$F_k = Cr^{n+1}(n \cos n\theta \sin\theta + m \sin n\theta \sin\theta - \sin n\theta \cos\theta) \tag{8.25}$$

The eigenvalues of n that satisfy Equation (8.25) must now be determined. For the corners to be considered, $\pi < \beta < 2\pi$ and there are two real roots, n_0 and n_1, and an infinite number of complex roots. The real roots can be found by direct substitution. The complex roots are found by letting $n_k = a_k + ib_k$ and determining a_k and b_k by trial and error. Thus, except for an arbitrary constant, C, the quantities in Equation (8.25) can be determined for the given corner angle β for each $k = 1,2,3\ldots$ Hence, as a final step, the corner stress function may be written as an infinite series in k, with each term multiplied by an unknown complex constant. Thus

$$F_\beta = R\sum_{k=1}^{\infty} C_k r^{n+1}(m \sin n\theta \sin\theta + n\cos n\theta \sin\theta - \sin n\theta \cos\theta) \quad (8.26)$$

in which, $m = \cot\beta - n\cot n\beta$, and $C_k = A_k + iB_k$. The constants A_k and B_k can then be found as indicated in Figure 8.14a from conditions at specific points on boundaries opposite from the corner or from other known stress stipulations necessary to provide a match with semi-infinite wedge results.

The stress field in the vicinity of the upstream and downstream corners given by this eigenfunction solution are shown in Figure 8.14b. The stress concentration for the upstream wedge angle $\beta = 260°$ is intense and is correct judging by the photoelastic isochromatics in Figure 8.12.

The actual procedure of combining the self-equilibrating corner effects from the eigenfunction with those from the actual wedge loading involves matching conditions and requires a computer.* However, it should be noted that since F_β is so constituted that it only redistributes stresses, it is useful not only in investigating corners and boundary lines different from those originally assumed, but also in considering the effects of cracks or zones of plasticity where limiting stresses can be specified as matching conditions.

8.9 Problems and Questions

P8.1 Show that the displacement components for the half-space with a concentrated normal load P, (Section 8.1) are:

$$u_r = -\frac{2P}{\pi E}\cos\theta \ln r - \frac{(1-\nu)P}{\pi E}\theta\sin\theta + C\cos\theta$$

$$v_\theta = \frac{2\nu P}{\pi E}\sin\theta + \frac{2P}{\pi E}\sin\theta \ln r - \frac{(1-\nu)}{\pi E}\theta\cos\theta + \frac{(1-\nu)P}{\pi E}\sin\theta - C\sin\theta$$

* See the first footnote for Section 8.7. Also the chapter problems provide details for determining the constants in the eigenfunction. Although this approach requires numerical evaluation on the computer it is nowhere near as complicated as a finite element solution and captures the stress concentration at a corner in much better detail.

and evaluate C assuming $u_r = 0$ at $x = 10r$, $y = 0$. Plot the displacement of the horizontal boundary $x = 0$ ($\theta = \pm\pi/2$).

P8.2 Derive the displacement field for a concentrated line shear, Q, on the half-space (Figure 8.2) corresponding to those given for P in P8.1.*

P8.3 Plot the flow net for the half-space loaded by a concentrated line shear, Q, (corresponding to Figure 4.2 for a normal load).

P8.4 For a concentrated load inclined at 30° from the x axis (℄ of the wedge), determine the stress field and plot the results for $\phi = 45°$ and $\phi = 90°$.

P8.5 Consider a wedge with a concentrated load P at its apex where $\phi = \pi$. Show that this is not a solution for Kelvin's problem of a point load within a large plate.

P8.6 Compare, at some cross-section $x = L$, the stresses for a 60° wedge ($\phi = 30°$) with transverse loading Q to a strength-of-materials approximation such as shown in Figure 8.2b.

P8.7 Compare at some vertical cross-section $x =$ const. the stresses for a wedge ($\phi = 30°$) with a concentrated moment at the apex to a strength-of-materials approximation, $\sigma_x = \dfrac{My}{I}$.

P8.8 For two equal and opposite concentrated loads of 1000 lb. on the half-space 2 inches apart (Figure P8.8), compute the stress distribution on a horizontal plane $x = 6$ inches beneath the surface by superimposing results from Equation (8.3). Compare to the results for a moment $M = 2000$ in. lbs. applied between them.

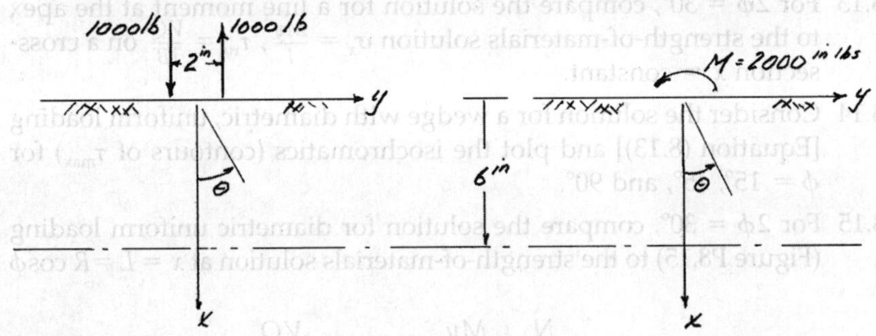

FIGURE P8.8

P8.9 Show that the stress solution for line moment on a wedge [Equation (8.8)], gives horizontal and vertical equilibrium when integrated over an arc $r = b$, $|\phi| < \pi/2$.

P8.10 For a line moment on the half-space Equation (8.9), plot the flow net for $(\sigma_r + \sigma_\theta)$ vs. $E\omega_z$ and also the isochromatics $(\sigma_1 - \sigma_2)/2$.

* A reference for both problems is Volterra and Gaines, *Advanced Strength of Materials*, 169–172. These results from P8.1 and P8.2 along with the stress field are basic functions for the boundary-element method in 2D.

P8.11 Combine a simple bending solution with that for a concentrated load on the half-space to deduce the isochromatic field in Figure P8.11 in the vicinity of the load on a deep beam (L/d = 6). Estimate the load that was applied if the stress fringe coefficient was $f = 46$ lb/fringe in. How could you further improve your analysis compared to the observed contours for τ_{max}?

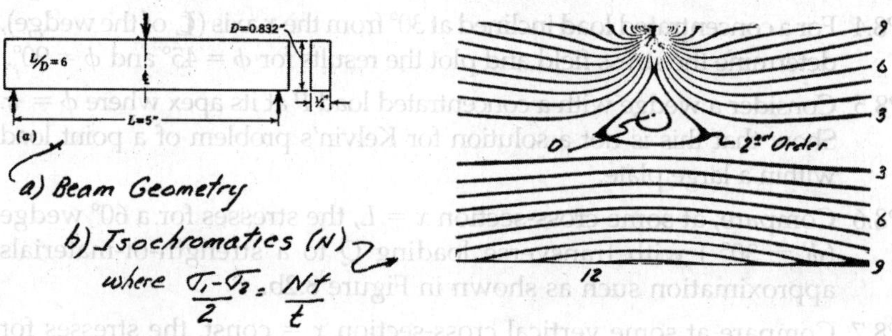

FIGURE P8.11
Deap beam loaded at its ℄. (Frocht)*

P8.12 For a wedge with $2\phi = 257.4°$, show from Equations (e) and (f) of Section 8.1 that the moment at the apex necessary for equilibrium is zero and thus the stress distribution given by the Stress Function Equation 8.7 is self-equilibrating. Does this resolve the paradox?

P8.13 For $2\phi = 30°$, compare the solution for a line moment at the apex to the strength-of-materials solution $\sigma_x = \frac{My}{I}$, $\tau_{xy} = \frac{VQ}{IB}$ on a cross-section x = constant.

P8.14 Consider the solution for a wedge with diametric, uniform loading [Equation (8.13)] and plot the isochromatics (contours of τ_{max}) for $\phi = 15°$, 45°, and 90°.

P8.15 For $2\phi = 30°$, compare the solution for diametric uniform loading (Figure P8.15) to the strength-of-materials solution at $x = L = R\cos\phi$

$$\sigma_x = \frac{N}{A} \pm \frac{My}{I}, \qquad \tau_{xy} = \frac{VQ}{Ib}$$

P8.16 For the wedge shown in Figure P8.16 (i.e, $2\phi = 30°$) plot the isopachic flow net (i.e., $\sigma_1 + \sigma_2$ and ω_z). What would be the deflection of the apex if the wedge has a fixed support at $x = L = R\cos\phi$.

P8.17 A three-hinged, haunched frame is loaded as shown (Figure P8.17). Complete the free-body diagram, consider equilibrium and boundary conditions, and estimate and plot stresses σ_n σ_t τ_{nt}

* Most of the photoelastic patterns used for Chapter 8 problems are from Frocht, M.M., *Photoelasticity*, Vol. 1, John Wiley & Sons, New York, 1941.

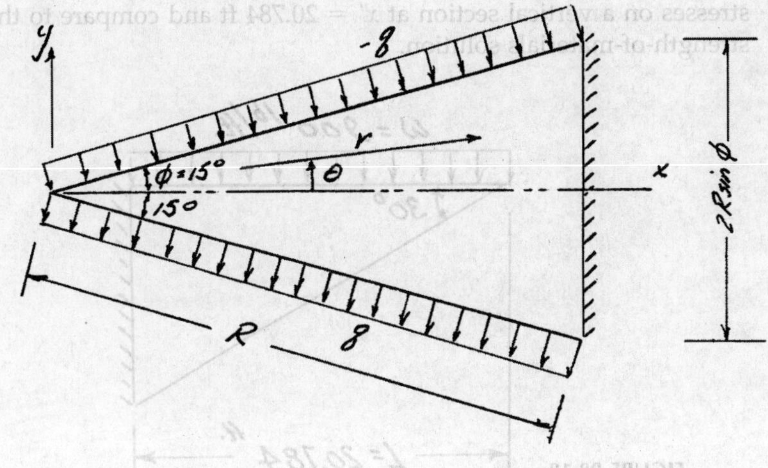

FIGURE P8.15

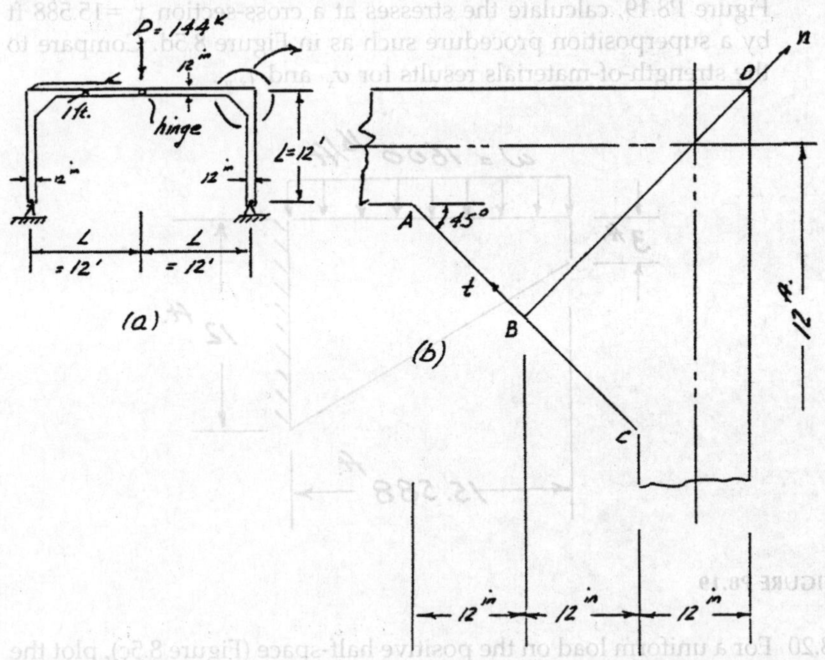

FIGURE P8.17

on sections AC and BD. Compare to a strength-of-materials approximation.

P8.18 A wedge, $2\phi = 30°$, has a uniform load $w = 900$ lb/ft on its upper face, which is horizontal as shown in Figure P8.18. Determine the stresses on a vertical section at $x' = 20.784$ ft and compare to the strength-of-materials solution.

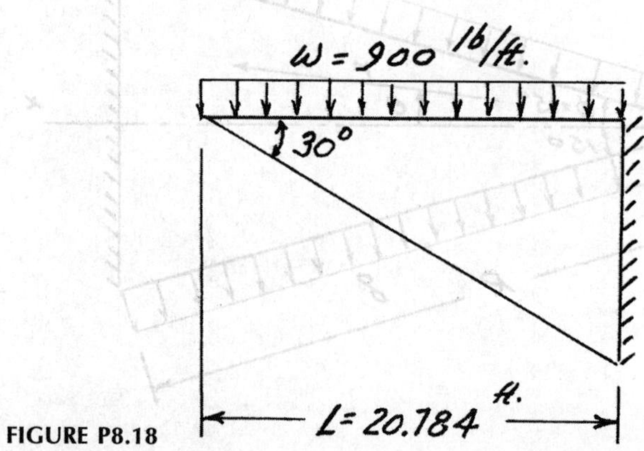

FIGURE P8.18

P8.19 For the haunched beam loaded by $w = 1600$ lb/ft as shown in Figure P8.19, calculate the stresses at a cross-section $x = 15.588$ ft by a superposition procedure such as in Figure 8.5d. Compare to the strength-of-materials results for $\sigma_{x'}$ and $\tau_{x'y'}$.

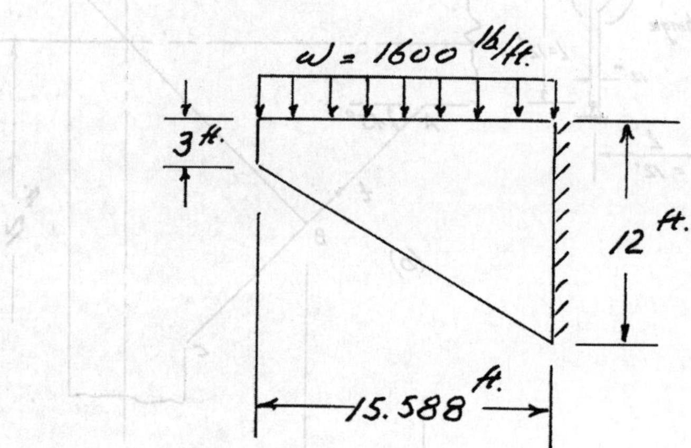

FIGURE P8.19

P8.20 For a uniform load on the positive half-space (Figure 8.5c), plot the isopachic flow net ($\sigma_1 + \sigma_2$ vs. ω_z) and separately, the isochromatics (τ_{max}). Plot the vertical normal stress σ_x and shear τ_{xy} on a section $x = constant = 6$.

P8.21 A 45° slope ($2\phi = 135$) is loaded along the top surface as shown in Figure P8.21. Do the results from the elasticity solution make sense near the apex? What support conditions are required for the elasticity solution? Plot the solution.

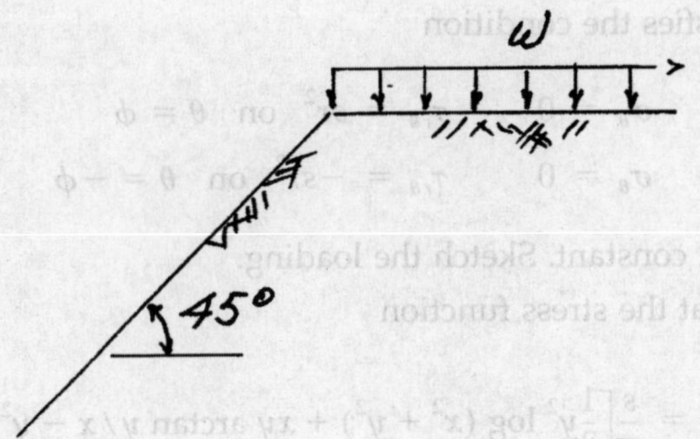

FIGURE P8.21

P8.22 Determine the value of the constant C in the stress function

$$F = Cr^2(\cos 2\theta - \cos 2\phi)$$

required to satisfy the conditions

$$\sigma_\theta = 0 \qquad \tau_{r\theta} = s \text{ on } \theta = \phi$$
$$\sigma_\theta = 0 \qquad \tau_{r\theta} = -s \text{ on } \theta = -\phi$$

corresponding to uniform shear loading on each edge of a wedge, directed away from the vertex. Verify that no concentrated force or couple acts on the vertex.

P8.23 Find the stress function of the type

$$ar^3 \cos 3\theta + br^3 \cos\theta$$

that satisfies the conditions

$$\sigma_\theta = 0 \qquad \tau_{r\theta} = sr \text{ on } \theta = \phi$$
$$\sigma_\theta = 0 \qquad \tau_{r\theta} = -sr \text{ on } \theta = -\phi$$

s being a constant. Sketch the loading.

P8.24 Find the stress function of the type

$$ar^4 \cos 4\theta + br^4 \cos 2\theta$$

that satisfies the condition

$$\sigma_\theta = 0 \qquad \tau_{r\theta} = sr^2 \quad \text{on} \quad \theta = \phi$$

$$\sigma_\theta = 0 \qquad \tau_{r\theta} = -sr^2 \quad \text{on} \quad \theta = -\phi$$

s being a constant. Sketch the loading.

P8.25 Show that the stress function

$$F = \frac{s}{\pi}\left[\frac{1}{2}y^2 \log\,(x^2 + y^2) + xy \arctan y/x - y^2\right]$$

satisfies the conditions on the edge $y = 0$ of the semi-infinite plate indicated in Figure P8.25 the uniform shear loading *s* extending from 0 indefinitely to the left.

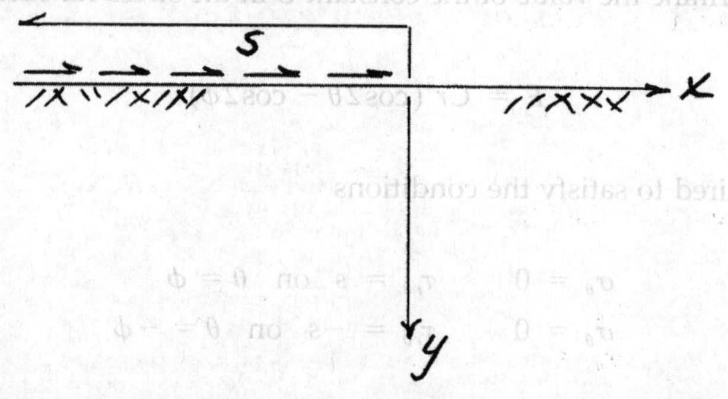

FIGURE P8.25

P8.26 Show the superposition procedure and derive Eq. (8.15) from Eq. (8.14) for the halfspace ($\phi = \pi/2$)

P8.27 Plot the influence curves for σ_x, σ_y, τ_{xy} for a concentrated tangential line load $Q = 1$ comparable to Figure 8.7 for a normal load.

P8.28 Using Figure 8.7 determine the stresses at point *P* in Figure P8.28 and compare to the exact result from Equation (8.15).

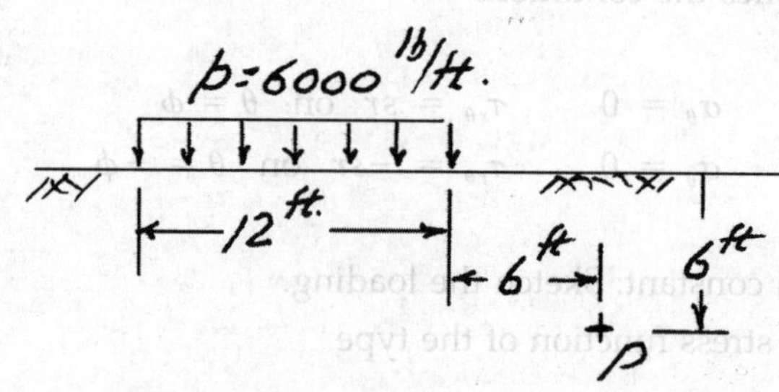

FIGURE P8.28

P8.29 The solution for a uniform tangential load, q, over a strip (Figure P8.29) is

$$\sigma_x = \frac{-4bqyx^2}{\pi[(y^2 + x^2 - b^2)^2 + 4b^2x^2]}$$

$$\sigma_y = \frac{-q}{\pi} \ln \frac{(y+b)^2 + x^2}{(y-b)^2 + x^2} - \sigma_x$$

$$\tau_{xy} = \frac{-q}{\pi}\left[\tan^{-1}\frac{x}{y-b} - \tan^{-1}\frac{x}{y+b} - \frac{2bx(y^2 - x^2 - b^2)}{(y^2 + x^2 - b^2)^2 + 4b^2x^2}\right]$$

Compare this solution at point P ($x = 6$ ft, $y = 12$ ft) to the result from the influence curves from P8.27 if 2b = 12 ft.

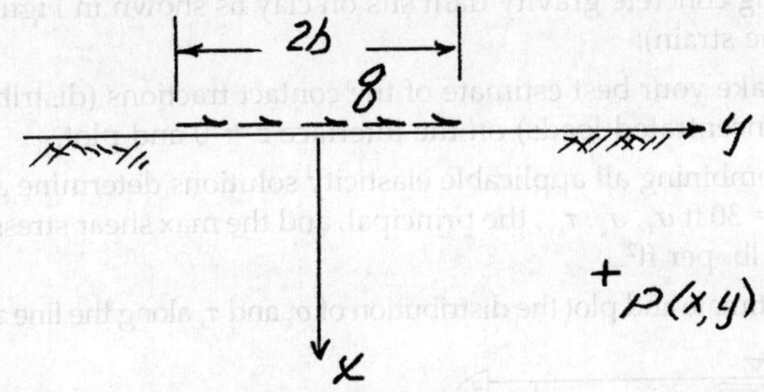

FIGURE P8.29

P8.30 The solution for a triangular normal loading over a strip (Figure P8.30) is:

$$\sigma_x = \frac{yp}{2\pi b}A(x, y) + \frac{px}{\pi} B(x, y)$$

$$\sigma_y = -\frac{qx}{2\pi b} \ln \frac{(y - 2b)^2 + x^2}{y^2 + x^2} + \frac{yp}{2\pi b}A(x, y) - \frac{px}{\pi}B(x, y)$$

$$\tau_{xy} = -\frac{px^2}{\pi(y - 2b)^2 + x^2} - \frac{qx}{2\pi b}A(x, y)$$

where: $A(x, y) = \tan^{-1}\frac{x}{y - 2b} - \tan^{-1}\frac{x}{y}$; $B(xy) = \frac{y - 2b}{(y - 2b)^2 + x^2}$

Compare this solution at point P ($x = 6$ ft, $y = 18$ ft) to the result from the influence curves in Figure 8.7 if $2b = 12$ ft and $p = 12{,}000$ lb/ft.

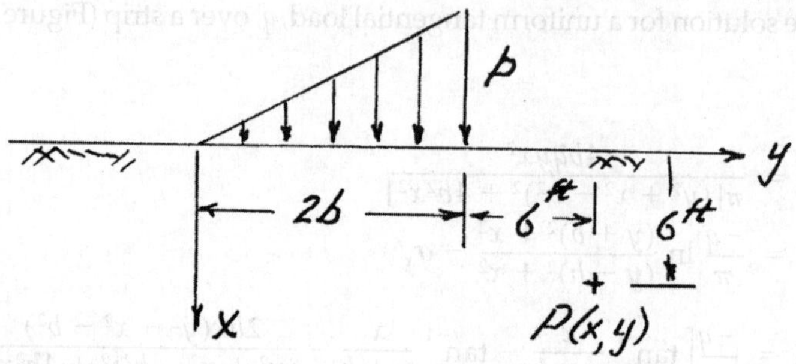

FIGURE P8.30

P8.31 For the triangular strip loading of P8.30, determine the vertical displacement of the surface ($x = 0$).

P8.32 A long concrete gravity dam sits on clay as shown in Figure P8.32 (plane strain):

a. Make your best estimate of the contact tractions (distributed or concentrated loads) on the interface $z = 0$ and plot.
b. Combining all applicable elasticity solutions determine at $x = 0$, $z = 30$ ft σ_x, σ_y, τ_{xy}, the principal, and the max shear stress values in lb. per ft^2.
c. Estimate and plot the distribution of σ_z and τ_x along the line $z = 30$ ft.

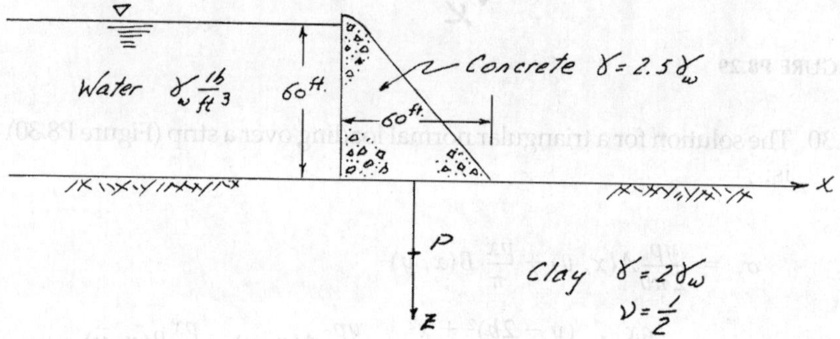

FIGURE P8.32

P8.33 For the gravity dam in P8.32, what is the factor of safety against yield at $x = 0$, $z = 60$ if for the clay $Y = 6000$ lb/ft^2, $\gamma_\omega = 62.4$ lb/ft^3. Where do you think the clay will yield first and what is the factor of safety there? Is the case where the reservoir is empty more critical and if so how much?

P8.34 For the vertical load within the half-space, determine the first stress invariant, $\sigma_x + \sigma_z$, from Equation (8.18) and plot the flow net for $\nu = 1/2$.

P8.35 For the horizontal load within the half-space determine the first stress invariant, $\sigma_x + \sigma_z$, from Equation (8.19) and plot the flow net for $\nu = 1/2$.

P8.36 Derive Equation (8.20) and plot the distribution of vertical stress on the horizontal diameter of a diametrically loaded disk giving values at increments of $r_3 = 0.2R$.

P8.37 Refer to Figure P8.37. Derive an expression for the horizontal stress on the vertical diameter of a diametrically loaded disk and discuss the split-cylinder test for the tensile strength of brittle materials (e.g., concrete, cast iron).

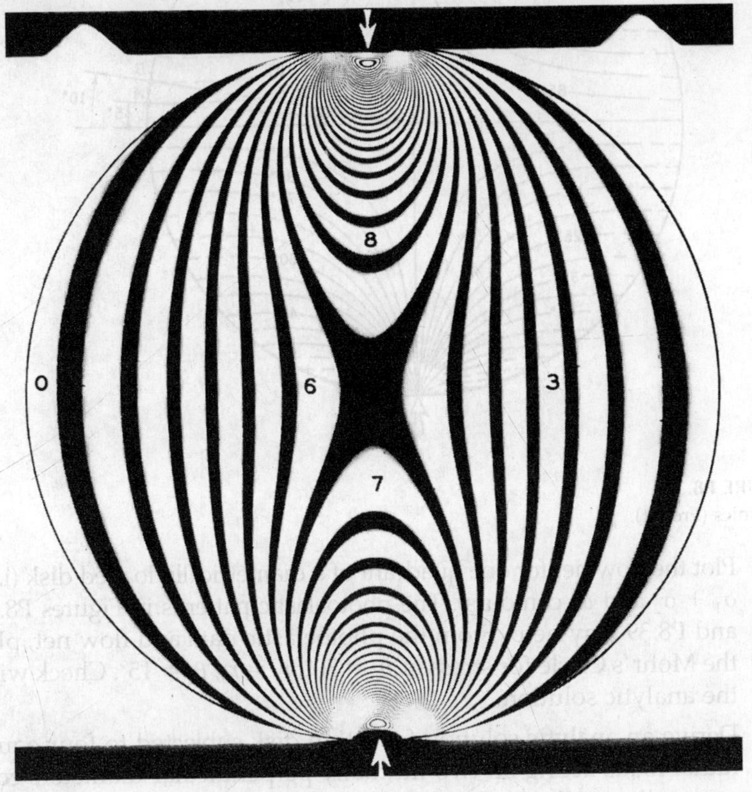

FIGURE P8.37
Isochromatics (Frocht).

P8.38 The half-order isochromatics in Figure P8.37 were obtained from a disk of diam = 1.25 in., thickness $t = 0.2$ in., and a shear-stress fringe coefficient $f = 43$ lb/fringe in. The fringe order at the center is $6\frac{1}{2}$. Determine the diametric load applied and compare the experimental distribution of $(\sigma_1 - \sigma_2)$ to the theoretical distribution on the horizontal and vertical diameters.*

P8.39 A sketch of isoclinics for a diametrically loaded disk is shown in Figure P8.39. From these draw an accurate sketch of the corresponding principal stress trajectories.

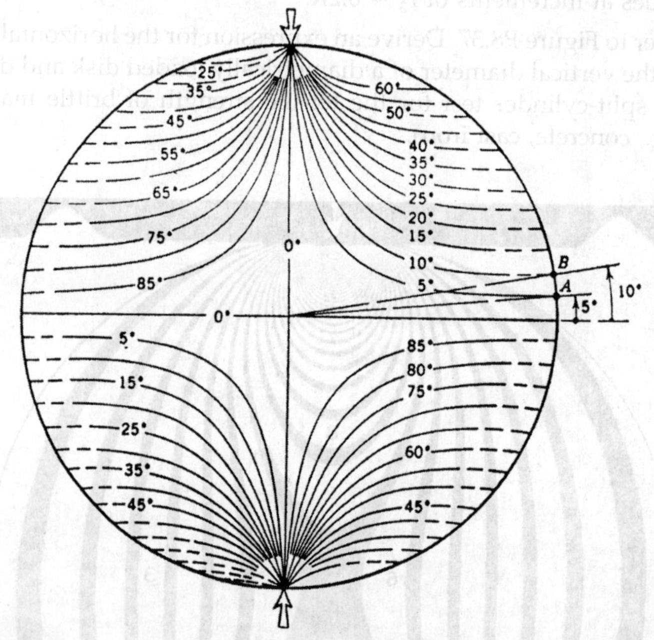

FIGURE P8.39
Isoclinics (Frocht).

P8.40 Plot the flow net for one quadrant of a diametrically loaded disk (i.e., $\sigma_1 + \sigma_2$ and ω_z contours). The photoelastic patterns in Figures P8.37 and P8.39 may help. From the photoelastic data and flow net, plot the Mohr's Circle for stress at $\theta_1 = 30°$, $r_1 = r_3$, $\theta_2 = 15°$. Check with the analytic solution.

P8.41 Derive an analytic solution for a thin disk subjected to four equal radial loads acting on two mutually perpendicular diameters corresponding to Figure P8.41. Plot the stress distribution on the horizontal and vertical planes and on the planes corresponding to the directions of the loads (at 45° to the horizontal and vertical). Compare the computed maximum shear stresses to those from Figure P8.41 where the diameter $D = 0.910$ in., thickness $t = 0.142$ in., and $f = 43$ psi shear stress for a 1-inch thickness. What must have been

* For P8.37 through P8.47 you may wish to program the analytic solution to give field values to compare to photoelastic and graphical results. To do this you must use the transformation equations to combine cases so you must be very careful to keep the signs straight.

the experimental loads? The isoclinics and corresponding stress trajectories are also shown in Figure P8.41.

a) Isochromatics (τ_{max} = Nf/t)

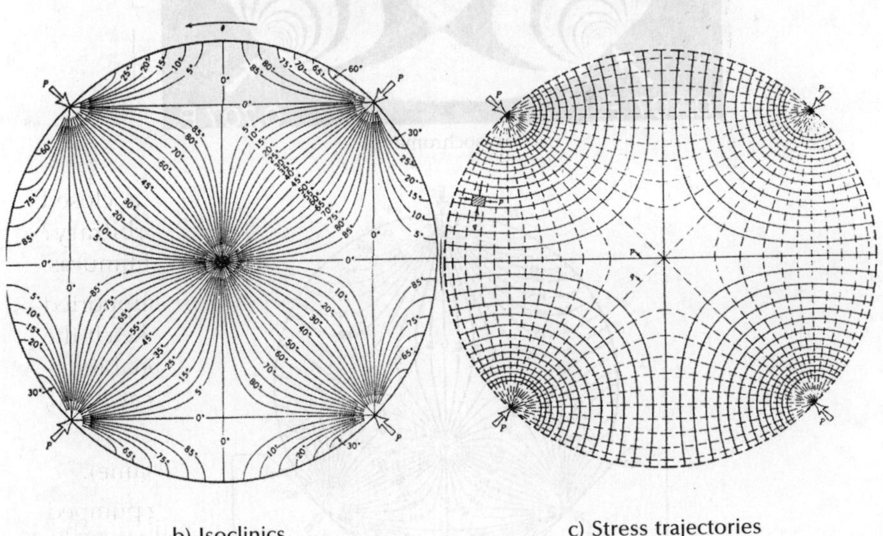
b) Isoclinics

c) Stress trajectories

FIGURE P8.41
Photoelastic patterns for a quadriradially loaded disk (Frocht).

P8.42 Plot the flow net (for $\sigma_1 + \sigma_2$ vs. ω_z) for one quadrant of the quadralaterally loaded disk of P8.41. From the information in P8.41 and this flow net, what is the stress distribution between two adjacent loads (i.e., along the zero horizontal and vertical isoclinics)? Check with the analytic solution if you wish.

P8.43 Isochromatics and isoclinics patterns for a disk with triradial loads are shown in Figure P8.43. Attempt an analytic solution for this stress field when: (a) the base angle $2\alpha = 90°$ as shown; and (b) the base angle $2\alpha = 120°$ giving an "equilateral" three-point loading.

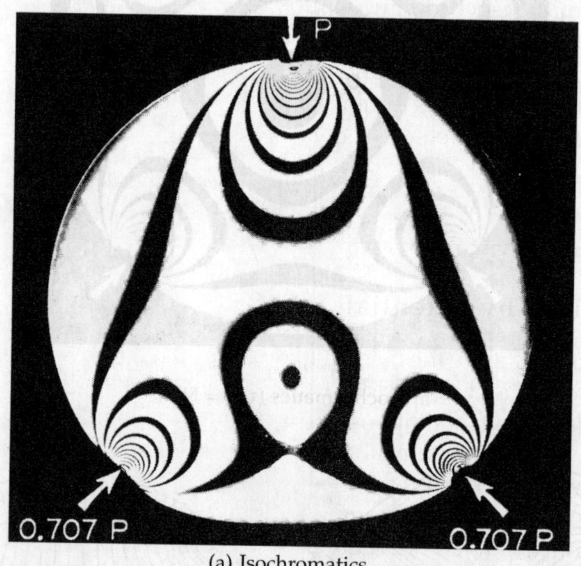

(a) Isochromatics

(b) Isoclinics

FIGURE P8.43
Disk with triradial loading (Frocht).

P8.44 From the isochromatic pattern on the boundary in Figure P8.43, plot the flow net ($\sigma_1 + \sigma_2$ vs. ω_z). Determine the state of stress and plot the Mohr's Circle at the center of the disk, and at the isotropic point. For the model: Diam = 1.205 in., t = .243 in., max shear stress f = 43 lb/fringe in., P = 92.5 lb.

P8.45 Isochromatics for a square loaded across a diagonal are shown in Figure P8.45. If the length of a diagonal = 1.885 in, thickness t = 0.255 in., and the shear fringe coefficient f = 36.2 lb/fringe in., estimate the load P. (Hint: consider the fringe pattern near the load points.)

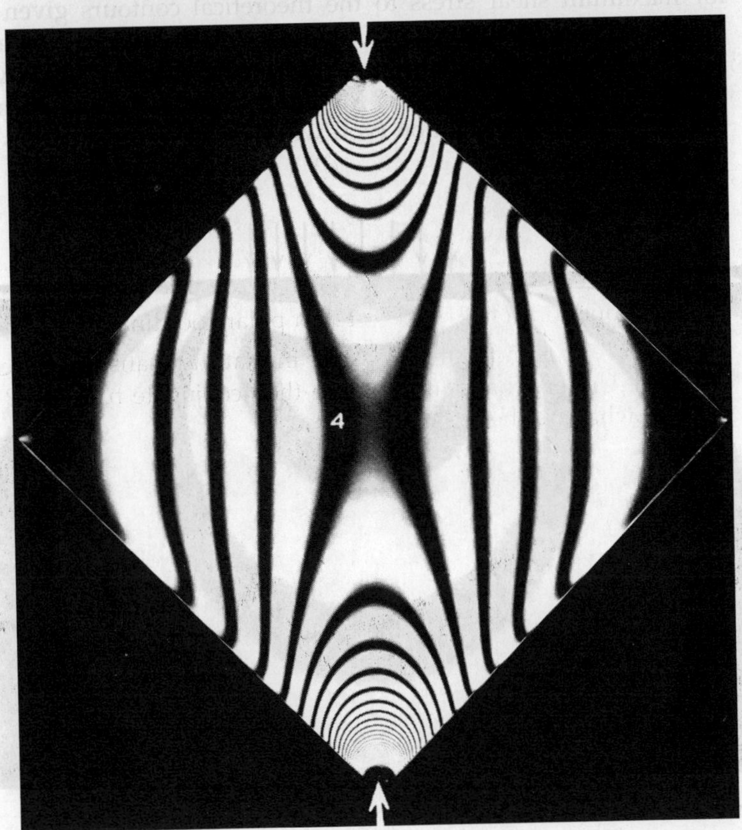

FIGURE P8.45
Diagonally loaded square (Frocht).

P8.46 From the photoelastic information (Figure P8.45) on the boundary, plot a flow net ($\sigma_1 - \sigma_2$ vs. ω_z). Note: check equilibrium at the horizontal and vertical cross-sections and keep adjusting the flow net accordingly. The fringe order at the center is 4.17.

P8.47 Suggest how you might try going about deriving an analytic solution for a square loaded across a diagonal as in Figure P8.45.

P8.48 Consider a square equally loaded across both diagonals. Is this more amenable to solution than a square loaded across only one diagonal? Attempt a solution.

P8.49 Shown in Figure P8.49 is the isochromatic fringe pattern for a strip loading on a gelatin plate. Gelatin has a density of 1.1 and a shear fringe coefficient of 0.0725 lb/fringe in. The gelatin plate is 4 inches thick, 11 inches deep, and 41 inches wide sitting on a table and the strip loading is 0.1878 psi over a width of 2 inches. The maximum fringe order is 3.5. Compare the observed contours (fringe pattern) for maximum shear stress to the theoretical contours given by Equation (8.16). Plot the experimental vs. analytic prediction as a function of depth beneath the center of the strip. What do you think is the cause of any error? Modify the theory accordingly and replot the corrected solution (you will now have three curves on the same plot).

FIGURE P8.49
Isochromatics for a strip load on a large gelatin plate (Frocht).

P8.50 Consider the polar coordinate system in Figure 8.10. Assume the stress function

$$F_\gamma = \frac{\gamma r^3}{6}\left[\frac{\cos\phi}{\tan\beta}\sin^3\theta + \frac{3\sin\phi\cos\theta}{\tan\beta}\sin^2\theta - \frac{2\sin\phi}{\tan^2\beta}\sin^3\theta \right.$$
$$\left. - \sin\phi\sin^3\theta + \cos\phi\sin^3\theta\right]$$

and the components γ_r and γ_θ of the body force per unit volume, γ, for a gravity field can be expressed in terms of the potential function $\Omega = \gamma \cos\psi$

$$\gamma_r = \gamma \cos\psi = \frac{\partial \Omega}{\partial r}; \qquad \gamma_\theta = -\gamma \sin\psi = \frac{1}{r}\frac{\partial \Omega}{\partial \theta}$$

a. Derive the stresses from Equation (6.16)

b. Show that they satisfy the boundary conditions at $\theta = 0$ and β.

c. Show that $\nabla^2(\sigma_r + \sigma_\theta) = 0$

d. Show that this solution transforms to that given by Equations (8.21) and (8.22) in Cartesian coordinates.

P8.51 For a 45° beam loaded by its own weight as shown in Figure P8.51, compare the elasticity solution, Equation (8.22), to the strength of materials solution on some vertical cross-section (x = const. = 1.0).

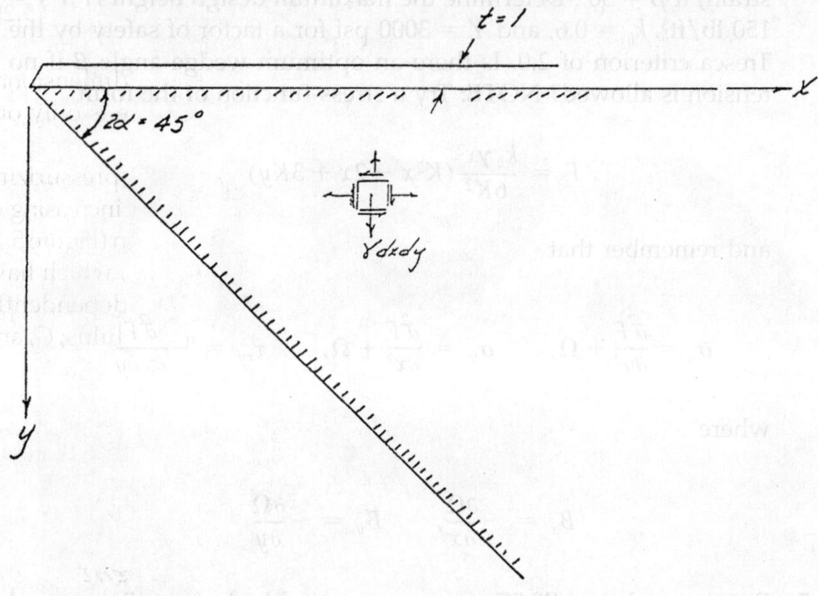

FIGURE P8.51

P8.52 Repeat the previous problem for a wedge angle $2\alpha = \beta = 30°$.

P8.53 For the haunched beam shown in Figure P8.53, derive an elasticity solution by superposition and compare to a strength of materials solution where self-weight is the load.

P8.54 For the 45° wedge (Figure P8.51)

a. Plot the flow net $\sigma_1 + \sigma_2$ vs. ω_z,

b. Determine the displacements u and v,

c. What support condition is implied?

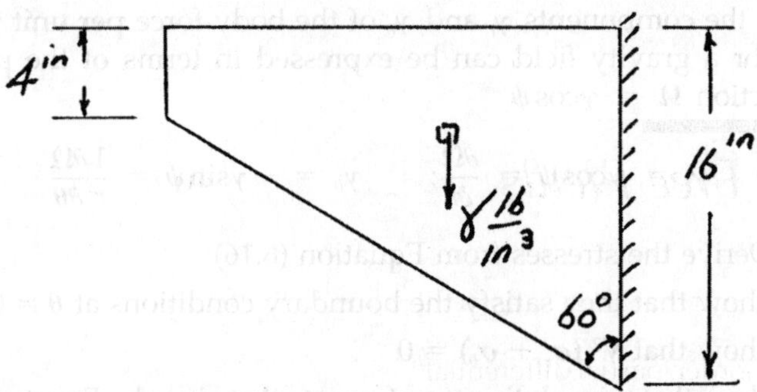

FIGURE P8.53

P8.55 Horizontal acceleration introduces a body force $B_x = k_h\,\gamma$ in a wedge as shown in Figure P8.55. Determine the stress field (assume plane strain) if $\beta = 30°$. Determine the maximum design height H if $\gamma = 150$ lb/ft^3, $k_h = 0.6$, and $Y = 3000$ psi for a factor of safety by the Tresca criterion of 2.0. Is there an optimum wedge angle β if no tension is allowed? NOTE: Try a stress-function of the form

$$F = \frac{k_h \gamma x^2}{6K^2}(K^2 x - 2x + 3Ky)$$

and remember that

$$\sigma_x = \frac{\partial^2 F}{\partial y^2} + \Omega, \qquad \sigma_y = \frac{\partial^2 F}{\partial x^2} + \Omega, \qquad \tau_{xy} = -\frac{\partial^2 F}{\partial x\,\partial y}$$

where

$$B_x = -\frac{\partial \Omega}{\partial x}, \qquad B_y = -\frac{\partial \Omega}{\partial y}$$

P8.56 For the wedge of P8.55, assume a vertical body force $B_y = \gamma$ and determine the design height. Is there an optimum wedge angle β if no tension is allowed?

P8.57 Consider the previous two problems and derive a general solution for a 30° wedge with a vertical upstream face subjected to seismic acceleration $a_v = k_v g$, $a_h = k_h g$ (either positive or negative).

P8.58 For a gravity dam ($B = 100\sqrt{2}$ lb/ft.^{3}) of height H $= 500\sqrt{2}$ ft as shown in Figure P8.58, the stress distribution is proposed to be:

$$\sigma_x = -B_x x = -100x \qquad \sigma_y = -B_y y = -100y$$
$$\sigma_z = v(\sigma_x + \sigma_y) \qquad \tau_{xy} = 0$$

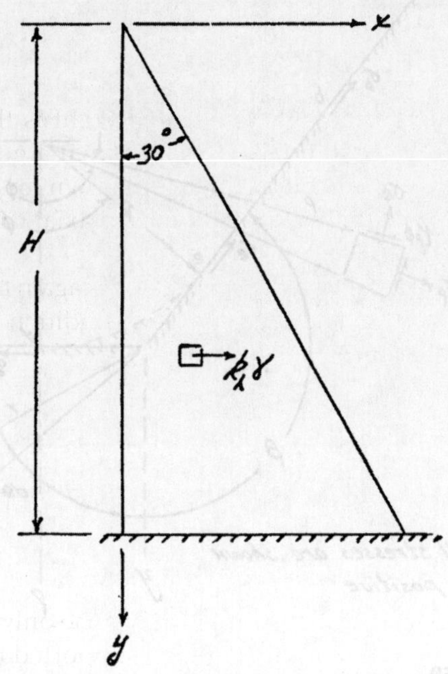

FIGURE P8.55

Is this the solution? Plot the normal and shear stresses on line AB and check overall equilibrium. What support conditions are implied by this stress field? Is it reasonable?

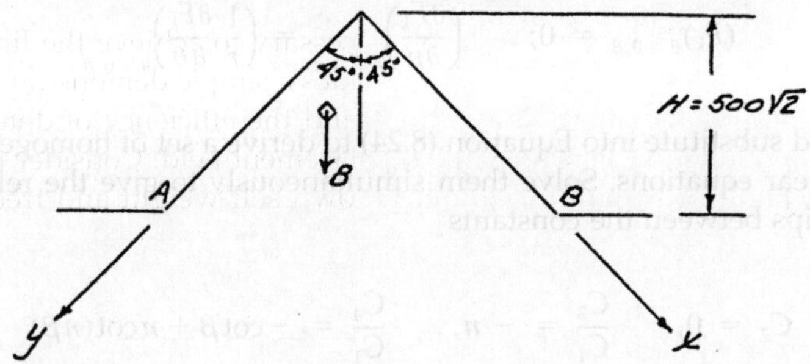

FIGURE P8.58

P8.59 Refer to Figure P8.59. Assuming the corner function only redistributes stresses and gives no stresses on the straight boundaries then $F_R = r^{n+1}f(\theta)$. Substitute this into $\nabla^4 F = 0$ to get the ordinary differential equation:

$$\frac{d^4 f}{d\theta^4} + 2(n^2+1)\frac{d^2 f}{d\theta^2} + (n^2-1)^2 f = 0$$

and thereby derive Equation (8.24).

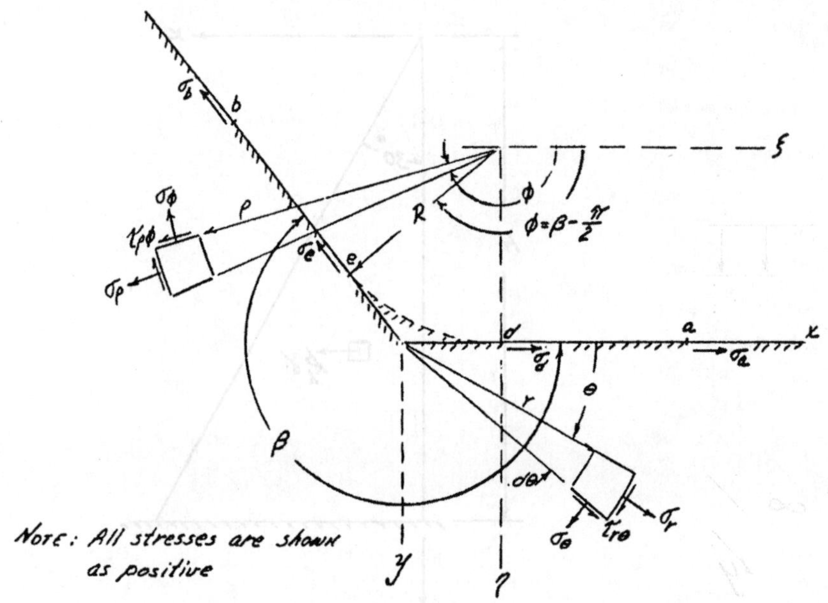

FIGURE P8.59

P8.60 Referring again to Figure P8.59, rewrite the boundary equations in terms of the stress-function as

$$(F_k)_{\theta = 0,\beta} = 0; \qquad \left(\frac{\partial F_k}{\partial n}\right)_{\theta = 0,\beta} = \left(\frac{1}{r}\frac{\partial F_k}{\partial \theta}\right)_{\theta = 0,\beta}$$

and substitute into Equation (8.24) to derive a set of homogeneous linear equations. Solve them simultaneously to give the relationships between the constants

$$C_2 = 0, \qquad \frac{C_3}{C_1} = -n, \qquad \frac{C_4}{C_1} = -\cot\beta + n\cot(n\beta)$$

and $n\sin\beta = \pm\sin n\beta$. Assume $C_1 = -1$ and derive Equation (8.25).

P8.61 Solve for the real eigenvalues of n referred to as "corner values" for $\pi < \beta < 2\pi$. There will be two real roots η_0 and η_1. Plot your results.

P8.62 Find the complex roots to the transcendental equation in P8.9 by substituting $n = a_k + ib_k$, equating real and imaginary parts, and solving by trial and error. Do this for $\beta = 235°$ and $\beta = 270°$.

9

Torsion

9.1 Elementary (Linear) Solution

The general problem of a shaft under torque is obviously of great practical importance. The linear solution, as presented in a first course in solid mechanics, assumes that for a circular cross-section, radical lines remain straight when the torque is applied. Thus, as shown in Figure 9.1, two neighboring cross-sections, dz apart rotate relative to each other by a small angle of twist $d\omega$. Thus, longitudinal straight lines become helixes at the angle of shear strain, γ, related to the angle of twist by the circumferential movement:

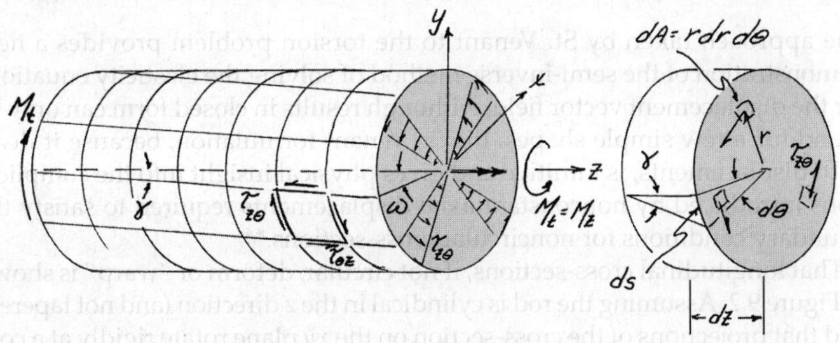

FIGURE 9.1
Elementary torsion (circular shaft).

$$ds = \gamma dz = r d\omega \quad \text{and} \quad \therefore \gamma = r\frac{d\omega}{dz} = r\theta \tag{9.1}$$

where the rate of twist or "unit angle of twist," $\frac{d\omega}{dz} = \theta$, is a constant. Thus, for an elastic material

$$\tau = \tau_{z\theta} = G\gamma = Gr\theta \tag{9.2}$$

and from equilibrium

$$M_t = \int_{\text{area}} r\tau dA = \int_{\text{area}} r^2 G\theta dA = G\theta \int_{\text{area}} r^2 dA = G\theta I_p \tag{9.3}$$

where $I_p = \int_A r^2 \, dA$ is the polar moment of inertia (the first invariant of the moment-of-inertia tensor). Thus

$$\theta = \frac{M_t}{GI_p} \quad \text{and} \quad \tau_{z\theta} = \tau_{\theta z} = \frac{M_t r}{I_p} \tag{9.4}$$

with all other stress and strain components zero.*

9.2 St. Venant's Formulation (Noncircular Cross-Sections)

The approach taken by St. Venant to the torsion problem provides a neat demonstration of the semi-inverse method of solving the elasticity equations for the displacement vector field. Although results in closed form can only be found for a few simple shapes, the St. Venant formulation, because it deals with displacements, is intuitive and gives physical insight into the complications introduced by nonconstant axial displacements required to satisfy the boundary conditions for noncircular cross-sections.**

That longitudinal cross-sections, if not circular, deform or "warp" is shown in Figure 9.2. Assuming the rod is cylindical in the z direction (and not tapered) and that projections of the cross-section on the xy plane rotate rigidly at a constant rate of twist, θ, St. Venant, pioneering the inverse method, postulated the displacement field (shown in Figure 9.3):

$$u = -\theta zy, \qquad v = \theta zx, \qquad w = w(x,y) \tag{9.5}$$

* Coulomb, in 1776, seems to have been the first to consider the torsional rigidity of rods (thin fibers) and deduce it was proportional to the moment of inertia. However, he considered shear stress and strain only at rupture and therefore did not introduce elastic stiffness and any consideration of the geometry of deformation. Coulomb must have understood torsional stiffness, however, since he used a torsional pendulum to measure electric charge, the units of which now bear his name.

** Although torsion is a 3D problem, its rigorous solution, as given by St. Venant in 1853, was the first such derived from formal elasticity theory. This convinced engineers who expected little from "theory" that the mathematical equations of Navier, Cauchy, et al., germinating in the literature for over 30 years, were really of great practical importance and would revolutionize stress analysis.

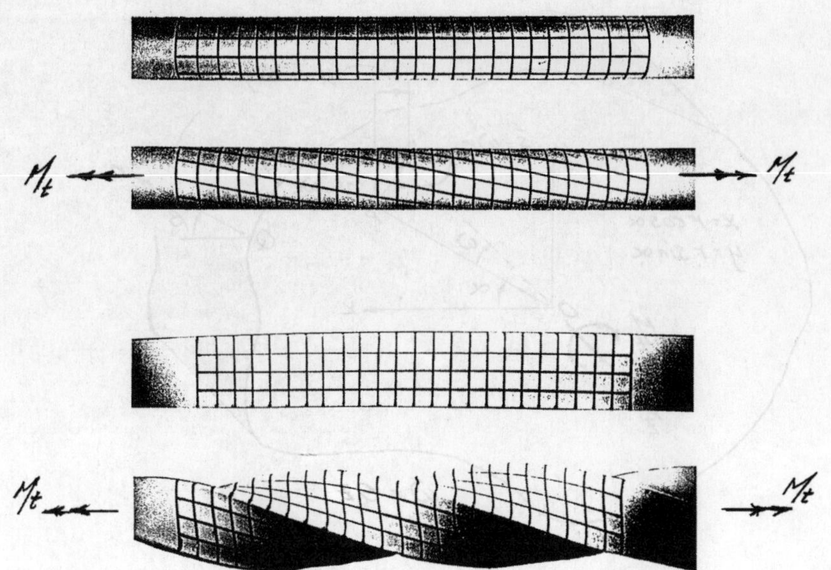

FIGURE 9.2
Rectangular shaft with warping.

as the solution (the answer). Thus the warp of the cross-section, w, is left as an unknown function (i.e., the warping function) to be adjusted to satisfy the boundary condition dictated by the shape. In terms of strains, he therefore postulated that:

$$\varepsilon_x = \varepsilon_y = \varepsilon_z = \gamma_{xy} = 0$$

$$\gamma_{yz} = \frac{\partial v}{\partial z} + \frac{\partial w}{\partial y} = \theta x + \frac{\partial w}{\partial y}; \qquad \text{and} \qquad \gamma_{xz} = \frac{\partial w}{\partial x} - \theta y \tag{9.6}$$

with the elastic rotation

$$\omega_z = \frac{1}{2}\left(\frac{\partial v}{\partial x} - \frac{\partial u}{\partial y}\right) = \frac{\theta z + \theta z}{2} = \theta z \tag{9.7}$$

being the total angle of twist of the cross-section. Geometric compatibility is automatically satisfied because this formulation assumes continuous and smooth displacements.

From the constitutive relationships:

$$\sigma_x = \sigma_y = \sigma_z = \tau_{xy} = 0 \tag{9.8}$$

$$\tau_{yz} = G\gamma_{yz} = G\left(\frac{\partial w}{\partial y} + \theta x\right) \tag{9.9a}$$

$$\tau_{xz} = G\gamma_{xz} = G\left(\frac{\partial w}{\partial x} - \theta y\right) \tag{9.9b}$$

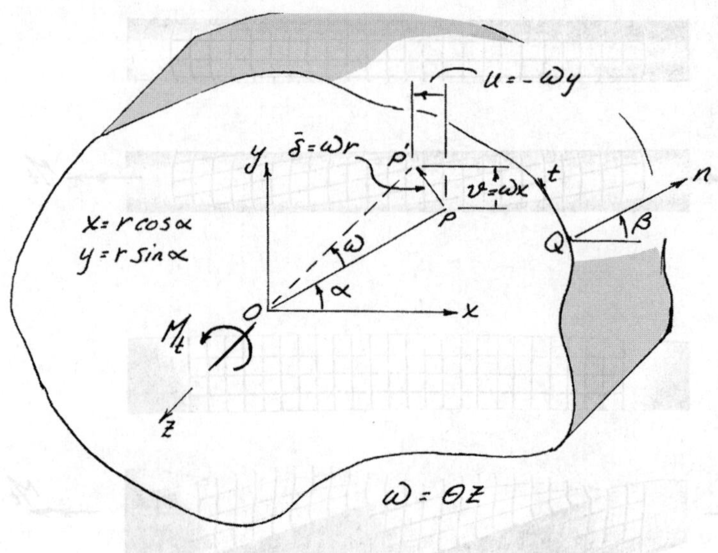

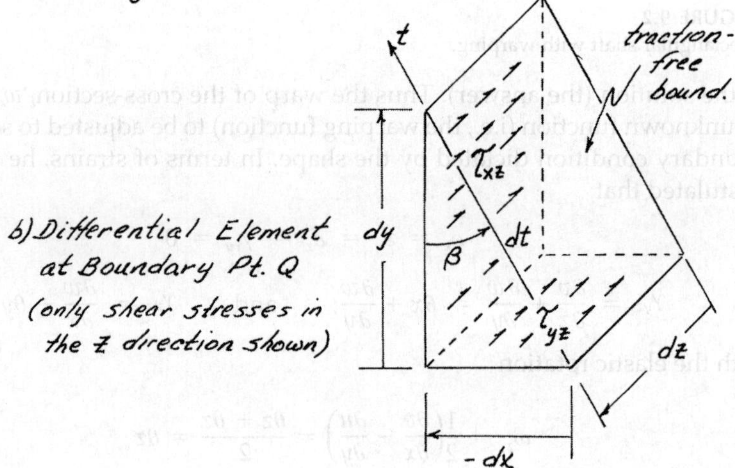

FIGURE 9.3
General torsion formulation.

The only equilibrium equation not already satisfied is

$$\frac{\partial \tau_{zx}}{\partial x} + \frac{\partial \tau_{zy}}{\partial y} = 0 \tag{9.10}$$

which, upon substitution becomes

$$\frac{\partial^2 w}{\partial x^2} + \frac{\partial^2 w}{\partial y^2} = \nabla^2 w = 0 \tag{9.11}$$

Thus any harmonic function for the longitudinal displacement or warping is a possible elastic field solution for the general torsion problem if we can find the shape of the cross-section where the traction-free boundary condition is satisfied.

The resultant twisting moment on the ends of the rod is:

$$\begin{aligned} M_t &= \iint_{\text{area}} (x\tau_{zy} - y\tau_{zx})dxdy \\ &= G\theta \iint_{\text{area}} \left(x^2 + y^2 + x\frac{\partial w}{\partial y} - y\frac{\partial w}{\partial x}\right)dxdy \end{aligned} \tag{9.12}$$

or

$$M_t = GJ\theta \tag{9.13}$$

where:

$$J = \iint_{\text{area}} \left(x^2 + y^2 + x\frac{\partial w}{\partial y} - y\frac{\partial w}{\partial x}\right)dxdy \tag{9.14}$$

is often called the "torsional constant" and GJ expresses the "torsional rigidity" of the bar.

Considering an arbitrary cross-section such as given in Figure 9.3a, there can be no shear τ_{nt} at the surface. Thus considering an element at any boundary point Q, longitudinal equilibrium requires that:

$$\tau_{xz}dydz + \tau_{yz}(-dxdz) = 0$$

where dx is negative in Figure 9.3b since a positive dt causes x to decrease. Therefore, the boundary condition becomes

$$\left(\frac{\tau_{zy}}{\tau_{zx}}\right)_{\text{bound}} = \left(\frac{dy}{dx}\right)_{\text{bound}} = \textit{slope of the edge} \tag{9.15}$$

which, on substitution of Equation (6.9) gives

$$\left[\frac{dy}{dx} = \frac{\dfrac{\partial w}{\partial y} + \theta x}{\dfrac{\partial w}{\partial x} - \theta y}\right]_{\text{bound}} \tag{9.16}$$

9.2.1 Solutions by St. Venant

The St. Venant formulation therefore leads to the search for harmonic functions $w(x,y)$ that satisfy Equation (9.16). In general this Neumann boundary condition is difficult to apply especially for nonsymmetric cross-sections where the center of rotation is not obvious. This illustrates the inherent difficulty, referred to in Chapter 4, in employing the displacement strategy for elasticity solutions where the boundary conditions involve stresses.

Nevertheless, a few simple harmonic functions for the warping are manageable leading to practical inverse solutions in closed form. The most obvious is to assume the simplest solution to the Laplace equation:

a. $w = const.$

Integrating the boundary Equation (9.15)

$$\int_s (xdx + ydy) = 0 \qquad \therefore x^2 + y^2 = \text{const.} = r^2$$

and we have a circular shaft and the elementary linear solution already discussed for plane sections is correct. From Equation (9.13), $J = I_p = \pi r^4/2$ and all the standard formulae follow.

b. $w = Kxy$

The hyperbolic function, $w = Kxy$, is harmonic and therefore satisfies the equilibrium requirement of Equation (9.6). The boundary condition [Equation (9.16)] gives: $x(K + \theta)dx = y\,(K - \theta)dy$ which, when integrated, gives the required boundary shape:

$$(K + \theta)x^2 + (\theta - K)y^2 = \text{const.} = C \tag{9.17}$$

which is an ellipse (Figure 9.4a) if $K < \theta$. If $K > \theta$, a hyperbola results which, not being a closed shape, can be disregarded.

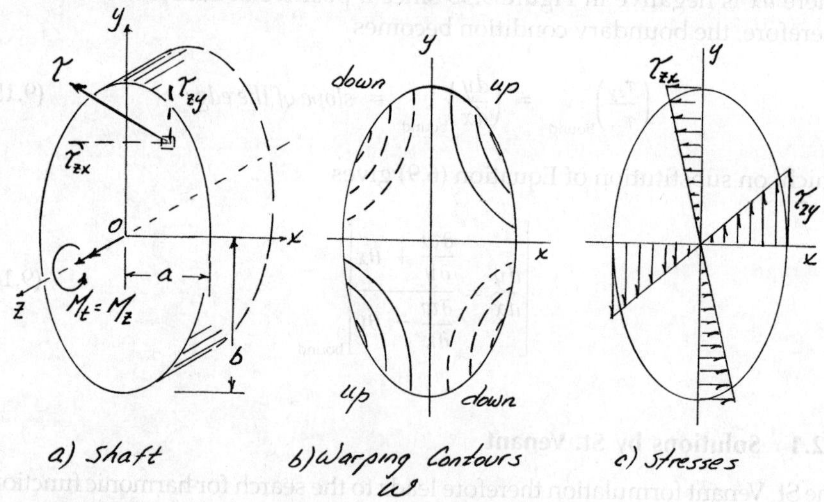

FIGURE 9.4
Torsion of a prismatic elliptic shaft.

Letting $a^2 = \frac{c}{\theta + K}$ and $b^2 = \frac{c}{\theta - K}$, then $\frac{x^2}{a^2} + \frac{y^2}{b^2} = 1$ and $\theta b^2 - Kb^2 = c = \theta a^2 + Ka^2$. Thus the warping coefficient of the elliptic cross-section is

$$K = \left(\frac{b^2 - a^2}{a^2 + b^2}\right)\theta \tag{9.18}$$

which is zero if circular ($a = b$). From Equation (9.14) for Region R

$$\begin{aligned} J &= \iint_R (x^2 + y^2 + Kx^2 - Ky^2)dA \\ &= (K+1)\iint_R x^2 dA + (1-K)\iint_R y^2 dA \\ &= (K+1)I_y + (1-K)I_x = \frac{\pi a^3 b^3}{a^2 + b^2} \end{aligned} \tag{9.19}$$

where I_x and I_y are the moments of inertia with respect to the x and y axes, respectively.

From Equation (9.9) the shear stress components are

$$\tau_{yz} = \frac{2M_t x}{\pi a^3 b}, \qquad \tau_{xz} = -\frac{2M_t y}{\pi a b^3}$$

and thus the maximum value

$$\tau_{max} = \frac{2M_t}{\pi a b^2}$$

occurs at the extremity of the minor axis ($x = 0$, $y = b$) which is somewhat counterintuitive but will be obvious from the analogies drawn later.

If the ends of the bar are not restrained from warping, there will be no normal stresses. However, if one or both ends are restrained such as for a built-in shaft, normal stresses will be induced similar to those that would result from equal and opposite bending moments and are, therefore, sometimes called the "bending stresses induced by torsion."

9.3 Prandtl's Stress Function

An alternative solution procedure that leads to a much simpler boundary condition, but a somewhat more difficult field equation, involves the introduction of a stress function $\Phi(x,y)$ defined as:

$$\begin{aligned} \frac{\partial \Phi}{\partial y} &= \tau_{zx} \\ -\frac{\partial \Phi}{\partial x} &= \tau_{zy} \end{aligned} \tag{9.20}$$

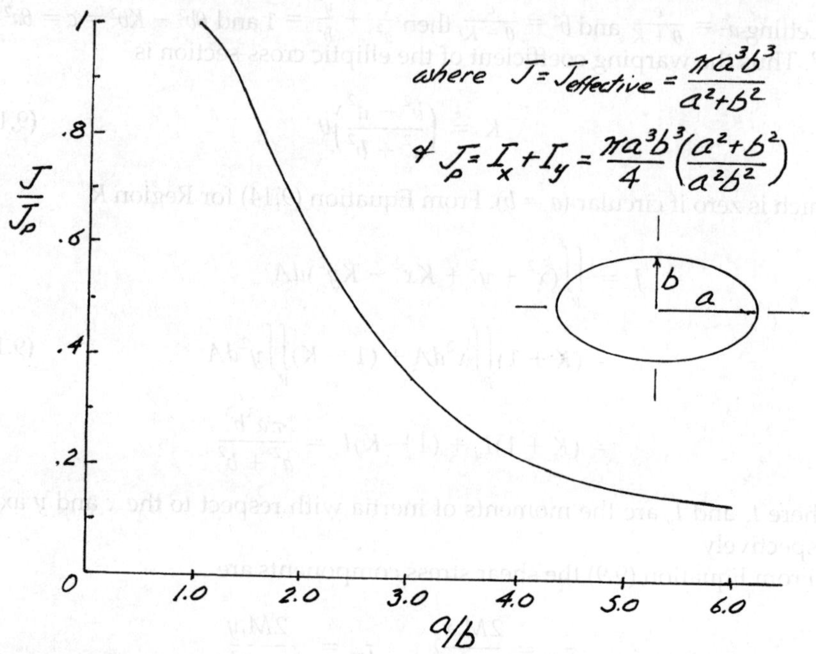

FIGURE 9.5
Ratio of J/J_P for ellipses.

By this definition, the equilibrium Equation (9.10) is automatically solved and the boundary condition Equation (9.15) becomes:

$$\frac{\partial \Phi}{\partial x}dx + \frac{\partial \Phi}{\partial y}dy = d\Phi = 0 \tag{9.21}$$

or Φ = constant (say zero) on the boundary.

Operating on Equation (9.9a) with $-\frac{\partial}{\partial x}$ and Equation (9.9b) with $\frac{\partial}{\partial y}$ and adding, the field equation insuring compatibility of strains becomes:

$$\frac{\partial^2 \Phi}{\partial x^2} + \frac{\partial^2 \Phi}{\partial y^2} = \nabla^2 \phi = -2G\theta \tag{9.22}$$

As noted in Chapter 4, this is a Poisson equation for which a solution can always be found and the boundary condition

$$\Phi = 0 \quad \text{on} \quad C \tag{9.23}$$

is particularly nice.

However, the power of the Prandtl stress function approach is in visualizing torsional shear fields. It is easy to show that the stress function Φ will transform in any direction as an invariant scalar function, and therefore in a new coordinate orientation x', y':

$$\tau_{zx'} = \frac{\partial \Phi}{\partial y'} \quad \text{and} \quad \tau_{zy'} = \frac{\partial \Phi}{\partial x'} \tag{9.24}$$

Thus at any point the shear stress in one direction is equal in magnitude to the slope of the Φ surface in the perpendicular direction, and the maximum shear is the maximum slope of the stress function curve and acts tangent (along a contour) to that curve. If we visualize the stress function plotted in the z direction above the cross-section, then contours (of equal Φ) can be plotted and:

a. The shear τ acts along these contours and is proportional to the shortest distance (slope normal) to the next contour;

b. τ_{max} occurs where the stress-hill contours become closest together, which will be on the boundary of the cross-section (generally where $r = \sqrt{x^2 + y^2}$ is a minimum).

From Figure 9.6 it is easy to show that the twisting moment M_t is proportional to the volume of this stress-function hill. The shear force dS on a differential element E will be

$$dS = \tau dA = \frac{\partial \Phi}{\partial \rho} d\rho \, dt \tag{a}$$

But $d\Phi = \frac{\partial \Phi}{\partial \rho} d\rho + \frac{\partial \Phi}{\partial \theta} d\theta$ where, on a contour line, $\frac{\partial \Phi}{\partial \theta} = 0$ and $d\Phi$ is a constant. Also

$$dS = d\Phi dt \tag{b}$$

Therefore, between two contour lines the resisting moment

$$dM_z = \int_{\text{ring}} dS\rho = d\Phi \int_{\text{ring}} \rho dt = d\Phi \int_{\text{ring}} 2dA' = 2d\Phi A' \tag{c}$$

But $dV = A' \, d\Phi$, and therefore

$$dM_z = 2dV \tag{d}$$

Thus for equilibrium:

$$M_t = \int_{\text{area}} dM_z = 2V = 2\int_{\text{area}} \Phi dx dy \tag{9.25}$$

which is twice the volume of the stress-function hill.

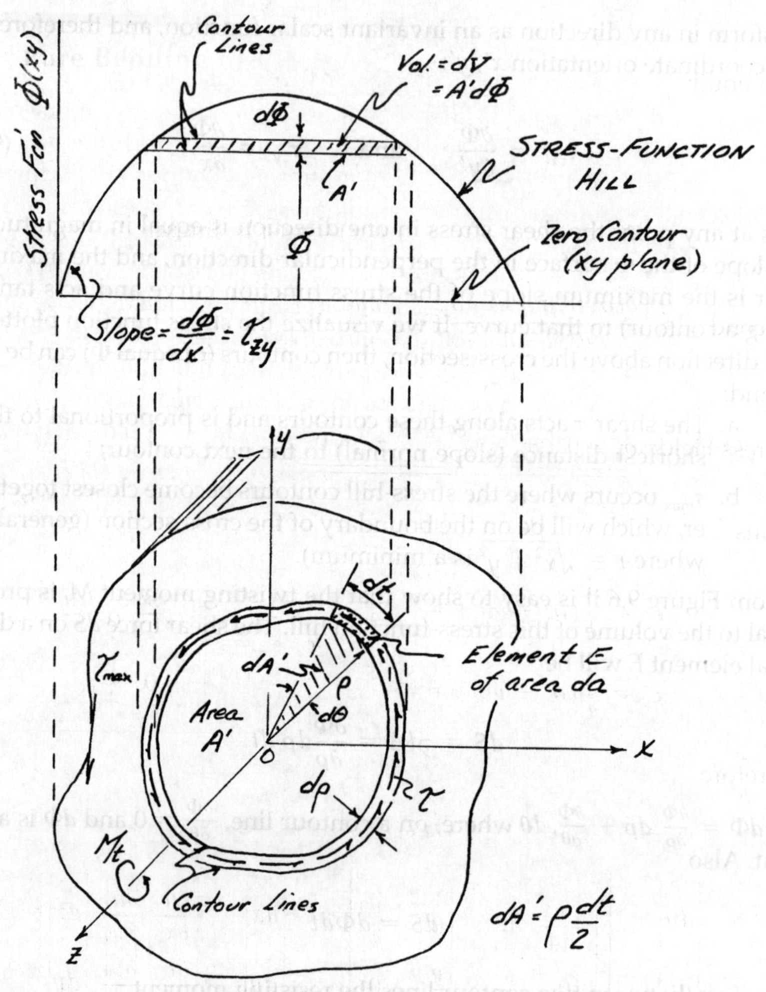

FIGURE 9.6
Stress function for a prismatic shaft.

For example, consider the elliptic cross-section discussed previously. Assume a stress function

$$\Phi(x,y) = h\left(1 - \frac{x^2}{a^2} + \frac{y^2}{b^2}\right) \quad \text{(a)}$$

which must satisfy Equation (9.22), and the boundary condition $\Phi_{edge} = 0$.

The volume of the stress-function hill (a paraboloid) is 0.5 π abh and therefore:

$$h = \frac{M}{\pi ab} \quad \text{and} \quad \Phi = \frac{M}{\pi ab}\left(1 - \frac{x^2}{a^2} + \frac{y^2}{b^2}\right) \quad \text{(b)}$$

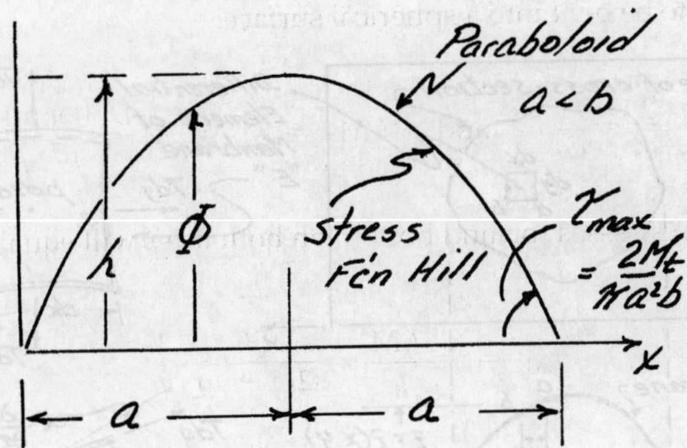

FIGURE 9.7
Φ for an ellipse.

Thus

$$\nabla^2\Phi = \frac{M_t}{\pi ab}\left(-\frac{2}{a^2}+\frac{2}{b^2}\right) \underset{\text{must}}{=} -2G\Theta \qquad \text{(c)}$$

or the unit angle of twist

$$\theta = \frac{M_t(a^2+b^2)}{G\pi a^3 b^3} \qquad (9.26)$$

and

$$\tau_{zx} = \frac{\partial\Phi}{\partial y} = -\frac{2yM_t}{\pi b^3 a}$$

$$\tau_{zy} = -\frac{\partial\Phi}{\partial x} = \frac{2xM_t}{\pi a^3 b}$$

which corresponds to the result obtained by the direct St. Venant approach* using the warping function.

9.4 Membrane Analogy

The crucial step in the stress-function approach, finding Φ, is not much easier than finding a warping function w. Its advantage comes from the simple

* The Prandtl stress function approach is also part of the general St. Venant theory of torsion since Equation (9.9) from Equation (9.5) are used to derive the compatibility equation $\nabla^2\Phi = -2G\theta$. It is also a semi-inverse strategy since Φ is assumed (guessed).

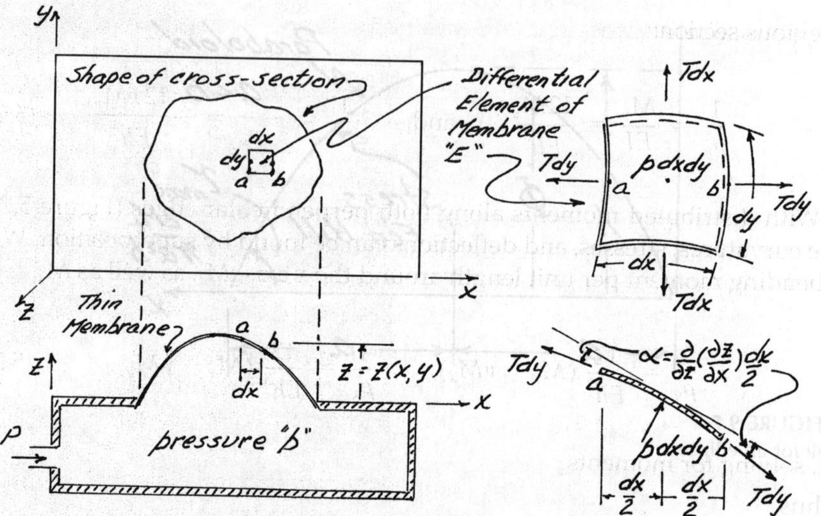

FIGURE 9.8
Membrane analogy setup.

boundary condition and, more importantly, the aid it gives in visualizing the stress-function hill by analogy. In his presentation in 1903, Prandtl* observed that the stress-function equation $\nabla^2\Phi = -2G\theta$ is the same Poisson type as that giving the elevation of an elastic membrane loaded by pressure on one side. While the computer with canned finite-difference or finite element solutions to Poisson's equation have replaced actual laboratory experiments with soap films, the membrane analogy is an extremely useful heuristic device to visualize the Φ surface and hence the stiffness of and the shear stresses on a cross-section under torque. In most cases, in fact, approximate calculations based on the membrane analogy are more than adequate for rough analysis and design.

As shown in Figure 9.8, imagine a flexible membrane stretched flat over a hole in a flat plate of the irregular shape of the cross-section to be analyzed. Air pressure is applied and the membrane is stretched a distance $z(x,y)$ above the plate. Summing forces in the z direction:

$$pdxdy + 2Tdy\frac{\partial^2 z}{\partial x^2}\frac{dx}{2} + 2Tdx\frac{\partial^2 z}{\partial y^2}\frac{dy}{2} = 0$$

* Ludwig Prandtl was born near Munich where he received his engineering degree at the Polytechnical Institute. His work is characterized by breadth, physical insight, and extreme originality. Practically everything he touched became the basis for a new fundamental field of research in mechanics from his first concept of the boundary layer in fluid flow to inelastic buckling and the collapse loads on the half-space, which consumed most of his time after 1925. Prandtl was a truly great teacher and student advisor at the Institute of Applied Mathematics and Mechanics at the University of Göttingen. Among his students who became active and well known in the U.S. are: Theodore von Karman (aerodynamics), A. Nadai (plasticity), and W. Prager who carried on Prandtl's development of mathematics for engineering at Brown University.

and therefore

$$\frac{\partial^2 z}{\partial x^2} + \frac{\partial^2 z}{\partial y^2} = \nabla^2 z = -\frac{p}{T} \tag{9.27}$$

This equation is the same as Equation (9.22) if $\frac{p}{T} = 2G\theta$. Thus we can visualize the stress-function, Φ, as the height of the membrane. On any contour of the membrane

$$\frac{dz}{dt} = \frac{d\Phi}{dt} = \left(\frac{\partial \Phi}{\partial x}\frac{dx}{dt} + \frac{\partial \Phi}{\partial y}\frac{dy}{dt}\right) = -\tau_{yz}\frac{dx}{dt} + \tau_{zx}\frac{dy}{dt} = 0$$

and similarly

$$\frac{dz}{dn} = -\frac{d\Phi}{dn} = -\left(\frac{\partial \Phi}{dy}\frac{dy}{dn} + \frac{\partial \Phi}{\partial x}\frac{dx}{dn}\right) = \tau$$

gives the resultant shearing stress at any point.

Thus the tangent to any membrane contour gives the direction of the resultant shear with a magnitude equal to the slope perpendicular to the contour line at that point. The twisting movement

$$M_t = 2\iint \Phi dxdy \sim 2\iint zdxdy \tag{9.28}$$

is twice the volume under the membrane. Finally, if we consider vertical equilibrium along a contour s,

$$\int_s T\frac{dz}{dn}ds = \int_s T\tau ds = pA$$

Since p, G, θ, and T are constants and $p/T = 2GA\theta$,

$$\int_s \tau ds = 2GA\theta \tag{9.29}$$

To summarize then, the following are analogous:

Membrane	Torsion
z	Φ
$1/T$	G
ρ	2θ
$-\frac{\partial z}{\partial x}, \frac{\partial z}{\partial y}$	τ_{zy}, τ_{zx}
Twice the volume	M_t

When there is a hole in the cross-section, the shear stress must be zero and the membrane must be flat, corresponding to a weightless plate "floating" on

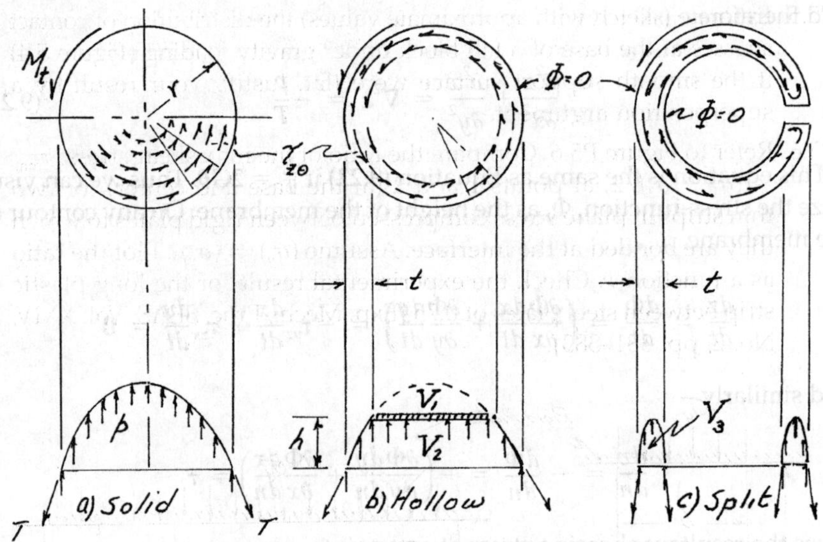

FIGURE 9.9
Membrane analogies for cylinders.

the surface. For example a solid cylinder is compared to hollow and split cylinders in Figure 9.9. Clearly, considering the weight reduction, removing the core of a solid cylinder is a good design choice since the reduction in torsional stiffness and moment capacity V_1 compared to V_2 (that remaining) is small.* For the split ring in Figure 9.9 with a moment capacity proportional to the volume of the small torus V_3, the reduction is dramatic. If the wall thickness is small, a cut reduces the capacity to where it is essentially that of a thin plate of thickness t and length equal to the perimeter $2\pi r$. In fact it is nearly correct for any thin-walled, open cross-section to calculate its moment capacity and stiffness as the sum of its elements considered as flat plates.

It is easy to solve for a flat plate under torsion from the membrane analog. If $b \gg l$ then, except near the ends, the shape of a nearly "flat" membrane under uniform vertical load on the horizontal projection is parabolic** as shown in Figure 9.10. Thus

$$z = \Phi = c\left(1 - 4\frac{y^2}{b^2}\right) \tag{9.30}$$

* Homework problems call for exact and approximate calculation.

** In deriving Equation (9.27), the pressure is assumed to act vertically and the tension T is assumed constant. Neither is true in an actual experiment but, if the membrane is pretensioned and the deflections z are small, then these assumptions are approximated. A long, pressurized membrane actually deforms into a circle with tensile force $T = pr$. Experiments and more exact analysis give results as much as 10% higher for both the rigidity and moment capacity (volume and maximum slope of the membrane). Thus assuming a parabolic shape is conservative as is neglecting the end effect.

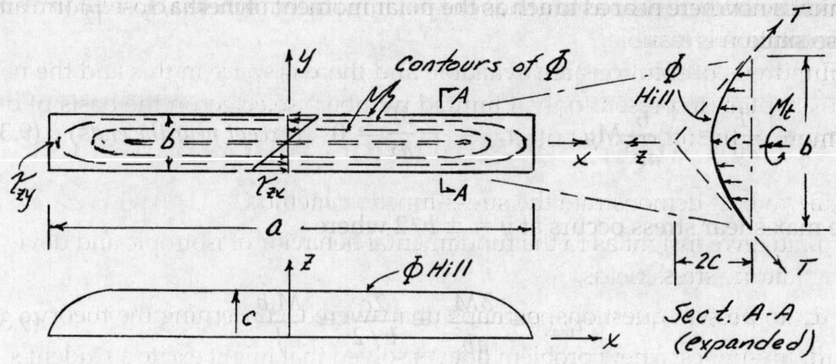

FIGURE 9.10
Torsion of a thin rectangular plate ($b \ll a$).

and

$$M_t = 2V = 2\int_{\text{area}} \Phi\, dxdy = 2cab - 16\frac{c}{b^2}\int_0^b\int_0^a y^2 dxdy = \frac{4}{3}abc$$

therefore

$$c = \frac{3M_t}{4ab} \quad \text{and} \quad \Phi = \frac{3}{4}\frac{M_t}{ab}\left(1 - \frac{4y^2}{b^2}\right)$$

since

$$\nabla^2\Phi = -2G\theta = -\frac{bM_t}{ab^3}$$

and

$$\theta = \frac{3}{Gab^3}M_t \tag{9.31}$$

evidently the "torsional constant" (i.e., the shape contribution to the torsional stiffness GJ) is

$$J = \frac{ab^3}{3} \quad (\text{for } a \gg b) \tag{9.32}$$

which is nowhere near as much as the polar moment of inertia, $I_p = \frac{ab}{12}(b^2 + a^2)$. Also since

$$\tau_{zx} = \frac{\partial F}{\partial y} = -\frac{6}{ab^3}My, \quad \tau_{zy} = -\frac{\partial F}{\partial x} = 0 \quad (\textit{except near the ends}) \tag{9.33}$$

the max shear stress occurs at $y = \pm b/2$ where

$$\tau_{\max} = \frac{3M_t}{ab^2} = \frac{2c}{b/2} = \frac{M_t b}{J} \tag{9.34}$$

This result can be generalized to any thin-walled, open section such as shown in Figure 9.11. It is approximately correct to say that the inflated membrane shape is parabolic throughout except near the ends and at any corner. Since θ must be the same for each part then:

$$J = \frac{1}{3}\sum_{i=1}^{n} a_i t_i \quad \text{or} \quad J = \frac{1}{3}\int t^3 ds \tag{9.35}$$

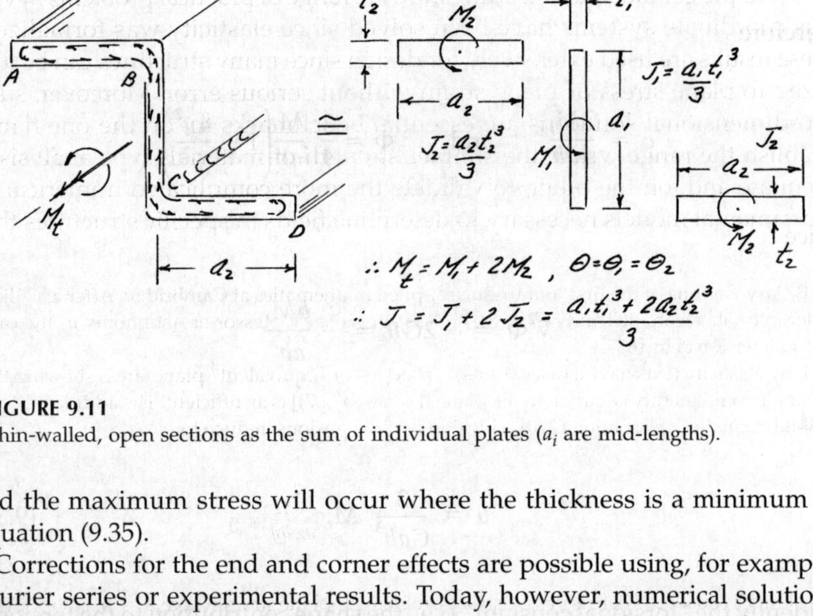

FIGURE 9.11
Thin-walled, open sections as the sum of individual plates (a_i are mid-lengths).

and the maximum stress will occur where the thickness is a minimum in Equation (9.35).

Corrections for the end and corner effects are possible using, for example, Fourier series or experimental results. Today, however, numerical solutions to Poisson's equation on the computer will generate results to any degree of precision. Such results are shown in Table 9.1. It is interesting that computing power and more exact solutions, if anything, only increase the necessity for the designer to understand the membrane analogy in order to visualize the results and then shape cross-sections for efficiency.

TABLE 9.1
Shear Stress and Unit Angle of Twist for Various Prismatic Sections

Cross-Section	τ_{max}	$\Theta = \frac{dw}{dz}$
Ellipse 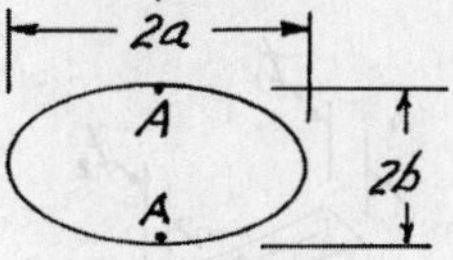 Area $= \pi ab$	$\tau_A = \frac{2M_t}{\pi a b^2}$ $a > b$	$\Theta = \frac{(a^2 + b^2)M_t}{\pi a^3 b^3 G}$
Equil. Triangle 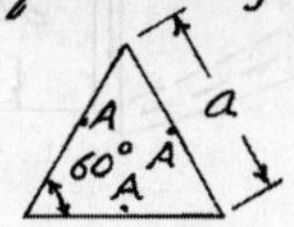 Area $= a^2\frac{\sqrt{3}}{4}$	$\tau_A = \frac{20M_t}{a^3}$	$\Theta = \frac{46\ M_t}{a^4 G}$
Reg. Hexagon 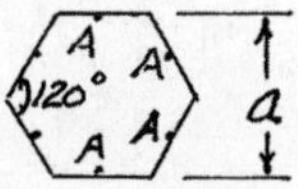 Area $= a^2\sqrt{3}/2$	$\tau_A = \frac{5.7M_t}{a^3}$	$\Theta = \frac{8.8\ M_t}{a^4 G}$
Rectangle 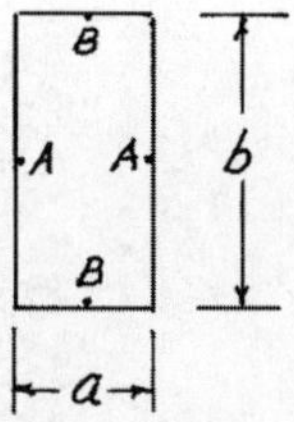 Area $= ab$	$\tau_A = \frac{M_t}{C_\alpha b a^2}$	$\Theta = \frac{M_t}{C_\beta b a^3 G}$

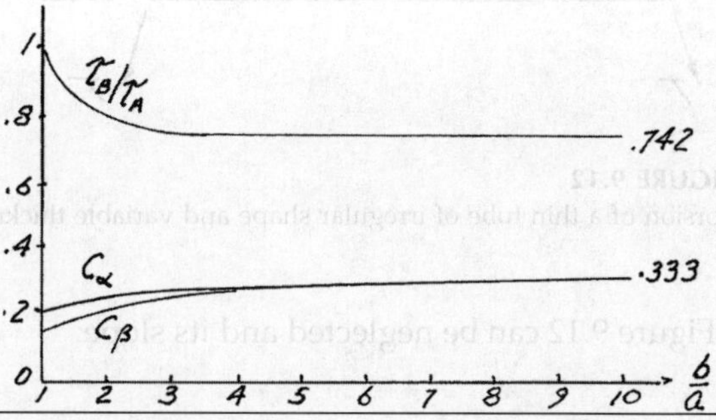

9.5 Thin-Walled Tubes of Arbitrary Shape

The membrane analogy is also very useful in analyzing torsion of thin tubes of arbitrary shape with variable thickness that is "small" in comparison to the overall dimensions. Because t is small, the curvature of the membrane

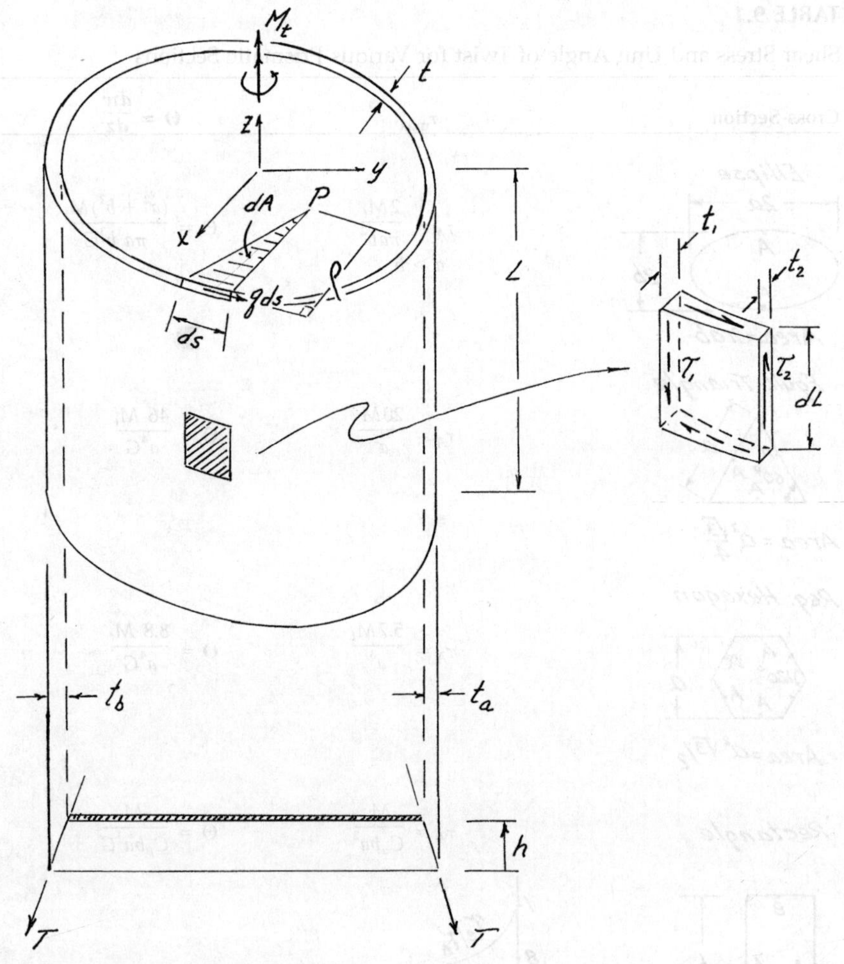

FIGURE 9.12
Torsion of a thin tube of irregular shape and variable thickness.

in Figure 9.12 can be neglected and its slope:

$$\tau = \frac{h}{t} = \text{constant} \qquad \text{or} \qquad \tau t = h = \text{const.} = q \tag{9.36}$$

where $q = h$ is often called the shear flow in units of force per unit circumferential length. Thus by the membrane analogy

$$M_t = 2 \text{ Volume} = 2Ah \qquad \text{or} \qquad q = \frac{M_t}{2A} \tag{9.37}$$

where A is the area enclosed by the average perimeter. Therefore, the shear stress

$$\tau = \frac{M_t}{2At} \tag{9.38}$$

will be a maximum at the minimum thickness. For the Tresca yield criterion for example ($\tau_{max} = Y/2$)

$$(M_t)_Y = AYt_{min} \tag{9.39}$$

and therefore, from a design standpoint, a circular tube of constant thickness is optimum since it has the maximum area for a given weight of material to sustain the constant shear stress.* To determine the angle of twist we can apply, Equation (9.29)

$$\int_s \tau\, ds = \frac{M_t}{2A}\int \frac{ds}{t} \tag{9.40}$$

and

$$\theta = \frac{M_t}{4A^2G}\int \frac{ds}{t} \tag{9.41}$$

For thin tubes of constant thickness, Equation (9.41) gives

$$\theta = \frac{M_t s}{4A^2Gt} \tag{9.42}$$

It is important to note that this analysis, in which we have used the membrane analogy can, just as well, be derived from statics alone. As shown in Figure 9.12, longitudinal equilibrium of a slice in the z direction requires that

$$L\tau_1 t_1 = L\tau_2 t_2 \quad \text{or} \quad \tau t = q = \text{constant}$$

Now taking moments about any arbitrary point P in the plane of the cross-section:

$$dM_t = q\,ds(\rho) = q(2dA)$$

* However, as long as the section is reasonably close to a circular shape, its area/volume ratio is relatively insensitive (P9.13) and a designer has great freedom. The designer must also be aware that torsional buckling often becomes critical for unstiffened thin tubes of any shape. This subject is covered in courses on structural instability.

and therefore

$$M_t = 2qA \quad \text{and} \quad \tau = \frac{M_t}{2At}$$

where A is the total area enclosed by the cross-section (as in the membrane analogy) and not the cross-sectional area of the wall of the tube.

Thus the special case of torsion of a thin tube is *statically determinate*. As we will see later, this allows us to use this solution as a building block to generate lower-bound estimates of the plastic behavior of all sorts of cross-sections in torsion.

Multicell thin tubes, which occur in box bridges, aircraft, cars, ships, and skyscrapers can be analyzed with a simple extension of the single-cell approach and the shear flow concept. An example with small wall thicknesses is shown in Figure 9.13. From the membrane analogy

$$\tau_1 = \frac{M_t h_1}{J t_1}, \quad \tau_2 = \frac{M_t h_2}{J t_2} \quad \text{and} \quad \tau_3 = \frac{M_t(h_1 - h_2)}{J} \frac{}{t_3} = \frac{\tau_1 t_1 - \tau_2 t_2}{t_3} \tag{9.43}$$

and the twisting moment equals twice the volume under the membrane or

$$M_t = 2V = 2[A_1 J h_1 + A_2 J h_2] = 2[A_1 t_1 \tau_1 + A_2 t_2 h_2] \tag{9.44}$$

From Equation (9.40) assuming t_1, t_2, and t_3 are constant

$$\tau_1 S_1 + \tau_3 S_3 = 2G\theta A_1, \quad \tau_2 S_2 - \tau_3 S_3 = 2G\theta A_2 \tag{9.45}$$

in which S_i are the length of the mid-section curves for each cell and we have assumed the direction of τ_3. Finally the shear stress in each wall τ_1, τ_2, and τ_3 and the rate of twist θ can be determined by solving Equations (9.44) and (9.45) simultaneously.

Again it should be noted that Equation (9.43) can, as shown in Figure 9.13, be thought of as the equilibrium condition at a junction in the longitudinal direction where:

$$\tau_1 t_1 L = \tau_z t_z L + \tau_3 t_3 L \quad \text{or} \quad q_1 = q_2 + q_3 \tag{9.46}$$

and $q = \tau t$ is a constant in any branch. This is also the continuity equation for fluid flow in a channel or pipe network stating that the amount of liquid coming into a junction must equal that flowing out. Hence the hydrodynamic analogy and the term "shear flow" for q.

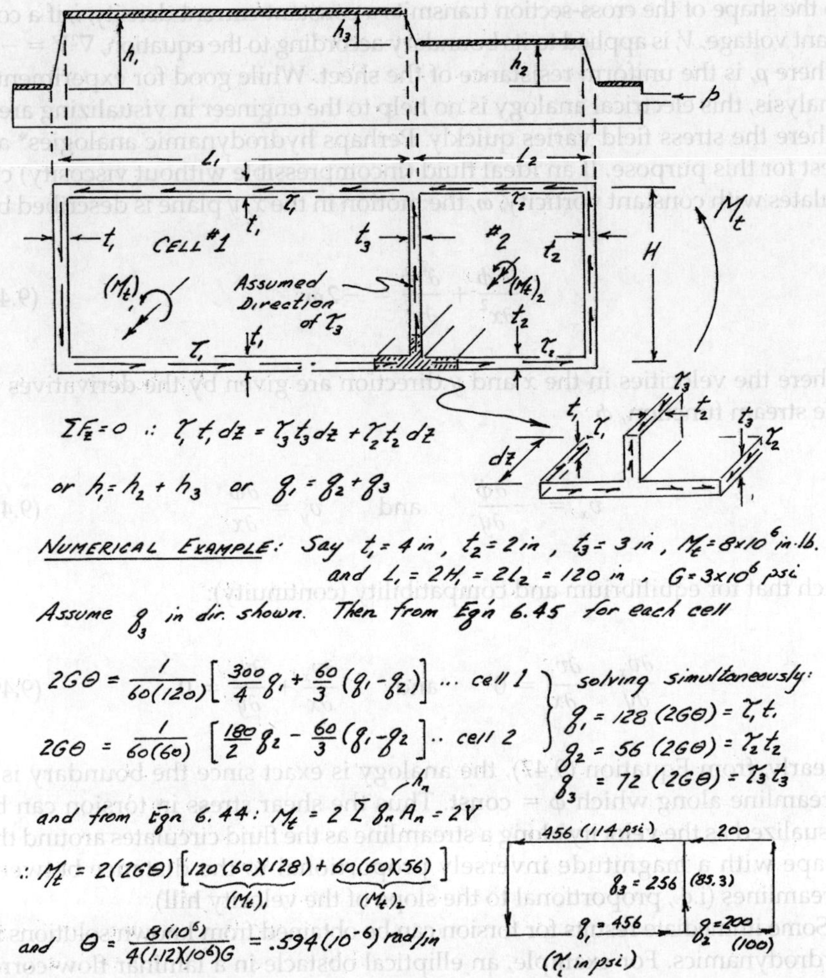

FIGURE 9.13
Torsion of multicell, thin-walled tubes.

9.6 Hydrodynamic Analogy and Stress Concentration

The membrane analogy is the most easily visualized and is, therefore, an excellent heuristic device for all cases. However, at corners or cutouts when there is stress concentration, the accuracy of any laboratory membrane model, such as a soap film experiment, is poor. The conducting sheet analogy mentioned in Chapter 4, used also to study ground water flow and temperature fields, is much better. An electrically conducting sheet (Telados Paper) cut to the shape of the cross-section transmits a constant current density, i, if a constant voltage, V, is applied to its boundary according to the equation, $\nabla^2 V = -\rho i$, where ρ, is the uniform resistance of the sheet. While good for experimental analysis, this electrical analogy is no help to the engineer in visualizing areas where the stress field varies quickly. Perhaps hydrodynamic analogies* are best for this purpose. If an ideal fluid (incompressible without viscosity) circulates with constant vorticity, ω, the motion in the x, y plane is described by:

$$\frac{\partial^2 \phi}{\partial x^2} + \frac{\partial^2 \phi}{\partial y^2} = -2\omega \tag{9.47}$$

where the velocities in the x and y direction are given by the derivatives of the stream function, ϕ

$$v_x = -\frac{\partial \phi}{\partial y} \quad \text{and} \quad v_y = \frac{\partial \phi}{\partial x} \tag{9.48}$$

such that for equilibrium and compatibility (continuity):

$$\frac{\partial v_x}{\partial y} - \frac{\partial v_y}{\partial x} = 0 \quad \text{and} \quad \frac{\partial v_x}{\partial x} + \frac{\partial v_y}{\partial y} = 0 \tag{9.49}$$

Clearly, from Equation (9.47), the analogy is exact since the boundary is a streamline along which ϕ = const. Thus the shear stress in torsion can be visualized as the velocity along a streamline as the fluid circulates around the shape with a magnitude inversely proportional to the distance between streamlines (i.e., proportional to the slope of the velocity hill).

Some immediate results for torsion can be obtained from known solutions in hydrodynamics. For example, an elliptical obstacle in a laminar flow corresponds to a hole or cutout in a cross-section resisting torque. Thus, in Figure 9.14a, the streamlines bunch up near the major diameter end points, A, and points B are stagnation points where $v = 0$ ($\therefore \tau = 0$). At points A where $x = \pm a$, the fluid velocity $v_y = v_o\,(1 + a/b)$. Thus for a small hole or cutout in a torsional stress field, as in Figure 9.14b, the shear stress at the root of a small elliptic cutout (point A) would be approximately

$$\tau = \tau_o\left(1 + \frac{a}{b}\right)$$

where τ_o is the stress at the point with no disturbance introduced. For a circular hole or notch such as point A or B, the stress concentration factor is 2. At

* Lord Kelvin, J. Boussinesq, and A.G. Greenhill all proposed versions. All involve laminar flow in a vessel of the same shape as the cross-section under torsion.

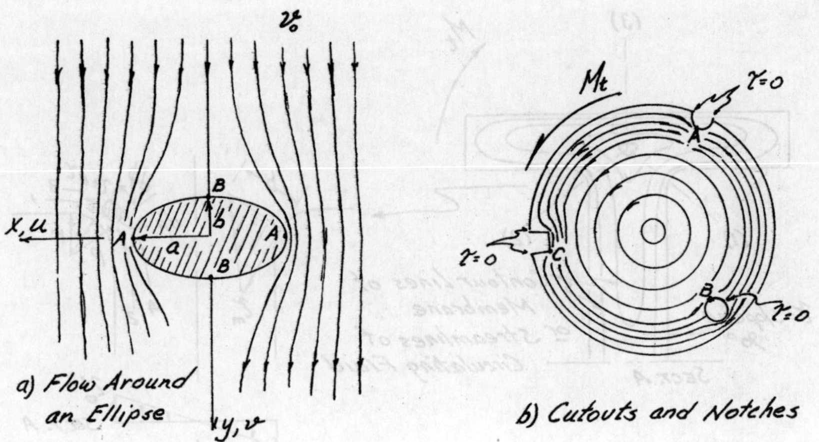

FIGURE 9.14
Inferences from the hydrodynamic analogy.

a very sharp corner such as C where a/b theoretically approaches infinity, the stress reaches yield quickly which explains the problems encountered at the corner of keyways in drive shafts and in plate girders constructed of flat plates.*

This question of determining stress concentration factors is particularly important for reentrant corners. Structural sections such as angles, wide flange beams, and channels (Figure 9.11), or rectangular tubes (Figure 9.13) are common in practice. If corners are perfectly sharp, the shear is theoretically infinite, so rounded corners or fillets are necessary so that the shear can flow smoothly around the corners and the streamlines pile up as little as possible. Take for example the wide flange shape shown in Figure 9.15. The maximum shears act at either sections (1), (2), or (3) and equal the slope, which is a maximum at the boundary. At (1) or (2) there is no shear along lines 1–3 or 2–4 perpendicular to the contour lines. From the membrane analogy

$$\int \tau ds = 2G\theta dA \quad \text{or} \quad \tau(\rho d\phi) + 0 - (\tau + d\tau)(\rho + d\rho)\, d\phi + 0 = 2G\theta\, r d\phi dr \tag{9.50}$$

and

$$\therefore \frac{d\tau}{dr} + \frac{\tau}{r} = -2G\theta$$

* Note that while small holes or notches may produce high local stress concentration they do not reduce the torsional stiffness significantly since the volume under the membrane is practically the same.

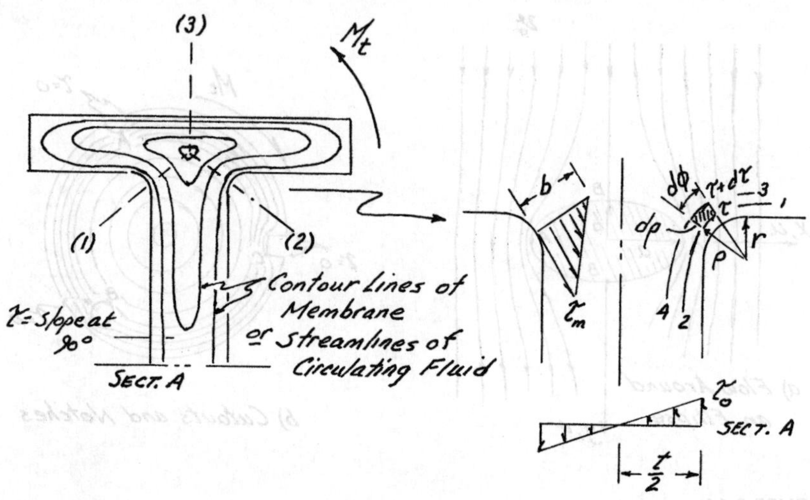

FIGURE 9.15
Stress concentration at fillet of radius r.

At Section A, $\frac{d\tau}{dr}$ = slope = $\frac{2\tau_0}{t}$ and = $\frac{\tau}{r}$ = 0 since $r = \infty$. Therefore from Equation (9.50):

$$\frac{-2\tau_0}{t} = -2G\theta$$

If we assume at Section (1) a uniform slope for the distribution of τ (a parabolic membrane shape), then from Equation (9.50):

$$\frac{-\tau_m}{b} + \frac{\tau_m}{r} = -2G\theta = -\frac{2\tau_o}{t}$$

Now from the hydrodynamic analogy, the quantity of flow through sections A and (1) must be equal so

$$b = \frac{\tau_o}{\tau_m}\frac{t}{2} \qquad \text{and} \qquad \left(\frac{\tau_m}{\tau_o}\right)^2 - \frac{t}{2r}\left(\frac{\tau_m}{\tau_o}\right) = 1$$

Thus, solving the quadratic

$$SCF = \frac{\tau_m}{\tau_o} = \frac{t}{4r}\left[1 + \sqrt{\left(\frac{4r}{t}\right)^2 + 1}\right] \tag{9.51}$$

This result is plotted as Figure 9.16 where Equation (9.51) is seen to be slightly conservative when compared to results from repeated finite difference solutions to Poisson's equation.

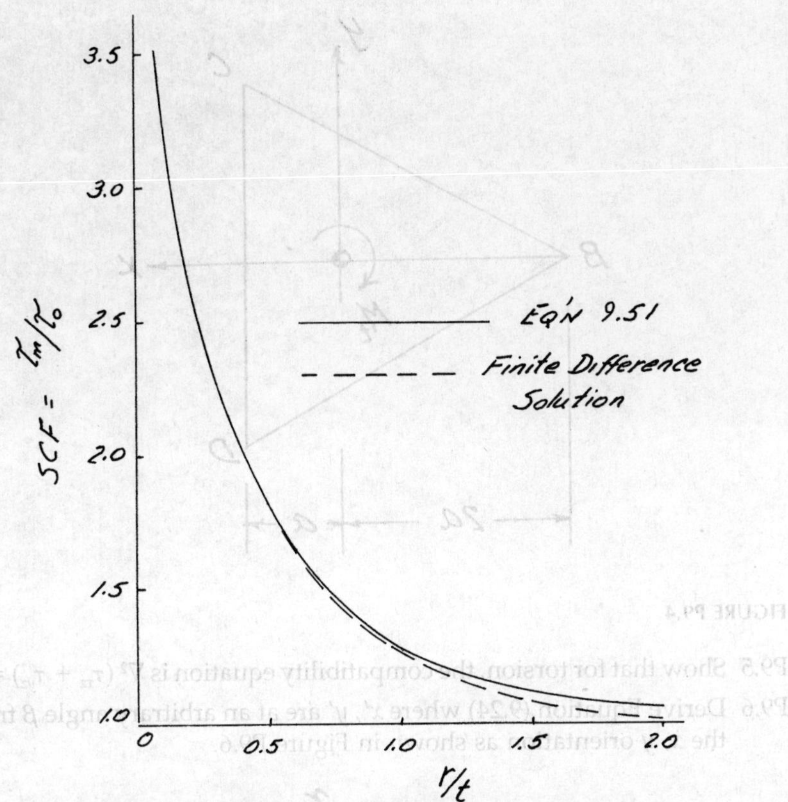

FIGURE 9.16
Stress concentration at reentrant corners of open sections.

9.7 Problems and Questions

P9.1 Derive Equation (9.11) from the Navier field Equation (5.2).

P9.2 What is the conjugate function to the harmonic warping function, w? Does it have physical meaning? What is the corresponding one-dimensional flow rule? Plot the flow net for an elliptic cross-section.

P9.3 Relate the harmonic warping function to the Prandtl stress function.

P9.4 Assume a warping function $w = A(y^3 - 3x^2y)$ is the solution for an equilateral triangle under torsion about its centroidal axis (Figure P9.4). Find the constant A, the torsional constant J, and the maximum shearing stress. Draw diagrams such as in Figures 9.4b and c for the warping and the stress distribution. Note: $x - a = 0$ on CD, $x + 2a - \sqrt{3}y = 0$ on BD, $x + 2a + \sqrt{3}y = 0$ on BD, and the value of A is most easily determined from the boundary condition on CD.

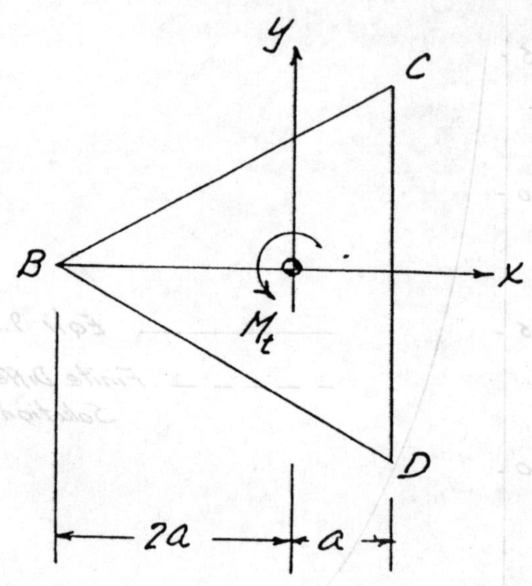

FIGURE P9.4

P9.5 Show that for torsion, the compatibility equation is $\nabla^2 (\tau_{xz} + \tau_{yz}) = 0$.

P9.6 Derive Equation (9.24) where x', y' are at an arbitrary angle β from the x, y orientation as shown in Figure P9.6.

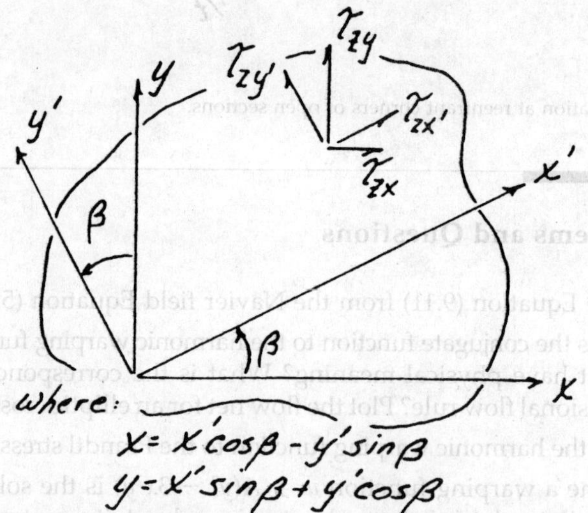

FIGURE P9.6

P9.7 Compare exact and approximate solutions for stiffness and τ_{max} for a hollow cylinder for $t = r/5, r/10, r/20$.

P9.8 Compare the results for the hollow cylinder in P9.7 to the comparable split cylinder using the membrane analogy neglecting corner effects.

P9.9 Revise the solution by membrane analogy for a long thin plate, Figure 9.10, in light of the footnote concerning a circle being the correct membrane shape for uniform pressure. What happens to the Poisson equation $\nabla^2 z = -p/T$? Compare results to the parabolic shape assumed.

P9.10 Compare the torsional stiffness, M_t/θ, for a split ring to a complete ring as shown in Figure 9.9 using a parabolic membrane shape and neglecting end effects at the cut. Plot for various r/t ratios. Show the solid cross-section on your plot.

P9.11 Do P9.7 for an elliptic ring with a 2:1 axes ratio (and some "rational" definition of thickness).

P9.12 Calculate the torque due to the linear distribution of τ_{zx} from Equation (9.33). Why is it only half that applied? Why do you think that τ_{max} is twice that of a circular bar of diameter, b, subjected to the same twist?

P9.13 Compare the yield torque, $(M_t)_Y$, for several thin-walled tubes all having the same thickness (constant) and perimeter, but different shapes. (For example: square, rectangle, triangle.) Nondimensionalize by $(M_t)_Y$ for a circle.

P9.14 Use the thin tube analysis Equations (9.36) through (9.41), to derive a torsion solution for: (a) an elliptic cross-section ($a/b = 2/1$); and (b) the equilateral triangle of P9.4, by assuming nested tubes touching, but not interacting with each other. Compare your results to the exact solution, Table 9.1, and comment on the reasons for any discrepancy.

P9.15 In the two-celled tube of Figure 9.13, what would be the increase in the largest shear stress and the unit angle of twist θ if: (a) the left vertical web (t = 4") was slit longitudinally; or (b) the central web (t = 3") was slit longitudinally; or (c) the right web (t = 2") was slit longitudinally.

P9.16 A section of a tall concrete building ($G = 1 \times 10^6$ psi) can be idealized as a three-cell tube as shown in Figure P9.16 (units in ft). Due to wind, a maximum torque of 2×10^6 ft lbs is applied. Find the rate of twist and average shear stresses. If the radius of fillets is 1 ft, estimate the maximum shear stress.

P9.17 A "bulbous" 3-cell airfoil of steel ($G = 12 \times 10^6$ psi) resists a torque of 100,000 in. lbs. The wall thicknesses are $t_1 = t_2 = t_3 = .250$ in. and $t_4 = t_5 = 0.125$ in. Determine the shearing stresses and the unit angle of twist. Refer to Figure P9.17.

P9.18 Calculate the stress concentration factor for a 8 × 8 × 1 inch angle, a 36 *WF* 300 and a 20 × 7 standard beam weighing 95 lb/ft using the dimensions in the AISC Manual.

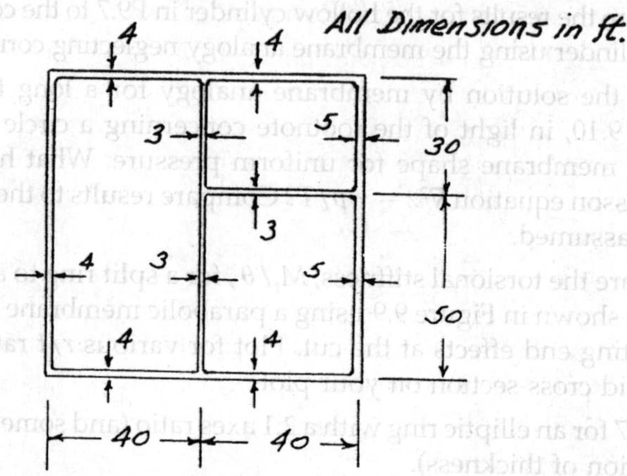

FIGURE P9.16
All dimensions in feet.

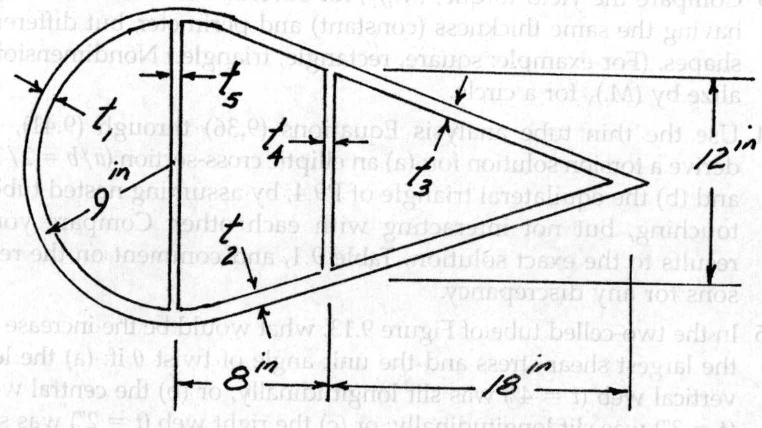

FIGURE P9.17

P9.19 Solve the torsion problem for a hollow elliptic cross-section where the elliptic hole is given by:

$$\frac{y^2}{(ak)^2} = \frac{x^2}{(bk)^2} = 1$$

Plot the results $1 < k < 0$ for θ and τ_{max} if $a = 2b$ and compare to the extreme cases of a solid and thin-walled tube.

P9.20 A thin isosceles triangle $a \gg b$ carries a torque M_t. Develop expressions for θ and τ_{max}. Refer to Figure P9.20.

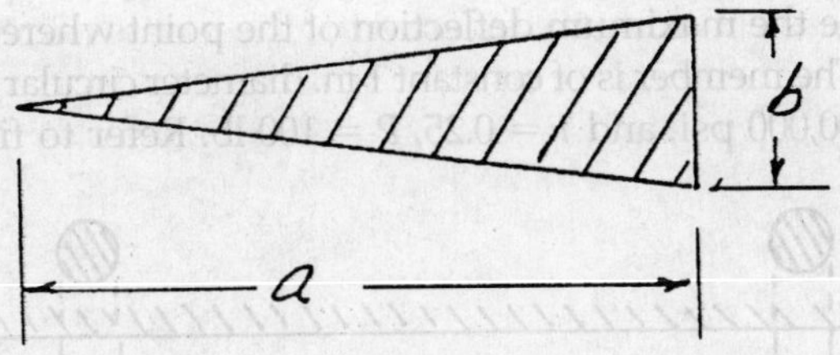

FIGURE P9.20

P9.21 For a thin-walled rectangular tube of constant thickness, t, plot r and θ as a function of the aspect ratio (length/width) for a given perimeter, M_t, t, and G. What shape is most efficient?

P9.22 A solid bar 300 in. long with a rectangular cross-section 10 × 10 in. is twisted until the ends have rotated 1.4° relative to each other. If $G = 10 \times 10^6$ psi what is τ_{max}?

P9.23 A thin circular ring is subjected to a distributed twisting moment M_θ per unit length where:

$$M_\theta = M\cos^2\theta$$

Find the angle of twist at points A and B. Also find the deformation pattern of the ring, i.e., its displacement out of its original plane. It may be assumed that the radius R is sufficiently large that the effect of curvature on the torsion formulas is negligible. Refer to Figure P9.23.

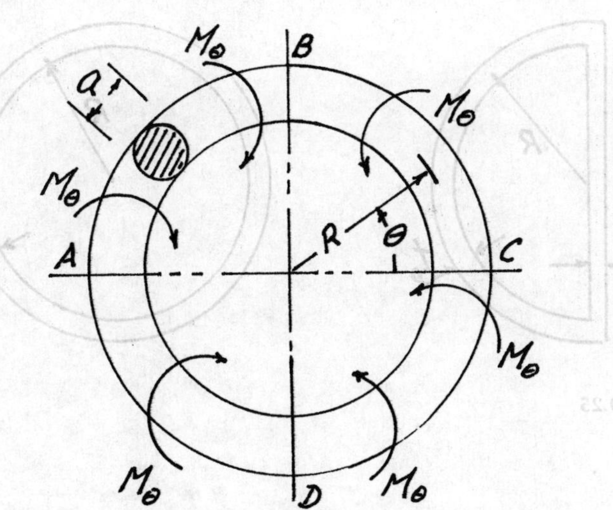

FIGURE P9.23

P9.24 Determine the maximum deflection of the point where the load P is applied. The member is of constant 1 in. diameter circular cross-section, $E = 30{,}000{,}000$ psi, and $\nu = 0.25$, $P = 100$ lb. Refer to figure P9.24.

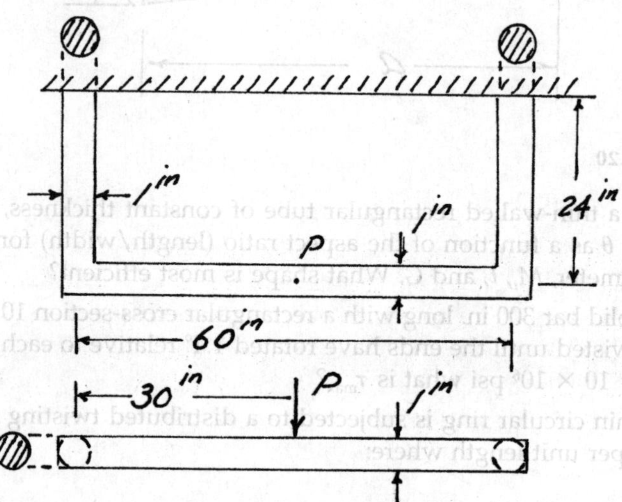

FIGURE P9.24

P9.25 Compare the stiffness of a thin, closed, semicircular (D-shaped) tube with that of a circular tube of the same radius. For equal stiffness determine the ratio of their material thicknesses. Refer to Figure P9.25.

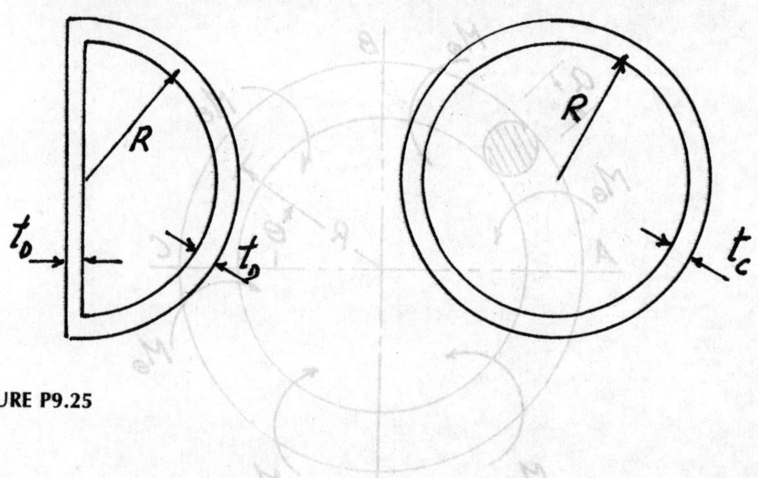

FIGURE P9.25

10

Concepts of Plasticity

10.1 Plastic Material Behavior

Behavior of a material beyond yield is called its plastic behavior. For this chapter, we will use the simplest idealization of a perfectly elastic-plastic solid (EPS) with the same yield stress (Y) in tension or compression able to deform at constant Y without strain hardening or strain softening. As can be seen from Figure 10.1, this assumption is not bad for typical metals strained $< 1\%$, but at greater strains most metals exhibit strain hardening. Mild steel is nearly a perfect EPS below 2% strain.

Since after yield our idealized material offers no further resistance to deviatoric stress, it essentially behaves as a viscous fluid and, without vorticity, must

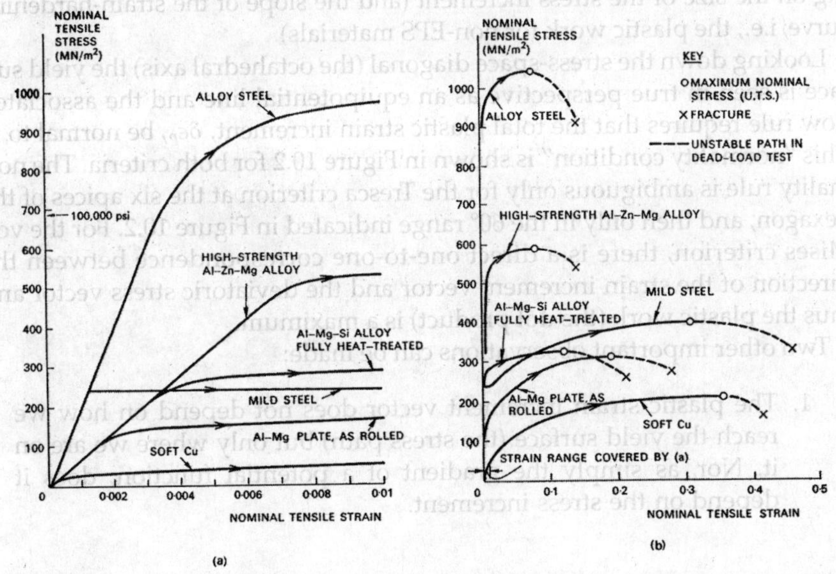

FIGURE 10.1
Tensile tests on metals (Calladine).

flow normal to lines of equipotential. For the classical Theory of Plasticity* with an EPS material, the yield function (either Tresca or von Mises) serves as the "plastic potential" hill, F, driving the flow in stress space down its steepest slope, which is the gradient of F. Thus the increments of the total plastic strain or deformation vector, $\delta\epsilon_i^p$, are found from the partial derivatives.

For the von Mises criterion, for example, where

$$F = (\sigma_2 - \sigma_3)^2 + (\sigma_3 - \sigma_1)^2 + (\sigma_1 - \sigma_2)^2 - 2Y^2 = 0 \tag{10.1}$$

then:

$$\begin{aligned}
\mu(\delta\epsilon_1^p) &= \frac{\partial F}{\partial \sigma_1} = 6\left(\sigma_1 - \frac{\sigma_1 + \sigma_2 + \sigma_3}{3}\right) = 6S_1 \\
\mu(\delta\epsilon_2^p) &= \frac{\partial F}{\partial \sigma_2} = 6\left(\sigma_2 - \frac{\sigma_1 + \sigma_2 + \sigma_3}{3}\right) = 6S_2 \\
\mu(\delta\epsilon_3^p) &= \frac{\partial F}{\partial \sigma_3} = 6\left(\sigma_3 - \frac{\sigma_1 + \sigma_2 + \sigma_3}{3}\right) = 6S_3
\end{aligned} \tag{10.2}$$

For given values of σ_1, σ_2, σ_3 at yield, these equations fix the *relative* magnitudes of the plastic strain increments but the scalar, μ, scales their actual values depending on the size of the stress increment (and the slope of the strain-hardening curve; i.e., the plastic work for non-EPS materials).

Looking down the stress-space diagonal (the octahedral axis) the yield surface is seen in true perspective as an equipotential line and the associated flow rule requires that the total plastic strain increment, $\delta\epsilon_P$, be normal to it. This "normality condition" is shown in Figure 10.2 for both criteria. The normality rule is ambiguous only for the Tresca criterion at the six apices of the hexagon, and then only in the 60° range indicated in Figure 10.2. For the von Mises criterion, there is a direct one-to-one correspondence between the direction of the strain increment vector and the deviatoric stress vector and thus the plastic work (the dot product) is a maximum.

Two other important observations can be made:

1. The plastic strain increment vector does not depend on how we reach the yield surface (the stress path) but only where we are on it. Nor, as simply the gradient of a potential function, does it depend on the stress increment.

---

* An outstanding text on classical plasticity for EPS materials is *Plasticity for Engineers* by C.R. Calladine, Ellis Horwood Ltd., West Sussex, England, 1985. All engineers should enjoy *The New Science of Strong Materials* by J.E. Gordon, 2nd Ed., Pitman Pub., London, 1976. For that matter Gordon's book *Structures or Why Things Don't Fall Down*, Plenum Press, New York, 1978 is somewhat less basic, but even more fun and equally illuminating.

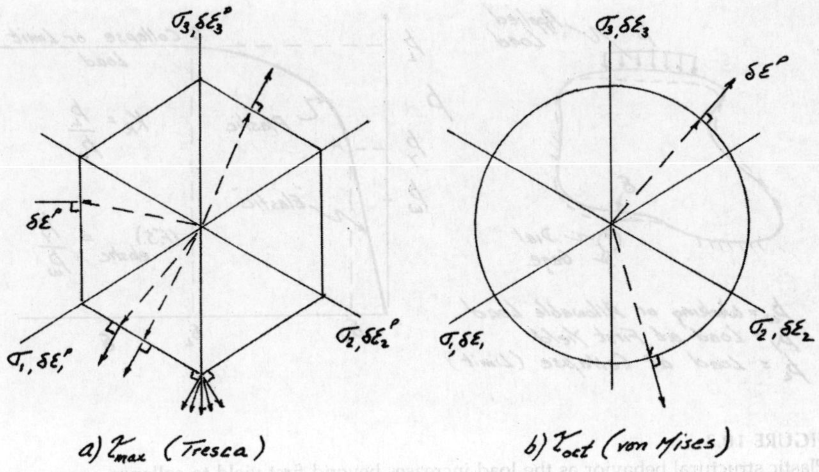

FIGURE 10.2
Yield surface in the octahedral plane and associated normality of the strain-increment vector.

2. Shear flow of an EPS material takes place at constant volume since, from Equation (10.1)

$$\delta\epsilon_1^P + \delta\epsilon_2^P + \delta\epsilon_3^P = 0 \tag{10.3}$$

Thus, beyond yield, $\nu^P = 1/2$ and the material becomes incompressible. Finally it should be noted that this "normality rule" is a more general concept in that it governs not only the plastic flow of EPS materials in stress space, but also applies, as we shall see, to structures made of EPS material when viewed in the appropriate "load space."

10.2 Plastic Structural Behavior

Yield and collapse will occur simultaneously only for structural shapes that carry constant stress fields (uniform distribution of tension, compression, or shear). In a general structure such as our standard blob (Figure 10.3), only a certain point or points reach yield and the load can be increased beyond that threshold. As the load increases beyond "first yield," the load-deformation response of the structure becomes nonlinear as plastic regions grow until a full "collapse mechanism" is formed. At this point, if there is no strain hardening or "geometric strengthening," the structure or parts of it become dynamic either suddenly (and violently, sometimes killing people) or gradually so there is time to get out of the way.

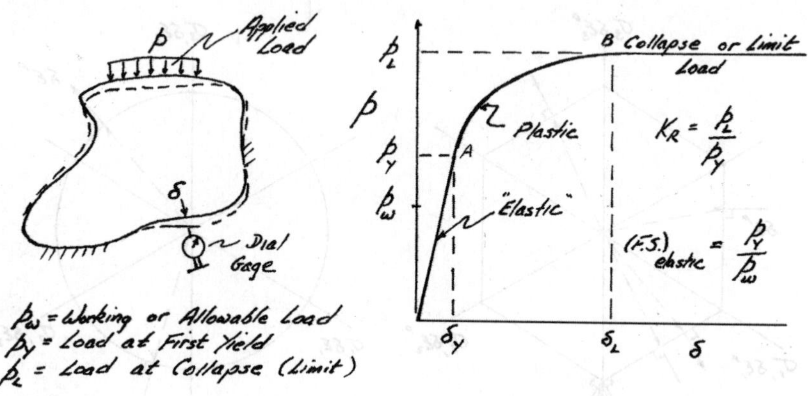

FIGURE 10.3
Plastic structural behavior as the load increases beyond first yield to collapse.

The plastic behavior of a structure must be clearly differentiated conceptually from the plastic behavior of the material from which it is made. For our EPS material, the stress–strain curve is made up of two straight lines and is, in that sense, piecewise linear. The load deformation structural response beyond first yield is generally not linear and is therefore much more difficult to determine than its behavior in the elastic range.

In fact only a handful of complete, formal, closed-form, plasticity solutions have ever been found. These solutions are all based on a strategy where the deformations are calculated from elastic strains in those portions of the structure that remain linear while the plastic zones spread with increasing load to generate the limit condition at collapse. Moreover, most of these are limited to two-dimensional problems with $\sigma_z = \sigma_2$ intermediate between σ_1 and σ_3 in the x, y, or r, θ plane.*

10.3 Plasticity Field Equations

Equilibrium must apply whether the material is behaving elastically or plastically or whatever. Moreover, in the plastic zone and normal and tangential to the elastic-plastic interface, the shear stress must equal the shear strength.** In the surrounding elastic regions (Figure 10.4) where the stress–strain relationship is still linear, the geometric compatibility requirement in terms of stresses still applies. Thus the field equations in terms of stress

* The same assumption will be made in this text. It must be checked, but it is a common case in engineering applications generally true for plane strain.

** Examples will assume the Tresca criterion. Corresponding results for the von Mises criterion will be treated in the chapter problems.

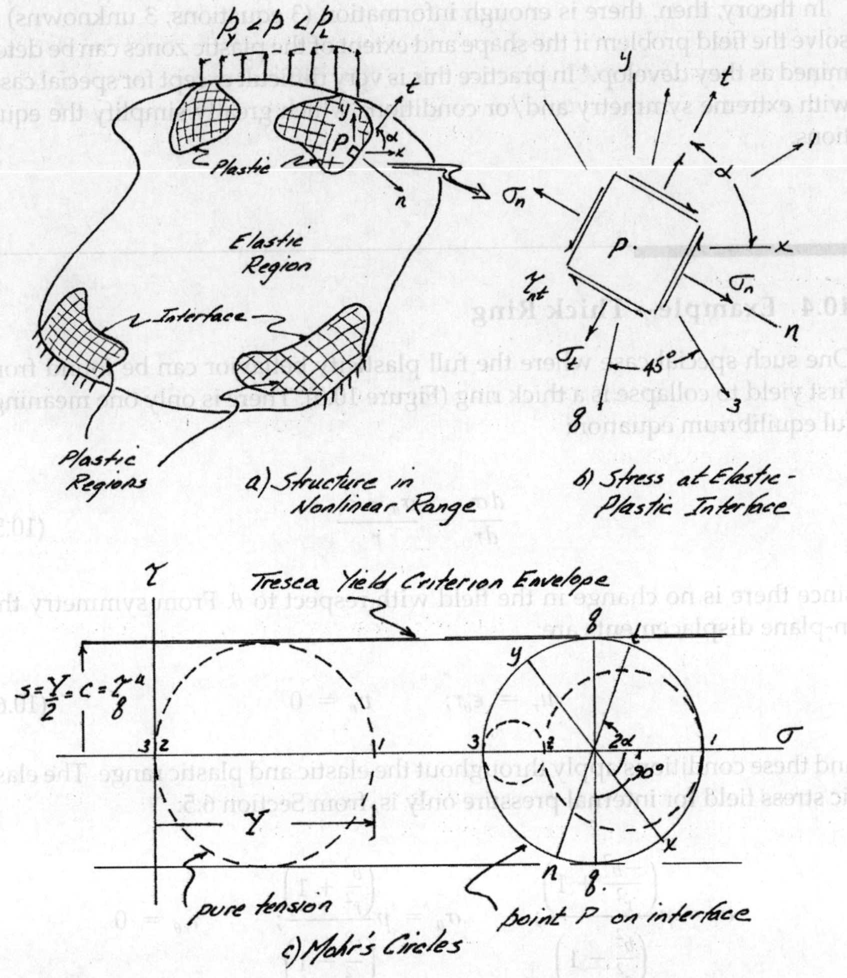

FIGURE 10.4
Relationships for plastic analysis (α = ?).

become:

Elastic Regions	Plastic Regions
$\dfrac{\partial\sigma_x}{\partial x} + \dfrac{\partial\tau_{xy}}{\partial y} + B_x = 0$	$\dfrac{\partial\sigma_x}{\partial x} + \dfrac{\partial\tau_{xy}}{\partial y} + B_x = 0$
$\dfrac{\partial\tau_{xy}}{\partial x} + \dfrac{\partial\sigma_y}{\partial y} + B_y = 0$	$\dfrac{\partial\tau_{xy}}{\partial x} + \dfrac{\partial\sigma_y}{\partial y} + B_y = 0$
$\nabla^2(\sigma_x + \sigma_y) = 0$	$\tau_{13} = s$

(10.4)

where s is the shear strength at yield. For the Tresca criterion for an EPS material, $s = \frac{Y}{2} = c$, which is constant and, as shown in Figure 10.4, is independent of the mean stress.

In theory, then, there is enough information (3 equations, 3 unknowns) to solve the field problem if the shape and extent of the plastic zones can be determined as they develop.* In practice this is very difficult except for special cases with extreme symmetry and/or conditions which greatly simplify the equations.

10.4 Example—Thick Ring

One such special case where the full plasticity behavior can be found from first yield to collapse is a thick ring (Figure 10.5). There is only one meaningful equilibrium equation:

$$\frac{d\sigma_r}{dr} = \frac{\sigma_\theta - \sigma_r}{r} \tag{10.5}$$

since there is no change in the field with respect to θ. From symmetry the in-plane displacements are:

$$u_r = \epsilon_\theta r; \qquad v_\theta = 0 \tag{10.6}$$

and these conditions apply throughout the elastic and plastic range. The elastic stress field for internal pressure only is, from Section 6.5:

$$\sigma_r = p\frac{\left(\frac{-b^2}{r^2} + 1\right)}{\left(\frac{b^2}{a^2} - 1\right)}; \qquad \sigma_\theta = p\frac{\left(\frac{b^2}{r^2} + 1\right)}{\left(\frac{b^2}{a^2} - 1\right)}; \qquad \tau_{r\theta} = 0$$

and, for plane stress: $\sigma_z = 0$

Thus the in-plane mean stress is:

$$\sigma_{\text{quad}} = \frac{\sigma_r + \sigma_\theta}{2} = p\frac{a^2}{b^2 - a^2} = \text{const!}$$

and the state of stress at any point is simply a state of pure shear superimposed on "hydrostatic tension." Since for plane stress, $\sigma_\theta < \sigma_z < \sigma_r$ the yield

* Although the plastic stress–strain relationship is nonlinear, plastic strains are continuous and displacements smooth *until* the limit load when slip lines and a failure mechanism finally develop. Thus the normality condition can, in theory, be used to solve for the plastic-elastic interfaces as they grow and migrate. This amounts to developing a relationship for α in Figure 10.4 as a function of load and position x, y.

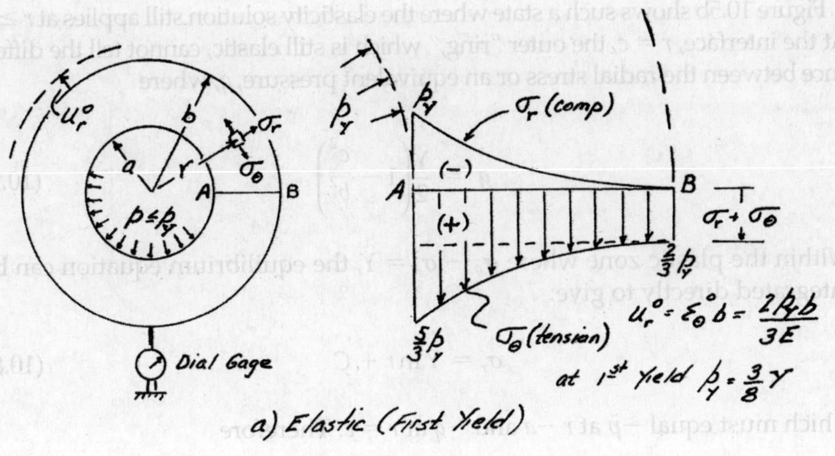

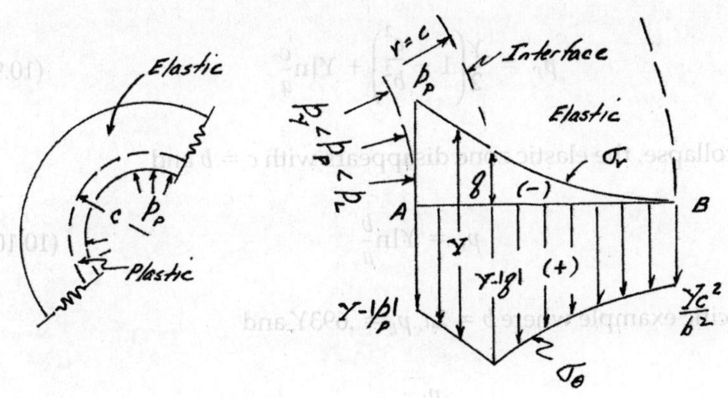

FIGURE 10.5
Plastification of a thick ring ($b = 2a$).

condition $\sigma_\theta - \sigma_r = Y$ is reached first on the inner fibers at $r = a$ at a value:

$$p_Y = \frac{Y}{2}\left(1 - \frac{a^2}{b^2}\right)$$

For the specific case $b = 2a$, $p_Y = \frac{3}{8}\,Y = .375Y$.

Although the ring yields at this pressure, it will not fail. As the pressure increases, the yield interface moves outward radially and, as the plastic zone enlarges, the stresses must adjust to provide equilibrium while satisfying the boundary conditions that at $r = a$, $\sigma_r = -p$, and at $r = b$, $\sigma_r = 0$.

Figure 10.5b shows such a state where the elasticity solution still applies at $r \geq c$. At the interface, $r = c$, the outer "ring," which is still elastic, cannot tell the difference between the radial stress or an equivalent pressure, q, where

$$q = \frac{Y}{2}\left(1 - \frac{c^2}{b^2}\right) \tag{10.7}$$

Within the plastic zone where $\sigma_\theta - \sigma_r = Y$, the equilibrium equation can be integrated directly to give:

$$\sigma_r = Y \ln r + C \tag{10.8}$$

which must equal $-p$ at $r -a$ and $-q$ at $r = c$. Therefore

$$p_P = \frac{Y}{2}\left(1 - \frac{c^2}{b^2}\right) + Y \ln\frac{c}{a} \tag{10.9}$$

Finally, at collapse, the elastic zone disappears with $c = b$ and

$$p_L = Y \ln\frac{b}{a} \tag{10.10}$$

For the specific example where $b = 2a$, $p_L = .693Y$ and

$$K_R = \frac{p_L}{p_Y} = 1.85$$

Thus, if a factor of safety of 1.5 was specified to determine the "allowable" pressure against yield (i.e., the "working stress")

$$F.S. = \frac{p_Y}{p_w} = 1.5$$

there would actually be a total safety margin* of

$$S.M. = (F.S.)(K_R) = \frac{p_L}{p_w} = 2.77$$

* Modern codes based on load factors (of ~1.8) and limit analysis to check collapse may be deficient for cyclic loadings. If yield occurs, *even once*, fatigue life can become a problem even if stress levels are maintained below yield thereafter. Older structures designed under codes based on working stress or working load concepts and elastic analysis usually have higher safety margins.

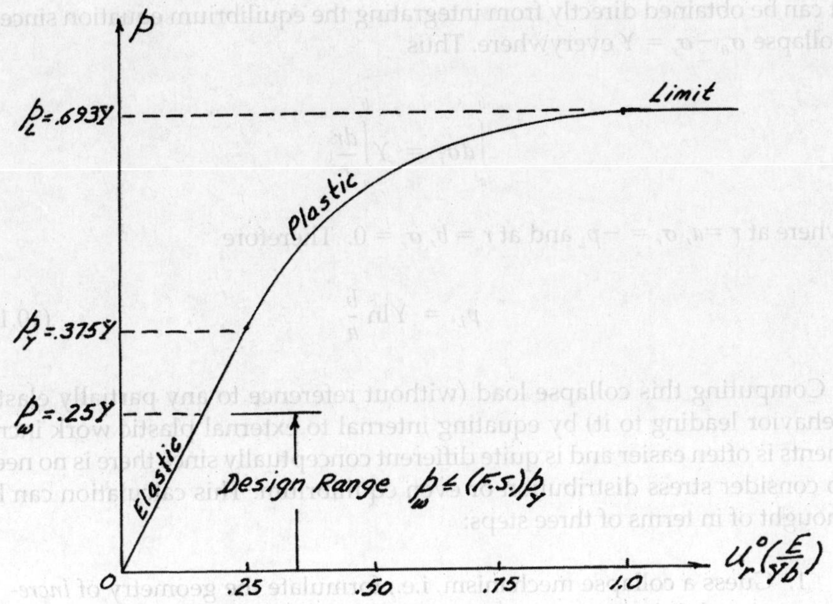

FIGURE 10.6
Load-deflection behavior, thick ring ($b = 2a$).

The elastic-plastic radial deformation of the outer circumference can be found directly since it remains elastic until the limit state is reached. At $r - b$ with c substituted for a and q for p in Equation (10.4)

$$\sigma_\theta = Y\frac{c^2}{b^2}, \qquad \sigma_r = \tau_{r\theta} = 0$$

and therefore

$$u_r^o = \epsilon_o b = \frac{Y}{E}\frac{c^2}{b^2}b$$

which can be evaluated for $p_y < p_p < p_L$ from Equation (10.8). At the limit load $c = b$, $p_L = Y \ln\frac{b}{a}$ and $u_L = \frac{Yb}{E}$. The load-deformation curve shown in Figure 10.6 illustrates the full-range, elastic-plastic behavior of a ring with $b = 2a$ made of EPS material.

10.5 Limit Load by a "Work" Calculation

The plastic collapse or limit load can be calculated from equilibrium or by a work calculation based on an assumed mechanism of collapse without first evaluating the elastic and plastic behavior of the structure. For the thick ring it can be obtained directly from integrating the equilibrium equation since at collapse $\sigma_\theta - \sigma_r = Y$ everywhere. Thus

$$\int_a^b d\sigma_r = Y\int_a^b \frac{dr}{r}$$

where at $r = a$, $\sigma_r = -p_L$ and at $r = b$, $\sigma_r = 0$. Therefore

$$p_L = Y\ln\frac{b}{a} \tag{10.10}$$

Computing this collapse load (without reference to any partially elastic behavior leading to it) by equating internal to external plastic work increments is often easier and is quite different conceptually since there is no need to consider stress distribution or even equilibrium. This calculation can be thought of in terms of three steps:

1. Guess a collapse mechanism, i.e., formulate the geometry of *incremental* deformation of a plausible failure mechanism.
2. For this incremental deformation, integrate the "work consumed" (i.e., energy) in plastic deformation over the body.
3. Equate this "internal plastic work" to the external work done by the limit load.

The first step is to describe the geometry of *incremental* deformation so we can use the flow rule for plastic deformation increments discussed in Section 10.1. For the thick ring:

$$\delta\epsilon_r = \dot{\epsilon}_r = \frac{d\dot{u}}{dr}; \qquad \delta\epsilon_\theta = \dot{\epsilon}_\theta = \frac{\dot{u}}{r}; \qquad \dot{\epsilon}_Z = \text{const} \tag{10.11}$$

where $\dot{\epsilon}$ rather than $\delta\epsilon$ is used to imply a derivative with respect to time and to avoid confusion.

Since plastic deformation occurs at constant volume $\nu = 1/2$

$$\dot{\epsilon}_r + \dot{\epsilon}_\theta + \dot{\epsilon}_z = 0$$

and by substitution

$$\frac{d\dot{u}}{dr} + \frac{\dot{u}}{r} = \text{constant} \tag{10.12}$$

If we assume no axial deformation ($\dot{\epsilon} = 0$), integration of Equation (10.12) (with the integration constant zero) gives $\dot{u} = c/r$. Since our external work is expressed in terms of the radial expansion of the inner circumference, let $c = a\dot{u}_a$ and therefore

$$\dot{u} = \frac{a\dot{u}_a}{r} \tag{10.13}$$

and

$$\dot{\epsilon}_r = -\frac{a\dot{u}_a}{r}; \quad \dot{\epsilon}_\theta = \frac{a\dot{u}_a}{r^2} \tag{10.14}$$

This is a state of pure shear deformation on planes at 45° to the radius.

For the Tresca yield condition, $\sigma_\theta - \sigma_r = Y$ with $\epsilon_r = -\epsilon_\theta$, the internal work dissipated

$$\dot{D} = \sigma_\theta\dot{\epsilon}_\theta + \sigma_r\dot{\epsilon}_r + \sigma_z\dot{\epsilon}_z \tag{10.15}$$

can be calculated as:

$$\dot{D} = (\sigma_\theta - \sigma_r)\dot{\epsilon}_\theta = Y\dot{\epsilon}_\theta \tag{10.16}$$

Thus the total dissipation of work per unit length will be:

$$\int_a^b 2\pi r\dot{D}dr = 2\pi Y\int_a^b r\dot{\epsilon}_\theta dr = 2\pi Ya\dot{u}_a\int_a^b \frac{dr}{r} = 2\pi Ya\dot{u}_a \ln\frac{b}{a} \tag{10.17}$$

Finally, equating this to the incremental external work done by p_L as the ring expands by $\dot{u}_a$

$$2\pi a\dot{u}_a p_L = 2\pi Ya\dot{u}_a \ln\frac{b}{a} \tag{10.18}$$

gives

$$p_L = Y\ln\frac{b}{a} \tag{10.10}$$

This is exactly the same result, but it is obtained by considering only the geometry of the deformation of the collapse mechanism, the normality rule, and the yield condition. In fact, no consideration of equilibrium was necessary or, for that matter, was the distribution of stress in the ring required.

10.6 Theorems of Limit Analysis

In Section 10.4, it was relatively easy to do a complete study of the plastic behavior of a thick ring because symmetry reduces the equilibrium requirement to one separable, first-order differential equation. However, most structures do not have symmetry, simple loadings, or ideal supports. In these situations even with modern computational power, a full range elastic-plastic analysis can be nearly impossible or of dubious value.

Fortunately, there is an alternative approach where we forget about obtaining information between first yield and collapse and concentrate on finding the collapse state directly. This strategy, called limit analysis, is reasonable if, as is usually the case, we are primarily interested in the strength of a structure. We may lose all information about deformations or even the stress distribution in the plastic range but, after all, most structures are, or should be, designed to operate under normal working loads in the elastic range well below yield. All we really need is a good estimate of the "collapse mechanism" and the load at which it occurs to allow design for safety under extreme loading (earthquake, hurricane, collision, etc). For these rare events, we usually want to also control the failure mode to prevent sudden collapse and to localize residual plastic deformations, whatever they may be, so that the structure can be repaired.

Application of the so-called "bound theorems" and in particular the "upper-bound theorem" of limit analysis provide a way to do this with a straightforward strategy. Limit analysis from either a lower-bound or upper-bound perspective is an intuitive and interactive process. We shall find that in most cases we are free to specify certain variables arbitrarily, make educated guesses as to failure modes or stress distribution, and use trial-and-error methods to converge on good answers. It is this freedom and the relatively simple calculations involved that makes limit analysis so valuable. In fact, graphical analysis is very often the best method giving the simplest approach and greatest insight to the interplay of variables. In an important sense, limit analysis is more of an art than elastic analysis. Limit analysis requires, and in turn develops, a physical feel for structural behavior that may be even more important for a creative engineer than the simplicity of the method itself.

10.7 The Lower-Bound Theorem

> The load corresponding to any stress distribution throughout the structure, which satisfies equilibrium without violating the yield condition, is a *safe* estimate of the limit load.

Another way of stating the theorem is:

> Of all internal stress distributions that satisfy equilibrium and do not violate the yield condition, the collapse (limit) load will be the *largest*.

When stated the first way, this theorem is sometimes called "the safe theorem." However, the limiting process as we approach the correct solution from below, emphasized in the second version, is the most important concept and therefore the lower-bound name.

The power of this theorem for applications is that we do not need to consider the geometry of deformations but use only stress conditions. There is a straightforward energy proof of the theorem, but the theorem is intuitively satisfying since the plasticity equations in Section 10.3 also involve only stress and essentially say the same thing. One important concept, however, which is used in the classic proof for physical understanding is that, in elastic- plastic theory *at the collapse load*, the stresses throughout do not change as the structure deforms. Thus there are no changes in strain in the "elastic" (or not fully plastic) zones during collapse and these regions act essentially as if they were rigid. Basically, the elastic energy is considered insignificant compared to the plastic energy as the individual elements move along the horizontal yield plateau. Similarly for the structure as a whole, the elastic-plastic energy used getting to the limit state can be disregarded in comparison to the fully plastic work as collapse progresses. Thus, localized areas of intense shear deformation, called "slip lines," develop during collapse with the surrounding materials, in comparison, acting as rigid blocks sliding relative to each other. Since the lower-bound theorem deals with stresses (or forces) and not with deformations (or the failure mechanism), this important concept, although used in the proof, is not needed in the application of the lower-bound approach. However, it becomes the crucial consideration in upper-bound analysis treated in the next section, which in turn, becomes the main focus of succeeding chapters.

To return to our example of a thick ring in plane stress under internal pressure, let us pretend we do not know the answer and approach the limit state by applying the lower-bound theorem. We, therefore, seek a stress distribution satisfying the equilibrium equation (on a differential scale):

$$\frac{d\sigma_r}{dr} = \frac{\sigma_\theta - \sigma_r}{r} \tag{10.19}$$

with stress boundary conditions:

$$\begin{aligned} &\text{at } r = a \qquad \sigma_r^a = -p \\ &\text{at } r = b \qquad \sigma_r^b = 0 \end{aligned} \tag{10.20}$$

that does not violate the yield condition:

$$|\sigma_\theta - \sigma_r| \preceq Y \tag{10.21}$$

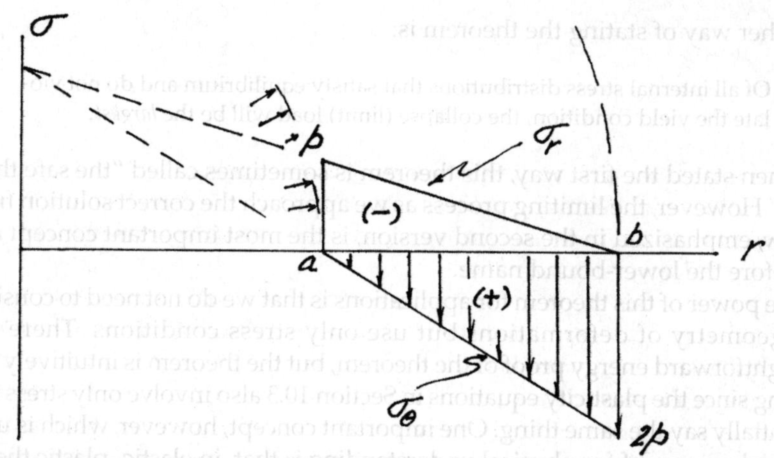

FIGURE 10.7
Lower-bound trial ($b = 2a$).

Cutting the ring through any diameter gives another overall equilibrium relationship in terms of stress resultants:

$$pa = \int_a^b \sigma_\theta \, dr \tag{10.22}$$

Clearly any number of possible stress distributions are possible. Our choice is wide open because any distribution of σ_r that satisfies the boundary conditions [Equation (10.20)] may be substituted into Equation (10.2) to give the corresponding (σ_θ) and then p_L determined so that Equation (10.21) is not violated.

An obvious and simple choice is a linear distribution as shown in Figure 10.7. For the ring with $b = 2a$:

$$\sigma_r = \frac{2pr}{b} - 2p$$

which satisfies the boundary conditions [Equation (10.20)]. Substituting into Equation (10.2):

$$\sigma_\theta = \frac{4pr}{b} - 2p$$

which is also linear and has a maximum value of $2p$ at $r = b$. Equation (10.22) is satisfied and since the yield criterion must not be violated anywhere

$$2p_L = Y \qquad \text{or} \qquad p_L = 0.5Y$$

This, as we know, is 28% lower than the correct value of $0.693Y$, but 33% higher than first yield.

Obviously to come closer to the exact value we must guess a distribution of σ_r and thereby σ_θ that come closer to the yield condition throughout the cross-section. There are obviously better strategies for making educated guesses. One method is to specify $\sigma_\theta(r)$ and determine $\sigma_r(r)$ from the equilibrium equation rather than the other way around. The elastic solution is a very useful clue as to how to proceed, but the lower-bound approach is in no way dependent on knowing the elastic solution beforehand.

10.8 The Upper-Bound Theorem

> If an estimate of the collapse load of a structure is made by equating internal rate of dissipation of energy to the rate external forces do work for *any postulated mechanism* of deformation (collapse mechanism), the estimate will either be high or correct.

This version of the upper-bound theorem emphasizes the important concept for its application of first guessing a failure mechanism. It also implies the importance of considering the plastic deformations associated with this collapse mechanism. Another way of stating the theorem is: Of all methods of collapse, the actual failure mechanism will require the least amount of force (or energy). This version couches the fundamental idea in terms of minimizing force. Either statement stipulates that we approach the correct solution from above not, as for the lower-bound theorem, from below.

Again this theorem is intuitively obvious and really needs no proof. We would expect, by definition, that a structure would naturally "choose to" collapse at the lowest possible load and if this were not the case, something would be wrong with the world.

To return to our example of a thick ring ($b = 2a$), the failure mechanism seems to be tensile yielding across a cross-section. However, since shear causes plastic flow, we would expect from the elastic solution that yield would develop at 45° to the $r - \theta$ orientation of principal stresses. For illustration, however, let us first guess the failure mechanism shown in Figure 10.8a. Along the shear surface, the shear stress is known to be the shear strength

$$\tau_Y = \tau_q = s = c = \frac{Y}{2}$$

but the normal stress is unknown. However, because we have anticipated this difficulty and chosen a failure mechanism with parallel slip surfaces, we

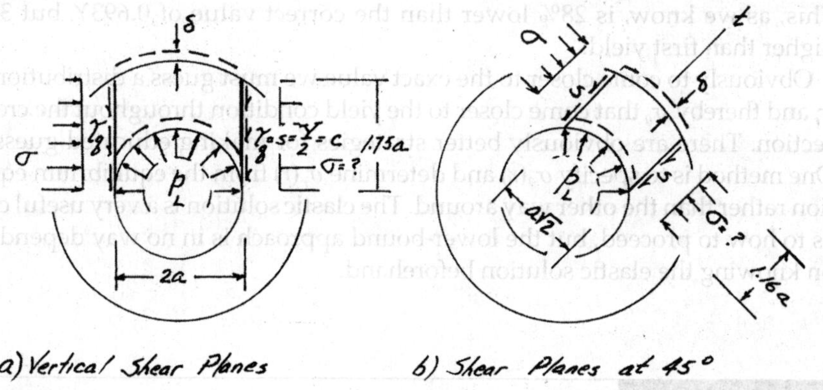

a) Vertical Shear Planes
b) Shear Planes at 45°

FIGURE 10.8
Two trial failure mechanisms.

can sum forces in this direction and the normal stresses on the failure planes do not enter into the equilibrium equation.* Equating vertical forces:

$$p_L(2a) = 2\frac{Y}{2}(1.75a)$$

$$\therefore p_L = 0.875Y$$

The length of the shearing surface, 1.75a, was scaled from the drawing rather than calculated exactly.

Similarly, for a second set of parallel slip planes at 45° to the vertical (Figure 10.8b).

$$\Sigma F_t = 0$$

$$\therefore p(a\sqrt{2}) = 2\frac{Y}{2}(1.16a)$$

$$p_L = 0.820Y$$

where again the slip length, 1.16a, is measured rather than computed. Note that rather than use an energy calculation, the limit load was obtained directly from equilibrium. Often this is easier and simpler to visualize but either a virtual work or an equilibrium calculation is equally valid.

As might be expected, the second choice of failure mechanism is somewhat better than the first (giving a lower estimate for p_L) since it at least starts off

* This trick will be used repeatedly in the following chapters for approximate analysis. Another extremely useful failure mechanism is circular for the same reason. In this case, normal stresses do not affect moment equilibrium.

from the inner boundary in the correct orientation of 45°. Neither is very good, however.

10.9 Example—the Bearing Capacity (Indentation) Problem

The contact problem of a foundation, bearing plate, or punch being driven into a body is encountered throughout engineering practice. As an introduction we will consider here only the simplest idealization by assuming:

1. The material being indented (the foundation) is an EPS material.
2. Plane strain (a long indenter) so that the out-of-plane stress is everywhere the intermediate principal stress.
3. The actual distribution of contact pressure is not important and can therefore be considered uniform.
4. There are no contact shears (a smooth indenter).
5. The weight of the foundation material can be neglected.
6. The Tresca yield condition; i.e., $\frac{\sigma_1 - \sigma_3}{2} = \frac{Y}{2} = c$.

In succeeding chapters a more thorough analysis for this general problem will be given and an "exact" solution obtained for this simple case using slip-line theory. However, it is extremely important to appreciate the simplicity and power of the trial-and-error approach using the upper-bound theorem before it is masked by the complexity of this theory.* For most bearing capacity problems and, for that matter, most two-dimensional problems in general, the assumptions above are only approximate at best and boundary conditions (geometry and loading) are complex. Thus the engineer usually faces problems for which "exact" answers are extremely difficult or impossible to obtain. Good approximate analysis using the upper-bound approach is, however, generally simple and always possible. Moreover, it develops that physical feel and sensitivity as to which variables are most significant, which is so essential for good design. Thus the upper-bound concept and its application will be emphasized in this section and hereafter.

10.9.1 Circular Mechanisms

The first and most crucial step in upper-bound analysis is to postulate a geometrically possible failure mode or "mechanism" of collapse. The closer this

* Simple lower-bound solutions for this idealized version of the bearing capacity problem are also possible (see chapter problems). However, in more general cases, they are very complex defeating the purpose of limit analysis. It is much more difficult to guess a stress distribution satisfying equilibrium while not violating the yield condition, than to guess a failure mechanism.

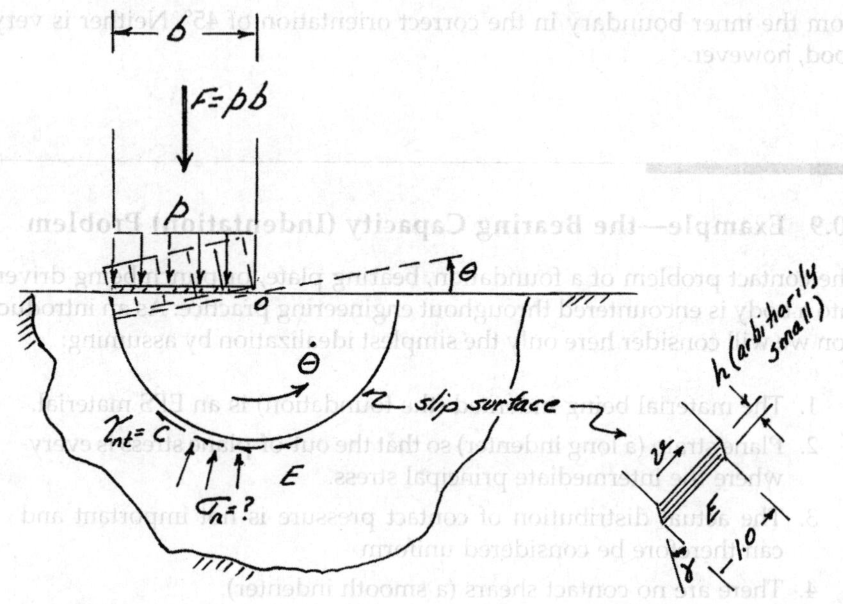

FIGURE 10.9
Simple circular mechanism.

is to the "correct" mechanism, the closer one approaches the "correct" answer, but there are trade-offs involved in that the subsequent calculation of the load to cause collapse must be reasonably easy. In fact, since after the mechanism is selected the problem of evaluating the corresponding failure load is simply a statics problem,* choosing and drawing a collapse mechanism is analogous to selecting the most advantageous free-body diagram to draw in structural analysis. Ninety-nine percent of thought and judgment should be spent on this first step since the remaining work is straightforward mechanical application of force equilibrium or its energy counterpart.

Perhaps the simplest possible mechanism is rotation about one corner as shown in Figure 10.9. The bearing, plate (indenter) rotates as a rigid body with angular velocity $\dot{\theta}$ around a corner and there is an intense shear zone (a slip surface) outside of which (region E) the EPS foundation material remains elastic. In this zone, which can be made arbitrarily narrow since the plastic plateau is assumed infinite, the shear stress must be $\tau_{\max} = c = \frac{Y}{2}$ and the shearing strain rate is uniform so that $\gamma \cong \frac{vt}{h}$ or $\dot{\gamma} = \frac{v}{h}$. Thus the rate of dissipation of energy is:

$$\dot{D} = c\left(\frac{v}{h}\right)h \text{ per unit volume } = cv \tag{10.23}$$

* And this is why upper-bound limit analysis and ultimate design are so popular. All the complexities of solving difficult simultaneous differential equations (often made impossible by complicated boundary conditions) are simply avoided.

where v = velocity of plastic deformation of one surface relative to the other. Therefore, equating external to internal work (in a small time)

$$F_L b \frac{\dot{\theta}}{2} = cb\dot{\theta}\pi b$$

and

$$F_L = 2\pi cb = 6.29cb \qquad \text{or} \qquad p_L = \frac{F_L}{b} = 6.29c$$

It is important to note that for this *circular* mechanism, the calculation could just as easily be done using equilibrium by taking moments about the corner at 0:

$$\Sigma M_0 = 0 = F\frac{b}{2} - c\pi b^2 \qquad \therefore p_L = \frac{F_L}{b} = 6.29c$$

However, this is only possible since the unknown normal stresses on the circular slip surface have no moment arm.

Thus we have found, very simply, an upper-bound *estimate* for the collapse load and it must be higher than the actual collapse load. The search for possible lower values for circular mechanisms in general can now be pursued by changing the origin and radius. This can be done as in the previous example, either by trial and error using a graphical minimization procedure, or formally (see P10.17). However, unlike the example of a loaded beam where the fundamental type of mechanism is known (i.e., plastic hinges), optimization of the location of the circular slip surface to reduce p_L to a minimum will not converge to the exact solution since this type of mechanism (circular) is not the correct shape to start off with (see P10.19).

10.9.2 Sliding Block Mechanisms

Another geometrically possible class of failure modes for this problem is made up of a number of plane, rigid blocks being pushed downward, sideways, and upward, such as is shown in Figure 10.10. To formulate the energy balance, the relative sliding velocities at each interface must be known.

The velocity (or displacement) diagram shown in Figure 10.10 does this directly.* Each sliding block is designated by a capital letter and the velocity

* Such diagrams (hodographs) where relative values in a vector field are found graphically are powerful conceptually and useful for analysis. The Williot and Williot-Mohr diagrams (developed in 1877) for determining displacements of all the joints of a truss are classic as is the Maxwell diagram for finding the bar forces. Graphic analysis used by so many great structural engineers in the last century is seldom even mentioned in this day of computers. This is too bad as graphic methods are less tiresome and quicker than numerical calculation. More important, there is no better way of understanding vectors and two-dimensional vector fields than to physically use vectors in their invariant form drawn as arrows in space on a piece of paper to get answers directly in visual form.

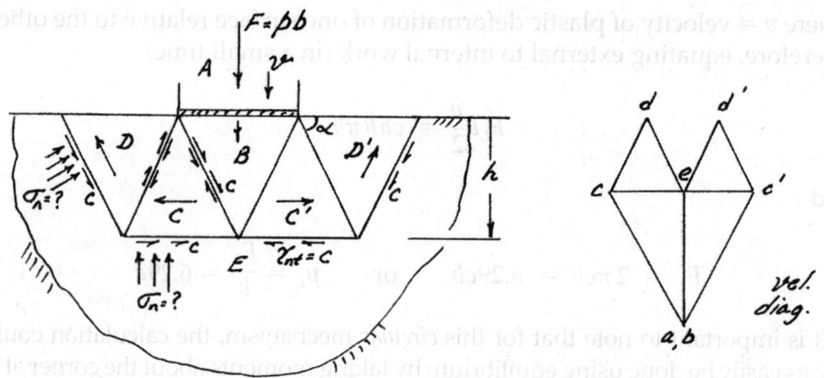

FIGURE 10.10
Simple "rigid-block" mechanism and associated velocity diagram.

of each block is represented by a single point. Thus the *relative* velocity between two blocks is that vector joining the two corresponding points.

To construct the diagram, one starts at point *e* as the zero reference since it represents the stationary elastic portion of the structure. The velocity of the bearing plate (indenter) is laid off to some arbitrary scale (since the magnitude is unknown). Blocks A and B move vertically the same amount (wedge B is an extension of A) so *a* and *b* must coincide. Then C is known since it must be horizontal to *e*, and at a given angle, to *b*. Similarly *d* is located graphically from *c* and *e* and the points *c'* and *d'* from symmetry.

The energy balance is therefore:

$$F_L v_{\text{ea}} = 2c(v_{\text{bc}}\ell_{BC} + v_{\text{ce}}\ell_{CE} + v_{\text{cd}}\ell_{CD} + v_{\text{de}}\ell_{DE})$$

where ℓ is the length of the interface and all energy terms are, of course, positive. Note also that the absolute size of the velocity diagram will have no effect on the calculation. For the mechanism shown, made up of equilateral triangles, all the lengths equal b and $v_{\text{bc}}/v_{\text{ca}} = \frac{1}{\sqrt{3}}$ etc. Therefore:

$$F_L = 2cb\left(\frac{2}{\sqrt{3}} + \frac{1}{\sqrt{3}} + \frac{1}{\sqrt{3}} + \frac{1}{\sqrt{3}}\right) = \frac{10}{\sqrt{3}}cb$$

or

$$p_L = \frac{F_L}{b} = 5.76c$$

This is 9% less than the simple slip circle mechanism ($p_L = 6.29c$) and it can be reduced slightly by optimizing the size and shape of the triangles (see P10.17). However, this is certainly not the correct family of slip surfaces and thus, no matter how hard we might try to find the optimum layout of rigid sliding blocks in this family of collapse modes, we shall never reach the "exact" value of p_L. As engineers we should not let this bother us too much.

However, for EPS materials and in some cases for more difficult Mohr-Coulomb materials (cast iron, concrete, soil), it is possible to derive from the basic equilibrium equations and yield condition, the geometric characteristics of the correct family of ship surfaces. We may even be able to solve for the most critical one and the corresponding exact limit load. This is the topic explored in Chapter 12.

10.10 Problems and Questions

P10.1 The full-range behavior of various metals is described in Figure 10.1 in terms of "normal" stress and strain for a dead load test. Compare the curves for mild steel and alloy steel to those you might obtain in terms of:

a. True stress and strain.

b. A controlled strain test (i.e., strain rather than stress is applied in controlled increments).

P10.2 Consider a stress block such that $\sigma_1 > \sigma_2 > \sigma_3$ during plastic deformation (Tresca condition). Refer to Figure P10.2.

a. Plot the planes on which the deformation increment (sliding) occurs.

b. What is the plastic deformation $\delta\epsilon_2$?

c. Show that Poisson's ratio is 1/2 (i.e., what is the relationship between $\delta\epsilon_1$ and $\delta\epsilon_3$)?

d. Show that the normality rule holds.

P10.3 The plastic stress–strain curve for a work-hardening material is idealized as shown in Figure P10.3. What must be the corresponding stress–strain curve for pure shear?

a. For a Tresca material?

b. For a von Mises material?

What are the values for plastic K, G, E, ν? (Hint: Neglect the elastic response and equate plastic work per unit volume.)

P10.4 A perfectly plastic material yields in three-dimensional states of stress, according to the Tresca yield condition and plastic deformation obeys the normality rule. Plot the yield condition for the material in σ_1, σ_2 space: (a) for a state of plane stress, i.e., $\sigma_3 = 0$, and (b) for a state of plane plastic strain, i.e., $\delta\epsilon_3 = 0$.

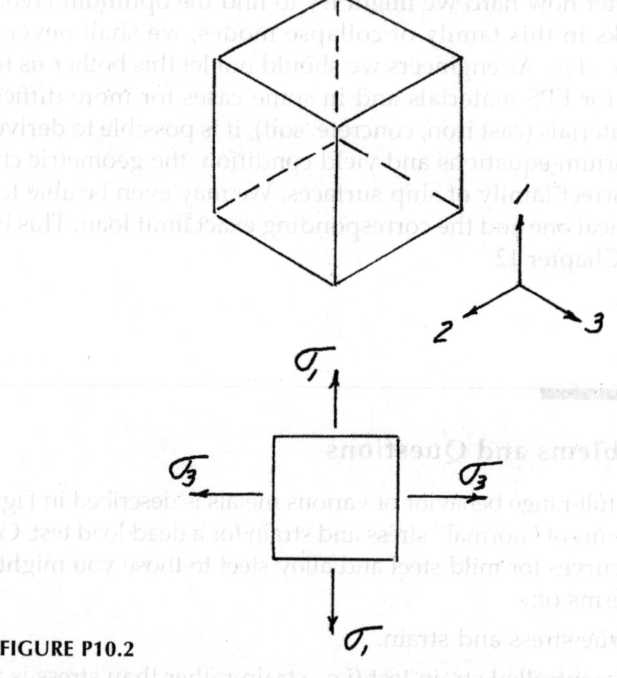

FIGURE P10.2

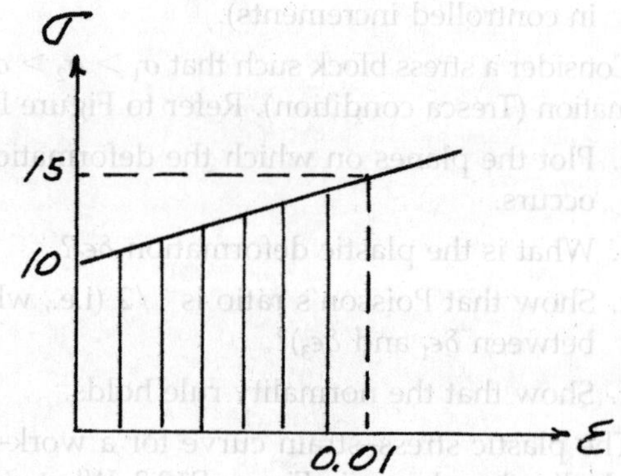

FIGURE P10.3

Relate the two yield conditions to the three-dimensional yield surface. Repeat for a von Mises material.

P10.5 Rather than a thick ring in plane stress, consider a thick-walled tube ($b = 2a$) closed at its ends under internal pressure, p. Revise the analysis (where appropriate) in Section 10.4 and *plot* the load-deflection curve similar to Figure 10.6. Note: Assume $\nu = 0.5$ and first show that $\sigma_z = \left(\frac{\sigma_\theta + \sigma_r}{2}\right)$.

P10.6 For the thick ring example of Section 10.4, *plot* "universal" elastic-plastic pressure-expansion curves for $b/a = 1.5$, $b/a = 2$, $b/a = 3$. To do this, plot p/Y downward vs. $(E/Y)\ \epsilon_\theta^b$ on the horizontal axis.

P10.7 Consider the thick-walled tube of P10.5, $b = 2a$ closed at its ends loaded with a pressure such that half the section is plasticized (i.e., $c = 1.5a$). Compute the residual stresses and residual deformation when the pressure is removed. Plot the full load deformation curve and show that your analytic results could be estimated graphically. (Hint: The easiest approach is to superimpose a suction $-p$ to remove the pressure.)

P10.8 Make a full-range, elastic-plastic analysis of a thick-walled sphere $b = 2a$ and compare the yield and limit pressures to those for the thick-walled tube of P10.5.

P10.9 How would the analysis of the thick-ring in plane stress (Section 10.4) be altered if the von Mises (octahedral shear) yield criterion had been used rather than the Tresca?

P10.10 Solve the compatability Equation (10.11) for a nonzero constant (say, $i_z = g$). Show that the resulting mode of deformation is equivalent to a superposition of the modes:

$$\dot{\epsilon}_\theta = \frac{a\dot{u}_a}{r^2}, \quad \dot{\epsilon}_r = -\frac{a\dot{u}_a}{r^2}, \quad \dot{\epsilon}_z = 0$$

and

$$\dot{\epsilon}_\theta = -\frac{g}{2}, \quad \dot{\epsilon}_r = -\frac{g}{2}, \quad \dot{\epsilon}_z = g$$

and that, in particular, the change in volume (per unit length $\Delta z = 1.0$) is independent of g.

P10.11 Make the lower-bound analysis of a thick ring ($b = 2a$) under internal pressure by putting $\sigma_\theta = const.$ in the equilibrium equation and solving the boundary equations without violating the yield condition. Note that Equation (10.2) can be rearranged as:

$$\sigma_\theta = \frac{d}{dr}(r\sigma_r)$$

P10.12 Figure P10.12 shows three distributions of $\sigma_\theta(r)$ for a thick ring ($b = 2a$). Compute the corresponding p_L from Equation (10.22). Which if any, are possible lower bound solutions; possible upper-bound solutions? Discuss.

P10.13 From the elastic solution for a thick ring, plot maximum shear stress trajectories (i.e., lines or planes along which the maximum shear stress, $\frac{\sigma_\theta - \sigma_r}{2}$, act).

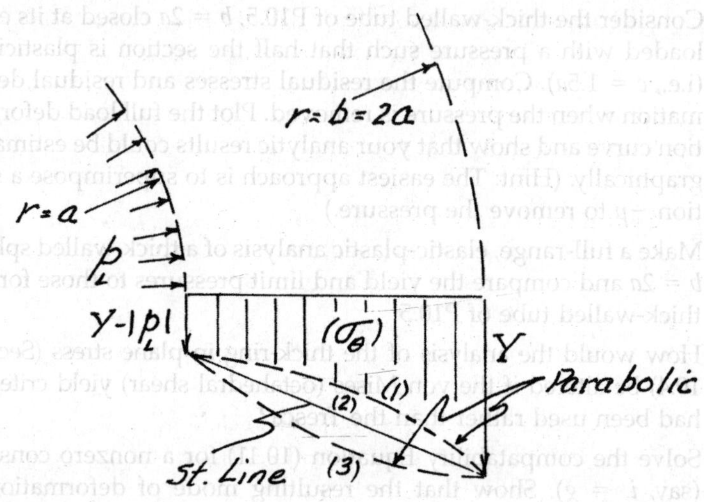

FIGURE P10.12

P10.14 Calculate the limit angular velocity, ω_L for a flat, rotating disc of EPS material (no central hole) and compare to the yield angular velocity ω_Y. Assume plane stress and the Tresca yield criterion.

P10.15 From the elastic solution for a uniform line load on the half space (Figure 8.6) plot (for $r < 4b$):

a. Lines of equal $\sigma_q = \frac{\sigma_x + \sigma_y}{2}$

b. Isochromatics (lines of equal $\tau_{max} = \tau_q = \frac{\sigma_x - \sigma_y}{2}$)

c. Trajectories of maximum shear stress (lines along which τ_{max} acts which will be 45° to principal stress trajectories) for the two loadings of the half-space.

In each case, determine the point or points where yield first occurs and the corresponding value of p_Y. If possible, develop some *simple* lower-bound solutions beyond yield. (Hint: Consider the vertical planes at $x = 0$ and/or $x = \pm\frac{b}{2}$.)

P10.16 Suppose you can guess a lower-bound for the limit load (i.e., a stress distribution that satisfies equilibrium and does not violate the yield condition). Does this lower-bound solution occur (i.e., does it correspond to a point on the load deflection curve on the way from first yield to ultimate collapse)? Why?

P10.17 Minimize the limit load p_L for the circular failure mechanism for the bearing capacity problem in Figure 10.8. You can do this formally or graphically following some sort of scheme (Fellenius problem).

P10.18 Minimize the limit load p_L for the family of rigid body mechanisms similar to Figure 10.9 with the vertical dimension h as the variable (using h and the Pythagorian theorem is easier than using an angle).

P10.19 Why do we know that both the circular and the block mechanisms are not correct for the bearing capacity problem? (Hint: Consider boundary conditions and "corners.")

P10.20 Calculate the result for the given sliding-block collapse mechanism (Figure 10.8) by equilibrium.

P10.21 Attempt a full plasticity solution for the rotating disc of P10.13. Is this of engineering significance?

11

One-Dimensional Plasticity for Design

In the previous chapter, a full-range plasticity solution for an EPS thick ring was developed in detail to illustrate the fundamental concepts of plastic analysis. This was possible only because a thick ring subjected to internal or external "pressure" is essentially a one-dimensional problem in that there is only one governing equilibrium equation since there is no variation of the field in the hoop direction. At any stage of plastic deformation, this equation can be integrated directly to determine the stress field. The strain and deformation field is then determined from the behavior of the elastic rim, which has not yet yielded.

The same is true for pure bending and the other one-dimensional linear elastic fields of Chapter 5. It is also possible to obtain full plasticity solutions in pure torsion for many cases considered in Chapter 9. Combinations of linear fields can also be considered. Since limit analysis and design* of beams, shafts, frames, plates, and many other structural systems are based on these one-dimensional plasticity solutions, it is important to study them.

11.1 Plastic Bending

The basic assumption of the theory of flexure as stated in Section 5.5 is that plane sections normal to the axis of the beam remain plane so that the beam under pure bending deforms into a circle with radius of curvature ρ. This assumption as to the geometry of deformations has nothing to do with material properties and requires that longitudinal strains, ε_x, in the fibers vary

* The concept is to apply "load factors" to working values and then design the structure to collapse. This philosophy today dominates the steel and concrete codes for structural design. Limit analysis is much easier than elastic analysis, especially for hand calculations and generally the load factors are established such that lighter structures result compared to those designed using the older working-stress philosophy. With experience it is becoming clear that an understanding of both elastic and plastic behavior should be incorporated in a good structural design.

directly as their distance, y, from the neutral axis as given by Equation (5.15).

$$\varepsilon_x = -\frac{y}{\rho} \quad \text{or} \quad \frac{1}{\rho} = \frac{d\omega_z}{dx} = -\frac{\varepsilon_x}{y}$$

Thus, from equilibrium requirements, a general theory can be developed for bending beyond initial yield.*

Any prismatic beam with a cross-section with symmetry about a vertical axis can be considered. The neutral axis, defined as where the strain is zero, goes through the centroid in the elastic range but may migrate as the yielding progresses such that horizontal equilibrium is maintained. That is $C = T$ or

$$\int \sigma_x dA = 0 \tag{11.1}$$

The moment is given by $M = Ca = Ta$ or

$$-\int \sigma_x y dA = M \tag{11.2}$$

The solution of the general case of inelastic bending requires that these equilibrium Equations (11.1) and (11.2) for the given stress–strain relationship be solved by trial and error. Two relatively simple cases are explored as chapter problems.

Let us consider only the case of a beam of an elastic, perfectly-plastic solid and start with a rectangular cross-section as shown in Figure 11.1. At yield of the outermost fibers, the moment $M_Y = YI/c = Ybh^2/6 = YS$. As more moment is applied, the plastic zone moves toward the neutral axis and, because of our idealization of the material as an EPS, there is a sharp separation leaving an elastic core at the center. If, for example, the strain at the outer surface is twice the yield strain, $\phi = 2\phi_Y$, the middle half of the beam remains elastic (Figure 11.1b) and the moment works out to be $\frac{11}{48}Ybh^2$. At very large rotations of the cross-section, as in Figure 11.1c, the linear distribution in the small elastic core can be neglected and the stress assumed as Y throughout. Thus, this ultimate or plastic moment is

$$M_P = C\left(\frac{h}{2}\right) = Yb\frac{h}{2}\left(\frac{h}{2}\right) = Y\frac{bh^2}{4} = YZ$$

* Or, for that matter, pure bending for materials with a nonlinear stress–strain behavior differing even in tension and compression. Actually, because $\nu = 1/2$ at yield, the anticlastic curvature relationship is more complicated, but is seldom of concern.

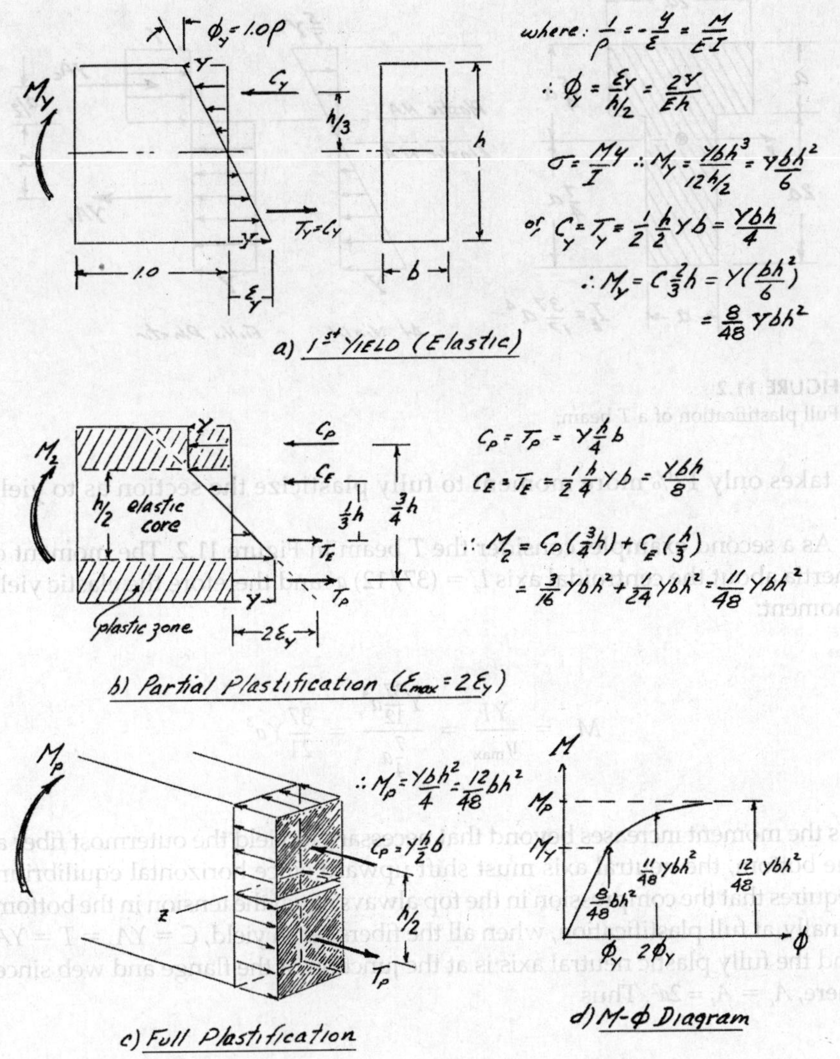

FIGURE 11.1
Plastification of an EPS rectangular beam.

Throughout, the neutral axis remains at mid-height because of the symmetry around the z axis coupled with the requirement that the compression force in the top equal the tension force in the bottom. The elastic section modulus, S, and plastic section modulus, Z, are given in various handbooks for standard shapes such as angles, channels, wideflange, and I beams. The ratio $Z/S = M_P/M_y$ is often called the shape factor, K_s. For a rectangular cross-section then, $K_s = 1.5$ indicating this shape has a large reserve capacity to take additional moment beyond yield because of its inefficiency in the elastic range. For WF and I beams, $K_s \cong 1.12$ and

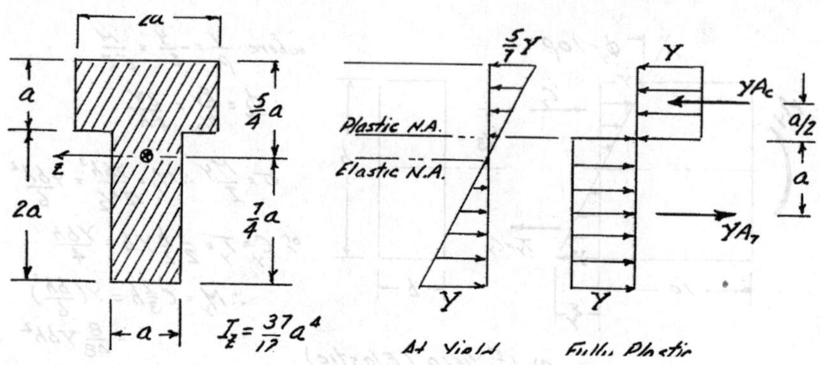

FIGURE 11.2
Full plastification of a *T* beam.

it takes only 12% more moment to fully plasticize the section as to yield it.

As a second example, consider the *T* beam in Figure 11.2. The moment of inertia about the centroidal axis $I_z = (37/12)\,a^4$ and therefore the elastic yield moment:

$$M_y = \frac{YI}{y_{\max}} = \frac{Y\frac{37}{12}a^4}{\frac{7}{4}a} = \frac{37}{21}Ya^3$$

As the moment increases beyond that necessary to yield the outermost fiber at the bottom, the neutral axis must shift upward since horizontal equilibrium requires that the compression in the top always equal the tension in the bottom. Finally, at full plastification, when all the fibers reach yield, $C = YA_c = T = YA_t$ and the fully plastic neutral axis is at the junction of the flange and web since, there, $A_c = A_t = 2a^2$. Thus

$$M_p = Y(2a^2)\frac{3}{2}a = 3Ya^3 \qquad \text{and} \qquad K_s = \frac{M_p}{M_y} = \frac{63}{37} = 1.71$$

This shape factor is roughly the same for any *T* beam showing they are even less efficient than rectangular cross-sections in the elastic range.

Example 11.1

A steel cantilever beam with a hole in it is loaded with a vertical shear force (distributed parabolically) at its free end as shown. From simple beam theory (plane sections) determine the yield load S_Y. Estimate the limit load from an upper-bound approach and thereby the plasticity factor S_L/S_Y.

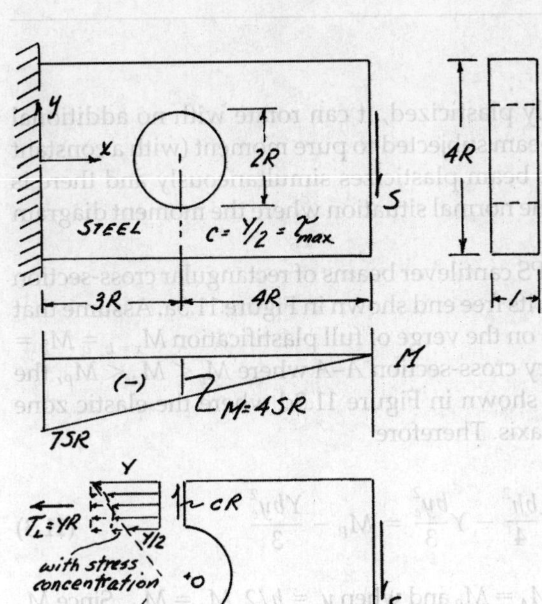

$\therefore$ at $x = 0$

$$I = \frac{b(4R)^3}{12} = \frac{16}{3}R^3$$

at $x = 3R$

$$I = \frac{16}{3}R^3 - \frac{(2R)^3}{12}$$

$$= \frac{14}{3}R^3$$

i) @ Fixed Support

$$\text{at } x = 0 \quad Y = \frac{7SR(2R)}{\frac{16}{3}R^3}$$

$$S_Y = \frac{8}{21}RY = \underline{\underline{0.381RY}}$$

$$S_L = K_s S_Y = 1.5S_y$$

$$= \underline{0.571RY}$$

ii) @ $x = 3R$

$$\text{elastic } Y \cong \frac{4SR(2R)}{\frac{14}{3}R^3}$$

$$\therefore S_Y \cong \frac{7}{12}RY \cong \underline{.583RY}_{\text{NOT CRIT.}}$$

at the limit $\Sigma F_y = 0 \quad \therefore 2CR = YR = S_L \quad \therefore S_L = \underline{1.0RY}_{\text{NOT CRIT.}}$

$$\Sigma M_0 = 0 \quad \therefore S(4R) = C_L(3R) = 3YR^2 \quad \therefore S_L = \underline{0.75RY}_{\text{NOT CRIT.}}$$

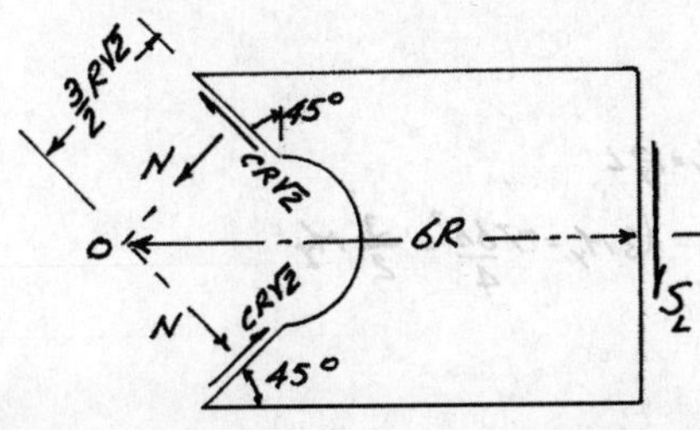

iii) Slip @ 45° (correct B.C.^{s})

$$\Sigma M_0 = 0$$

$$\therefore S_L(6R) = 2(CR\sqrt{2})\frac{3}{2}R\sqrt{2}$$

$$S_L = CR = \underline{\underline{0.5RY}} \text{ CRIT.!}$$

$$\therefore K_p = \frac{S_L}{S_Y} = \frac{.5}{.381} = \underline{\underline{1.31}} \text{ ANS.}$$

11.2 Plastic "Hinges"

Once the cross-section is fully plasticized, it can rotate with no additional moment applied. Thus for a beam subjected to pure moment (with a constant moment diagram), the entire beam plasticizes simultaneously and there is unrestrained plastic flow. In the normal situation where the moment diagram varies, this is not the case.

Consider for example the EPS cantilever beams of rectangular cross-section with a concentrated load, P, at its free end shown in Figure 11.3a. Assume that the cross-section at the wall is on the verge of full plastification $M_{x=L} = M_P = K_s M_Y = 1.5 P_Y L$. At an arbitrary cross-section A–A where $M_y < M_A < M_P$, the stress distribution will be as shown in Figure 11.3d where the elastic zone extends $\pm y_e$ from the neutral axis. Therefore

$$M_A = Y\frac{bh^2}{4} - Y\frac{by_e^2}{3} = M_P - \frac{Yby_e^2}{3} \tag{11.3}$$

and we see that when $y_e = 0$, $M_A = M_P$ and when $y_e = h/2$, $M_A = M_Y$. Since M_A is a known function of x or x^* for a given loading, the elastic-plastic interface can be determined from Equation (11.3). In this case, $M_A = M_P - M_P(x - L)/L$ so

$$y_e^2 = \frac{3h^2}{4L}(x - L), \qquad L_P = L(1 - M_Y/M_P) = \frac{L}{3}$$

and the interface is parabolic as shown.

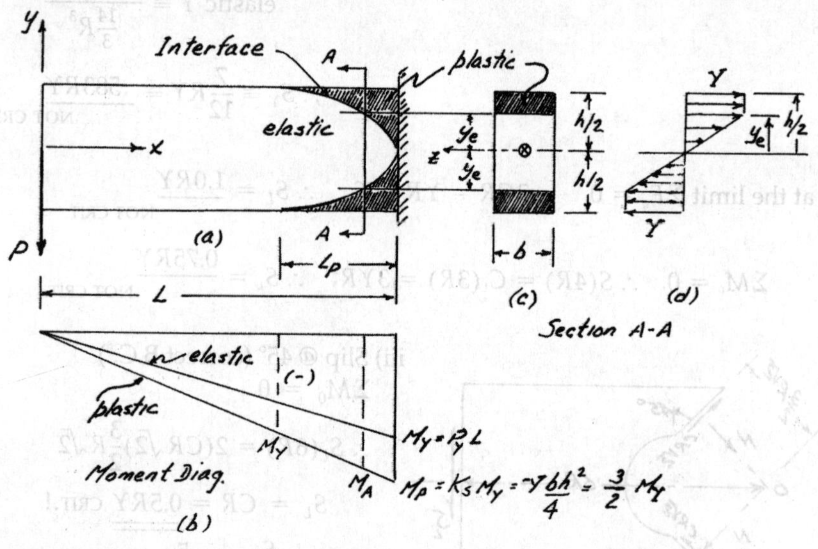

FIGURE 11.3
Contained "plastic hinge."

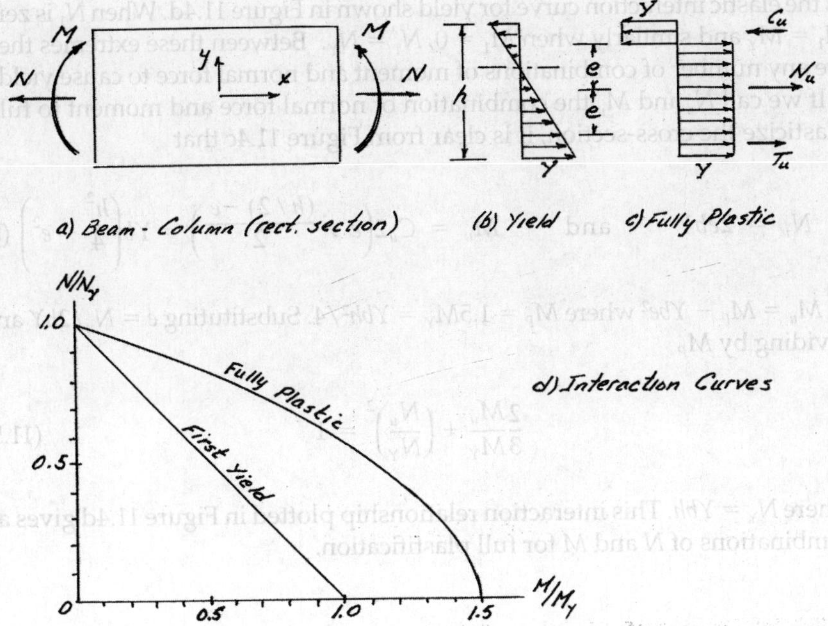

FIGURE 11.4
Combined axial load and bending moment.

As long as the elastic core remains, plastic flow is constrained. However, when P increases such that $PL = M_P$, the cross-section becomes fully plastic and the curvature can increase without limit but contained within the plastic zone. The beam rotates like a rigid body about a "plastic hinge" holding the ultimate plastic moment M_P. If, as in this case, the structure is determinate, it collapses and $P_L = M_P/L = K_s M_Y/L$. It should be noted that no provision is made in this simple strength-of-materials approach for the effect of any shear stresses, τ_{xy}, or vertical stress, σ_y, which may be of some significance in short beams. The chapter problems explore these questions in a preliminary way.

On the other hand, the effect of axial load N combined with bending moment M in the plastic range is straightforward. Suppose N and M combine to give the stress state at yield shown in Figure 11.4b. The maximum stress in the bottom fiber is

$$Y = \frac{N_1}{bh} + \frac{6M_1}{bh^2} \tag{a}$$

The limits on $N(M = 0)$ and $M(N = 0)$ when each act alone are $N_Y = Ybh$ and $M_Y = Ybh^2/6$ which, upon substitution, gives

$$\frac{N_e}{N_Y} + \frac{M_e}{M_Y} = 1 \tag{11.4}$$

as the elastic interaction curve for yield shown in Figure 11.4d. When N_1 is zero, $M_1 = M_Y$ and similarly when $M_1 = 0$, $N_1 = N_Y$. Between these extremes there are any number of combinations of moment and normal force to cause yield.

If we call N_u and M_u the combination of normal force and moment to fully plasticize the cross-section, it is clear from Figure 11.4c that

$$N_u = 2ebY \quad \text{and} \quad M_u = C_u 2\left(e + \frac{(h/2) - e}{2}\right) = Yb\left(\frac{h^2}{4} - e^2\right) \quad \text{(b)}$$

or $M_u = M_P - Ybe^2$ where $M_P = 1.5M_Y = Ybh^2/4$. Substituting $e = N_u/2bY$ and dividing by M_P

$$\frac{2M_u}{3M_Y} + \left(\frac{N_u}{N_Y}\right)^2 = 1 \tag{11.5}$$

where $N_Y = Ybh$. This interaction relationship plotted in Figure 11.4d gives all combinations of N and M for full plastification.

11.3 Limit Load (Collapse) of Beams

With the concept of a plastic hinge it is possible to determine the collapse or limit load of structures by the upper-bound theorem (as was done for thick rings and the bearing capacity problem in Chapter 10). Any type of structure: beams, frames, arches, plates, even shells, are susceptible to this approach no matter how complicated or irregular the geometry. Various collapse "mechanisms" are assumed and the limit or collapse load computed for each from equilibrium or conservation of energy. That mechanism corresponding to the lowest load is critical. For linear structures (one-dimensional from the standpoint of equilibrium) this minimization procedure by analytic analysis or by trial and error leads to an "exact" upper-bound solution since only the number and longitudinal position of the discrete plastic hinges are unknown for a limited number of possible types of collapse mechanisms. For plates and shells, the exact mechanism is much more difficult to determine, but it is often not really necessary for a good approximation since the type of mechanism is seldom in doubt.

Only a few examples will be given in this and the following section to illustrate the fundamental idea and to show how similar it is to the yield surface concept used for upper-bound analysis for two-dimensional stress fields in the previous chapter.* Take first the simple case of a once-indeterminate

* Since the upper-bound procedure is so simple and intuitively satisfying, it is quite fun. All sorts of lovely examples permeate the literature and textbooks on structural mechanics. Moreover, structures are easily tested to failure in the laboratory and, unfortunately, sometimes in the field so observed failure mechanisms are readily available to verify the theory.

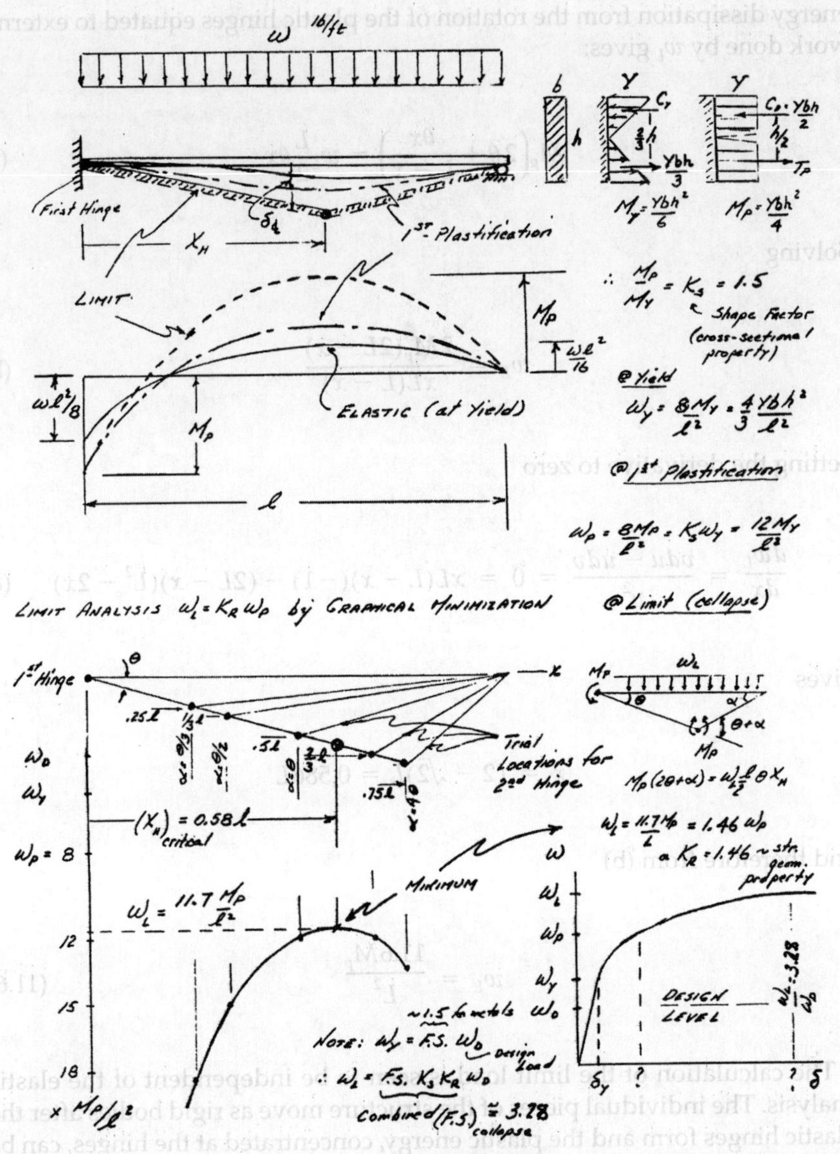

FIGURE 11.5
Upper-bound limit analysis of an indeterminate beam by trial and error minimization.

beam with a uniform load shown in Figure 11.5. Two plastic hinges, one at the wall and one near midspan, is the only possible collapse mechanism. The second hinge will be, by the upper-bound theorem, located such that the limit load w_L is a minimum. The minimization procedure can be done graphically by trial and error as shown or formally. Since $(L - x)\alpha = \theta x$ the internal energy dissipation from the rotation of the plastic hinges equated to external work done by w_L gives:

$$M_p\left(2\theta + \frac{\theta x}{L-x}\right) = w_L \frac{L}{2}\theta x \tag{a}$$

Solving

$$w_L = \frac{2M_p(2L-x)}{xL(L-x)} \tag{b}$$

Setting the derivative to zero

$$\frac{dw_L}{dx} = \frac{vdu - udv}{v^2} = 0 = xL(L-x)(-1) - (2L-x)(L^2 - 2x) \tag{c}$$

gives

$$x = (2 - \sqrt{2})L = 0.586L$$

and therefore from (b)

$$w_L = \frac{11.6M_p}{L^2} \tag{11.6}$$

The calculation of the limit load is seen to be independent of the elastic analysis. The individual pieces of the structure move as rigid bodies after the plastic hinges form and the plastic energy, concentrated at the hinges, can be made arbitrarily large for an EPS material. Thus the elastic contribution can be neglected.

With the formation of the first plastic hinge, the one-degree indeterminate beam becomes determinate and when the second one forms, the structure is unstable (on the verge of becoming dynamic). This gives the extra equation to calculate the limit load. Since the most complicated indeterminate structure is reduced to a simple computation in statics, limit analysis is very popular. Moreover, small support settlements or rotations will not affect the limit load while they have a great influence on the elastic forces and moments in indeterminate structures.

The price paid for this simplification is the loss of any straightforward calculation of displacements beyond first yield.* As outlined in Figure 11.5, two interacting nonlinear effects are occurring. The plastification of the first hinge redistributes the moment diagram which, in turn, influences the location of the next plastic hinge. The increase in moment, and therefore load to cause the first plastic hinge is the shape factor, K_s, which, as we have seen, is a geometric property of the shape of the cross-section. The added strength introduced by the indeterminacy and geometry of the structure can be called the redistribution factor, K_R, defined as

$$K_R = \frac{w_L}{w_P} = \frac{w_L}{K_s w_Y} \tag{11.7}$$

For the beam in Figure 11.5, $K_R = 1.46$ and K_s for a rectangular cross-section is 1.5 giving a combined plasticity factor $K_P = K_s\, K_R = w_L/w_Y = 2.19$.** Thus, if the beam were designed to give a factor of safety of 1.5 against yield, it would have a factor of safety against collapse of (1.5)(2.19) = 3.28. Even in an *H*-beam with $K_s = 1.12$ and a working load of two-thirds, the yield load would have a factor of safety against collapse of 2.46 if supported as in Figure 11.5. Such large reserves against flexural failure are one reason indeterminate bridges and other structures built in the first half of this century using the working-stress design code are still functioning well, even though trucks and trains are now so much heavier.

11.4 Limit Analysis of Frames and Arches

The upper-bound analysis of frames and arches follows much the same strategy as for beams. However, in addition to the location of enough plastic hinges to bring about collapse, there may be a variety of ways they can be arranged into different "failure mechanisms" and, of course, the interaction effect of axial forces may reduce the moment capacity in a nonlinear way (Figure 11.4) as the loads increase. We will consider a few simple examples to illustrate these concepts.***

* Since in theory it takes an infinite rotation to fully develop the plastic moment, the "hinges" experience constrained plastic flow. Therefore a numerical analysis for the complete load deformation curve is possible to any degree of accuracy by the piecewise-linear approximation of increasing the load in small increments and calculating the corresponding increment of displacement. This can become very complicated quickly.

** For the thick ring in Chapter 10, $K_P = P_L/P_Y = 1.85$. We shall see in Chapter 12 that the correct value for P_L for the punch problem is $5.14\ Y/2$ so that $K_P = 5.14/3.14 = 1.64$.

*** A number of good textbooks on limit analysis of structures of all types are available. One of the earliest and still one of the best is *Plastic Design Analysis of Structures* by P.G. Hodge, McGraw-Hill, New York, 1963.

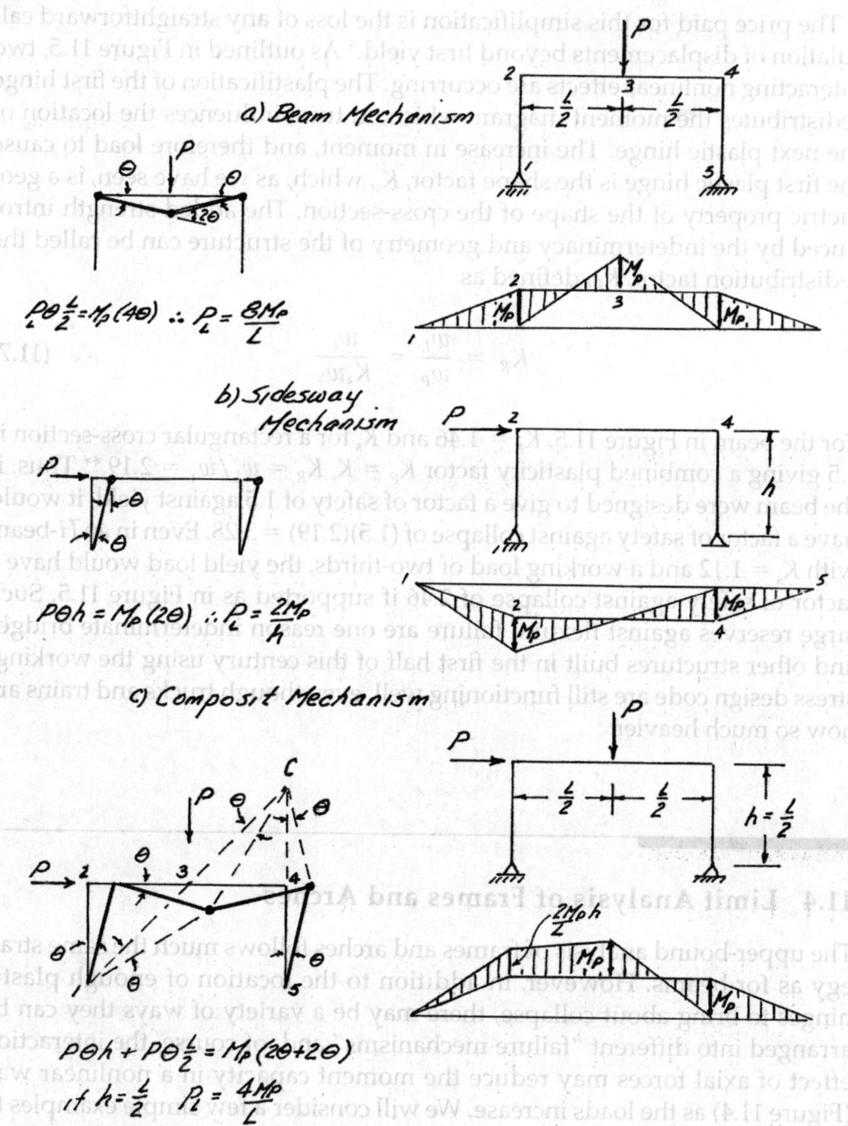

FIGURE 11.6
Limit analysis of a portal frame.

Consider first a basic single-story portal frame (Figure 11.6). For a concentrated load at midspan, a local beam mechanism would occur at $P_L = 8M_P/L$. Similarly, if the wind loading is approximated by a concentrated lateral force P at the top, the local "panel" or sidesway mechanism would occur at $P_L = 2M_P/h$. If, for example, $h = L/2$ then $P_L = 4M_P/L$ and the panel mechanism would occur first. Finally, with both loads acting, a composite mechanism is possible. For the particular case where $h = L/2$, this mode of collapse occurs simultaneously with the panel mechanism at $P_L = 4M_P/L$. Also shown in Figure 11.6 is the instantaneous center approach to the virtual work calculation. There are three movable parts that rotate as rigid bodies. Part 123 rotates about 1, part 45 rotates about 5 and part 34 must therefore rotate about a point C, located as shown (since point 3 rotates about both A and C, point 4 rotates about both 5 and C). The instantaneous-center method is particularly useful for gable frames and arches.

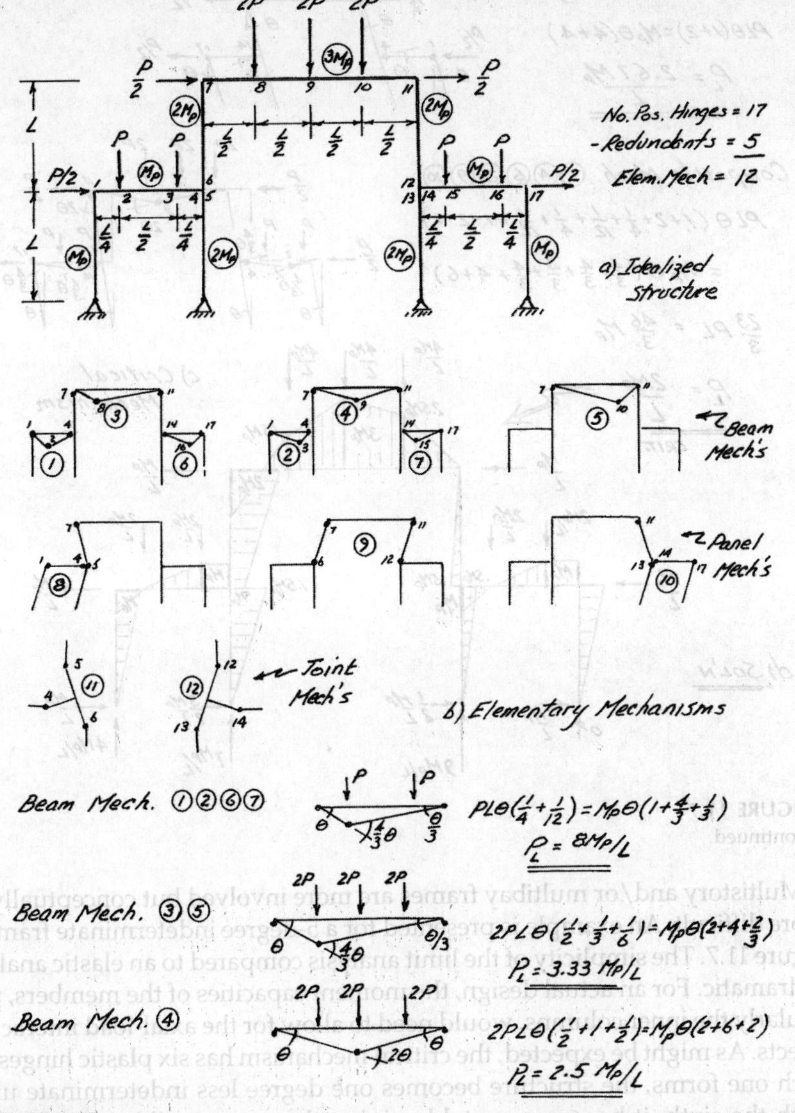

FIGURE 11.7
Limit analysis of an industrial frame.

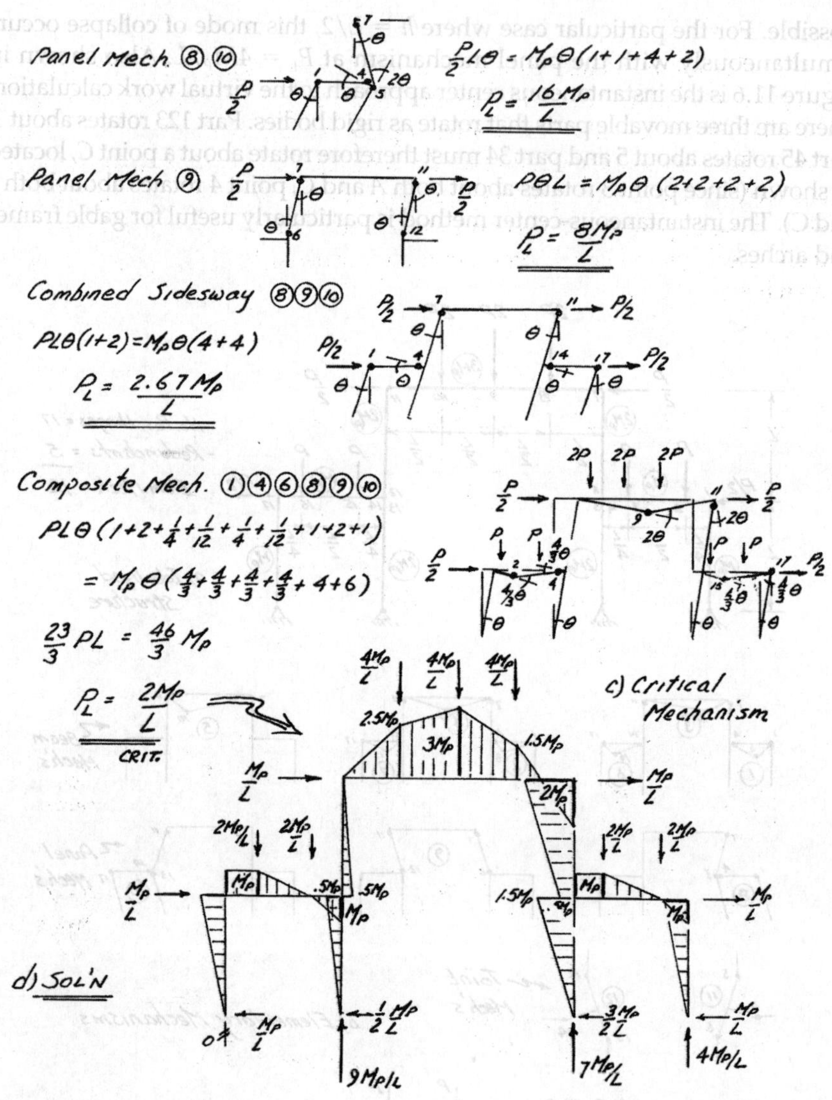

FIGURE 11.7
Continued.

Multistory and/or multibay frames are more involved but conceptually no more difficult. An example is presented for a 5-degree indeterminate frame in Figure 11.7. The simplicity of the limit analysis compared to an elastic analysis is dramatic. For an actual design, the moment capacities of the members, particularly the inner columns, would need to allow for the axial-load interaction effects. As might be expected, the critical mechanism has six plastic hinges. As each one forms, the structure becomes one degree less indeterminate until, with the sixth, it becomes unstable giving the extra equilibrium equation needed to solve for the critical load.

The minimization strategy becomes clear as you work with various trial mechanisms. Obviously, you want to reduce the number of hinges and the angles through which they rotate to keep the internal plastic energy low while at the same time getting all the loads to move to increase the external work. Moreover, from a design perspective it quickly becomes apparent where increasing moment capacity can help and where it will not. Since the calculations are so simple so is a design by successive approximation. Usually for structures, determining the collapse load for each possible mechanism by equilibrium equations is much more difficult than by using a work calculation. However, when you have found what you think is the upper-bound minimum, in this case $P_L = 2M_P/L$, you must use the equilibrium equations to solve for the reactions and plot the moment diagram (Figure 11.7d) to make sure that at no point is $M > M_P$ and thereby check that you have, in fact, determined the correct solution.

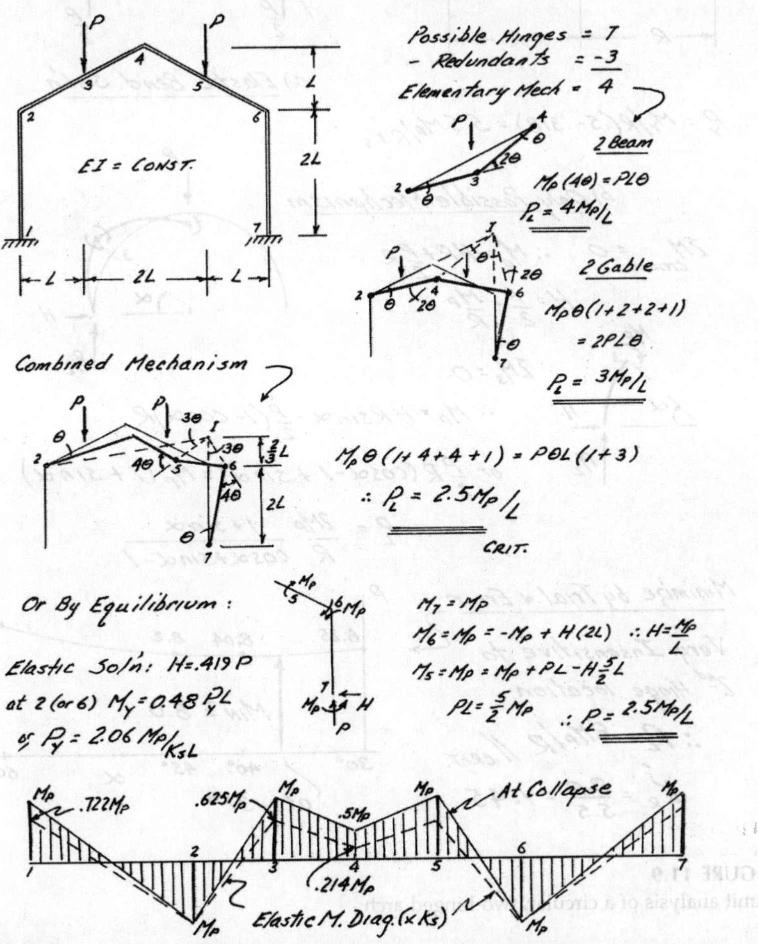

FIGURE 11.8
Limit analysis of a gable frame.

Two further examples of simple limit analysis, a gable frame and a circular arch each with constant cross-section, are shown in Figures 11.8 and 11.9. In both cases the analysis is made easier by assuming concentrated loads. The elastic bending solution is shown for the gable frame to compare to the collapse bending moment diagram. The redistribution factor is simply the ratio of the areas of the two.

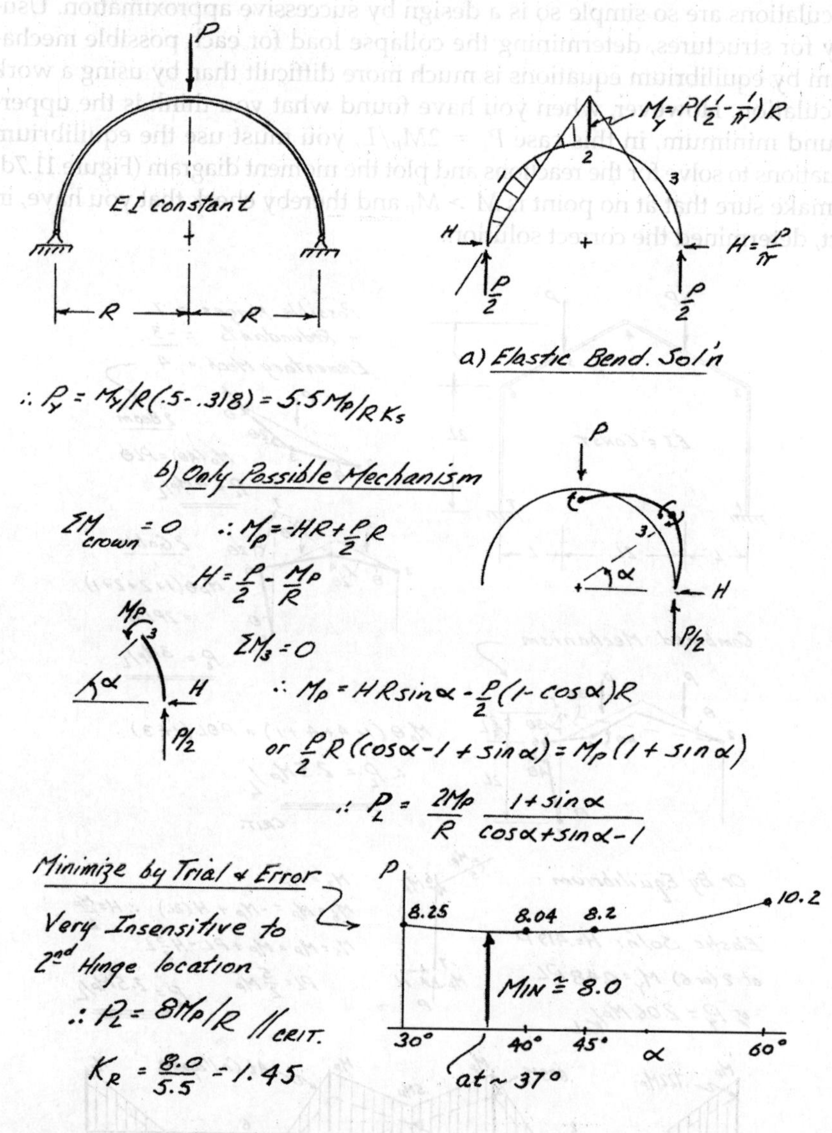

FIGURE 11.9
Limit analysis of a circular, two-hinged arch.

Even with a concentrated load, one possible hinge location is uncertain for the arch and therefore must be determined by application of the upper-bound theorem. A formal equation could be derived, but minimization by trial and error is perfectly adequate. The analysis at these two structures with uniform loads are called for in the chapter problems where determining the critical hinge locations is more difficult. For more complicated structural geometry or loading, graphical analysis for the kinematics of collapse is perfectly adequate and is commonly used in practice. Since real arches are designed to transmit loads primarily by axial force, the bending solution for the collapse load may need to be reduced to include the interaction effect discussed in Section 11.2.

11.5 Limit Analysis of Plates

The elastic analysis of plates with a distributed lateral load "q," while straightforward in concept, can be extremely complicated because of difficulties introduced by the boundary conditions. The fundamental governing equation in terms of the lateral deflection, w,

$$\frac{\partial^4 w}{dx^4} + 2\frac{\partial^4 w}{\partial^2 x \partial^2 y} + \frac{\partial^4 w}{\partial y^4} = \frac{q}{D}$$

is essentially a "two-way" beam equation where the flexural rigidity of the plate

$$D = \frac{Eh^2}{12(1 - \nu^2)} \tag{5.28}$$

replaces the simple beam rigidity EI. As with a beam, slopes are proportional to the first derivatives, moments to the second derivatives, and shears to the third derivatives of the deflection. Thus the boundary conditions of a free edge, simply-supported edge, and fixed edge can be expressed. In theory, the elastic problem can be solved directly for circular plates, by using Fourier series for rectangular shapes, and by numerical methods for irregular boundaries. Elastic solutions in the form of extensive tables of coefficients are given in the literature for rectangular and circular plates.*

* A complete development of the theory with solution by various analytic methods is given in *Theory of Plates and Shells* by Timoshensko and Woinowsky-Krieger, McGraw Hill, New York, 1959. Since 1960, finite flexural plate elements have developed with ever-increasing sophistication and power to the point where Fourier series or other such solutions are seldom used anymore. Since the details of specifying the element mesh, the boundary condition, geometry, and variable material properties can now be automated on the computer, the finite element strategy has become extremely versatile and now dominates in the elastic analysis of plates and shells.

In contrast, the limit analysis of plates using the upper-bound theorem, is straightforward and much easier than elastic analysis. Simply assume a collapse mechanism and calculate the corresponding collapse load by the energy balance or by equilibrium. While an elastic analysis may require hours of intensive calculation, a good estimate of the limit load can be determined in a few minutes. Equally important, a good physical feel for the structural behavior of the plate is obtained in the process.

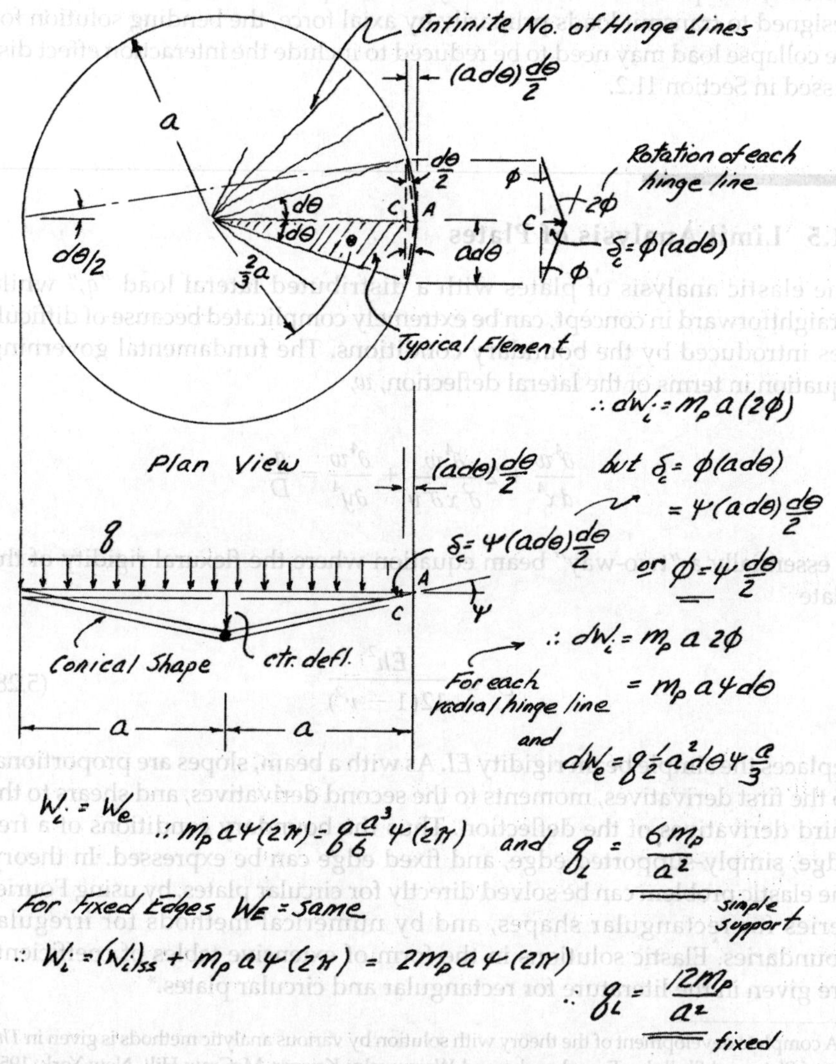

FIGURE 11.10
Limit analysis of a circular plate.

Take, for example, the simple case shown in Figure 11.10 of a circular plate of radius, a, and thickness, h, with a uniform vertical load, q, over the entire surface. If the edge is simply supported (free to rotate) the maximum elastic moment at the center is $m_r = m_t = (3+\nu)qa^2/16$. If the edge is built-in (clamped), the moment at the center is reduced to $(1+\nu)qa^2/16$ while at the edge the radial moment is $m_r = -qa^2/8$ and there is a tangential or twisting moment $m_t = -\nu qa^2/8$. The corresponding maximum stresses are, for a simply supported plate:

$$(\sigma_r)_{\max} = \frac{My}{I} = \frac{6m_r}{h^2} = \frac{3(3+\nu)qa^2}{8h^2} \quad \text{at} \quad r = 0$$

and for the edge built-in

$$(\sigma_r)_{\max} = \frac{6m_r}{h^2} = -\frac{3}{4}\frac{qa^2}{h^2} \quad \text{at} \quad r = a$$

Thus, to cause yield,

$$(q_Y)_{s.s.} = \frac{Y8h^2}{3(3+\nu)a^2} \quad \text{and} \quad (q_Y)_{\text{fixed}} = \frac{4}{3}\frac{Yh^2}{a^2}$$

If we assume $\nu = \frac{1}{3}$ then for the simply supported circular plate

$$(q_Y)_{ss} = \frac{4}{5}\frac{Yh^2}{a^2} = 3.2\frac{m_P}{a^2} \quad \text{at} \quad r = 0$$

or for a fixed boundary

$$(q_Y)_{\text{fixed}} = \frac{5.33m_P}{a^2} \quad \text{at} \quad r = a$$

The plastic analysis follows the procedure developed for linear structures, except that instead of elastic hinge points with a moment capacity $M_p = K_s M_y$, "hinge" lines develop with a moment capacity per unit length.

$$m_p = K_s m_Y = 1.5\frac{Yh^2}{6} = \frac{Yh^2}{4}$$

Because of the symmetry, the only possible failure mechanism is that, at collapse, the surface is pushed downward into a cone as shown in Figure 11.10. The radial hinge lines, along which $m_t = m_p$, rotate 2ϕ doing internal work while q over the area of a typical differential slice does external work moving downward $4a/3$. For simply supported edges, the energy balance gives $q_L = 6m_P/a^2$ and a reserve strength from first yield to collapse of $K_P = K_s K_R = 6/3.2 = 1.875$. For the case where the edge is fixed, the edge itself of length $2\pi a$ becomes a negative hinge line adding $\Delta W_i = 2\pi a m_p \psi$ to the total internal work. Thus $q_L = 12m_P/a^2$ and the reserve strength $q_L/q_Y = K_P = 12/5.33 = 2.25$. Thus, not only do plates take greater loads in the elastic range by beam action in two directions rather than one, but they have greater plastic load-carrying capacity than beams for the same reason.

Yield-line limit analysis of plates is the flexural counterpart to yield line or slip-surface analysis for two-dimensional planar structures. Like the punch problem, the determination of the critical yield line can be accomplished by trial and error or by a formal minimization procedure for a particular type of failure mechanism. Only in special cases like circular plates, is the determination of the exact pattern of yield surfaces straightforward. In general, however, for plates of uniform thickness:

i. Hinge lines are straight with a constant $m_P = Yh^2/4$,
ii. Axes of rotation are along supports,
iii. Axes of rotation are over columns , and
iv. Hinge lines pass through axes of rotation of adjacent slab segments.

These rules of thumb, which help guess the critical hinge line patterns, are illustrated with a few examples in Figure 11.11.

As an example of refining the pattern of hinge lines to converge by the upper-bound theorem on the critical limit load, consider a square plate with either pinned or fixed supports shown in Figure 11.12. The elastic solution gives:

$$(q_Y)_{ss} = \frac{13.9m_P}{L^2} \quad \text{and} \quad (q_Y)_{\text{fixed}} = \frac{12.9m_P}{L^2}$$

The simplest assumption for the failure mechanism is diagonal hinges from the corners shown as pattern #1. The corresponding collapse loads for the simply supported and fixed boundary cases are calculated by equilibrium to illustrate that this approach can be used rather than enforcing the energy balance.

However, the corners of a rectangular plate actually would like to lift off the edge support if not restrained. Thus the moments are negative near the corners, which act as cantilevers. To capture this, a second hinge line pattern (#2 in

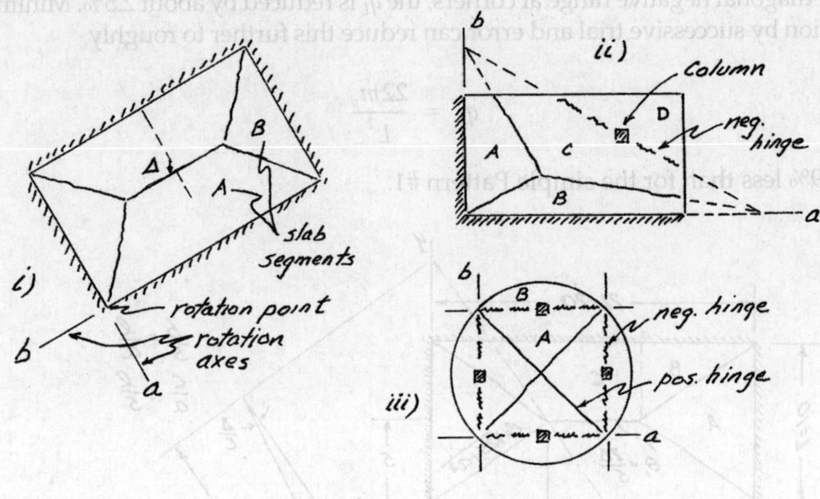

FIGURE 11.11
Hinge line patterns.

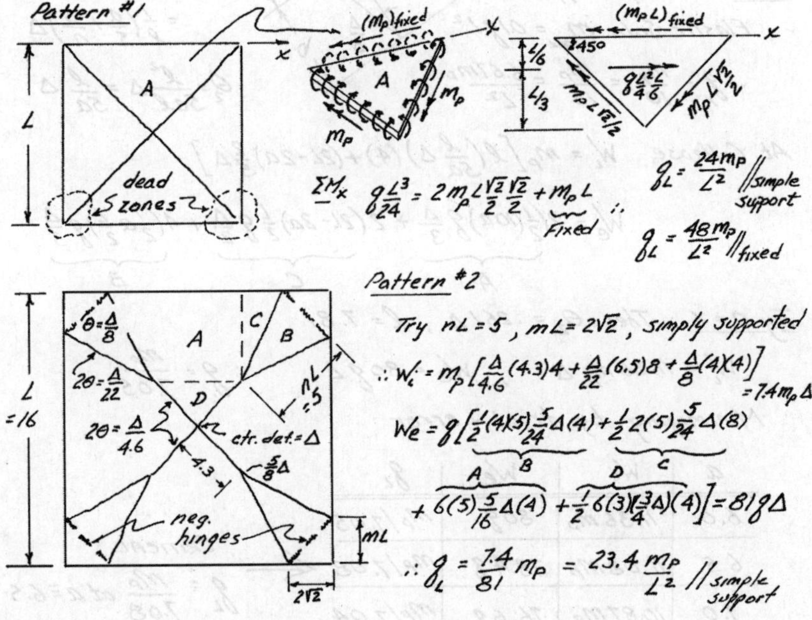

FIGURE 11.12
Limit analysis of a square plate.

Figure 11.12) with "corner cantilevers" or "corner levers" should be investigated. For a particular case we see, for the simply supported plate, that by including the diagonal negative hinge at corners, the q_L is reduced by about 2.5%. Minimization by successive trial and error can reduce this further to roughly

$$q_L = \frac{22m_p}{L^2}$$

or 9% less than for the simple Pattern #1.

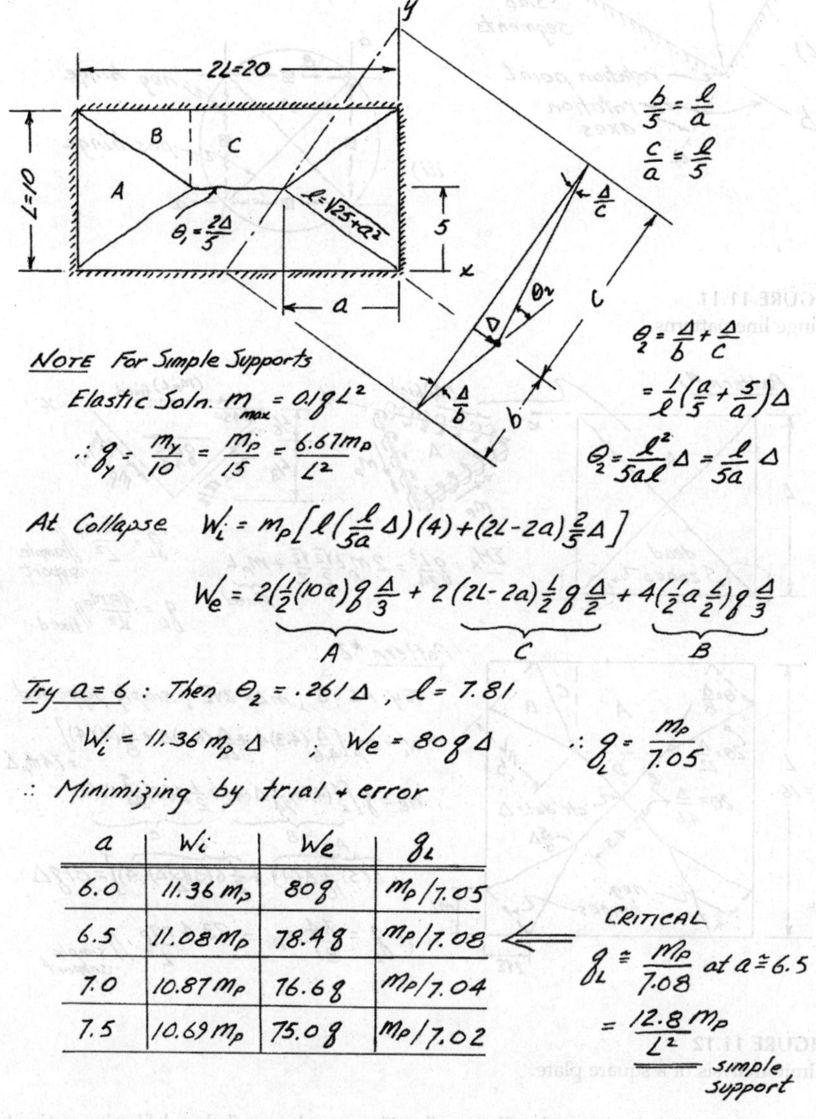

FIGURE 11.13
Limit analysis of a rectangular plate.

As a final example, consider a 2:1 rectangular plate shown in Figure 11.11. Assuming the simple mechanism without including the corner cantilevers, the limit load works out to be $q_L = m_P/7.08 = .1412m_P$. Including the corner cantilevers might reduce this 10% to $q_L = .127m_P$. From the elastic solution

$$q_Y = \frac{m_P}{15} = .0677m_P \quad \text{so} \quad K_P \cong 1.9$$

We could now consider what happens if the edges are fixed or if one edge were free, but these and other examples for rectangular plates are left as chapter problems.

11.6 Plastic Torsion

The elastic solution for pure torsion (Chapter 9) predicts that the maximum shear stress, τ_{zt}, is always a maximum on the boundary and is equal to the slope of the pressurized elastic membrane (or ϕ hill) used in Prandtl's Membrane Analogy presented in Section 9.4. Using the Prandtl stress function, the shear stresses are

$$\tau_{zx} = -\frac{\partial \phi}{\partial y}; \qquad \tau_{zy} = \frac{\partial \phi}{\partial x} \tag{9.20}$$

and the equilibrium Equation (9.10) is automatically solved. Now, following the strategy of plasticity, we use the yield condition rather than the compatibility equation. Writing the Tresca Criterion

$$\tau_{\max} = \tau_Y = (\tau_{zx}^2 + \tau_{zy}^2)^{\frac{1}{2}} = \frac{Y}{2}$$

in terms of ϕ gives:

$$\left(\frac{d\phi}{dy}\right)^2 + \left(\frac{d\phi}{dx}\right)^2 = \tau_Y^2 \tag{11.8}$$

Since our boundary condition, $\phi = 0$, was also based only on equilibrium, it remains applicable in the plastic range. Therefore, the theory of plastic torsion, formulated in terms of Prandtl's stress function with no reference to strains or displacements, is complete.

Equation (11.8) tells us that the square of the gradient of ϕ, which is the square of the maximum slope of the "ϕ hill" at any point, is equal to the square of the yield stress. Therefore the "ϕ Hill" in regions of the cross-section that have gone plastic have a constant maximum slope equal to τ_Y.

11.6.1 Sand-Hill and Roof Analogies

The solution, that in plastic zones the "ϕ Hill" has a constant maximum slope, suggested to Nadai* in 1923 the so-called sand-heap analogy for full plastification. A slope of dry sand has a natural angle of repose slightly greater than the friction angle which, if exceeded, will cause the particles to roll down the hill. Thus if we take a flat horizontal plate of the same shape as the cross-section under torsion and pour sand on it until it will take no more, the resulting surface satisfies the conditions on the plastic "ϕ Hill" that its maximum slope be constant everywhere and its height be zero at the external boundary. Moreover, the expression for the total torque

$$M_t = 2\int_{\text{area}} \phi dA$$

remains valid since it again is based only on equilibrium. Thus the torque applied to the cross-section is twice the volume of sand on the plate (easily determined in the laboratory by weighing it).

We can also think of the sand-hill analogy as a roof of straight sections of constant slope, τ_Y, in the shape of the sand-hill surface erected over the cross-section. This allows us to better visualize stages of partial plastification by thinking of the pressurized membrane inside representing the elastic "ϕ Hill" being pressed flat against the limiting roof in the plastic zones and still curving beneath the roof where the cross-section remains elastic. This is illustrated in Figure 11.14a for a circular and a square cross-section. At initial yield the membrane first touches the roof at some point or points at the boundary. With further increase in air pressure, which is directly proportional to the rate of twist and therefore the torque, the membrane gradually touches more and more of the roof until, at full plastification, there is no space between them.

These analogies give us both a heuristic way to visualize solutions and, for fairly simple shapes, a device to actually calculate exact values for the fully plastic torque $(M_t)_P$. Since τ_Y flows along contour lines, it abruptly changes direction at "ridge lines" which, since the slope is constant, must bisect corner angles. Also, since the slope is constant, the contours are equidistant.

---

* A lifetime worth of work on plasticity is presented in the classic text *Theory of Flow and Fracture of Solids* by A. Nadai, McGraw Hill, New York, 1950. Chapter 35 presents the torsion solution and the Sand-Hill Analogy in detail.

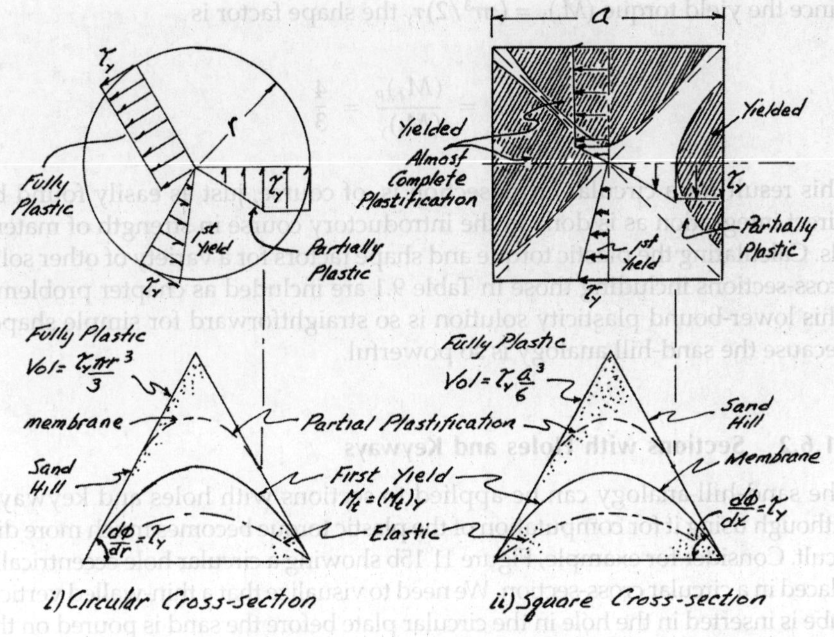

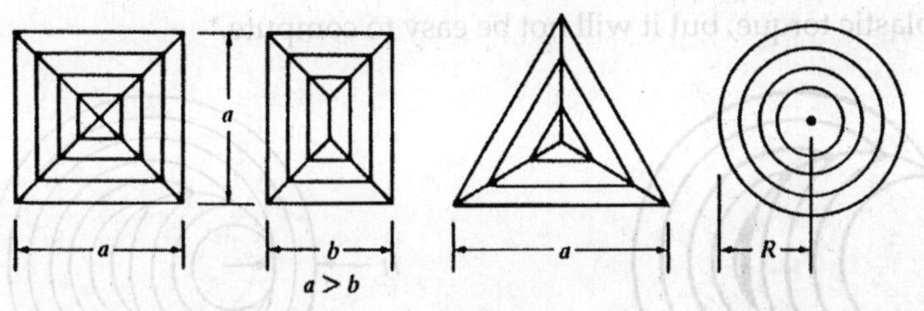

FIGURE 11.14
Sand-hill analogy for fully plastic torsion.

For example, contour lines for some simple shapes are shown in Figure 11.14b. The sand hill for a circular cross-section is a cone of height $\tau_Y R/2$. Thus the ultimate torque capacity is:

$$(M_t)_P = 2\int_{\text{area}} \phi dA = \frac{2}{3}\pi R^3 \tau_Y$$

Since the yield torque $(M_t)_Y = (\pi r^3/2)\tau_Y$ the shape factor is

$$K_s = \frac{(M_t)_P}{(M_t)_Y} = \frac{4}{3}$$

This result for a circular cross-section is, of course, just as easily found by direct integration as is done in the introductory course in strength of materials. Calculating the plastic torque and shape factors for a variety of other solid cross-sections including those in Table 9.1 are included as chapter problems. This lower-bound plasticity solution is so straightforward for simple shapes because the sand-hill analogy is so powerful.

11.6.2 Sections with Holes and Keyways

The sand-hill analogy can be applied to sections with holes and keyways, although using it for computation of the plastic torque becomes much more difficult. Consider for example, Figure 11.15b showing a circular hole eccentrically placed in a circular cross-section. We need to visualize that a thin-walled vertical tube is inserted in the hole in the circular plate before the sand is poured on the plate. We then lower the tube allowing the sand to fall down it until its surface area is covered. In Figure 11.15b the top of the tube would be $\tau_Y h$ above the plate. Therefore the boundary conditions, $\phi = 0$ at the edge of the plate and $\phi = \tau_Y h$ (constant) on the edge of the hole, are satisfied. Twice the volume of the sand hill is the plastic torque, but it will not be easy to compute.*

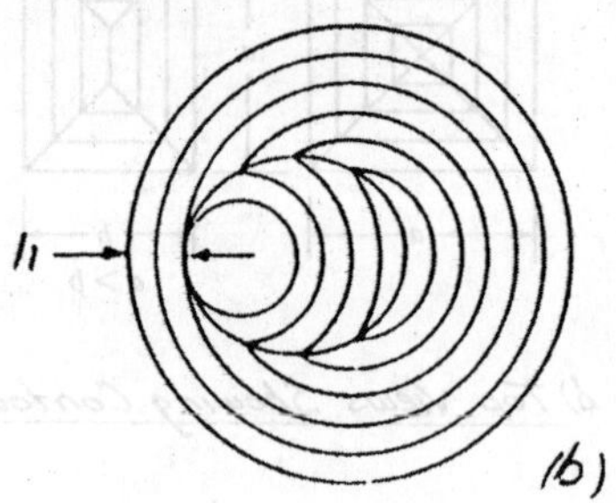

FIGURE 11.15
ϕ Contours for nonsymmetric sections.

In cases without symmetry such as shown in Figure 11.15, it is easier to draw the contours and compute the plastic torque using the "onion" analogy or "nesting tube" lower-bound approximation presented by Calladine.

* For very complicated cross-sections with cutouts and holes of irregular shape, actually doing the experiment in the laboratory may be easier and give more accurate results than some numerical scheme to evaluate the volume integral. Weighing the sand is so simple.

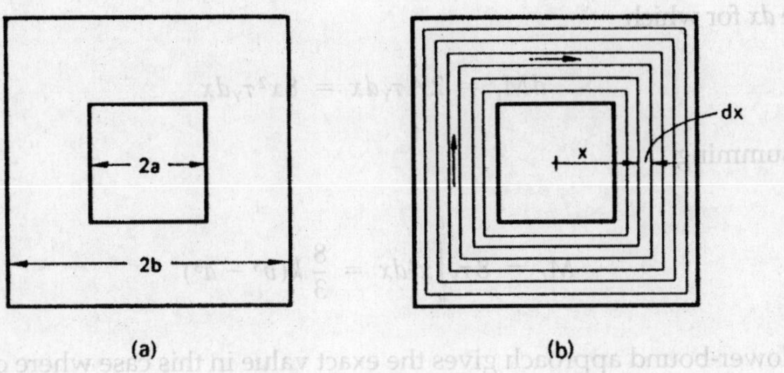

FIGURE 11.16
Subdivision of a square cross-section into nesting thin-walled tubes. (Calladine)

We have seen in Section 9.5 (Figure 9.12) from equilibrium considerations only that for thin-walled tubes of arbitrary shape

$$\tau = \frac{M_t}{2At} \tag{9.38}$$

where A is not the area of the cross-section but the area enclosed by the average perimeter. If we make t a constant, then at yield

$$M_t = 2At\tau_Y \tag{11.9}$$

Thus we can obtain a lower-bound solution for the plastic torque by dividing the cross-section into a set of nesting thin-walled tubes where each tube has the same constant thickness and encloses the largest possible area.

The procedure is, starting at the outer edge, to cut or strip off complete thin tubes each of the same constant thickness (so $\tau = \tau_Y$) until the cross-section is used up. We then sum the torques from each nested tube from Equation (11.9) to give the total plastic torque, which will be a lower bound. Calladine* uses the term "onion" analogy and it is clear that, in the limit, it is the same as the sand-hill analogy. However, it is much clearer conceptually, particularly for complicated shapes, and much easier to draw and compute.

Take, for example, the hollow square tube in Figure 11.16. This strategy of graphical analysis gives a set of nested square tubes. Because of the regular geometry we can reduce the thickness of each tube, $2x$ on a side, to a differential

* *Plasticity for Engineers* by C.R. Calladine, John Wiley & Sons (Halsed Press), New York, Chapter 6, 1985. This is possibly the best modern text on simple plasticity. It is beautifully written and follows a long British tradition of achieving clarity by precise, graphic prose unencumbered by unnecessary mathematics.

value dx for which

$$dM_t = 2A\tau_Y dx = 8x^2\tau_Y dx \tag{a}$$

and summing

$$M_t = 8\tau_Y \int_a^b x^2 dx = \frac{8}{3}k(b^3 - a^3) \tag{11.10}$$

This lower-bound approach gives the exact value in this case where calculus can be used.

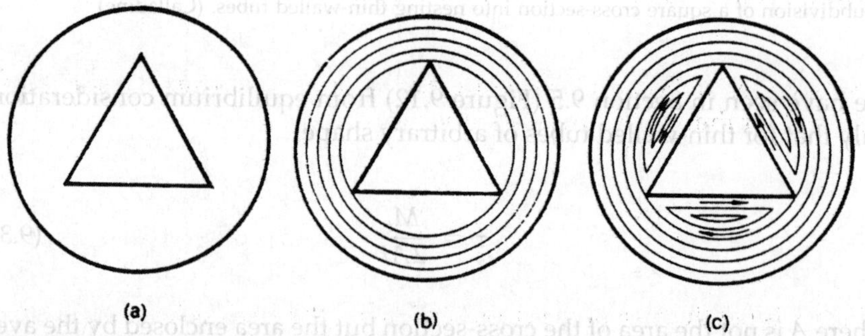

FIGURE 11.17
Nesting thin-walled tubes for a circular section with a triangular hole. (Calladine)

For cross-sections with holes, the calculation is more difficult but manageable, and the conceptual visualization is clear. Take for example the circular cross-section with a triangular hole shown in Figure 11.17. Starting from the outside we get four complete nested tubes, but then we are left with three irregular zones. Our striping procedure must produce complete tubes so we continue on as shown. With a ruler, protractor, and some poetic license we can determine the area circumscribed by each of the three onion slices in each of these three areas and their contribution to the total plastic torque. They turn out to be small and neglecting them and using only the four circular tubes gives a satisfactory lower-bound value

$$(M_t)_P = \frac{2\pi}{3}\tau_Y(b^3 - a^3)$$

It is clear that this "stripping" procedure is perfectly general and that by using it we can develop a satisfactory lower-bound estimate for the plastic torque for any cross-section, no matter how irregular. Starting from the outside, rings of uniform cross-section are "removed" treating separately any areas that may become separated in the process. Areas can be evaluated by graphical approximations with or without the aid of a planimeter. Figure 11.18 shows

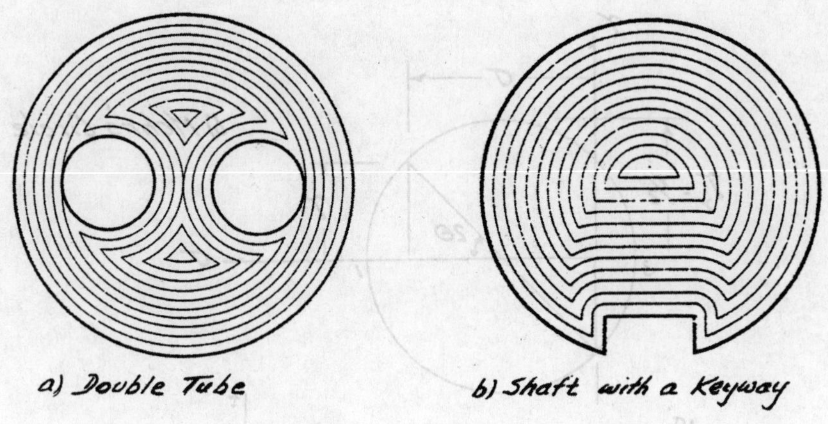

FIGURE 11.18
Nested tubes for irregular sections. (Calladine)

two more examples that emphasize the relative simplicity of the nested-tube approach compared to the sand-hill analogy in visualizing the contours of the "ϕ hill" for complicated sections.

11.7 Combined Torsion with Tension and/or Bending

Interaction diagrams for a safe estimate of the effect of combined loadings of prismatic members can be determined from our lower-bound approach to plastic bending and torsion. Consider first the case of torsion and axial force acting simultaneously on a cross-section. Reviewing our yield condition for combined tension and shear shown in Figure 11.19b, for the Tresca criterion:

$$\sin 2\theta = \frac{\tau}{\tau_Y}, \quad \cos 2\theta = \frac{\sigma/2}{\tau_Y} \quad \text{and} \quad \therefore \left(\frac{\tau}{\tau_Y}\right)^2 + \left(\frac{\sigma/2}{\tau_Y}\right)^2 = 1$$

Thus our yield criterion is:

$$\left(\frac{\tau}{\tau_Y}\right)^2 + \left(\frac{\sigma}{Y}\right)^2 = 1 \tag{11.11}$$

If we say that the stresses are some fraction of the yield $\tau = \lambda\tau_Y$ and $\sigma = \mu Y$ then by Equation (11.11)

$$\lambda^2 + \mu^2 = 1$$

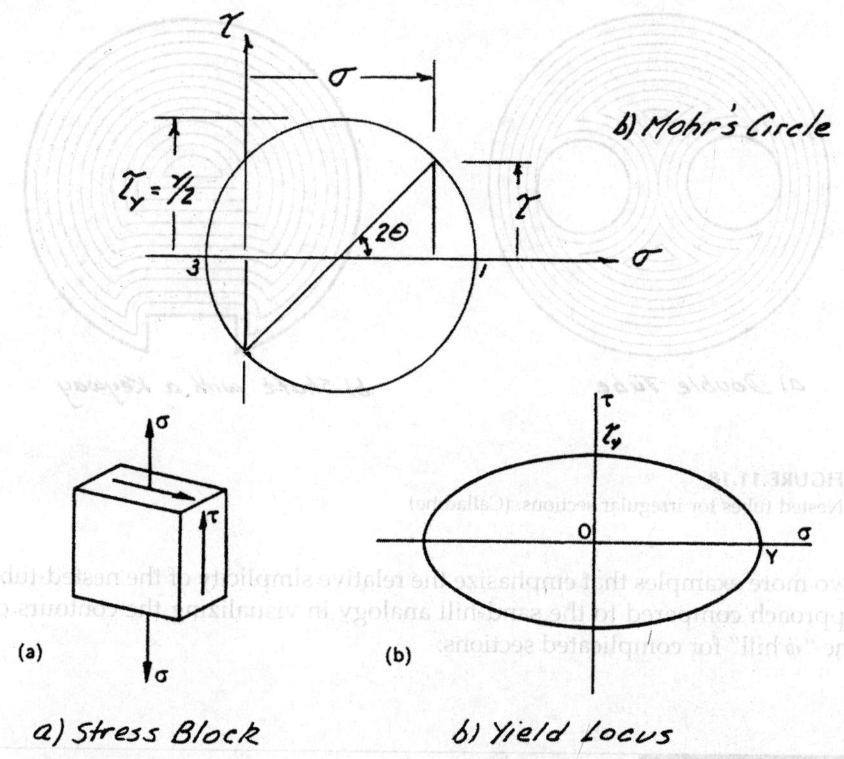

FIGURE 11.19
Combined tension and shear.

But the stresses are proportional to the torsion and the axial force. That is, with some combined loading, $\lambda\tau_Y$ is in equilibrium with $\lambda\,(M_t)_P$ and μY is in equilibrium with μN_Y so that

$$\left(\frac{M_t}{(M_t)_P}\right)^2 + \left(\frac{N}{N_Y}\right)^2 = 1 \tag{11.12}$$

represents the safe combinations. This plots as an ellipse as shown in Figure 11.20a.

It is easy to show with the same arguments, that if $M_p = K_s\,M_y$ is the fully plastic bending moment, then:

$$\left(\frac{M_t}{(M_t)_P}\right)^2 + \left(\frac{M}{M_P}\right)^2 = 1.0 \tag{11.13}$$

represents safe combinations of torque and bending (Figure 11.20b).

However, as discussed in Section 11.2, such a simple interaction equation does not represent safe combinations of bending moment and axial force. This is because both produce normal stresses that combine in a different way

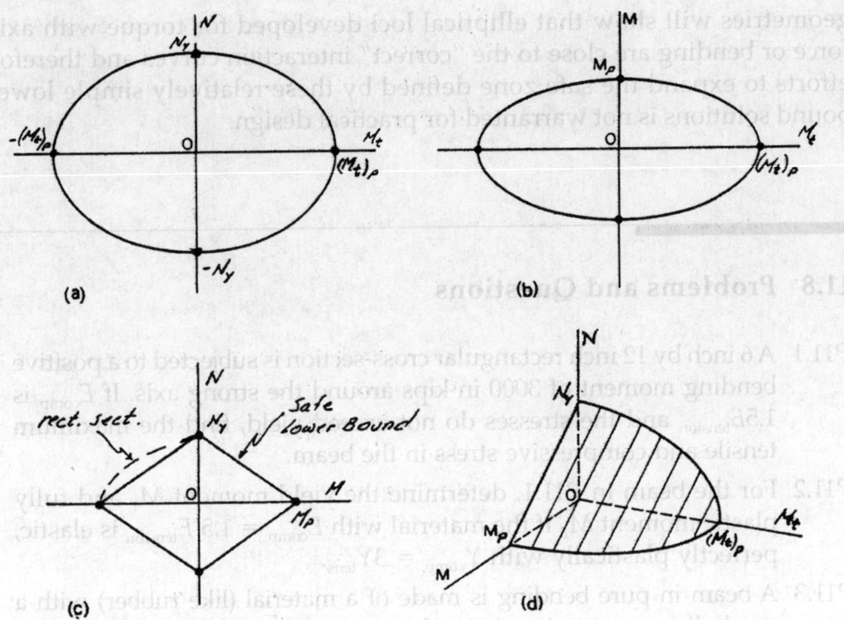

FIGURE 11.20
Safe combinations of tension (N), torsion (M_t), and bending (M).

to progressively yield the cross-section. A safe assumption is a straight line connecting $N_P = N_Y = YA$ and $M_p = K_s M_Y$

$$\frac{N}{N_P} + \frac{M}{M_P} = 1.0 \tag{11.14}$$

This would be nearly true for an I beam but, as shown in Figure 11.4 and replotted in Figure 11.20c, the safe region can be extended considerably for a rectangular beam.

Nevertheless, using the safe lower-bound straight line relationship [Equation (11.14)], an excellent general safe load locus is:

$$\left(\frac{N}{N_P} \pm \frac{M}{M_P}\right)^2 + \left(\frac{M_t}{(M_t)_P}\right)^2 = 1.0 \tag{11.15}$$

The positive octant of these two intersecting surfaces is shown in Figure 11.20d derived from the three, two-dimensional loci a, b, and c.

It should be noted that in this discussion of combined loading we have limited ourselves to so-called "compact" cross-sections where buckling instability leading to premature collapse is not a concern. Nor have we considered other than the perfect EPS material. Within those limitations however, we have been able to define safe combinations of torque, axial force, and bending moment independent of the dimensions other than that $(M_t)_P$, N_Y, and M_P depend on the shape of the cross-section. A review of detailed solutions for specific geometries will show that elliptical loci developed for torque with axial force or bending are close to the "correct" interaction curves and therefore efforts to expand the safe zone defined by these relatively simple lower-bound solutions is not warranted for practical design.

11.8 Problems and Questions

P11.1 A 6 inch by 12 inch rectangular cross-section is subjected to a positive bending moment of 3000 in-kips around the strong axis. If E_{comp} is $1.5E_{tension}$ and the stresses do not exceed yield, find the maximum tensile and compressive stress in the beam.

P11.2 For the beam in P11.1, determine the yield moment M_Y and fully plastic moment M_P if the material with $E_{comp} = 1.5\,E_{tension}$ is elastic, perfectly plastically with $Y_{comp} = 3Y_{tens}$.

P11.3 A beam in pure bending is made of a material (like rubber) with a parabolic stress–strain relationship $\sigma = 10^4\sqrt{\varepsilon}$. Plot the $M - \phi$ diagram for cross-section (a) and for cross-section (b) in Figure P11.3 to a maximum strain of $\pm 9\%$. ϕ is the relative rotation of the cross-section *per unit length* or $\phi = \frac{-\varepsilon_x}{y}(1.0) = \Delta\omega_z$ for small angles.

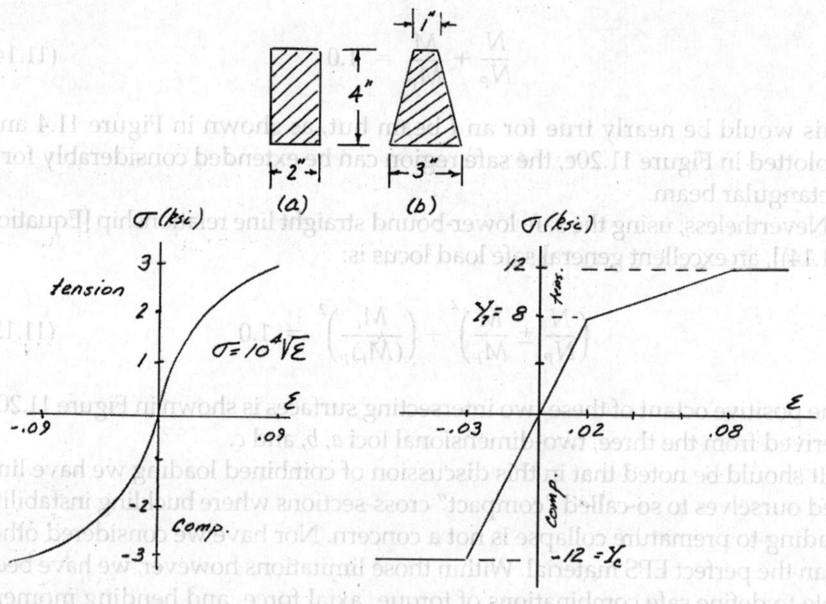

FIGURE P11.3 FIGURE P11.4

P11.4 Plot the $M - \phi$ diagram for cross-sections (a) and (b) as in P11.3, but for the elastic-plastic material shown in Figure P11.4.

P11.5 Find the shape factor $K_s = M_P/M_Y$ for EPS beams having the cross-sections given in Figure P11.5.

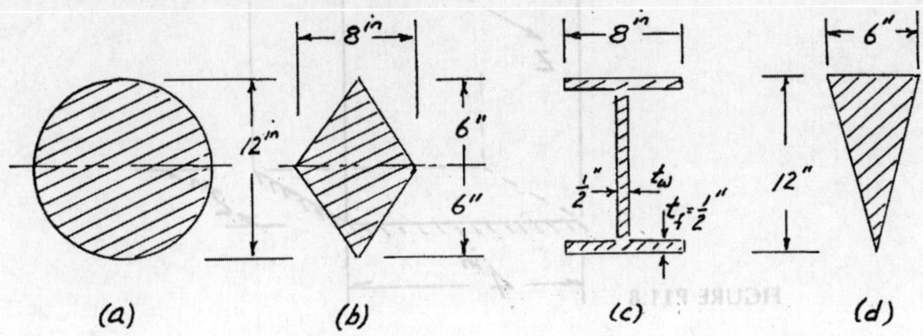

FIGURE P11.5

P11.6a Plot a $M - \phi$ diagram similar to that in Figure 11.1 for the EPS *T*-Beam of Figure 11.2.

P11.6b Revise the analysis for the plastic hinge profile (i.e., the elastic-plastic interface) in Figure 11.3 for a uniform load q lb/ft on the cantilever beam (rather than a concentrated load at the end) and compare.

P11.6c Consider the short ($L = 3h$) beam of Figure 11.3 and estimate the reduction in ultimate moment capacity M_P due to the flexural shear stresses $\tau_{xy} = (VQ)/(\text{Ib})$. Can you present your result in a general way that applies to any rectangular EPS beam?

P1.7 Now consider the cantilever beam with a uniform loading q as shown in Figure 6.1 and, as in P11.6c, estimate the reduction in ultimate moment capacity, M_P, due to *both* the flexural shear stress τ_{xy} and the vertical normal stress σ_y. Again present your result in a general way if possible. For your analysis, assume the 1–3 orientation is in the *x-y* plane, but point out where this might not be true (i.e., where $\sigma_z = 0 = \sigma_3$) and estimate the effect. Compare the resulting elastic-plastic interface profile to that for the simple strength-of-material result called for in P11.6.

P11.8 A short steel column of rectangular cross-section is loaded at its edge as shown in Figure P11.8. Assume $Y = 40{,}000$ psi and determine P_Y and P_L to check the values given by the interaction curves in Figure 11.4d.

P11.9 An EPS beam of constant ultimate moment capacity, M_P, has a concentrated load, P, at its quarter point and a uniform load $w = 3P/L$ as shown in Figure P11.9. Determine the critical collapse mechanism and corresponding limit load P_L. Do this by the graphical trial-and-error approach and by a formal minimization procedure.

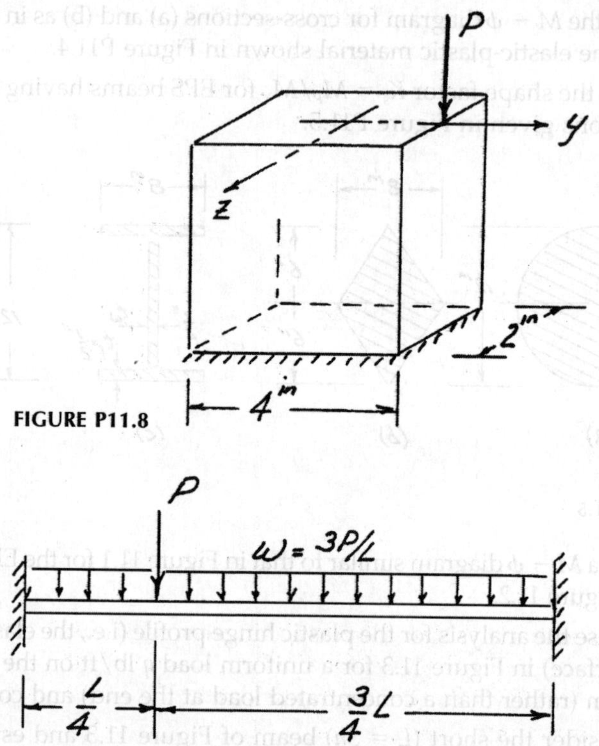

FIGURE P11.8

FIGURE P11.9

P11.10 The 12-foot long beam shown (Figure P11.10) built in at *A* is constructed by welding together two 6-foot long square bars of the same size (4 in. × 4 in.). Bar (1) is steel ($Y = 60$ ksi) and bar (2) is aluminum ($Y = 30$ ksi). Estimate the limit load P_L to be applied at the quarter point if:

a. The beam is simply supported at B.
b. The beam is built in at B.

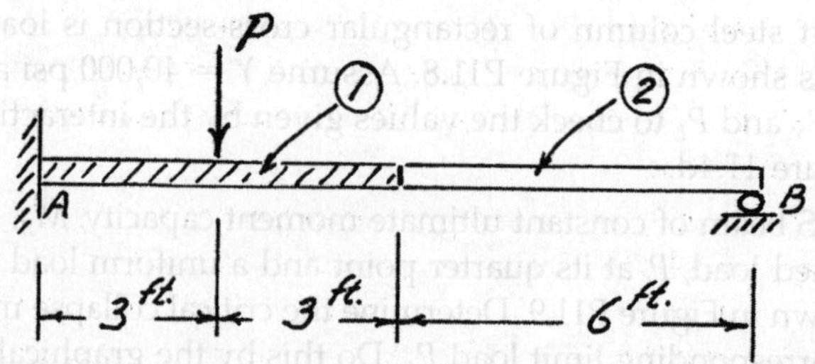

FIGURE P11.10

P11.11 For an optimum design, where should the transition point between the steel and aluminum be located to minimize the weight of the beam for P_L in each case in P11.10.

P11.12 An EPS beam of constant ultimate moment capacity, M_P, has a triangular load as shown in Figure P11.12. Determine the critical collapse mechanism and corresponding limit load w_L.

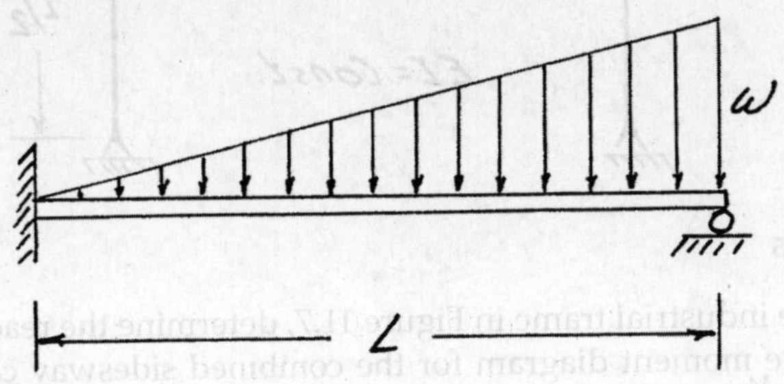

FIGURE P11.12

P11.13 For the beam of P11.12, determine the critical collapse mechanism and corresponding w_L if the right end is built in rather than simply supported.

P11.14 Determine the collapse load for the portal frame load as shown in Figure P11.14 if the moment capacity, M_P, of the beam BC is 1.2 times that of the columns AB and CD. Plot the limit moment diagram and compute the reactions at collapse. Check that at no point is $M > M_P$.

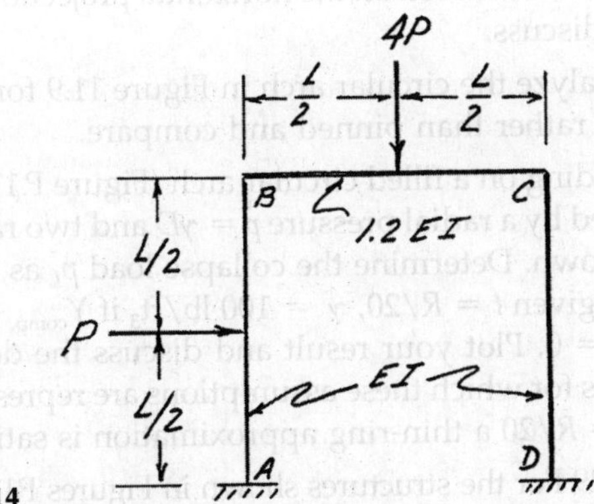

FIGURE P11.14

P11.15 Determine the collapse load for the frame loaded as shown in Figure P11.15 assuming EI is a constant (M_P the same for the columns and beam). Plot the limit moment diagram and reactions to check that at no point is $M > M_P$. Compare to an elastic computer solution to determine K_R.

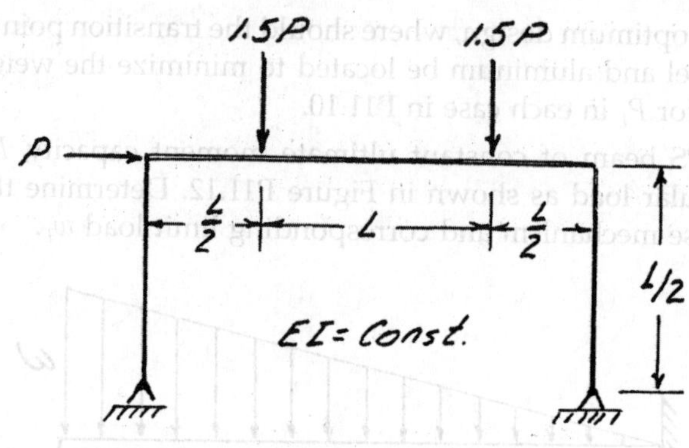

FIGURE P11.15

P11.16 For the industrial frame in Figure 11.7, determine the reactions and plot the moment diagram for the combined sidesway case where $P_L = \frac{8}{3}\frac{M_P}{L}$. Discuss your result in relation to the upper-bound and lower-bound theorems.

P11.17 Reanalyze the industrial frame in Figure 11.7 to determine P_L if

a. The inner columns are fixed at their base.

b. All four columns are fixed at their base.

P11.18 Reanalyze the gable frame in Figure 11.8 for the limit load, w_L, if, rather than concentrated loads, the frame is subjected to a uniform vertical load of w lb/ft (distributed on the horizontal projection of the roof beams). Compare to the solution in Figure 11.8 if $P = 2wL$. Discuss.

P11.19 Reanalyze the circular arch in Figure 11.9 for P_L if the concentrated load is distributed on the horizontal projection, $w = P/2R$. Compare and discuss.

P11.20 Reanalyze the circular arch in Figure 11.9 for P_L if the supports are fixed rather than pinned and compare.

P11.21 A loading on a filled circular arch (Figure P.11.21) might be approximated by a radial pressure $p = \gamma D$ and two radical loads $P - \gamma R^2/4$ as shown. Determine the collapse load p_L as a function of D and R for a given $t = R/20$, $\gamma = 100$ lb/ft$_3$ if $Y_{comp.} = f_c' = 200{,}000$ lb/ft.^{2}, $Y_{tens} = 0$. Plot your result and discuss the design of Roman stone arches for which these assumptions are representative values. Note: at $t = R/20$ a thin-ring approximation is satisfactory.

P11.22–P11.29 For the structures shown in Figures P11.22 through P11.29:

a. Determine the critical collapse load using the upper-bound theorem.

b. Plot the corresponding limit moment diagram to be sure that nowhere does the moment exceed the capacity.

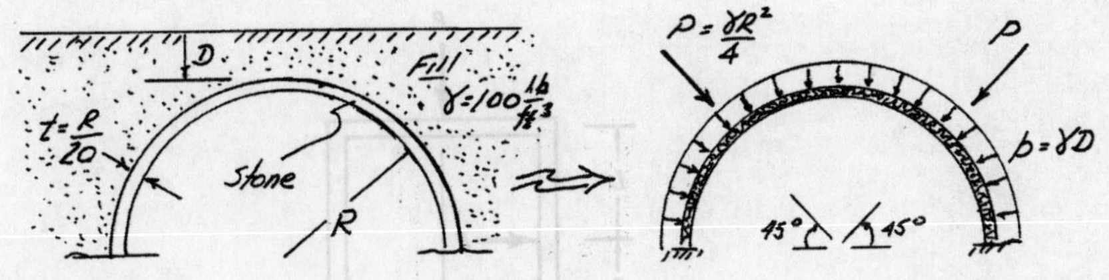

FIGURE P11.21

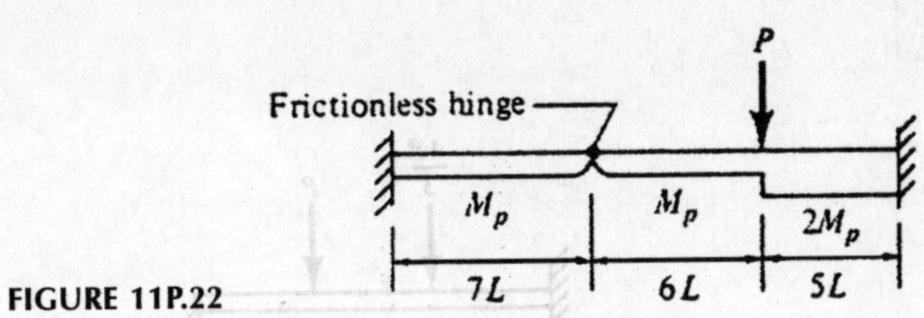

FIGURE 11P.22

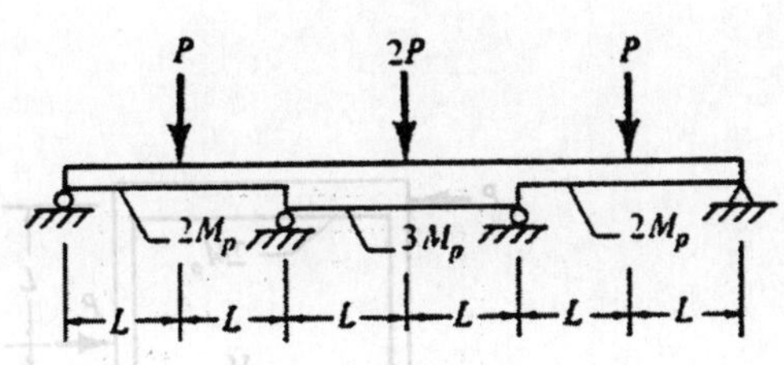

FIGURE P11.23

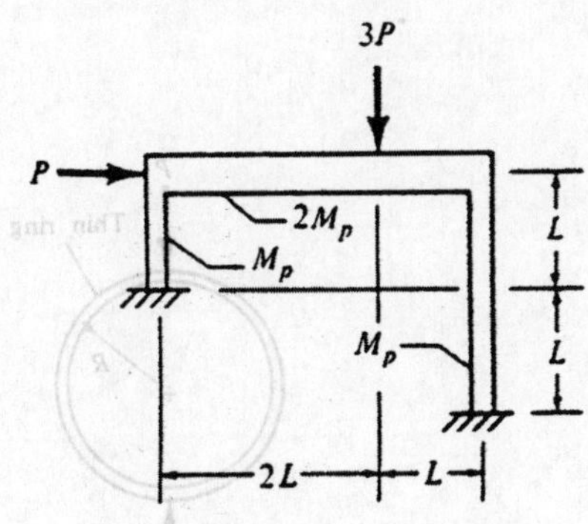

FIGURE P11.24

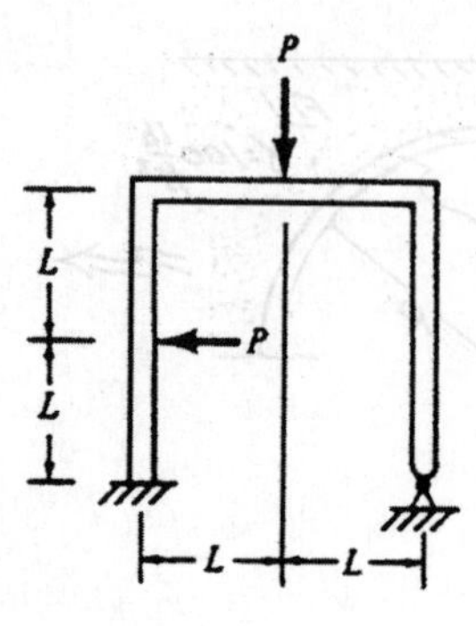

FIGURE P11.25

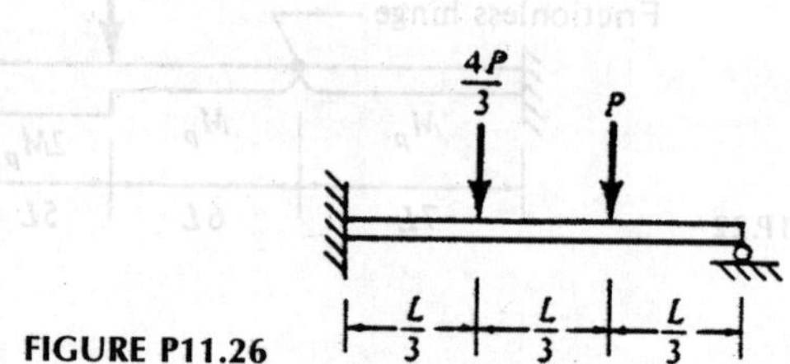

FIGURE P11.26

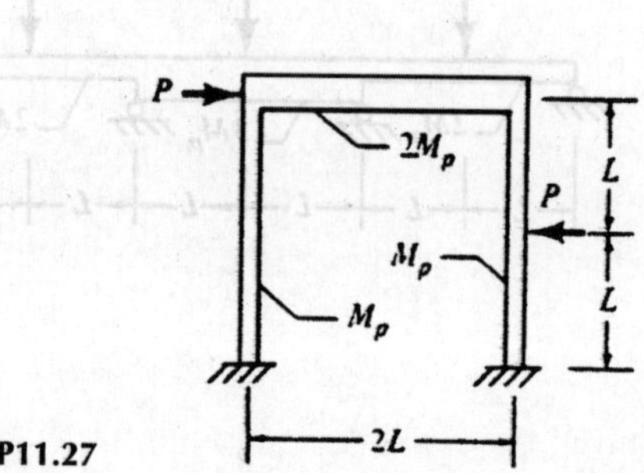

FIGURE P11.27

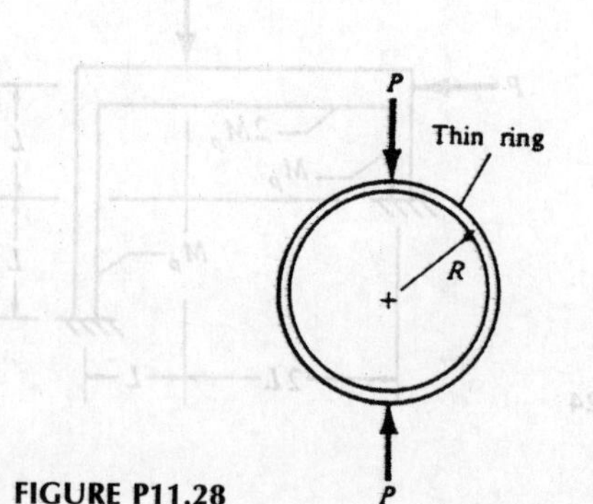

FIGURE P11.28

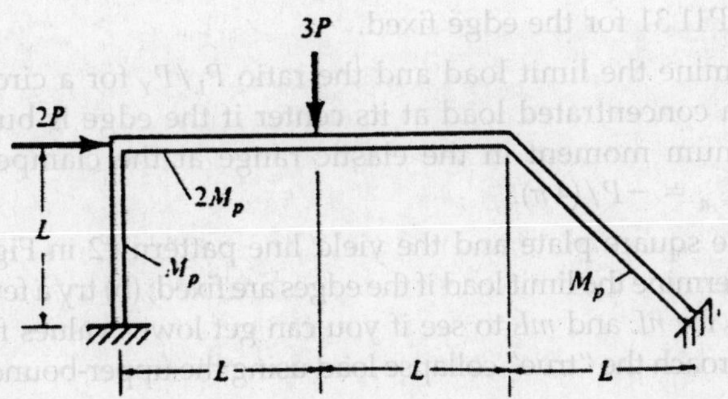

FIGURE P11.29

P11.30 For the circular plate in Figure 11.10 the elastic moments are for fixed edges:

$$m_r = \frac{q}{16}[a^2(1+v) - r^2(3+v)]; \qquad m_t = \frac{q}{16}[a^2(1+v) - r^2(1+3v)]$$

and for simply supported edges:

$$m_r = \frac{q}{16}(3+v)(a^2 - r^2); \qquad m_t = \frac{q}{16}[a^2(3+v) - r^2(1+3v)]$$

Compare the elastic to the limit moment for each case to determine $K_p = K_s K_R$.

P11.31 A circular plate with a uniform load q to a radius from the center, c, is shown in Figure P11.31. The elastic solution gives a maximum moment at the center:

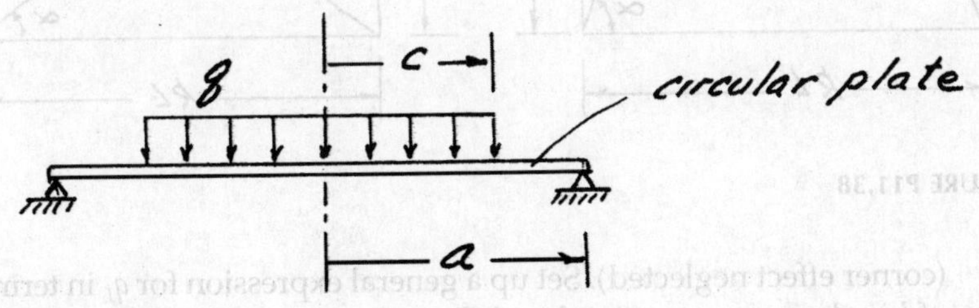

FIGURE P11.31

$$m_r = m_t = \frac{c^2 q}{4}\left[(1+v)\ln\frac{a}{c} + 1 - \frac{(1-v)c^2}{4a^2}\right]$$

Determine q_L and the reserve strength $q_L/q_Y = K_s K_R$. Plot q_L and q_L/q_Y as a function of c. Does the result make sense for small c?

P11.32 Redo P11.31 for the edge fixed.

P11.33 Determine the limit load and the ratio P_L/P_Y for a circular plate with a concentrated load at its center if the edge is built in. The maximum moment in the elastic range at the clamped edge is $(M_r)_{r=a} = -P/(4\pi)$.

P11.34 For the square plate and the yield line pattern #2 in Figure 11.12, (a) determine the limit load if the edges are fixed; (b) try a few different values for nL and mL to see if you can get lower values for q_L so as to approach the "true" collapse load using the upper-bound theorem.

P11.35 For the rectangular plate and yield-line pattern in Figure 11.11 determine the limit load if the edges are fixed. Is the critical value of the dimension, a, the same?

P11.36 Develop a general limit load solution for a rectangular plate simply supported as in Figure 11.11 but where the length is kL compared to the width L. Plot q_L as a function of k showing the limiting cases $k = 1.0$ and $k = \infty$ are the solutions for a square plate and a one-way slab. (Use the simple yield line pattern of Figure 11.11 where corner effects are neglected.)

P11.37 Repeat P11.36, but with the edges fixed.

P11.38 Consider the rectangular plate L by kL simply supported on three sides and the two simple yield line patterns shown in Figure P11.38

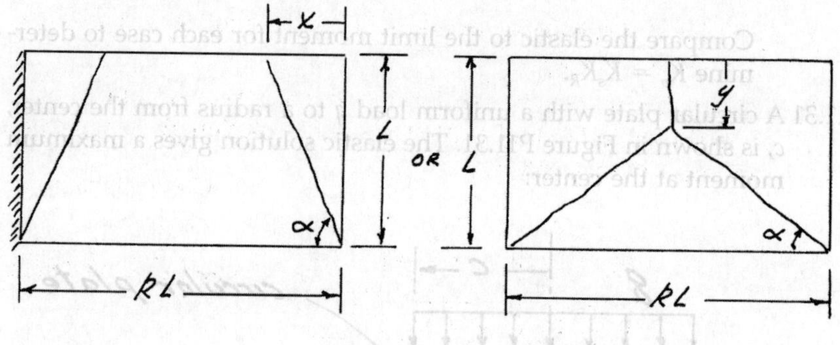

FIGURE P11.38

(corner effect neglected). Set up a general expression for q_L in terms of k and some parameter that defines the geometry of each pattern (an angle or distance). Minimize and plot q_L vs. k for each pattern on the same graph. Discuss from the standpoint of design.

P11.39 Repeat P11.38, but with the edges fixed.

P11.40 Determine the collapse load for the plate shown in Figure 11.11(ii) if the length is twice the width, L, and the column is one-quarter of the length from the free corner along the diagonal. Assume the two edges are simply supported.

P11.41 In P11.40 position the column along the diagonal to give the highest collapse load.

P11.42 Repeat P11.40, but with the two supporting edges fixed.

P11.43 Repeat P11.41, but for P11.42 (with the edges fixed).

P11.44 Make a limit analysis for the circular plate of radius R shown in Figure 11.11 (iii) and then determine the optimum placement of the four columns to maximize the collapse load q_L.

P11.45 Determine the fully plastic torque capacity and the shape factor $K_s = (M_t)_P/(M_t)_Y$ for each of the four cross-sections shown in Table 10.1. For the rectangle use $b = 2a$. Compare their efficiency.

P11.46 For the hollow square cross-section in Figure 11.16 show how the sand-hill analogy leads to the same result as obtained by the nested-tube strategy.

P11.47 Calculate by some graphical procedure, the amount the three irregular areas (Figure 11.17) contribute to the total plastic torque compared to the contribution of the outer tubes.

P11.48 Determine the total plastic torque for the cross-sections shown in Figure 11.18a and compare to a solid circular cross-section of radius r. Scale the dimensions from the figure.

P11.49 For the square shaft $4a$ by $4a$ with a keyway in one side (a by a) shown in Figure P11.49, estimate the reduction in the plastic torque due to cutting the keyway.

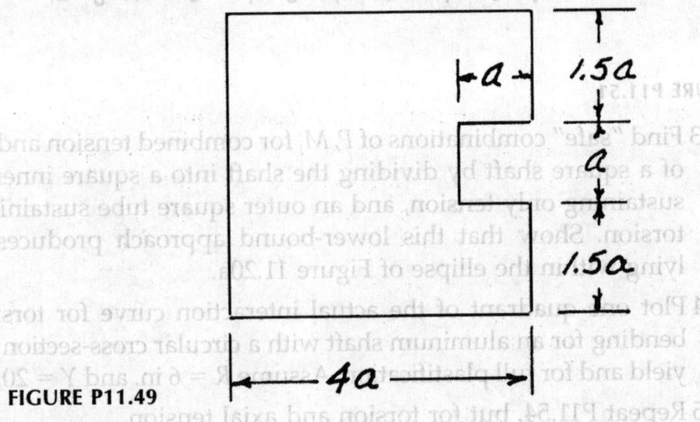

FIGURE P11.49

P11.50 Find $(M_t)_P$ for the T Beam section shown in Figure P11.50.

P11.51 Find $(M_t)_P$ for the hollow box section shown in Figure P11.51 and assess the contribution of the bridge area between the holes to the total.

P11.52 Show that the Tresca yield criterion is the same as the von Mises condition and therefore the interaction Equation (11.12) is general.

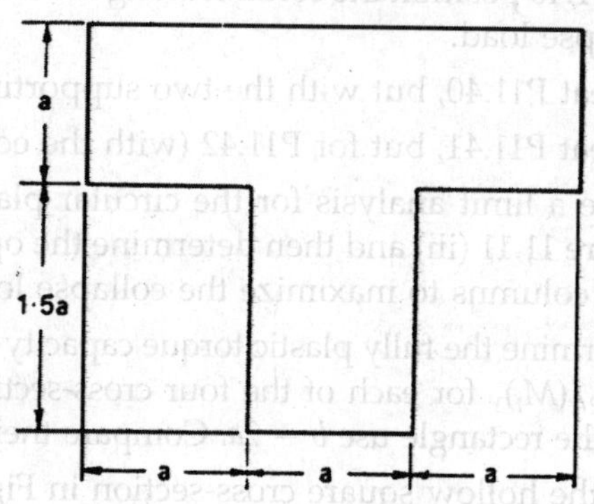

FIGURE P11.50

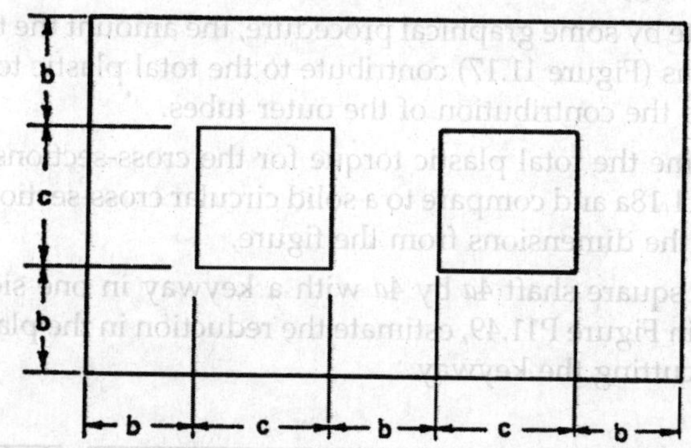

FIGURE P11.51

P11.53 Find "safe" combinations of P, M_t for combined tension and torsion of a square shaft by dividing the shaft into a square inner region sustaining only tension, and an outer square tube sustaining only torsion. Show that this lower-bound approach produces points lying within the ellipse of Figure 11.20a.

P11.54 Plot one quadrant of the actual interaction curve for torsion and bending for an aluminum shaft with a circular cross-section for first yield and for full plastification. Assume $R = 6$ in. and $Y = 20{,}000$ psi.

P11.55 Repeat P11.54, but for torsion and axial tension.

12

Slip-Line Analysis

Either for a full solution or to aid in the trial-and-error approach to upper-bound limit analysis, further knowledge of the shape of preferred mechanisms is crucial. Nature will select the type and location of slip surface that requires the least amount of energy (or force) to bring it about. In Chapter 10, this minimization was demonstrated without knowledge of the correct "type" of mechanism and, by the upper-bound theorem, values obtained for the collapse load must necessarily be incorrect on the high side (unconservative).

The question then is: how sensitive is the collapse load to the choice of failure mechanism or, put differently, how close must we estimate the shape and the position of the slip surface to get good approximations?

To address this question and better understand the physics of collapse, the geometric characteristics of correct slip mechanisms must be investigated and "exact" solutions found for some basic problems to compare to the more easily obtained approximate answers. This can be done by looking at the basic differential field equations of equilibrium coupled with the yield condition. In this chapter we will do this not only for an ideal EPS material but also for the more general Mohr-Coulomb (M-C) material.

12.1 Mohr-Coulomb Criterion (Revisited)

All materials have some component of frictional strength and therefore are stronger in compression than tension. For EPS materials (e.g., steel), this difference is insignificant, but for cast iron, rock, concrete, or soil, frictional strength predominates and cannot be neglected. The Mohr-Coulomb criterion introduced in Chapter 3 has the fundamental form:

$$s = c + \sigma_n \tan \phi \tag{3.16}$$

which plots as a straight line in stress space. Thus the shear strength, s, is a linear function of the normal stress σ_n on the potential slip plane where $\tan \phi$ is the coefficient of friction.

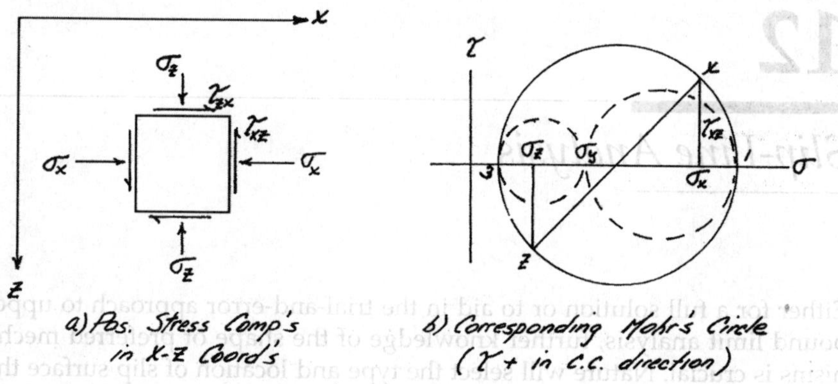

FIGURE 12.1
Sign convention and Mohr's Circle for soil mechanics (reversed from elasticity).

Since very "brittle" materials such as concrete have little tensile strength and granular materials like sand have none, we will, in this chapter, reverse the arbitrary sign conventions of elasticity. Thus, compression will be positive and shear stresses will be positive if the signs of the subscripts are mixed. The convention for plotting Mohr's Circle is shown in Figure 12.1. Also, throughout, it will be assumed that we have a plane-strain situation where the out-of-plane normal stress, σ_y, is intermediate. That is:

$$\sigma_3 \leq \sigma_y = \sigma_2 \leq \sigma_1 \qquad \text{and} \qquad \tau_{yx} = \tau_{yz} = 0$$

Therefore the slip planes on which $\tau' = s$ are in the x–z plane and only the largest Mohr's Circle need be plotted in future figures.

One fundamental difference introduced by the M-C criterion is the possibility of two potential failure states and two sets of slip surfaces, "active" and "passive" at a point, even when one principal stress component is the same. This is easily seen in the triaxial "test" situation shown in Figure 12.2. Assuming this M-C material is initially at constant hydrostatic pressure $\sigma_x^i = \sigma_z^i = p$, the specimen can be brought to failure (Mohr's Circle touching the strength envelope) by either increasing σ_x (Figure 12.2a) or decreasing σ_x (Figure 12.2b) resulting in different sized and oriented Mohr Circles as shown in Figure 12.2c.

It is useful to go through a formal plastic analysis for either the active or passive case (although the result is obtained by simply drawing Mohr's Circle touching the strength envelope as in Figure 12.2c). Consider, for example, the passive case illustrated again in Figure 12.3 where the stress field is constant (b any size). Note that the active case would give the same stressed element rotated 90° where σ_x would now be the minimum principal stress and σ_z the major. Thus, in terms of the orientation of the principal stress components to the slip surfaces, there is no difference between the active and passive case.

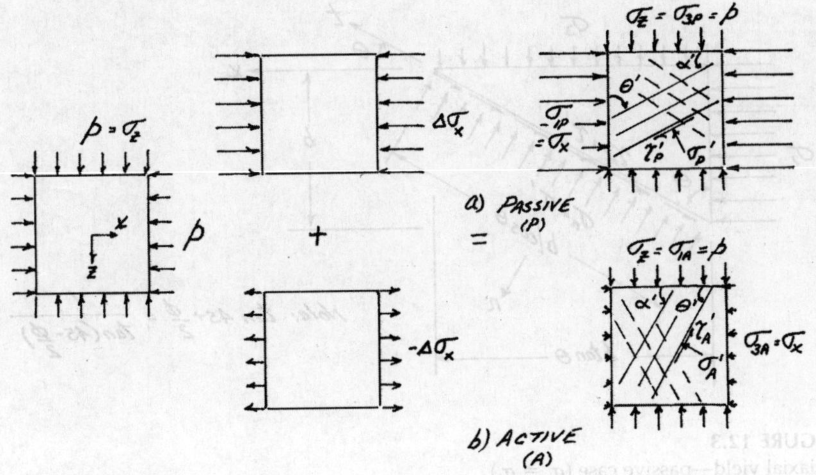

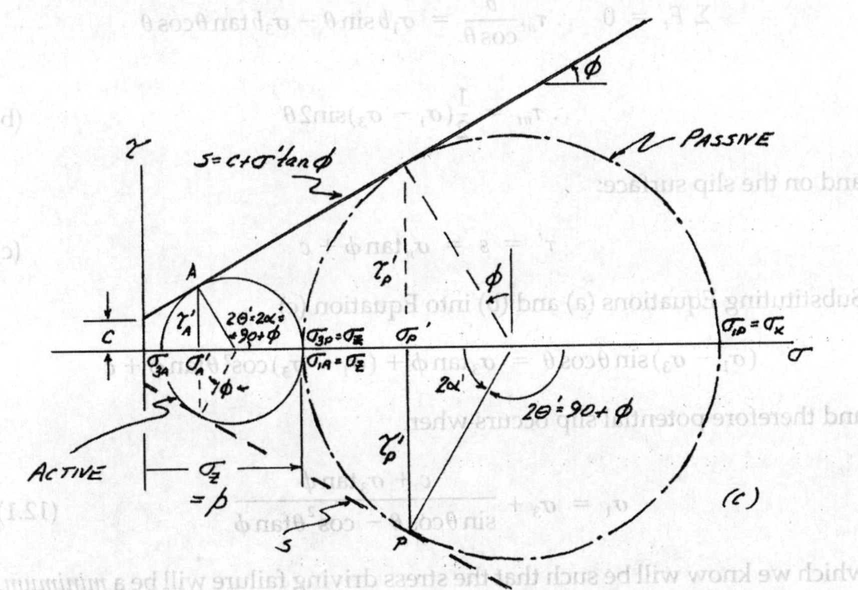

FIGURE 12.2
Triaxial test to failure of a M-C material.

For equilibrium:

$$\Sigma F_n = 0 \quad \therefore \sigma_n \frac{b}{\cos\theta} = \sigma_3 b \tan\theta \sin\theta + \sigma_1 b \cos\theta$$

and

$$\therefore \sigma_n = \sigma_3 + \cos^2\theta(\sigma_1 - \sigma_3) = \sigma_1 - \sin^2\theta(\sigma_1 - \sigma_3) \qquad \text{(a)}$$

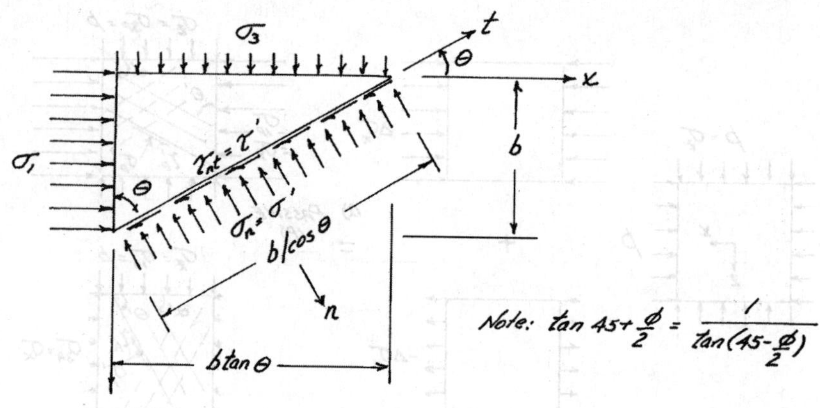

FIGURE 12.3
Triaxial yield—passive case ($\sigma_x = \sigma_1$).

$$\Sigma F_t = 0 \quad \therefore \tau_{nt}\frac{b}{\cos\theta} = \sigma_1 b\sin\theta - \sigma_3 b\tan\theta\cos\theta$$

$$\therefore \tau_{nt} = \frac{1}{2}(\sigma_1 - \sigma_3)\sin 2\theta \tag{b}$$

and on the slip surface:

$$\tau' = s = \sigma_n\tan\phi + c \tag{c}$$

Substituting Equations (a) and (b) into Equation (c)

$$(\sigma_1 - \sigma_3)\sin\theta\cos\theta = \sigma_3\tan\phi + (\sigma_1 - \sigma_3)\cos^2\theta\tan\phi + c$$

and therefore potential slip occurs when

$$\sigma_1 = \sigma_3 + \frac{c + \sigma_3\tan\phi}{\sin\theta\cos\theta - \cos^2\theta\tan\phi} \tag{12.1}$$

which we know will be such that the stress driving failure will be a *minimum*. Therefore:

$$\frac{d}{d\theta}(\sin\theta\cos\theta - \cos^2\theta\tan\phi) = 0 = \cos 2\theta' + \tan\phi\sin 2\theta'$$

defining the failure mechanism as the straight slip surfaces at

$$\theta' = \pm\left(45 + \frac{\phi}{2}\right) \tag{12.2}$$

from the major principal stress direction.

Therefore, substituting this result into Equation (12.1):

$$\sigma_1' = \sigma_3' \tan^2\left(45 + \frac{\phi}{2}\right) + 2c\,\tan\left(45 + \frac{\phi}{2}\right) \tag{12.3a}$$

or

$$\sigma_3' = \sigma_1' \tan\left(45 - \frac{\phi}{2}\right) - 2c\,\tan\left(45 - \frac{\phi}{2}\right) \tag{12.3b}$$

which apply in the principal stress orientation for either the active or passive case. Thus only when σ_x or σ_z is specified does the question arise of which "type of failure" occurs.

Finally, it is important to reemphasize that the block in Figure 12.3 can be thought of as a differential element ($b = db$) and therefore the analysis and equations are general for a point. However, they only apply on a finite scale in a stress field where the orientation of the principal stresses (the isoclinic field) is constant. This is not the normal situation. Thus a straight slip surface as an overall failure mechanism for limit analysis, while it may be a nice guess, can seldom be exact.

Example 12.1

A 6-inch diameter concrete ($f_c' = 4000$psi) column is to be reinforced by confining it in a thin steel ($Y = 30^{\text{ksi}}$) cylinder as shown below. Compare the confined to unconfined *ultimate* load capacity of the column $\frac{P_{\text{conf}}}{P_{\text{unconf}}}$ as the thickness of the jacket increases. ($0 \le t \le \frac{2}{5}$ in.) . ϕ for concrete is ~30°.

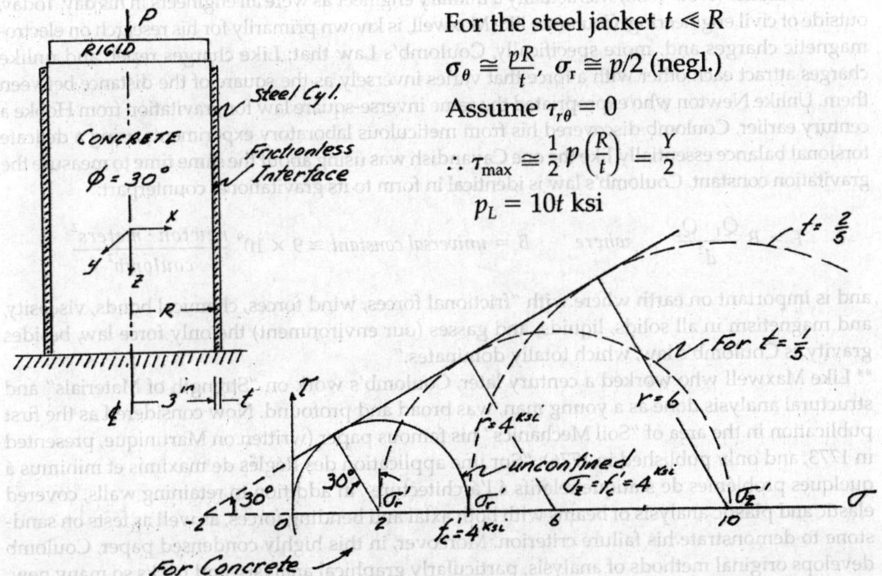

For the steel jacket $t \ll R$

$\sigma_\theta \cong \frac{pR}{t}$, $\sigma_r \cong p/2$ (negl.)

Assume $\tau_{r\theta} = 0$

$\therefore \tau_{\max} \cong \frac{1}{2} p\left(\frac{R}{t}\right) = \frac{Y}{2}$

$p_L = 10t$ ksi

t	$\sigma'_r = p_L$	$\sigma'_z = f'_c$
0	0	4
1/5	2	10
2/5	4	16

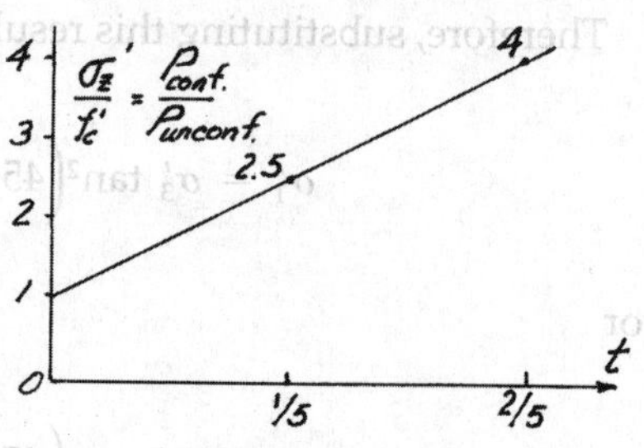

in general from Mohr's Circle

$$\sigma'_z = 3\sigma'_r + 4$$

Demonstrating why, because of frictional strength, encased concrete columns are so much stronger.

12.2 Lateral "Pressures" and the Retaining Wall Problem

To further set the stage, let us look at the same retaining wall problem that Coulomb* considered as a young civil engineer in charge of various construction projects on the island of Martinique in the 1760s. In order to address this problem, Coulomb first had to develop from scratch the concepts of slip surfaces and limit analysis along with his law of dry frictional resistance and cohesion.** Unfortunately, the information in his 1776 paper, while not obscure, is so condensed that it must be studied with great attention and expanded to be appreciated. Coulomb was elected to the academy in 1781 and never returned to soil mechanics and the retaining wall problem, so his

* C.A. Coulomb (1736–1806) was actually a military engineer as were all engineers in his day. Today, outside of civil engineering, Coulomb, like Maxwell, is known primarily for his research on electromagnetic charges and, more specifically, Coulomb's Law that: Like charges repel, and unlike charges attract each other with a force that varies inversely as the square of the distance between them. Unlike Newton who expropriated the same inverse-square law for gravitation from Hooke a century earlier, Coulomb discovered his from meticulous laboratory experiments using a delicate torsional balance essentially like the one Cavandish was using about the same time to measure the gravitation constant. Coulomb's law is identical in form to its gravitational counterpart:

$$F = B\,\frac{Q_1 \cdot Q_2}{d^2} \qquad where \qquad B = universal\ constant \cong 9 \times 10^9\ \frac{newton \cdot meters^2}{coulomb^2}$$

and is important on earth where with "frictional forces, wind forces, chemical bonds, viscosity, and magnetism in all solids, liquids, and gasses (our environment) the only force law, besides gravity, is Coulomb's law, which totally dominates."

** Like Maxwell who worked a century later, Coulomb's work on "Strength of Materials" and structural analysis done as a young man, was broad and profound. Now considered as the first publication in the area of "Soil Mechanics" his famous paper (written on Martinique, presented in 1773, and only published in 1776): "Sur une application des Régles de maximis et minimus á quelques problémes de statique relatifs á l'árchitecture" in addition to retaining walls, covered elastic and plastic analysis of beams with both axial and bending forces, as well as tests on sandstone to demonstrate his failure criterion. Moreover, in this highly condensed paper, Coulomb develops original methods of analysis, particularly graphical analysis, and packs so many new solution comes down to us as a few algebraic equations and two small figures* with a minimum of explanation. It took nearly 200 years and thousands of pages of text to recognize that his work includes all the concepts and methodology needed for analysis and design of gravity retaining walls.

If we consider a vertical line (the wall) embedded in a M-C halfspace (Figure 12.4a), the weight of the soil, $B_z = \gamma$, introduces the elastic stress field:

$$\sigma_z = \gamma z, \qquad \sigma_x = \frac{\nu}{1-\nu}\sigma_z = K_o \gamma z, \qquad \tau_{xz} = 0 \tag{12.4}$$

The lateral pressure increases linearly with depth at some proportion, K_o, of the vertical stress (a standard value for K_o for sand is 0.5).

If the wall, now considered rigid, is to exactly replace the material on one side of it (Figure 12.b), a force $P_o = \frac{1}{2} K_o \gamma H^2$ must be provided at the third point, perhaps by a jack or prestressed strut. Coulomb now visualized the

---

insights into such a short space that, according to Poncelet, "during the next 40 years the attention of engineers seemed not to focus on any of them." Perhaps because the presentation of this paper in 1773 was his introduction to the French Academy of Sciences, Coulomb felt intimidated as "only a practicing engineer" and kept throwing in fundamental insight after insight with less and less explanation to the point where the power of the whole obscures the beauty of the individual parts. Certainly his introduction in which he states:

> If I dare to present it (this paper) to this Academy it is only because the feeblest endeavors are kindly welcomed by it when they have useful objective . . .

would suggest he suffered at this time from the same "engineering inferiority syndrome" that Hooke developed a century earlier trying to deal with Newton's intolerable egocentricity and wholesale theft of Hooke's ideas in the name of science. In fact, Coulomb amplifies this paranoia (probably justified) with further self-denigration noting that

> ... while great men will be carried to the top of the edifice (science) where they can mark out and construct the upper stories, ordinary artisans (engineers) who are scattered through the lower stories .... should seek only to perfect that which cleverer hands have created.

Unfortunately, such nonsense is as prevalent a view today as it was in Hooke's and Coulomb's time.
* Coulomb's figure (1776).

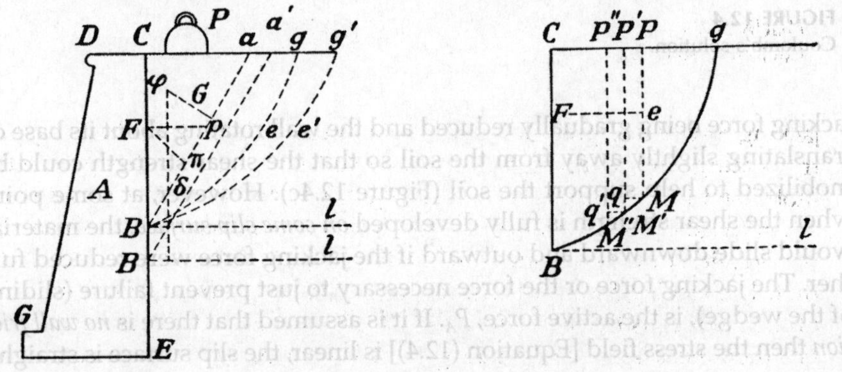

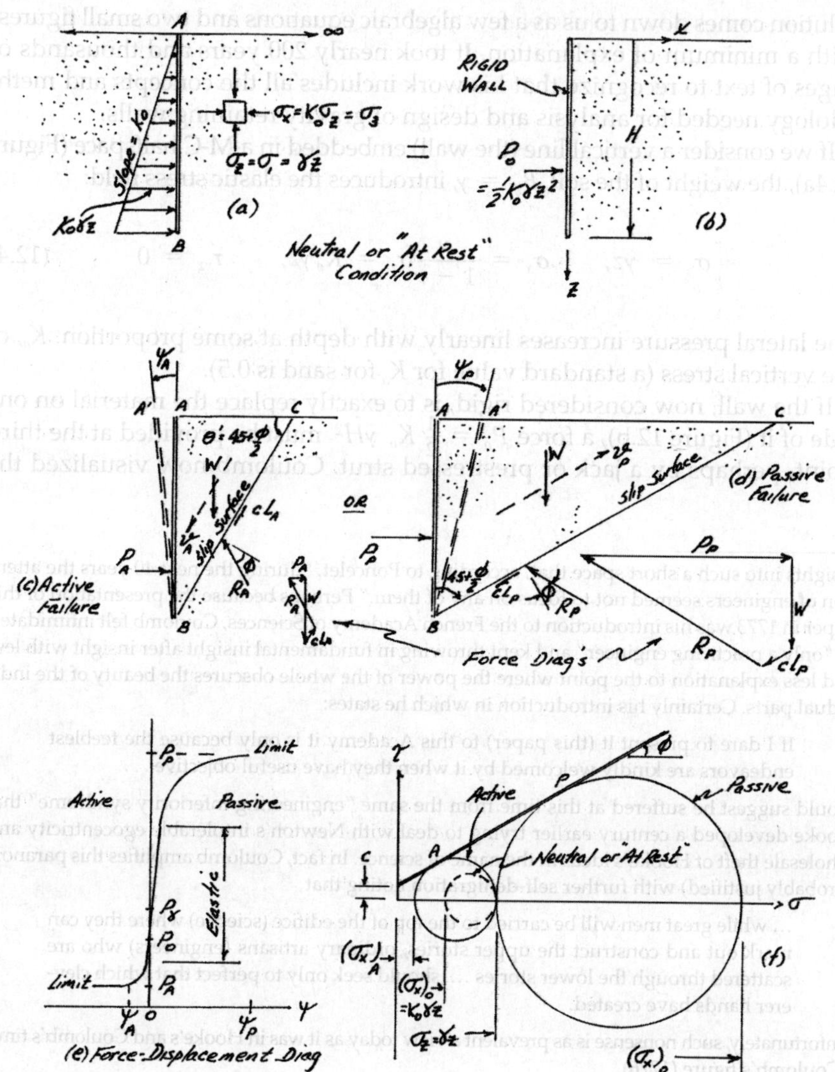

FIGURE 12.4
Coulomb's solution.

jacking force being gradually reduced and the wall rotating about its base or translating slightly away from the soil so that the shear strength could be mobilized to help support the soil (Figure 12.4c). However, at some point when the shear strength is fully developed *on some slip surface*, the material would slide downward and outward if the jacking force were reduced further. The jacking force or the force necessary to just prevent failure (sliding of the wedge), is the active force, P_A. If it is assumed that there is *no wall friction* then the stress field [Equation (12.4)] is linear, the slip surface is straight, and Equation (12.3b) does apply on a global scale with $\sigma_x = \sigma_3$, $\sigma_z = \sigma_1 = \gamma z$. The lateral pressure σ_3 is linear as shown in Figure 12.5 (line fd) for $c > 0$, $\phi > 0$. Integrating:

$$P_A = \int_0^H \sigma_x dz = \frac{1}{2}\gamma H^2 \tan^2\left(45 - \frac{\phi}{2}\right) - 2cH\tan\left(45 - \frac{\phi}{2}\right) \qquad (12.5b)$$

If instead the wall is jacked into the soil, Figure 12.4d, the passive case develops where the friction mobilized in the soil again resists the motion but, in this case, adds to the jacking force necessary to cause failure. Again, if there is *no wall friction*, the slip surface will be straight and Equation (12.3a) applies with $\sigma_3 = \gamma z = \sigma_z$ (line ml in Figure 12.5 for $c > 0, \phi > 0$). Thus the passive

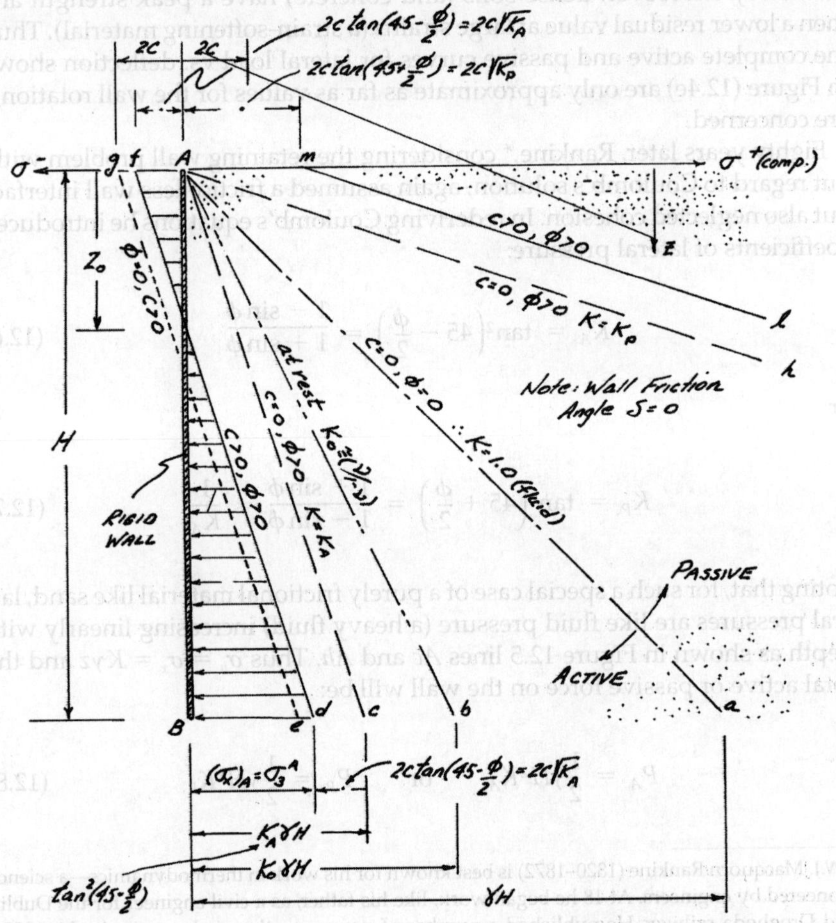

FIGURE 12.5
Coulomb solution for lateral pressures.

force P_P just necessary to cause the failure wedge to slide upward is

$$P_P = \int_0^H \sigma_x dz = \frac{1}{2}\gamma H^2 \tan^2\left(45 + \frac{\phi}{2}\right) + 2cH\tan\left(45 + \frac{\phi}{2}\right) \qquad (12.5a)$$

The amount of plastic strain necessary to fully mobilize the shear strength and fully develop the slip surface corresponding to either P_A or P_P is uncertain. Yield may occur first at some point on the eventual slip surface and then spread with continuing small movements of the wall until the complete mechanism develops. In soils (to some extent rock and concrete) the issue is further complicated in that not only does the shear strength depend on the confining pressure, but also on the soil's density, the water content, and even its history. Moreover, dense soils (and concrete) have a peak strength and then a lower residual value at large strains (a strain-softening material). Thus, the complete active and passive curves for lateral load vs. deflection shown in Figure (12.4e) are only approximate as far as values for the wall rotation ψ are concerned.

Eighty years later, Rankine,* considering the retaining wall problem without regard to Coulomb's solution, again assumed a frictionless wall interface but also neglected cohesion. In rederiving Coulomb's equations he introduced coefficients of lateral pressure:

$$K_A = \tan^2\left(45 - \frac{\phi}{2}\right) = \frac{1 - \sin\phi}{1 + \sin\phi} \qquad (12.6)$$

or

$$K_P = \tan^2\left(45 + \frac{\phi}{2}\right) = \frac{1 + \sin\phi}{1 - \sin\phi} = \frac{1}{K_A} \qquad (12.7)$$

noting that, for such a special case of a purely frictional material like sand, lateral pressures are like fluid pressure (a heavy fluid) increasing linearly with depth as shown in Figure 12.5 lines Ac and Ah. Thus $\sigma_h = \sigma_x = K\gamma z$ and the total active or passive force on the wall will be:

$$P_A = \frac{1}{2}\gamma H^2 K_A \quad \text{or} \quad P_P = \frac{1}{2}\gamma H^2 K_P \qquad (12.8)$$

* W.J. Macquorn Rankine (1820–1872) is best known for his work in thermodynamics—a science pioneered by engineers. At 18 he began work, like his father, as a civil engineer for the Dublin and Drogheda railway. He published a number of papers on thermodynamics in the 1850s, was elected to the Royal Society in 1853, and was appointed as professor of engineering at the University of Glasgow in 1855. Glasgow may be the first British university to have engineering education per se.

12.3 Graphic Analysis and Minimization

In his paper, Coulomb only derives the active pressure Equation (12.3b). However, although not stated explicitly, this implies both the basic Mohr-Coulomb strength criterion, Equation (3.16) and the passive pressure Equation (12.3a). The actual method Coulomb used to obtain the minimum active and maximum passive forces was graphical. Developed in greater detail by Culmann*, Example 12.2 presents the method for the active case. For any straight slip-line trial,** a force diagram can be drawn as in Figures 12.4c and d. By plotting a number of trials superimposed, the maximum can be seen directly. This is similar in every respect to the one-dimensional problem of the collapse of an indeterminate beam as in Figure 11.5 where a graphical minimization is used to determine the limit load and position of the plastic hinges (critical collapse mechanism) simultaneously. For retaining walls, however, the computation of the thrust for each trial is also done graphically.

For homogeneous cases with $c = 0$, straight boundaries, and no external loading, the forces can be expressed in terms of the geometric variables β, i, and friction angles δ, ϕ to obtain a closed form expression for P. This can be formally minimized or maximized to obtain formulas for the active and passive thrusts and the critical straight-line slip surfaces associated with them. These will be given in the concluding section of this chapter.

However, the graphical method for any particular case is very quick and much more general. With the basic assumption of a straight slip surface, any shaped wall, wall roughness, backfill geometry, layered backfill with

---

* Karl Culmann (1821–1881) at seventeen entered the foremost German engineering school, Karlshrue Polytechnicum and, upon graduation, worked on the design of important railway structures. His study of English and American bridges is filled with important criticism derived from his excellent theoretical training. Particularly intriguing is his discussion of the Burr Truss where the proportion of the load taken by the arch vs. the truss is indeterminate and his approximate analysis is clearly a prelude to the better understanding of this design by his student, N. Ritter, and final flowering in the deck-stiffened, concrete, bridge forms of Ritter's famous student, R. Maillart.

Culmann was appointed as the first Professor of Structures at the Zürich Polytechnicum in 1855 where his work on graphic statics became preeminent. His application of graphical analysis to all sorts of structures was published in *Die graphische statik* in 1866 (and subsequently improved upon by Ritter). Among other wonderful constructions, Culmann demonstrates that normal and tangential components of stress acting on any plane are given by coordinates on a "circle of stress." This construction allows him to find principal values and directions, maximum shears at 45°, and the plane tensor components at any other orientation. Mohr's Circle should probably be called Culmann's circle.

Culmann believed that graphic analysis was the way to understand solid mechanics physically and should be a major component in engineering education. However, only vestiges of this powerful approach are even mentioned in today's curriculum.

** It is clear from the figures in Coulomb's paper (see footnote on page 395) that Coulomb knew perfectly well that, with wall friction, the critical slip surface might well be curved and even anticipated the "method of slices" now used for this situation for nonhomogeneous cases.

discontinuous c, ϕ, γ, condition of the water table, or practically any other complexity can be included with very little additional work. Moreover, it is a design as much as an analysis approach. Thus Coulomb's method of analysis, relying on successive applications of graphic statics, is as much of an accomplishment, in its way, as are his equations.

Example 12.2

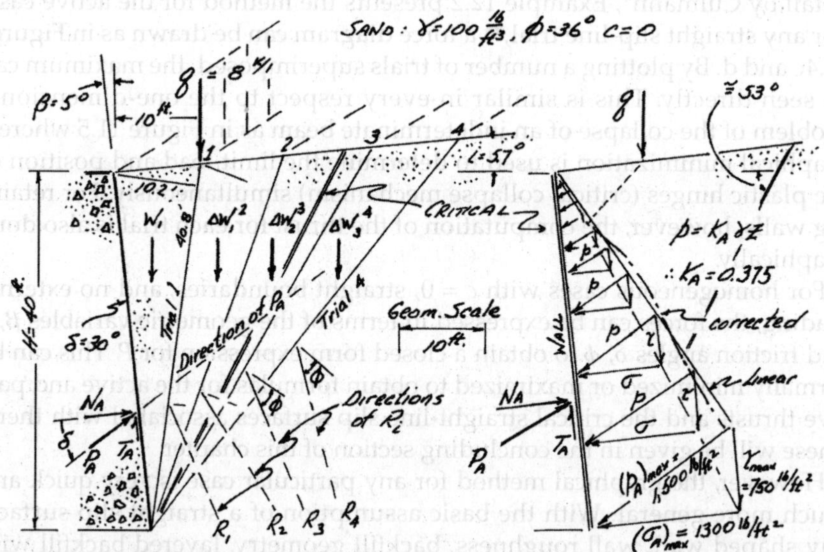

1) Select trial slip surfaces 1, 2, 3, & 4

2) Calculate W for each Wedge:

$$W_i = W_{i-1} + W_{i-1}^i$$

$$W_1 = \frac{1}{2}40.8(10.2)(100) = 20.8^K$$

$$W_1^2 = 21.1^K, \quad W_2^3 = 21.5^K, \quad W_3^4 = 21.8^K$$

$$\therefore W_2 = 41.9^K, \quad W_3 = 63.4^K, \quad W_4 = 65.2^K$$

3) Lay out force polygons (Superimposed)

4) Draw P Line

5) Scale off $P_{max} = P_A$ (also gives critical slip surface) $P_A = 1.2\ (25) = 30^K$

$$\therefore N_A = 26^K; \quad T_A = 15^K$$

6) Draw linearized distributions p_A, σ_A, τ_A (estimate correction for q)

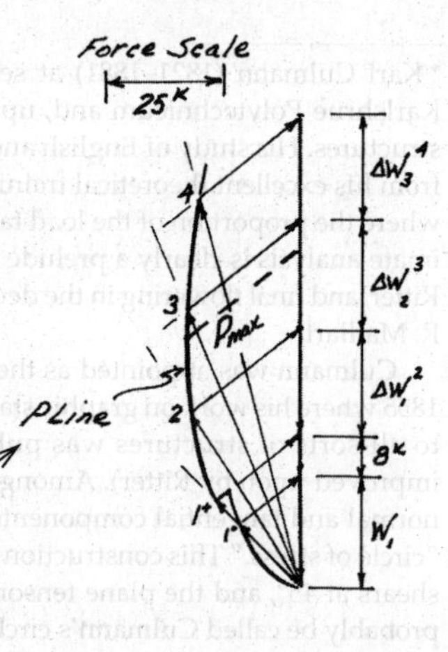

Example 12.3

a. An anchor for a bridge tower is designed as shown below. Assume no wall friction and neglecting any active pressures behind the wall, determine the ultimate (limiting) static passive resistance T_L of the anchor.

b. Modify your analysis to include inertial forces due to an earthquake with a peak horizontal acceleration $a_h = 0.3g$. Show the difference between the seismic and static value for T_L and the critical seismic and static slip surfaces.

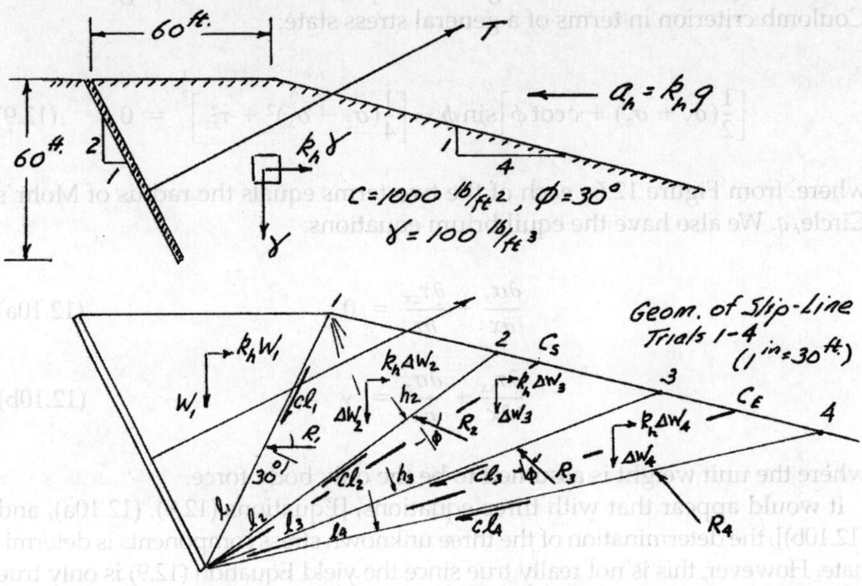

	ΔW_i (k)	W_i (k)	cl_i (k)	h_i (ft.)	$k_h W_i$ (k)
1		180	60		54
2	120	300	84	30	90
3	250	430	111	45	129
4	380	560	150	51	168

$$\therefore (T_{\text{min}})_{\text{static}} = 440^K$$

$$(T_{\text{min}})_{\text{e'quake}} = 380^K$$

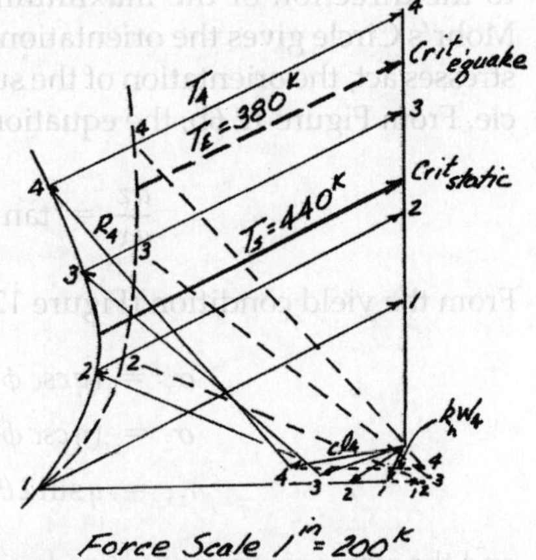

12.4 Slip-Line Theory

Only at points where we know the orientation of the principal stresses do we know the orientation of the slip surface to be ± (45 − ϕ/2) to the direction of σ_1. We often know the principal orientation at boundaries, but seldom anywhere else. As the stresses change throughout the field in some fashion so do the isoclinics and the slip surface must, therefore, in general be curved.

To determine the shape or characteristic of the slip-line field we must solve the basic equations in terms of geometric parameters. Rewriting the Mohr-Coulomb criterion in terms of a general stress state:

$$\left[\frac{1}{2}(\sigma_x + \sigma_z) + c\cot\phi\right]\sin\phi - \left[\frac{1}{4}(\sigma_z - \sigma_x)^2 + \tau_{xz}^2\right]^{\frac{1}{2}} = 0 \qquad (12.9)$$

where, from Figure 12.6a each of the two terms equals the radius of Mohr's Circle, q. We also have the equilibrium equations:

$$\frac{\partial\sigma_x}{\partial x} + \frac{\partial\tau_{zx}}{\partial z} = 0 \qquad (12.10a)$$

$$\frac{\partial\tau_{zx}}{\partial x} + \frac{\partial\sigma_z}{\partial z} = \gamma \qquad (12.10b)$$

where the unit weight is assumed to be the only body force.

It would appear that with three equations, [Equations (12.9), (12.10a), and (12.10b)], the determination of the three unknown stress components is determinate. However, this is not really true since the yield Equation (12.9) is only true on the slip surfaces for which we do not know the shape or critical location.

The two slip lines (or characteristic lines) are inclined at an angle $\mu = \pm\left(45 - \frac{\phi}{2}\right)$ to the direction of the maximum principal compressive stress σ_1. Since Mohr's Circle gives the orientation of the normal to the surface on which the stresses act, the orientation of the surface itself is at 90° or 180° on Mohr's Circle. From Figure 12.6b, the equations for the slip lines are then:

$$\frac{dz}{dx} = \tan\alpha = \tan(\theta \pm \mu) \qquad (12.11)$$

From the yield condition (Figure 12.6a) where $\sigma_m = q\csc\phi - c\cot\phi$:

$$\begin{aligned}\sigma_x &= (q\csc\phi - c\cot\phi) + q\cos 2\theta \\ \sigma_z &= (q\csc\phi - \cot\phi) - q\cos 2\theta \\ \tau_{zx} &= q\sin 2\theta\end{aligned} \qquad (12.12)$$

and the stress components producing potential slip are expressed in terms of the invariant q and the isoclinic angle θ.

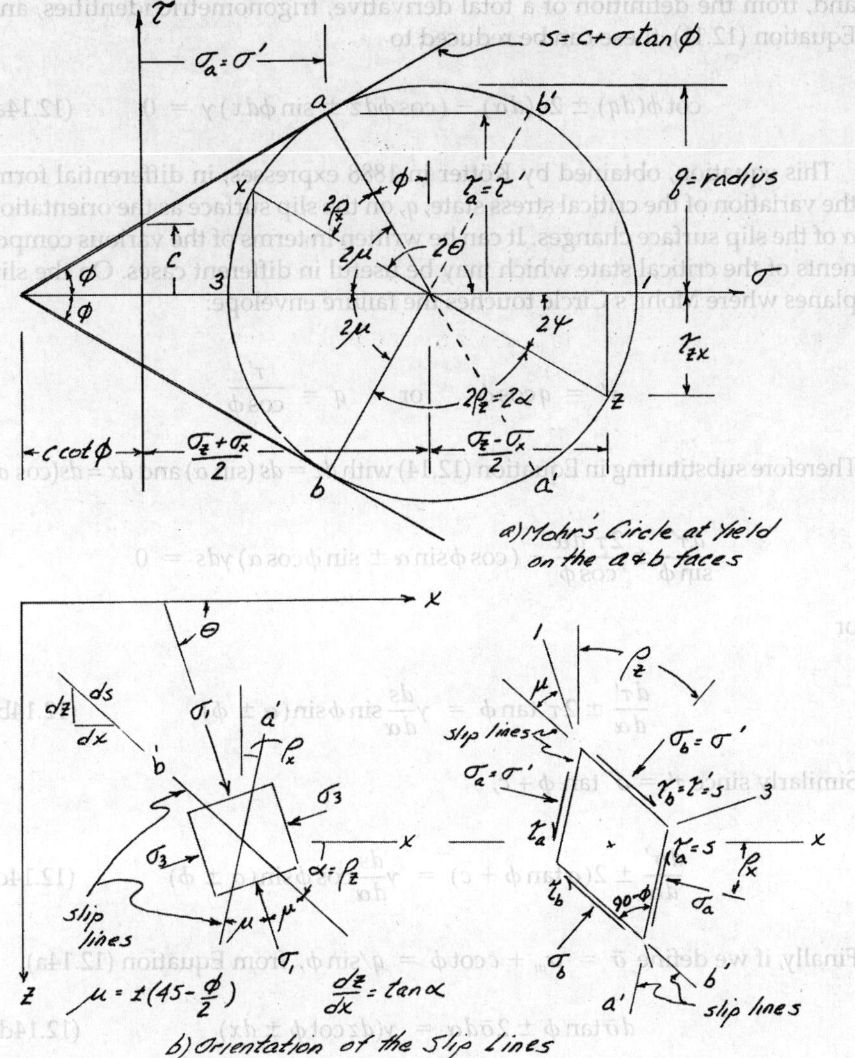

FIGURE 12.6
Stress components and slip lines.

Substituting the stress components [Equation (12.12)] into the equilibrium Equations (12.10):

$$\frac{\partial q}{\partial x}(\csc\phi + \cos 2\theta) + \frac{\partial q}{\partial z}\sin 2\theta - 2q\left(\sin 2\theta\frac{\partial\theta}{\partial x} - \cos 2\theta\frac{\partial\theta}{\partial z}\right) = 0$$

$$\frac{\partial q}{\partial z}(\csc\phi - \cos 2\theta) + \frac{\partial q}{\partial x}\sin 2\theta + 2q\left(\sin 2\theta\frac{\partial\theta}{\partial z} + \cos 2\theta\frac{\partial\theta}{\partial x}\right) = \gamma \tag{12.13}$$

and, from the definition of a total derivative, trigonometric identities, and Equation (12.11), these can be reduced to

$$\cot\phi(dq) \pm 2q(d\alpha) - (\cos\phi dz \pm \sin\phi dx)\gamma = 0 \qquad (12.14a)$$

This equation, obtained by Kötter in 1888 expresses, in differential form, the variation of the critical stress state, q, on the slip surface as the orientation α of the slip surface changes. It can be written in terms of the various components of the critical state which may be useful in different cases. On the slip planes where Mohr's Circle touches the failure envelope:

$$\tau' = q\cos\phi \qquad \text{or} \qquad q = \frac{\tau'}{\cos\phi}$$

Therefore substituting in Equation (12.14) with $dz = ds(\sin\alpha)$ and $dx = ds(\cos\alpha)$

$$\frac{d\tau'}{\sin\phi} \pm \frac{2\tau' d\alpha}{\cos\phi} - (\cos\phi\sin\alpha \pm \sin\phi\cos\alpha)\gamma ds = 0$$

or

$$\frac{d\tau'}{d\alpha} \pm 2\tau'\tan\phi = \gamma\frac{ds}{d\alpha}\sin\phi\sin(\alpha \pm \phi) \qquad (12.14b)$$

Similarly since $\tau' = \sigma'\tan\phi + c$,

$$\frac{d\sigma'}{d\alpha} \pm 2(\sigma\tan\phi + c) = \gamma\frac{ds}{d\alpha}\cos\phi\sin(\alpha \pm \phi) \qquad (12.14c)$$

Finally, if we define $\bar{\sigma} = \sigma_m + c\cot\phi = q/\sin\phi$, from Equation (12.14a)

$$d\bar{\sigma}\tan\phi \pm 2\bar{\sigma}d\alpha = \gamma(dz\cot\phi \pm dx) \qquad (12.14d)$$

which is the version often used for integration along a slip line.

Various combinations of the three material properties used to characterize yield of a Mohr-Coulomb material are summarized in Table 12.1. If we assume a yield plateau and thus eliminate strain effects, these properties will not change as the yield area spreads to form a collapse mechanism (the slip surface) at the limit load.

In the next sections we will discuss a few simple cases where Equation (12.14) can be integrated. With the stress boundary conditions, it may be possible then to determine the shape of the critical slip surface and the stresses on the slip surface at collapse. Solutions for difficult boundary loads and for the general case where ϕ, c, or ψ are not neglected can sometimes be obtained by a shooting technique called "the method of characteristics" where Equation (12.14)

TABLE 12.1

Strength Characterization of Materials

Type	γ	ϕ	c	Material
1	0	0	0	Ideal gas } No strength
2		0	0	Ideal liquid } $\therefore$ Flow at $q > 0$(Fluids)
3	0		0	Sand without weight
4	0	0		EPS without weight-Material of classic elasticity and plasticity
5	0			General material without weight-Usual limit analysis of brittle materials
6			0	Sand
7		0		Clay or EPS with weight
8				GENERAL-Soil, confined concrete, rock.

is numerically integrated by finite differences starting from one boundary and working in increments, which can be arbitrarily small, to the other boundary. Results from this method will be referred to but not derived. Some observations on the "method of characteristics" or "the slip-line method" are included in the final section.

12.5 Purely Cohesive Materials ($\phi = 0$)

Materials with no significant frictional strength (cases 4 and 7 in Table 12.1) range from ductile metals to overconsolidated clay. They are the elastic, perfectly-plastic solids (EPS) of classical elasticity and plasticity (at least in compression).

With $\phi = 0$, Mohr's Circle at yield is of constant radius $q = \frac{Y}{2} = c = \tau'$ and the slip planes are at 45° to the principal orientation. Equation (12.14c) can be integrated directly to give the normal stress

$$\sigma' = \gamma z \pm 2c\alpha + K \tag{12.15}$$

where K is the constant of integration.

The pair of slip lines at a point are at right angles. Let us consider the slip surface to have the counterclockwise angle α from the x axis so that the normal to it (the other slip line) is at the clockwise angle ρ_z in Figure 12.7. Then from Mohr's Circle

$$\begin{aligned} \sigma_z &= \sigma' + c\sin 2\alpha = z\gamma + c(2\alpha + \sin 2\alpha) + K \\ \sigma_x &= \sigma' - c\sin 2\alpha = z\gamma + c(2\alpha - \sin 2\alpha) + K \\ \tau_{xz} &= -c\cos 2\alpha \end{aligned} \tag{a}$$

where z and α are independent variables.

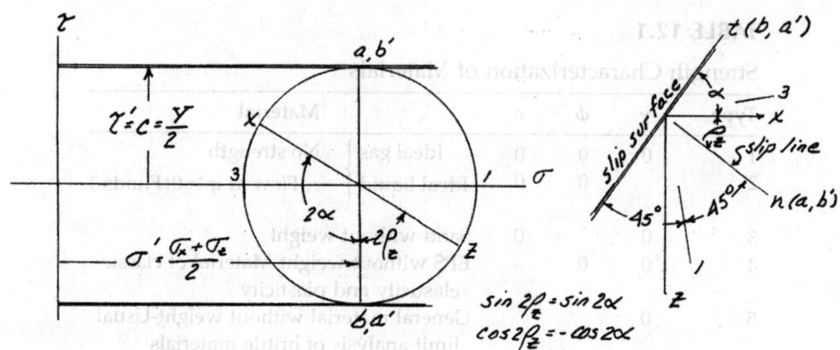

FIGURE 12.7
Stress components and orientation for $\phi = 0$.

Introducing the equilibrium Equations (12.10):

$$\frac{\partial \sigma_z}{\partial z} + \frac{\partial \tau_{zx}}{\partial x} = \gamma = \gamma + 2c\frac{\partial \alpha}{\partial z} + 2c\cos 2\alpha\frac{\partial \alpha}{\partial z} + 2c\sin 2\alpha\frac{\partial \alpha}{\partial x}$$

$$\frac{\partial \sigma_x}{\partial x} + \frac{\partial \tau_{zx}}{\partial z} = 0 = 2c\frac{\partial \alpha}{\partial x} - 2c\cos 2\alpha\frac{\partial \gamma}{\partial x} + 2c\sin 2\gamma\frac{\partial \gamma}{\partial z} \tag{b}$$

and adding

$$(1 + \cos 2\alpha + \sin 2\alpha)\frac{\partial \alpha}{\partial z} + (1 - \cos 2\alpha + \sin 2\alpha)\frac{\partial \alpha}{\partial x} = 0 \tag{c}$$

leads to the basic differential equation

$$\frac{\partial \alpha}{\partial z} = \tan \alpha \frac{\partial \alpha}{\partial x} \tag{12.16}$$

which describes the shape of the failure surface.

One obvious solution is

$$\alpha = \text{const. (a straight failure surface)} \tag{12.17}$$

While mathematically trivial, this solution is physically profound. We know it is correct in regions where τ_{zx} = const and therefore the orientation of the principal stresses do not change.

A general solution can be obtained by assuming

$$\tan \alpha = f(x)g(z) \tag{d}$$

It can be shown then that:

$$f(x) = kx + a, \qquad g(z) = -\frac{1}{b + kz} \tag{e}$$

and therefore:

$$\frac{dz}{dx} = \tan\alpha = -\frac{kx + a}{kz + b} \tag{f}$$

Finally, multiplying this out and integrating term by term we get the equation of a *circle*:

$$x^2 + z^2 + \frac{2ax}{k} + \frac{2bz}{k} - \frac{2c}{k} = 0 \tag{12.18}$$

where a, b, and c are constants.

Thus, for a material with $\phi = 0$, the failure surface in Cartesian coordinates must be *made up of straight lines and/or circles* combined in some way to satisfy the stress boundary conditions.

12.6 Weightless Material ($\gamma = 0$)

A weightless Coulomb material provides another case where Kötter's equation can be integrated directly. Such an idealization is often used for concrete, cast iron, or other brittle building materials and occasionally for soil such as in the bearing capacity problems when self weight adds only to the resistance and not the driving force causing the collapse.

For $\gamma = 0$, Equation (12.14b) reduces to

$$\frac{\partial\tau'}{\partial\alpha} = 2\tau'\tan\phi \qquad \text{or} \qquad \frac{d\tau'}{\tau'} = 2\tan\phi d\alpha \tag{12.19}$$

where again α is measured counterclockwise from the x axis to the failure surface. As before there is the important "trivial" solution

$$\alpha = \text{constant}$$

giving a *straight failure surface* on which the stresses are constant.

In general for a curved surface

$$\tau' = Ke^{2\tan\phi(\alpha)} \tag{12.20a}$$

or in terms of the maximum shear

$$\frac{1}{2}\cot\phi \ln q + \alpha = \text{constant} \tag{12.20b}$$

The shape of the curved slip surface on which these stresses act is easily found by considering Figure 12.8 in conjunction with Figure 12.6 where the

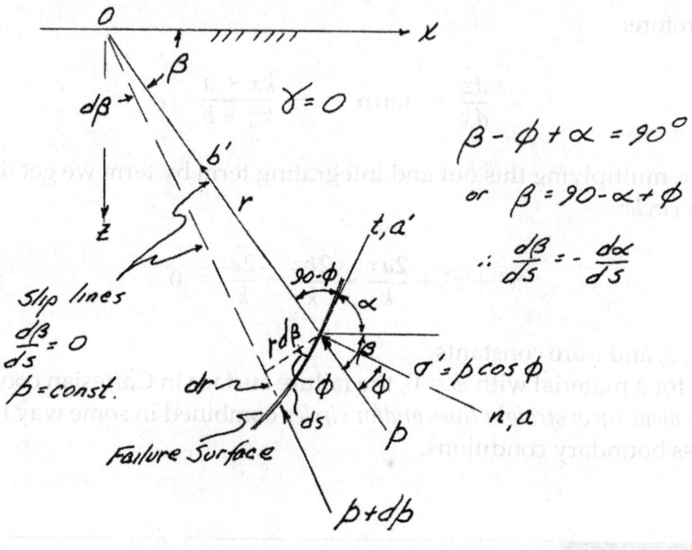

FIGURE 12.8
Slip-line field and relationships for $\gamma = 0$.

slip surface a' is the failure surface on which $p = \sigma' \csc \phi$ is the stress resultant of the frictional component of strength. For $\gamma = 0$, p acts through 0 in the direction of the secondary slip surface at $90 - \phi$ to the failure surface α from the horizontal. Therefore, in polar coordinates

$$\tan\phi = \frac{dr}{rd\beta} \quad \text{and} \quad \therefore r = r_o e^{\beta \tan\theta} \tag{12.20c}$$

Thus with a frictional component of strength but neglecting weight, the failure surface is made up of *straight line* and/or *log spiral* sections combined in some way to satisfy the boundary conditions.

12.7 Retaining Wall Solution for $\phi = 0$ (EPS Material)

First let us consider a wedge as shown in Figure 12.9 where for simplicity we will assume the top surface, BC, is free of boundary tractions and there is a rigid retaining wall supporting the EPS material with unit weight γ on the sloping face AB. If the wall moves outward in the direction of the v arrows, the active state of limiting equilibrium develops. A general failure surface, which by our analysis is a combination of straight and circular segments, is assumed. This is a combination of an upper wedge (slip field I) with plane surfaces of sliding, a circular transition section, II, and a third wedge, III, adjoining the wall.

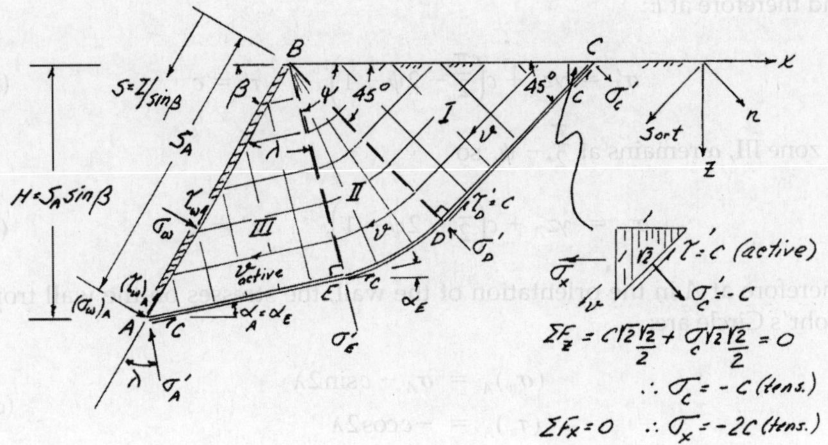

FIGURE 12.9
Slip field for $\phi = 0$, BC unloaded.

Since we do not know the extent of slip fields II and III or the stresses on the wall, we cannot solve directly but must work our way along the slip surface* from the known boundary, BC, to the unknown boundary, AB, using the solution [Equation (12.15)] to Kötter's equation for $\phi = 0$. Starting at point C we know the complete state of stress. Since $\tau_{xz} = \sigma_z = 0$, x and z are principal directions and therefore $\alpha = 45°$. From Mohr's Circle and the direction of τ' for the active case, $\sigma' = -c$, $\tau' = c$ and from Equation (12.15)

$$\sigma' = \gamma z + 2c\frac{\pi}{4} + K = c\frac{\pi}{2} + K = -c \tag{a}$$

Therefore:

$$K = -c\left(1 + \frac{\pi}{2}\right) \tag{12.21}$$

As we move to point D, x and z remain principal so $\sigma'_D = \gamma z_o - c$, $\tau' = c$. However, from D to E on the circular portion of the failure surface, α changes as well as the depth so

$$\sigma_E = \gamma z_E + 2c\alpha_E + K = \gamma z_E + 2c\left(\frac{\pi}{4} - \psi\right) + K$$

* This so-called "shooting technique" is a direct consequence of the field Equations (12.13), which are of the hyperbolic type. Thus, as discussed in Chapter 4, the boundary tractions cannot be specified over the entire boundary, for a solution to exist. For a unique and stable solution, (Table 4.2), we must specify both the value *and* slope of the function (i.e., the complete state of stress) on one boundary, and then, when we get to the other boundary, the stress state will be whatever it turns out to be. This requirement can lead to difficulties that are often not appreciated.

and therefore at E:

$$\sigma'_E = \gamma z_E + c\left(\frac{\pi}{2} - 2\psi - 1\right), \qquad \tau' = c \tag{b}$$

In zone III, α remains at $\frac{\pi}{2} - \psi$, so

$$\sigma_A = \gamma z_A + c\left(\frac{\pi}{2} - 2\psi - 1\right), \qquad \tau' = c \tag{c}$$

Therefore at A in the orientation of the wall, the stresses on the wall from Mohr's Circle are:

$$\begin{aligned}(\sigma_w)_A &= \sigma_A - c\sin 2\lambda \\ (\tau_w)_A &= -c\cos 2\lambda\end{aligned} \tag{d}$$

Since this "shooting" procedure could be used along any slip line, $C'D'E'A'$, the normal stress on the wall (or any radial line from B) must be linear with depth. Therefore, from Equation (d), noting that $\sin 2\lambda = -\sin(2\beta + 2\psi)$ and $\cos 2\lambda = \cos(2\beta + 2\psi)$, the solution for the unknown stresses on the wall for $\phi = 0$ becomes

$$\begin{aligned}\sigma_w &= \gamma z + c\left(\frac{\pi}{2} - 2\psi - 1\right) + c\sin(2\beta + 2\psi) \\ \tau_w &= -c\cos(2\beta + 2\psi)\end{aligned} \tag{12.22}$$

It is important to note that the actual values of the wall stresses depend on the choice of ψ, which specifies the extent of the circular zone II. The distribution of stress on the wall however, is determined by the solution to be a constant shear and a normal stress linear with depth. Any other distribution cannot be specified for a unique solution to this Cauchy problem and maintain the unloaded boundary conditions on the surface BC.*

The direction of the *resultant stress* on the wall varies since the normal stress increases with depth while the shear stress is constant. Therefore, to talk about a constant "angle of wall friction" δ is a misnomer. However, we can compute the components of the force resultants on the wall by integrating the stresses, i.e.,

$$\begin{aligned}N_A &= \int_0^s \sigma_w ds = \frac{\gamma H^2}{2\sin\beta} + \frac{cH}{\sin\beta}\left[\left(\frac{\pi}{2} - 2\psi - 1\right) + \sin(2\beta + 2\psi)\right] \\ T_A &= \int_0^s \tau_w ds = -\frac{cH}{\sin\beta}\cos(2\beta + 2\psi)\end{aligned} \tag{12.23}$$

* See the previous footnote on hyperbolic equations and the shooting techniques to solve them. Various other conditions on surface BC are reserved for the chapter problems. Also, the wall stress distribution for the passive case for various specified conditions on BC are left as problems.

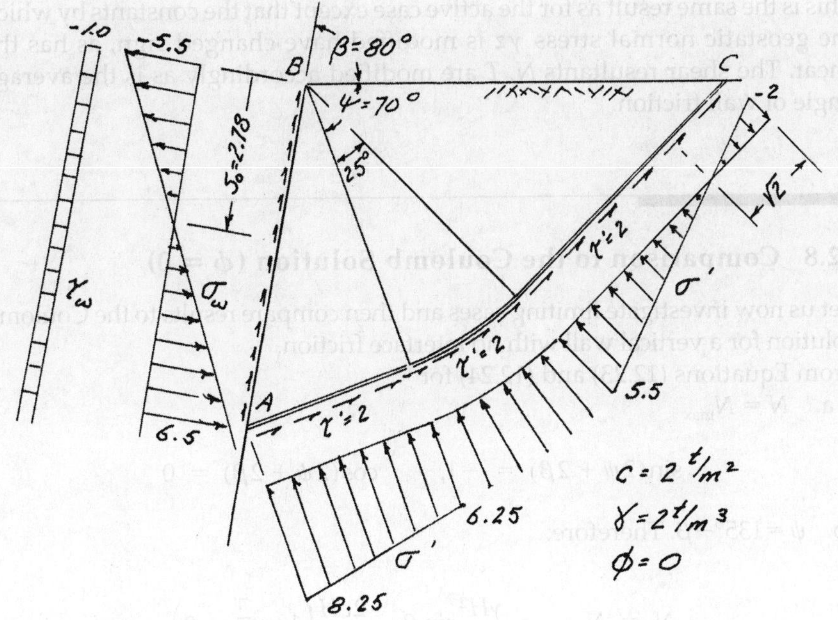

FIGURE 12.10
Limit equilibrium with assumed $\psi = 70°$.

and the direction of the total thrust $P_A = \sqrt{N^2 + T^2}$ is at an "average" angle of wall friction δ given by

$$\tan \delta_A = \frac{T}{N} = \frac{\cos(2\beta + 2\psi)}{\frac{\gamma H}{2c} + \left[\left(\frac{\pi}{2} - 2\psi - 1\right) + \sin(2\beta + 2\psi)\right]} \tag{12.24}$$

A specific example to illustrate this complete stress solution at limit equilibrium for the case $\psi = 70°$ is shown in Figure 12.10.

For the passive case where the wall is forced into the material, the failure surface is the same but the direction of impending motion is revised. Thus the shear stress on the failure surface acts in the opposite direction to resist the motion so that at C, $\tau' = -c$, $\sigma' = +c$ (compression), $\sigma_x = 2c$ and α is now $-45°$. Thus $K = c(1 + \frac{\pi}{2})$, $\alpha_E = -(\frac{\pi}{4} + \psi)$ and, since the direction of the shear on the wall has changed for the passive case,

$$\begin{aligned} \sigma_w &= \sigma_A + c\sin 2\lambda \\ (\tau_w)_A &= c\cos 2\lambda \end{aligned} \tag{e}$$

Therefore, the stresses on the wall for the passive case are:

$$\begin{aligned} \sigma_w &= \gamma z - c\left(\frac{\pi}{2} - 2\psi - 1\right) - c\sin(2\beta + 2\psi) \\ \tau_w &= c\cos(2\beta + 2\psi) \end{aligned} \tag{12.25}$$

This is the same result as for the active case except that the constants by which the geostatic normal stress γz is modified have changed sign, as has the shear. The shear resultants N, T are modified accordingly as is the average angle of wall friction.

12.8 Comparison to the Coulomb Solution ($\phi = 0$)

Let us now investigate limiting cases and then compare results to the Coulomb solution for a vertical wall with no interface friction.

From Equations (12.23) and (12.24) for

a. $N = N_{max}$

$$\sin(2\psi + 2\beta) = -1, \qquad \cos(2\psi + 2\beta) = 0$$

so $\psi = 135° - \beta$. Therefore:

$$N = N_{max} = \frac{\gamma H^2}{2}\sin\beta - \frac{2cH}{\sin\beta}\left(1 + \frac{\pi}{2} - \beta\right)$$

$$T = T_{min} = 0$$

For a vertical wall ($\beta = 90°$), $\psi = 45°$ and the circular zone II disappears (Figure 12.11 a). Therefore

$$N = N_{max} = \frac{\gamma H^2}{2} - 2cH, \qquad T = \delta = 0$$

and theoretically no wall friction develops. Thus we obtain the failure surface and stress distribution identical to Coulomb's solution, $\sigma_w = \gamma z - 2c$ and $\tau_w = 0$, plotted in Figure 12.5 as line ge.

At the other extreme for:

b. $N = N_{min}$; $T = T_{max}$

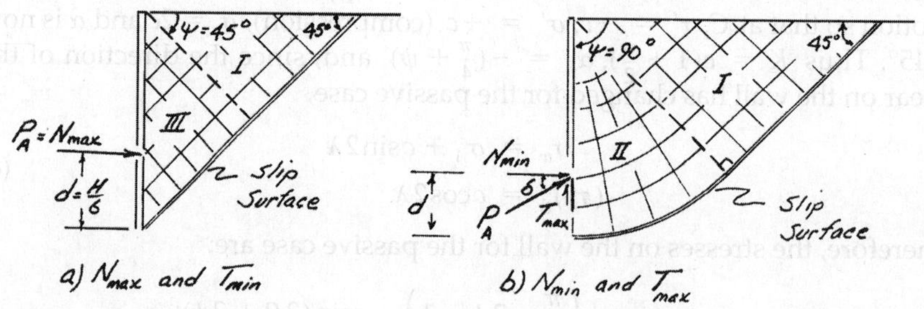

FIGURE 12.11
Limits for a vertical wall.

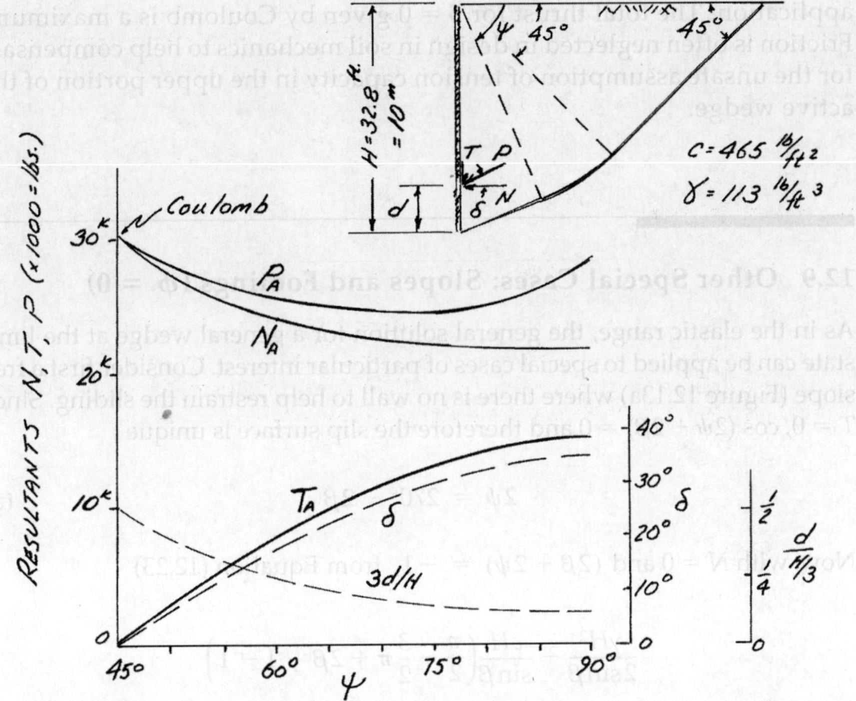

FIGURE 12.12
Solution for a vertical wall—active case.

$\sin(2\psi - 2\beta) = 0$ so $\psi = 180° - \beta$. Therefore:

$$N = N_{\min} = \frac{\gamma H^2}{2}\sin\beta - \frac{cH}{\sin\beta}\left(1 + \frac{3}{2}\pi - 2\beta\right)$$

$$T = T_{\max} = \frac{cH}{\sin\beta}$$

For a vertical wall, $\psi = 90°$ and zone III disappears (Figure 12.11b). Therefore

$$N_{\min} = \frac{\gamma H^2}{2} - 2.57cH, \qquad T_{\max} = cH$$

for a very rough wall along which the shear strength is fully mobilized.

Thus T, N, and δ are functions of ψ, which defines the failure surface. In practice, however, the average angle of wall friction, δ, is usually selected on the basis of the possibilities of wall movement and the relative roughness of the wall (e.g., steel is relatively smooth compared to concrete). For a known δ, T, N, P, the angle ψ and the point of application of the resultant thrust are uniquely defined. The complete solution for a vertical wall is shown in Figure 12.12 where it can be seen that wall friction always helps both by reducing P_A and lowering the point of its application. The total thrust for $\delta = 0$ given by Coulomb is a maximum. Friction is often neglected in design in soil mechanics to help compensate for the unsafe assumption of tension capacity in the upper portion of the active wedge.

12.9 Other Special Cases: Slopes and Footings ($\phi = 0$)

As in the elastic range, the general solution for a general wedge at the limit state can be applied to special cases of particular interest. Consider first a free slope (Figure 12.13a) where there is no wall to help restrain the sliding. Since $T = 0$, cos $(2\psi + 2\beta) = 0$ and therefore the slip surface is unique

$$2\psi = 270° - 2\beta \tag{a}$$

Now, with $N = 0$ and $(2\beta + 2\psi) = -1$, from Equation (12.23)

$$\frac{\gamma H^2}{2\sin\beta} + \frac{cH}{\sin\beta}\left(\frac{\pi}{2} - \frac{3}{2}\pi + 2\beta - 1 - 1\right)$$

and

$$H_{cr} = \frac{4c}{\gamma}\left(1 + \frac{\pi}{2} - \beta\right)$$

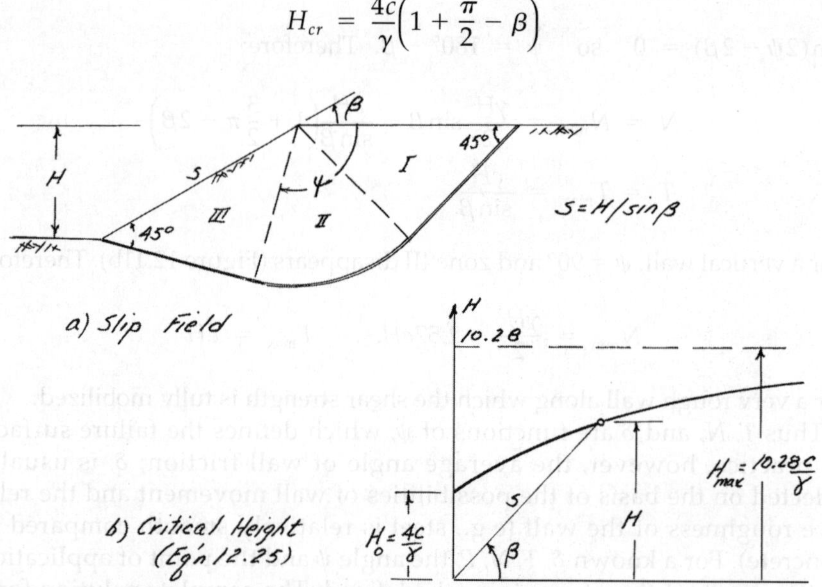

FIGURE 12.13
Solution for any slope with $\phi = 0$ (Equation 14.25).

Since there is no closed-form elasticity solution for a slope loaded by its own weight, this result is particularly important in that at least the stress field at collapse is known to the designer.

The result is plotted in Figure 12.13b. For a vertical cut $\beta = \pi/2$), $H_{cr} = 4c\Upsilon$ which is the same given by Coulomb for a plane surface of sliding. The elasticity solution for a uniform load, $p = \gamma H$, gives $\tau_{max} = \rho/\pi$. Therefore setting τ_{max} equal to c at yield

$$H_{yield} = \frac{\pi c}{\gamma} = \frac{3.14c}{\gamma}$$

and, for a vertical slope, there is little load capacity beyond first yield.

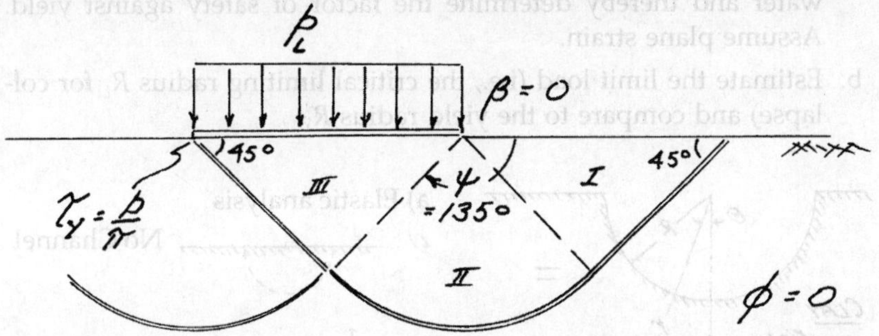

FIGURE 12.14
"Exact" limit analysis for the punch problem.

At the other extreme, as β approaches zero (a very gentle slope), ψ approaches 135° and $H_{cr} = 10.28/\gamma$.*

We have also, in fact, finally derived the exact limit-load solution for the footing problem for which upper-bound estimates were made using circular and sliding block mechanisms. The solution for the passive case for a wall at $\beta = 0$ gives it to us directly. If we assume a smooth footing (the wall) then from Equation (12.25),

$$\tau_w = 0 = c\cos(2\beta + 2\psi)$$

and $\psi = 135°$ giving the failure mechanism as shown in Figure 12.14. The normal stress on the wall is the uniform distributed limit load p_L so

$$\sigma_w = P_L = \gamma z - c\left(\frac{\pi}{2} - 2\psi - 1\right) - c\sin(2\beta + 2\psi)$$

and since $z = 0$ and $\sin(2\beta + 2\psi) = -1$.

$$P_L = -c\left(\frac{\pi}{2} - \frac{3}{2}\pi - 1 - 1\right) = c(\pi + 2) = 5.14c \qquad (12.26)$$

* However, as covered in the chapter problems, the critical failure surface does *not* have to pass through the toe for shallow slopes! This illustrates the unsafe aspect of using the upper-bound theorem. Equation (14.25) is, in fact, limited to a range of β greater than some value left for the student to explore.

This is considered the correct solution. The "best" circular and sliding-block mechanisms from Chapter 12 give $p_L \cong 5.7c$ illustrating the upper-bound theorem. The symmetry of the correct slip surface shows why the unit weight drops out of the expression for p_L when $\phi = 0$.

Example 12.4

A long semi-circular channel with radius R is to be cut from an elastic, perfectly-plastic, incompressible half-space.

a. Solve for the elastic stress field before the channel is filled with water and thereby determine the factor of safety against yield. Assume plane strain.
b. Estimate the limit load (i.e., the critical limiting radius R_L for collapse) and compare to the yield radius R_Y.

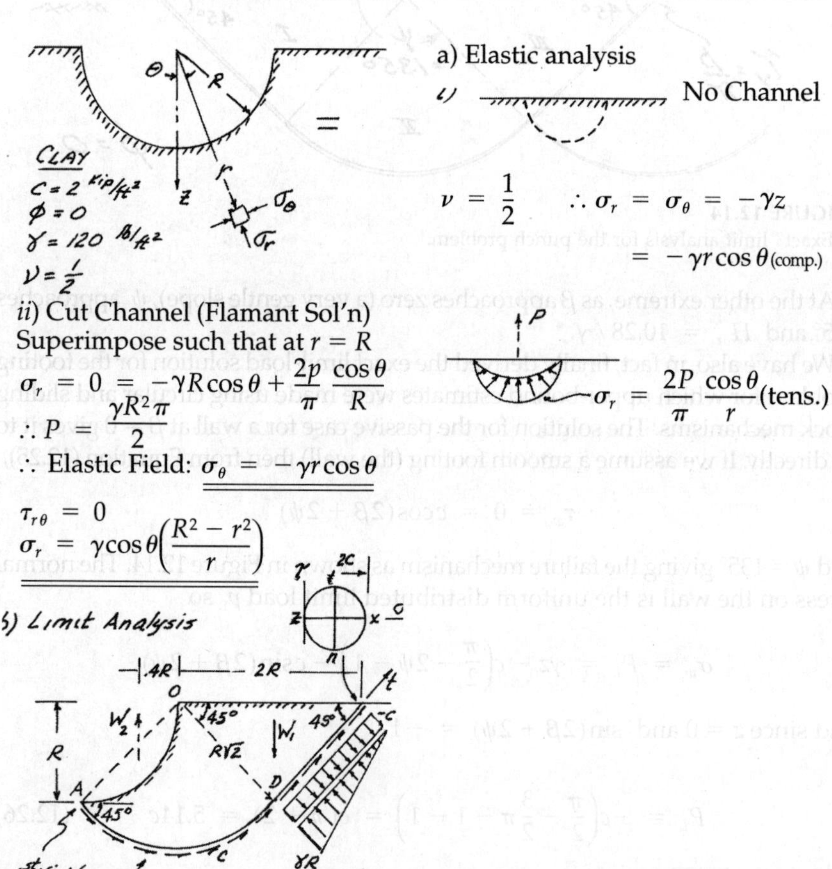

∴ Since we know the point of first yield at A and orientation of slip surface, only one combination of circle and straight line is possible and the slip surface is defined.

∴ First yield at $r = R$, $\theta = 0$ where $\sigma_\theta = -\gamma R = Y = 2c$

$$\therefore\ R_Y = \frac{2c}{\gamma} = \frac{4000}{120} = \underline{\underline{33.3\text{ ft.}}}$$

$$W_1 = \frac{1}{2}(R\sqrt{2})^2\gamma = \gamma R^2, \qquad W_2 = \frac{\gamma\pi R^2}{4} - \frac{R^2}{2}\gamma = \gamma\frac{\pi - 2}{4}R^2 = \frac{1.14}{4}R\gamma$$

ΣM_o

$$\underbrace{W_1(R) + W_2(.4R)}_{\text{driving}} = \underbrace{c(R\sqrt{2})^2 + c\frac{\pi}{2}(R\sqrt{2})^2 - c\frac{(R\sqrt{2})^2}{2} + \frac{1}{2}\gamma R\frac{(R\sqrt{2})^2}{3}}_{\text{resisting}}$$

$$\therefore\ .781\gamma R = 4.14c \qquad \therefore\ R_L = \frac{4.14}{.781}\frac{c}{\gamma} = \underline{\underline{\frac{5.3c}{\gamma}}}$$

$$\therefore\ R_L = \underline{\underline{88\text{ft}}} \qquad K_P = \frac{R_L}{R_Y} = \frac{8}{3} = \underline{\underline{2.67}}$$

12.10 Solutions for Weightless Mohr-Coulomb Materials

When there is a frictional component of shear strength (i.e., $\phi > 0$), but no body forces ($\gamma = 0$), we have shown in Section 12.6 that for uniform surface loadings the slip lines or "characteristics" must be straight or log spirals. For wedges then, the combined failure mechanism will be similar to that for EPS material, but with the circular portion replaced by a log-spiral fan as in Figure 12.15.

We could now work our way along the slip surface from the known boundary point at C on BC to the unknown boundary point A on BA using the solution [Equation (12.20)] to Kötter's equation for $\gamma = 0$. Just as for the previous case for a heavy EPS with $\phi = 0$, this shooting technique would give a solution dependent on the extent of zone II (the angle ψ) which, in turn, is related to the angle of wall friction. Unfortunately this general development and the resulting expressions are complicated and cumbersome. For example, even if we assume a smooth wall so that the stress on the wall is a constant, p_o, the solution is not simple. For this case shown in Figure 12.15 since $\sigma_3 = p_o$, on AB, $\alpha_A = 45° - \beta + \frac{\phi}{2}$ and therefore ψ defining the critical slip surface is known for a given load inclination ξ and slope angle β. The critical limit load p_L is given by Kezdi* as

* Kezdi, Á; *Erddrucktheorien*, Springer Verlag, Verlin, Göttingen, Heidelberg, 1962.

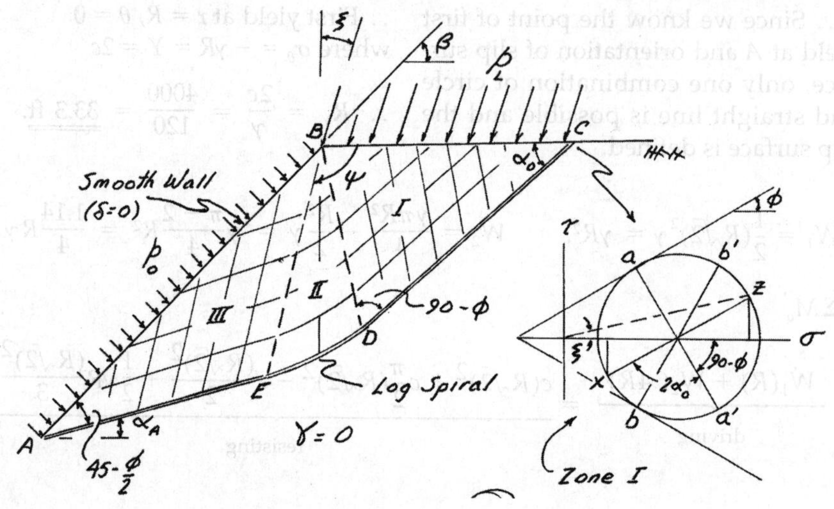

FIGURE 12.15
Slip field for $\gamma = 0$ (shown for $\delta = 0$).

$$
\begin{aligned}
P_L \cos\xi &= P_o\left[\frac{1+\sin\phi\sin(2\alpha_o-\phi)}{1-\sin\phi}\right]e^{(2\alpha_o-2\beta+90^\circ-\phi)\tan\phi} \\
&\quad + c\cot\phi\left[\frac{1+\sin\phi\sin(2\alpha_o-\phi)}{1-\sin\phi}\right]e^{(2\alpha_o-2\beta+90^\circ-\phi)\tan\phi - 1}
\end{aligned}
\tag{12.27}
$$

where the angle of the slip surface in zone I can be determined from the boundary condition at C by Mohr's Circle (Figure 12.15) or the expression

$$\frac{p_L}{c} = \frac{\cos\phi\cos(2\alpha_o-\phi)}{\sin\xi - \sin\phi\cos(2\alpha_o-\phi+\xi)} \tag{12.28}$$

Let us instead work out the specific case of the classic punch or strip footing problem shown in Figure 12.16. If p_L is the major principal stress and slip surface for the active wedge (zone I) must be at $45+\frac{\phi}{2}$. Therefore at D (in zone I and zone II) from Mohr's Circle

$$\sigma' = p_L - \frac{\tau'}{\cos\phi} - \tau'\tan\phi = (\tau'-c)\cot\phi \tag{a}$$

or

$$p_L = \tau'\left(\cot\phi + \tan\phi + \frac{1}{\cos\phi}\right) - c\cot\phi \tag{b}$$

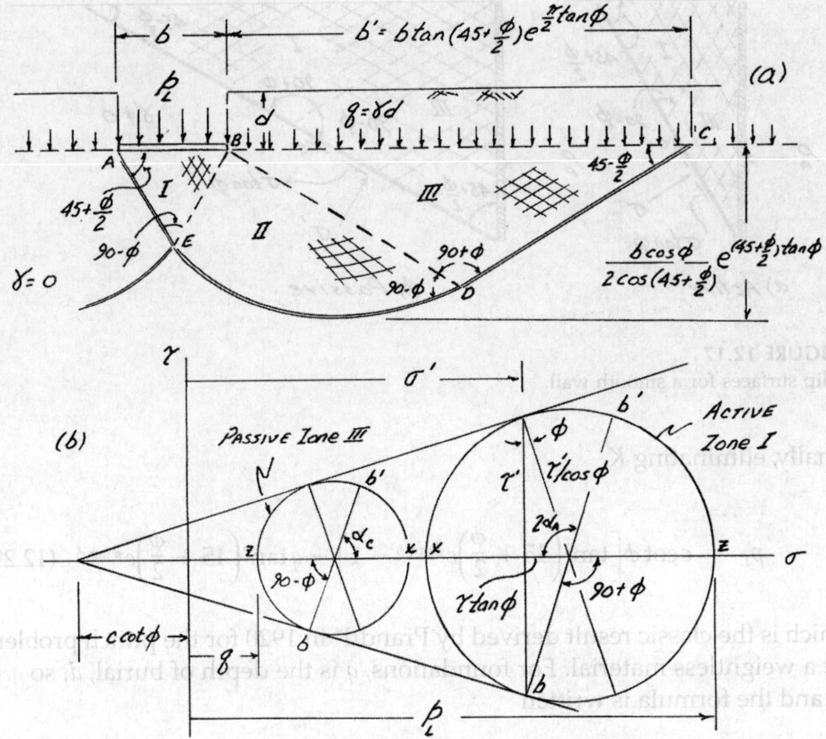

FIGURE 12.16
Solution to the "Punch Problem" for a $\phi - c$ material a) Prandtl mechanism (slip field); b) stresses in the active and passive wedges.

But since D is also on the curved log-spiral, the solution [Equation (12.20a)] to Kötter's equation applies and

$$\tau' = Ke^{2\alpha\tan\phi} = Ke^{-2\left(45+\frac{\phi}{2}\right)\tan\phi} \tag{c}$$

Therefore

$$p_L = \left[Ke^{-2\tan\phi\left(45+\frac{\phi}{2}\right)}\right]\left[\frac{1+\sin\phi}{\sin\phi\cos\phi}\right] - c\cot\phi \tag{d}$$

The same procedure can now be used at point E, which is on the log-spiral fan and wedge III. In this case $\sigma_3 = q$, so

$$q = \left[Ke^{-2\tan\phi\left(\frac{3}{4}\pi+\frac{\phi}{2}\right)}\right]\left[\frac{1-\sin\phi}{\sin\phi\cos\phi}\right] - c\cot\phi$$

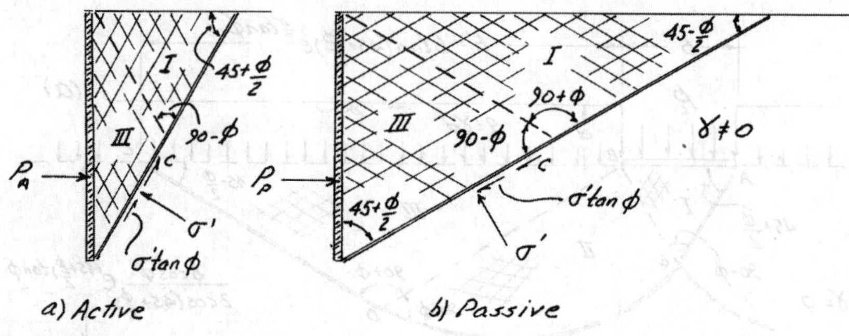

FIGURE 12.17
Slip surfaces for a smooth wall.

Finally, eliminating K,

$$p_L = c\cot\phi\left[\tan^2\left(45+\frac{\phi}{2}\right)e^{\pi\tan\phi}-1\right]+q\tan^2\left(45+\frac{\phi}{2}\right)e^{\pi\tan\phi} \quad (12.29)$$

which is the classic result derived by Prandtl* in 1920 for the punch problem for a weightless material. For foundations, q is the depth of burial, d, so $q = \gamma d$ and the formula is written

$$P_L = cN_c + qN_q \quad (12.30a)$$

where

$$N_q = \tan^2\left(45+\frac{\phi}{2}\right)e^{\pi\tan\phi} \quad (12.31a)$$

$$N_c = (N_q - 1)\cot\phi \quad (12.31b)$$

As a final example, consider the case of a vertical retaining wall. If the wall is smooth, the normal stress on it is principal and the Coulomb solution, Equation (12.3) or (12.5), must be correct since zone II disappears (Figure 12.17) and the slip surface is straight. It is important to note that when the slip lines are straight it is not necessary to assume the material weightless since α is a constant and Kötter's equation can be integrated directly giving the Coulomb solution.

* We encountered Prandtl in Chapter 9 on torsion where there is a brief footnote concerning his life as an engineer and "founder," perhaps, of what today is called applied mathematics. His derivation of Equation (12.29) as presented in Über die Härte plastischer Körper, Nach.r.K.Ges. Wiss, Gött., Math.-Phys. Kl., 1920, 74–85 is, however, not the same as ours.

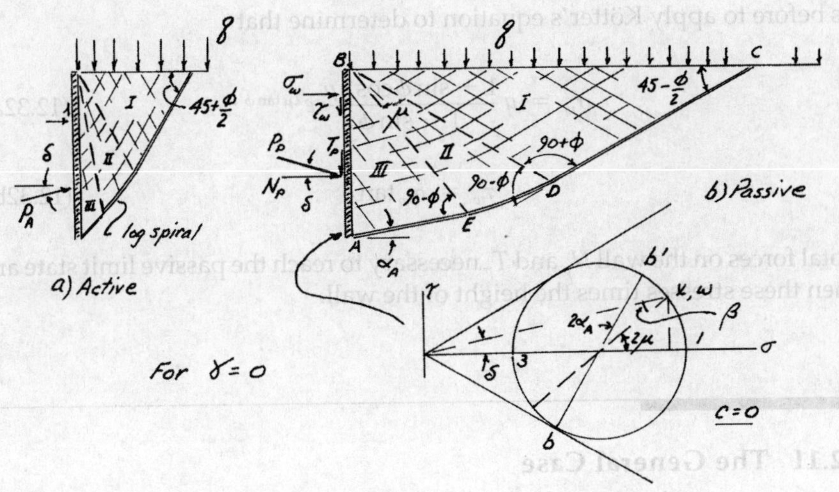

FIGURE 12.18
Slip surfaces for a rough wall ($\gamma = 0$).

However, if the wall is rough ($\delta > 0$) we must assume $\gamma = 0$ to integrate along the log-spiral portion of the slip surfaces for either the active or passive cases as shown in Figure 12.18. It will turn out that the effect of roughness is not very significant for the active case while it is very important for passive failure. Therefore, let us solve the passive case and leave the active case for a chapter problem. To simplify the situation further, let us assume that $c = 0$.

The stresses on the wall are constant so that at any depth the Mohr's Circle is as shown in Figure 12.18. Therefore

$$2\mu + 2\alpha_w = 90 - \phi \tag{f}$$

where μ is the angle from the x axis to the major principal stress. But it can be shown from Mohr's circle (Figure 12.18) that

$$2\mu = 2\beta - (90 - \phi) \tag{g}$$

and

$$\sin\beta = \frac{\sin\delta}{\sin\phi} \tag{h}$$

so that μ, which is the change in angle from D to E as we move along the log-spiral fan, can be found in terms of any angle of wall friction. Since we now know the critical shear stress on the slip surface at D and E, we can proceed as before to apply Kötter's equation to determine that

$$\sigma_w = q\frac{1+\sin\phi\cos 2\mu}{1-\sin\phi}e^{2\mu\tan\phi} \tag{12.32a}$$

$$\tau_w = \sigma_w \tan\delta \tag{12.32b}$$

Total forces on the wall N_p and T_p necessary to reach the passive limit state are then these stresses times the height of the wall.

12.11 The General Case

Unfortunately Kötter's equation cannot be integrated directly for a closed-form solution in the general case of a Coulomb material with weight or for nonconstant boundary loads. The most common numerical procedure for integrating Kötter's equation was developed by V.V. Sokolovski* where the integration proceeds from one boundary to the other in the direction of the characteristic slip lines using a finite-difference approximation. This shooting technique called "the method of characteristics" is necessary since, as already discussed, the governing equations are hyperbolic and, if the load is specified completely on the entire boundary, a solution will not normally exist.

The result then may be an "exact" solution for the slip field but only approximate the load prescribed. Take for example the footing problem shown again in Figure 12.19. We would like to solve for the *uniform* load intensity p_L that will cause collapse by forcing the material to shear under the punch and bulge upward lifting the *uniform* surcharge q upward with the passive wedge of region I. The solution starts from the boundary BC on which $\sigma_z = q =$ constant, $\tau_{xz} = 0$, and b' is selected to some convenient scale. However, in this case region I is the passive Coulomb solution with $\sigma_3 = q$ so the stresses on BD are known and we can actually start from there. Recurrence formulas for piecewise integration of Kötter's equation along the characteristics in the fan-shaped Region II and the Active Region III are derived by Sokolovski. The accuracy depends on how dense we make the finite-element grid of characteristics (or slip lines).

The resulting slip-line field and limit loading $p(x)$ are shown in Figure 12.19b for a smooth, surface footing ($q = 0$) with $c = \gamma = 1$ and $\phi = 30°$. Neither the width of the footing, b, nor the distribution of the limit load p_L can be specified in advance and we see that p_L is far from uniform. Thus we have a

* The second edition of his book *Statics of Soil Media* as translated by Jones and Schofield was published by Butterworths, London, 1960. A more modern version of the method is presented in *Foundations of Theoretical Soil Mechanics* by M.E. Harr, McGraw-Hill, 1966.

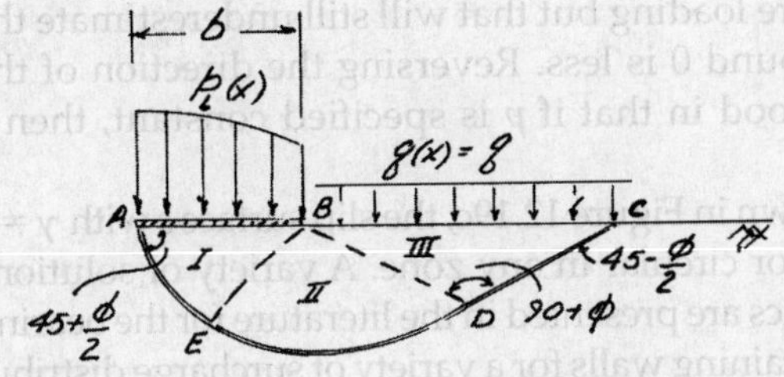

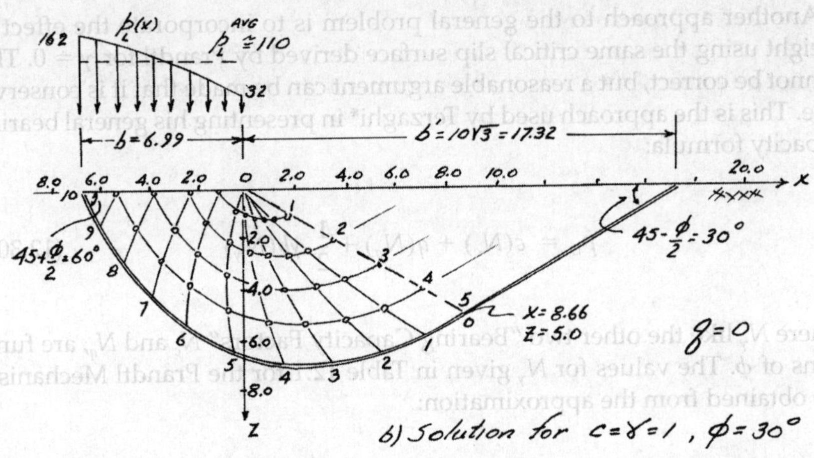

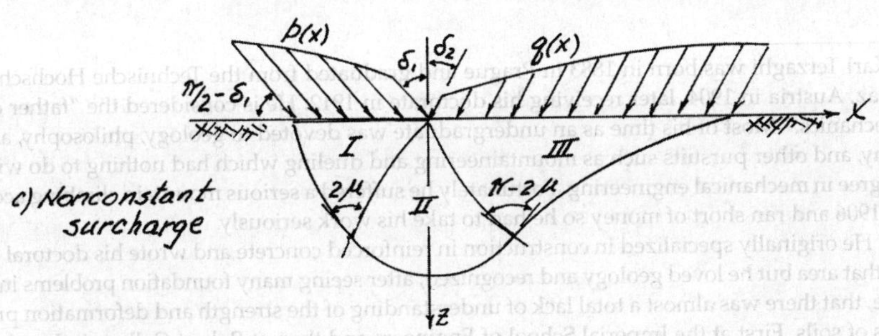

FIGURE 12.19
Solution by method of characteristics ($\gamma \neq o$).

method that we can make as accurate as we want to solve for the wrong load! One can use an equivalent limit load intensity

$$(p_L)_{\text{equiv.}} = \int_{o}^{-b} \frac{p_L(x)}{b}$$

to approximate a uniform failure loading but that will still underestimate the capacity since the moment around 0 is less. Reversing the direction of the shooting technique does no good in that if p is specified constant, then q becomes a function of x.

If q is not constant, then as shown in Figure 12.19c, the slip surface s with $\gamma \neq 0$ is neither straight, a log spiral, or circular in any zone. A variety of solutions using the method of characteristics are presented in the literature for the bearing capacity problem, slopes, and retaining walls for a variety of surcharge distributions. Since these seldom correspond to realistic-boundary conditions they are primarily of academic interest.

Another approach to the general problem is to incorporate the effect of weight using the same critical slip surface derived by Prandtl for $\gamma = 0$. This cannot be correct, but a reasonable argument can be made that it is conservative. This is the approach used by Terzaghi* in presenting his general bearing capacity formula:

$$p_L = c(N_c) + q(N_q) + \frac{1}{2}\gamma b(N_\gamma) \tag{12.30b}$$

where N_γ like the other two "Bearing Capacity Factors" N_c and N_q, are functions of ϕ. The values for N_γ given in Table 12.1 for the Prandtl Mechanism are obtained from the approximation:

$$N_\gamma = 2(N_q + 1)\tan\phi \tag{12.31c}$$

---

* Karl Terzaghi was born in 1883 in Prague and graduated from the Technische Hochschule in Graz, Austria in 1904, later receiving his doctorate in 1912. He is considered the "father of soil mechanics." Most of his time as an undergraduate was devoted to geology, philosophy, astronomy, and other pursuits such as mountaineering and dueling which had nothing to do with his degree in mechanical engineering. Fortunately he suffered a serious mountain climbing accident in 1906 and ran short of money so he had to take his work seriously.

He originally specialized in construction in reinforced concrete and wrote his doctoral thesis in that area but he loved geology and recognized, after seeing many foundation problems in practice, that there was almost a total lack of understanding of the strength and deformation properties of soils. First at the Imperial School of Engineers and then at Robert College in Istanbul, he developed the equipment and procedures to test soils—particularly rich clays in compression. This, in turn, led to his basic insight into the transfer of applied pressure from pore water to grain structure and the consequent deformation of clay modeled as a Kelvin Body (Chap. 3) through the defusion equation. His paper in 1924 on the consolidation process and his book *Erdbaumechanik auf Bodenphysikalischer Grundlage* published in 1925 ushered in a new field of human knowledge still being explored at a basic level. Of all the practitioners of rational mechanics there is no other more original, diversified, ingenious, and practical than Terzaghi. All engineers can enjoy and profit from reading *From Theory to Practice in Soil Mechanics*, John Wiley & Sons, 1960 which is a selection of the writings of Terzaghi (including the early classics translated to English) with an introduction on Terzaghi's life and achievements by four of the leading pioneers in the field who followed in his footsteps.

The power of the upper-bound approach to limit analysis, however, is supposed to be the simplicity of the method. To return to this theme, let us conclude by introducing a simple failure mechanism based, for the bearing capacity problem, on Coulomb wedges, which is more versatile, easy for computation, and is intuitively satisfying.

12.12 An Approximate "Coulomb Mechanism"

For the Coulomb wedge in the active case, the free body diagram for no cohesion is as shown in Figure 12.20a. Let i be the angle at which the surface of the backfill rises above the horizontal, β be the angle between the wall and the plane, which is vertical to the horizontal surface of the ground in front of the wall, and ρ_A be the angle between the failure surface and the horizontal.

From the force equilibrium of the failure wedge, the maximum value of P_A, the total force acting on the wall, is:

$$P_A = \frac{1}{2}\gamma H^2 K_A \tag{12.8}$$

wherein

$$K_A = \frac{\cos^2(\phi - \beta)}{\cos^2\beta\cos(\delta + \beta)\left[1 + \sqrt{\frac{\sin(\phi + \delta)\sin(\phi - i)}{\cos(\delta + \beta)\cos(i - \beta)}}\right]^2} \tag{12.32a}$$

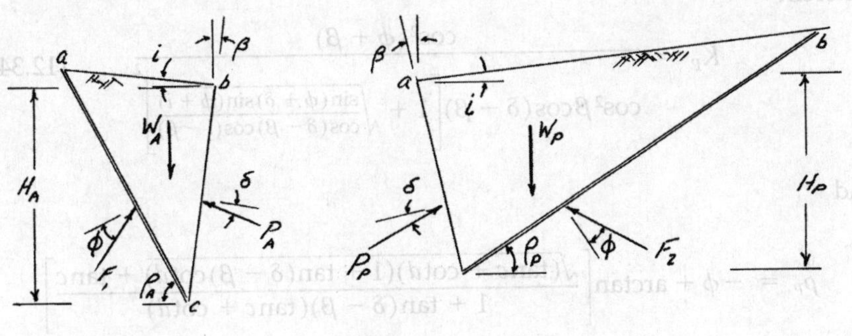

FIGURE 12.20
Equilibrium of Coulomb wedges.

Also

$$\rho_A = \phi + \arctan\left[\frac{\sqrt{\tan a(\tan a + \cot b)(1 + \tan(\delta+\beta)\cot b)} - \tan a}{1 + \tan(\delta+\beta)(\tan a + \cot b)}\right] \tag{12.33a}$$

in which

$$a = \phi - i \qquad b = \phi - \beta$$

For $i = 0$ and $\beta = 0$:

$$K_A = \frac{\cos^2\phi}{\cos\delta\left[1 + \sqrt{\frac{\sin(\phi+\delta)\sin\phi}{\cos\delta}}\right]} \tag{12.33b}$$

$$\rho_A = \phi + \arctan\left[\frac{\sqrt{\tan\phi(\tan\phi + \cot\phi)(1 + \tan\delta\cot\phi)} - \tan\phi}{1 + \tan\delta(\tan\phi + \cot\phi)}\right] \tag{12.33c}$$

Similarly, the corresponding expression for the passive thrust (the free body diagram is shown in Figure 12.20b) is:

$$P_P = \frac{1}{2}\gamma H^2 K_P \tag{12.8}$$

wherein

$$K_P = \frac{\cos^2(\phi+\beta)}{\cos^2\beta\cos(\delta-\beta)\left[1 + \sqrt{\frac{\sin(\phi+\delta)\sin(\phi+i)}{\cos(\delta-\beta)\cos(i-\beta)}}\right]^2} \tag{12.34a}$$

and

$$\rho_P = -\phi + \arctan\left[\frac{\sqrt{(\tan c + \cot d)(1 + \tan(\delta-\beta)\cot d)} + \tan c}{1 + \tan(\delta-\beta)(\tan c + \cot d)}\right] \tag{12.34b}$$

in which

$$c = \phi + i, \qquad d = \phi + \beta$$

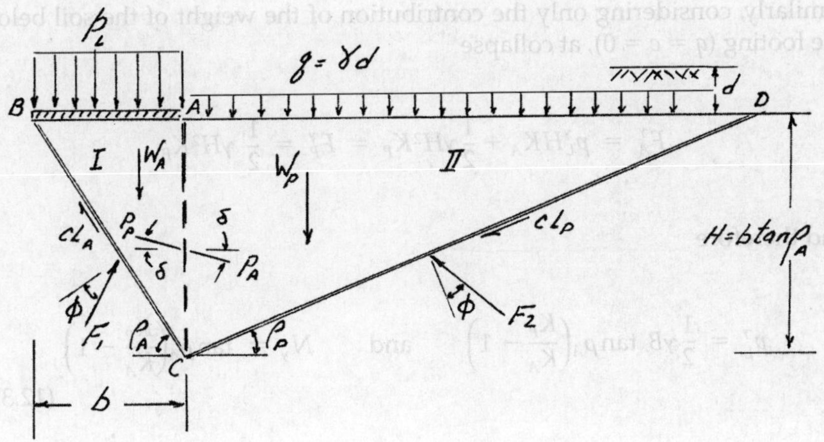

FIGURE 12.21
Coulomb mechanism for bearing capacity.

For $i = 0$ and $\beta = 0$,

$$K_P = \frac{\cos^2\phi}{\cos\delta\left[1 - \sqrt{\frac{\sin(\phi+\delta)\sin\phi}{\cos\delta}}\right]^2} \tag{12.35a}$$

$$\rho_P = -\phi + \arctan\left[\frac{\sqrt{\tan\phi(\tan\phi + \cot\phi)(1 + \tan\delta\cot\phi)} + \tan\phi}{1 + \tan\delta(\tan\phi + \cot\phi)}\right] \tag{12.35b}$$

A very useful simplification of the Prandtl or Sokolovski failure surfaces is, as shown in Figure 12.21, to eliminate the fan-shaped transition zone and thereby average its effect concentrating the shear transfer at AC. Line AC can then be thought of as a retaining wall with the active lateral thrust P_A from region I pushing against the passive resistance P_P from region III. From equilibrium of the two Coulomb wedges involved, bearing-capacity factors for each component of strength can be derived. Considering first the surcharge ($c = \gamma = 0$)

$$F_A = p_L^q HK_A = F_P = qHK_P$$

where P_L^q equals the load component supported by q. Thus

$$p_L^q = \frac{qK_P}{K_A} \quad \text{and} \quad N_q = \frac{K_P}{K_A} \tag{12.36}$$

Similarly, considering only the contribution of the weight of the soil below the footing ($q = c = 0$), at collapse

$$F_A^\gamma = p_L^\gamma H K_A + \frac{1}{2}\gamma H^2 K_P = F_P^\gamma = \frac{1}{2}\gamma H^2 K_P$$

and therefore

$$p_L^\gamma = \frac{1}{2}\gamma B \tan\rho_A\left(\frac{K_P}{K_A} - 1\right) \quad \text{and} \quad N_\gamma = \tan\rho_A\left(\frac{K_P}{K_A} - 1\right) \tag{12.37}$$

Finally, it might be noted that for only cohesion ($\phi = 0$), equilibrium considerations of the Coulomb mechanism gives a value of $N_c = 6.0$, which is not unreasonable compared to the exact value of 5.14.

The corresponding static-bearing capacity factors N_q and N_γ for a granular soil for various values of ϕ and δ can now be calculated for the Coulomb mechanism from Equations (12.36) and (12.37) with the expressions for K_A, K_P, and ρ_A. The results are shown in Table 12.2 compared to the corresponding values for the Prandtl-type mechanism and solution by the method of characteristics. For the correct kinematic Coulomb mechanism, the wedges must slip relative to each other and $\delta = \phi$. This, by the upper-bound theorem of limit analysis, must overestimate the bearing capacity, which it does. However, since the fictitious wall replaces the fan-shaped transition zone, it should average the shear transfer, and setting $\delta = \phi/2$ seems more reasonable as an approximation. As shown in Table 12.1, such an assumption for δ does in fact give a very good correlation (± ~ 10%) between the mechanisms for all values of ϕ. The values for N_γ by the method of characteristics in Table 12.2 are those given by Sokolovski. His values seem to have been computed by simply taking the average

$$(P_L) = \frac{b}{2}(p_{x=0} + p_{x=-b}) = b\left(N_c c + N_q q + \frac{1}{2} b N_\gamma\right)$$

rather than computing the limit load P_L using the integral of the limit pressure distribution. Therefore, they are unnecessarily low since both the curvatures of the actual critical load intensity computed and its eccentricity are neglected. If we assume the distribution is parabolic, the value is ~15% higher, which is a better representation but still conservative. This modification is reflected in Figure 12.22 which gives the standard accepted value for bearing capacity factors for design of shallow strip footings.

TABLE 12.2

Comparison of Bearing Capacity Factors by Different Mechanisms

	ϕ	0°	10°			20°			30°			40°		
	δ	0	0	5°	10°	0	10°	20°	0	15°	30°	0	20°	40°
	K_A	1	.704	.662	.635	.49	.447	.427	1/3	.301	.297	.217	.199	.210
	ρ_a (deg.)	45°	50	45.2	41.8	55	51.1	48.1	60	56.9	54.3	65	62.6	60.4
	K_P	1	1.42	1.57	1.73	2.04	2.635	3.525	3	4.978	10.09	4.60	11.77	92.6
	ρ_P (deg.)	45	40	33.2	28.4	35	27.0	21.2	30	20.7	13.4	25	14.1	4.8
N_q	Sokolovski	1	2.47			6.40			18.40			64.20		
	Prandtl	1	2.47			6.40			18.40			64.20		
	Coulomb	1	2.02	2.37	2.73	4.16	5.90	8.26	9.0	16.5	34.0	21.1	59.0	144.1
N_γ	Sokolovski	0	.46			3.16			15.32			86.5		
	Prandtl	0	1.22			5.39			22.40			109.41		
	Coulomb	0	1.21	1.38	1.54	4.51	6.06	8.10	13.8	23.7	46.0	43.2	114	774
N_c	Sokolovski	5.14	8.34			14.8			30.1			75.3		
	Prandtl	5.14	8.34			14.8			30.1			75.3		
	Coulomb[a]	6.0												

[a] Chapter Problem P12.36.

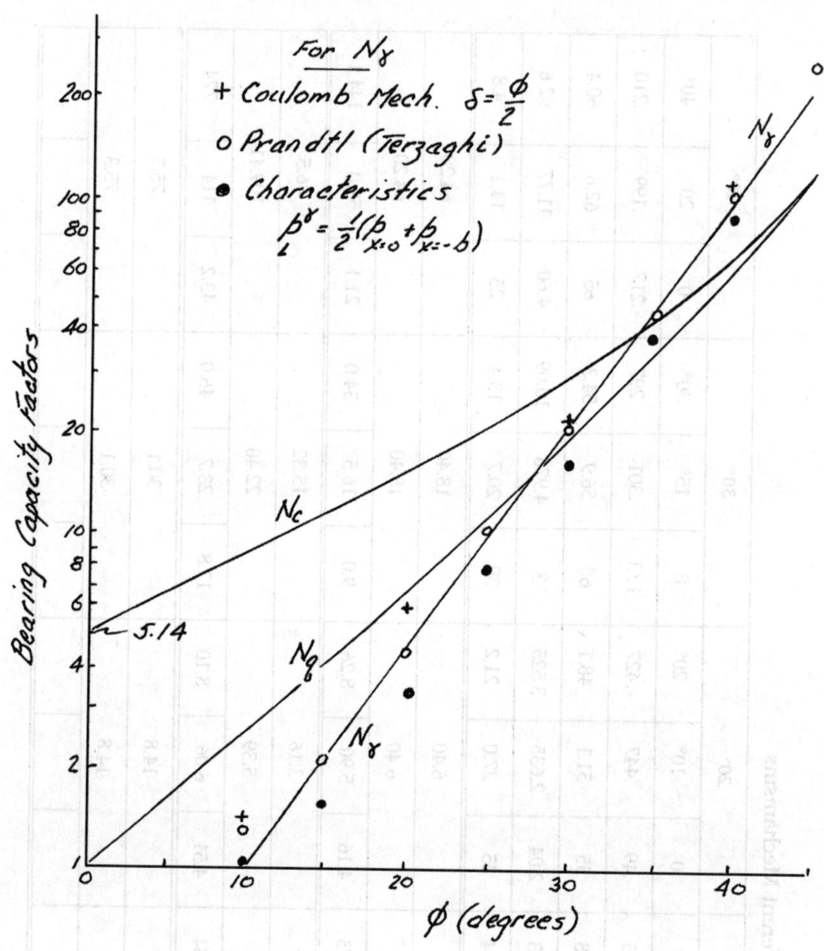

FIGURE 12.22
Bearing capacity factors for shallow strip-footings.

12.13 Problems and Questions

P12.1 Determine from either the equations of Section 12.1 or Figure 12.2 the normal and tangential stresses on the passive and active failure surfaces in terms of σ_1 and σ_3. Derive Equations (12.3).

P12.2 Determine the position (height above the base of the wall) of the resultant active and passive forces P_A and P_P from the Coulomb analysis [Equation (12.5) and Figures 12.4 and 12.5].

P12.3 Derive the free field stress solution for the "heavy half-space" given by Equations (12.4). Is this solution sensible in terms of a geologic deposit of soil, a monolithic pouring of concrete, a block of steel?

P12.4 Redo Example 12.1 for $c = 400\ \text{lb/ft}^2$. Compare P_A and the critical slip surface.

P12.5 Redo Example 12.1 for no concentrated load q and compare to the results from Equations (12.8), (12.32a), and (12.33a).

P12.6 Derive Equation (12.14) from Equation (12.13) as referred to in the text. Note that

$$du = \frac{\partial u}{\partial x}dx + \frac{\partial u}{\partial z}dz, \quad dz = dx \tan\alpha \quad \text{where} \quad \alpha = \theta \pm \left(45 - \frac{\phi}{2}\right).$$

P12.7 Assume uniform vertical and horizontal accelerations $a_v = k_v g$ and $a_h = k_h g$ so that the body forces are $B_x = k_h\gamma$, $B_z = -(1 - k_v)\gamma$. If we define $\bar{\sigma} = \frac{\sigma_x + \sigma_z}{2} + c\cot\phi = \frac{q}{\sin\phi}$ derive the more general Kötter equation:

$$d\sigma \pm 2\sigma\tan\phi(d\theta) =$$
$$[-k_h\gamma \pm (1 - k_v)\tan\phi]\,dx + [(1 - k_v)\gamma \pm k_h\tan\phi]\,dz.$$

P12.8 Obtain Equation (12.16) from Equation (c) above it. Is there an easier way to deduce this equation directly from Kötter's equation?

P12.9 Fill in the missing steps in deriving Equation (12.18) from Equation (12.16) only outlined in *d*, *e*, and *f* in Section 12.5.

P12.10 For the case $\gamma = 0$ in Section 12.6, it is stated that the frictional stress resultant p at the angle ϕ from the failure surface acts everywhere through an origin 0 in polar coordinates. Why (and under what conditions) is this true?

P12.11 Derive Equation (12.20b).

P12.12 Draw (on one set of stress axes) Mohr's Circle for the state of stress at points *C*, *D*, *E*, *A*, and *B* in Figure 12.9. Be sure to show the orientation x, z and n, t in each case. Scale the relative depths from the figure and show how the solution works by getting the same Mohr's Circle at *B* that you started with at *C*.

P12.13 Determine for the active case the failure surface and stresses on the wall (for a $\phi = 0$ material as in Figure 12.9) if there is a *constant* traction on the top surface BC as shown in Figure P12.13.

a. surcharge, q, acting downward. Plot for $q - c$,

b. shear, s, acting toward *B*. Plot for $s = -c/2$,

c. shear, s, acting toward C. Plot for $s = c/2$

P12.14 Go through the steps in deriving Equation (12.25) only outlined in the text. In particular justify $\alpha = -45°$ for the passive case rather

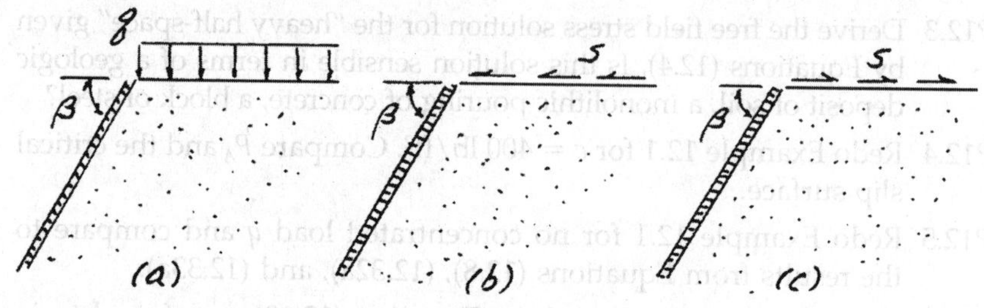

FIGURE P12.13

than +45° for the active case. Is there an alternative way to obtain the same result?

P12.15 As in P12.13, but for the passive case. Note P12.14 and refer to Figure P12.13.

P12.16 Do the example in Figure 12.10 for the passive case and compare.

P12.17 Make a rough equilibrium check for the example shown in Figure 12.10 with two force equations and $\Sigma M_B = 0$. Calculate and show the thrust on the wall, its orientation, δ, and its location.

P12.18 Replot the stresses for the example shown in Figure 12.10 for the passive case and check overall equilibrium. Calculate and show the thrust on the wall, its orientation, and its location.

P12.19 Solve and plot the limit–equilibrium stress state for a wall with $\beta = 75°$, $\psi = 90°$ and a $\phi = 0$ backfill for:

a. The active state

b. The passive state

P12.20 For the example in Figure 12.10 compare the thrust on the wall from the "exact" solution shown to that for a straight failure surface AC and circular failure surfaces R = BA, R = BC, and $R = \frac{1}{2}$ (BA + BC) all at the same average angle of wall friction δ given by Equation (12.24). Discuss the results in relationship to the upper-bound theorem.

P12.21 For a vertical wall, determine the maximum and minimum thrust components for the *passive* case (as in Section 12.8 for the active case). Compare the total thrust, P_p, to the Coulomb solution for the extreme cases. Is the assumption $\delta = 0$ safe or unsafe for design?

P12.22 For a vertical wall, derive a general expression for the position of the force resultant in the active case (i.e., an expression for d in Figure 12.12). Note that $d \lessdot H/6$.

P12.23 Derive Figure 12.12 for a vertical wall for the passive case.

P12.24 In Figure 12.13b where is ψ? What is the horizontal ordinate? The limiting case $H_{cr} = 10.28c/\gamma$ corresponds to what bearing capacity problem?

P.12.25 Work out the solution for the critical height of a 30° slope (i.e., β = 30°) showing the normal and shear stresses on the failure surface and that overall equilibrium is satisfied.

P.12.26 The assumption of our solution for slopes [Equation (12.25)] is that the failure surface goes through the toe. Is this necessarily true? (Hint: For the 30° slope of the previous problem try a deep circular failure surface.)

P.12.27 Referring to P12.26. What is going on here? Try to derive a general expression for "Deep" vs. "Toe"-type slope failure or at least determine at what slope angle β the solution given by Eqaution (12.25) no longer applies.

P12.28 Determine the limit values $\xi = \xi_L$ for the inclination of the collapse load p_L on a surface footing using Equations (12.27) and (12.28). (Hint: This is the special case $p_o = \beta = 0$ for the failure mechanism of surface sliding $\alpha_o = 0$.) Construct the graph of α_o (vertically) vs. the angle of incination for $\phi = 0, 10, 20, 30, 40°$. Note the value for $\alpha_o = 0$.

P12.29 Plot the solution [Equation (12.27)] for $\frac{P_L}{c}$ (vertical ordinate) for a surface footing ($\beta = 0, p_o = 0$) for $\phi = 0, 10, 20, 30, 40°$ vs. tan ξ. You may want to look at P12.28 first. How do the values for $\xi = 0$ compare to the Prandtl solution?

P12.30 To what value does N_c in Equation (12.30) converge as ϕ goes to zero? To what value should it converge?

P12.31 Derive Coulomb's solution [Equation (12.3)] for a smooth vertical wall by integrating Kötter's equation [Equation (12.14)] for both the active and passive case (Figure 12.17).

P12.32 Solve the active case for a rough vertical wall as shown in Figure 12.18. Compare the result for $\delta = \phi = 30°$ to the smooth wall.

P12.33 Consider the solution for a rough wall and $\gamma = 0$ given by Equation (12.32) for Figure 12.18. Calculate the passive thrust components and plot the slip surfaces (in one figure) for $\delta = 0$, $\delta = \frac{\phi}{2}$ and $\delta = \phi$.

P12.34 Give a full derivation of Equation (12.32) calculating τ' at points D and E using Kötter's equation. Can this solution be rewritten in terms of δ rather than μ? What happens if $c \neq 0$?

P12.35 Derive a solution for a rough, vertical wall for the active case (Figure 12.18a) with $\gamma = \phi = 0$. Calculate the active thrust components N_A and T_A for $\delta = 0$, $\delta = \phi/2$, and $\delta = \phi$ and compare the slip surfaces.

P12.36 Using the Coulomb mechanism for bearing capacity (Figure 12.21), derive an expression for N_c as a function of ϕ and δ. Compare the result to those given in Table 12.2 by the Prandtl and Sokolovski solutions.

P12.37 The prismatic (square $a \times a$ cross-section) block $\ell = 10a$ of Figure 5.4 is accelerated in the z direction. If the shearing strength is a constant, c, what is the critical acceleration for failure (i.e., $a_{cr} = k_{cr}g$) by:

a. Formal plasticity (lower-bound solution).

b. Trial and error (upper-bound solution).

Attempt an approximate solution for the problem if the material also has frictional strength ϕ. Work out the numbers for $c = 500$ psi, $\phi = 30°$, $\gamma = 144$ lb/ft³, and $\ell = 10a = 120$ ft.

P12.38 A 10-foot square plate is buried 20 ft deep to anchor a vertical pull of 60,000 lbs as shown. Circulate the factor of safety. Refer to Figure P12.38.

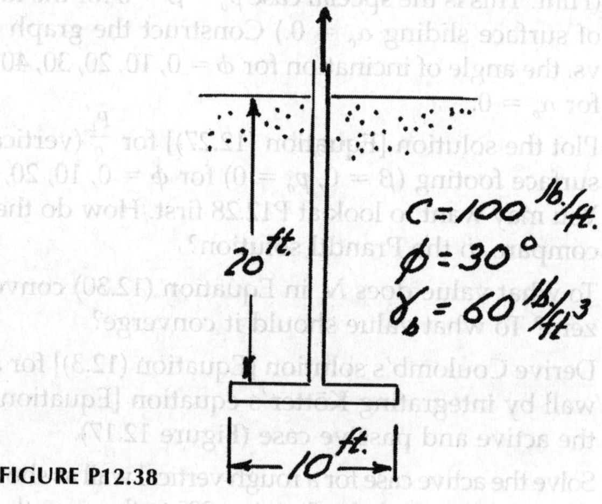

FIGURE P12.38

Plot your results in terms of a stability number $N = \gamma d/c$ vs. the aspect ratio D/r. What would happen if the strength increased with depth (i.e., $c = c_o + Az$)? (This is usually the case in soils.)

P12.39 At what internal pressure p_L will a thick ring $r_o = 2r_i$ in plane stress burst if it is made of steel $Y = 40{,}000$ psi.

Redo the problem for an external pressure if the ring is made of:

a. Steel: $Y = 40{,}000$ psi, $\phi = 0$

b. Concrete: $c = 300$ psi, $\phi = 30°$.

P12.40 A concrete ($c = 300$ psi, $\phi = 30°$, $\gamma = 150$ lb/ft³) gravity dam is in plane strain as shown in Figure P12.40. Estimate the factor of safety against collapse.

a. neglecting ϕ

b. including ϕ

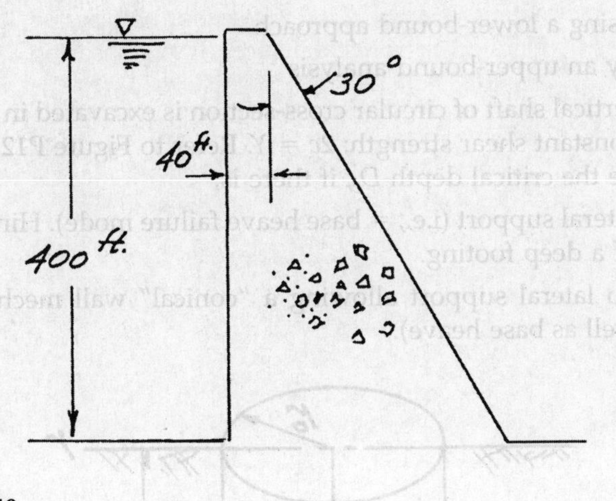

FIGURE P12.40

P12.41 The half-space (in plane strain) is loaded as shown in Figure P12.41. Determine the limiting load k_L for

a. c only material

b. c, ϕ, and γ material

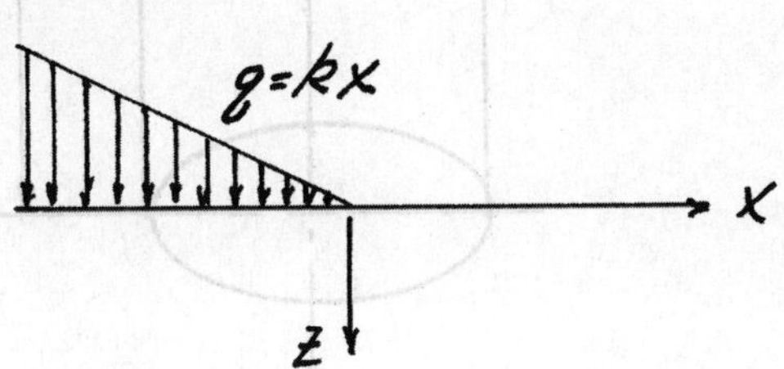

FIGURE P12.41

P12.42 A 30° tapered beam of steel ($\gamma = \phi = 0, Y = 2c$) is loaded uniformly on one edge as shown in Figure P12.42. Find the limit load w_L by:

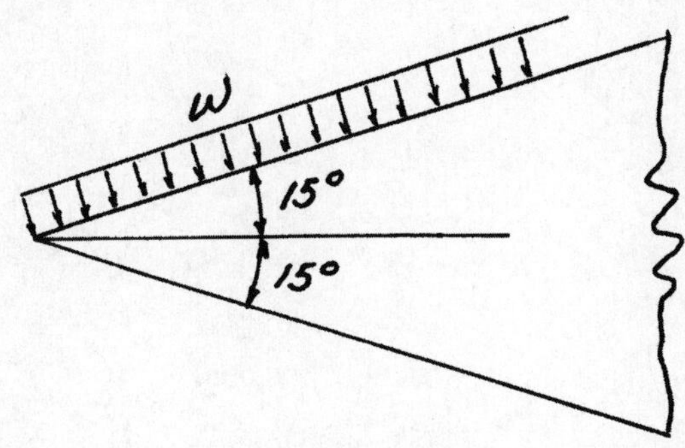

FIGURE P12.42

a. using a lower-bound approach

b. by an upper-bound analysis

P12.43 A vertical shaft of circular cross-section is excavated in a material of constant shear strength: $2c = Y$. Refer to Figure P12.43. Determine the critical depth D_{cr} if there is:

a. lateral support (i.e., = base heave failure mode). Hint: Reverse of a deep footing.

b. no lateral support allowing a "conical" wall mechanism (as well as base heave).

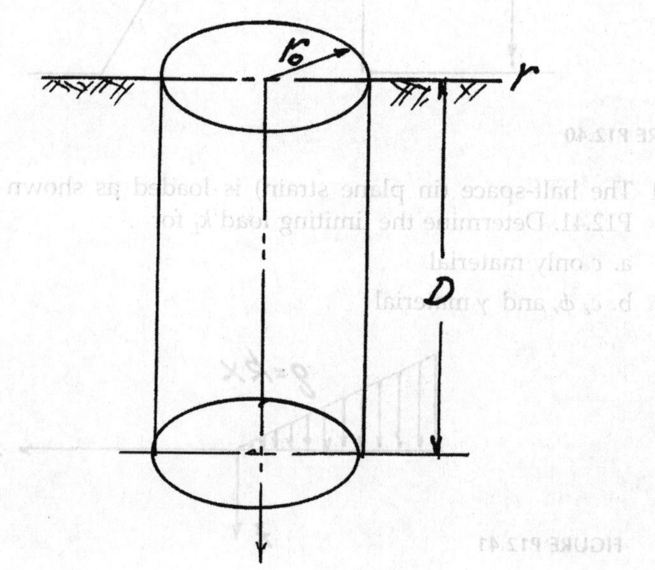

FIGURE P12.43

Index

Page numbers in italic indicate figures; page numbers with an "n" indicate footnotes.